MATHEMATICAL MORPHOLOGY IN IMAGE PROCESSING

OPTICAL ENGINEERING

Series Editor

Brian J. Thompson
Provost
University of Rochester
Rochester, New York

1. Electron and Ion Microscopy and Microanalysis: Principles and Applications, *Lawrence E. Murr*
2. Acousto-Optic Signal Processing: Theory and Implementation, *edited by Norman J. Berg and John N. Lee*
3. Electro-Optic and Acousto-Optic Scanning and Deflection, *Milton Gottlieb, Clive L. M. Ireland, and John Martin Ley*
4. Single-Mode Fiber Optics: Principles and Applications, *Luc B. Jeunhomme*
5. Pulse Code Formats for Fiber Optical Data Communication: Basic Principles and Applications, *David J. Morris*
6. Optical Materials: An Introduction to Selection and Application, *Solomon Musikant*
7. Infrared Methods for Gaseous Measurements: Theory and Practice, *edited by Joda Wormhoudt*
8. Laser Beam Scanning: Opto-Mechanical Devices, Systems, and Data Storage Optics, *edited by Gerald F. Marshall*
9. Opto-Mechanical Systems Design, *Paul R. Yoder, Jr.*
10. Optical Fiber Splices and Connectors: Theory and Methods, *Calvin M. Miller with Stephen C. Mettler and Ian A. White*
11. Laser Spectroscopy and Its Applications, *edited by Leon J. Radziemski, Richard W. Solarz and Jeffrey A. Paisner*
12. Infrared Optoelectronics: Devices and Applications, *William Nunley and J. Scott Bechtel*
13. Integrated Optical Circuits and Components: Design and Applications, *edited by Lynn D. Hutcheson*
14. Handbook of Molecular Lasers, *edited by Peter K. Cheo*
15. Handbook of Optical Fibers and Cables, *Hiroshi Murata*
16. Acousto-Optics, *Adrian Korpel*
17. Procedures in Applied Optics, *John Strong*
18. Handbook of Solid-State Lasers, *edited by Peter K. Cheo*
19. Optical Computing: Digital and Symbolic, *edited by Raymond Arrathoon*
20. Laser Applications in Physical Chemistry, *edited by D. K. Evans*
21. Laser-Induced Plasmas: Physical, Chemical, and Biological Applications, *edited by Leon J. Radziemski and David A. Cremers*

22. Infrared Technology Fundamentals, *Irving J. Spiro and Monroe Schlessinger*
23. Single-Mode Fiber Optics: Principles and Applications, Second Edition, Revised and Expanded, *Luc B. Jeunhomme*
24. Image Analysis Applications, *edited by Rangachar Kasturi and Mohan M. Trivedi*
25. Photoconductivity: Art, Science, and Technology, *N. V. Joshi*
26. Principles of Optical Circuit Engineering, *Mark A. Mentzer*
27. Lens Design, *Milton Laikin*
28. Optical Components, Systems, and Measurement Techniques, *Rajpal S. Sirohi and M. P. Kothiyal*
29. Electron and Ion Microscopy and Microanalysis: Principles and Applications, Second Edition, Revised and Expanded, *Lawrence E. Murr*
30. Handbook of Infrared Optical Materials, *edited by Paul Klocek*
31. Optical Scanning, *edited by Gerald F. Marshall*
32. Polymers for Lightwave and Integrated Optics: Technology and Applications, *edited by Lawrence A. Hornak*
33. Electro-Optical Displays, *edited by Mohammad A. Karim*
34. Mathematical Morphology in Image Processing, *edited by Edward R. Dougherty*
35. Opto-Mechanical Systems Design: Second Edition, Revised and Expanded, *Paul R. Yoder, Jr.*
36. Polarized Light: Fundamentals and Applications, *Edward Collett*

Additional Volumes in Preparation

Rare Earth Doped Fiber Lasers and Amplifiers, *edited by Michel Digonnet*

Speckle Metrology, *edited by Rajpal S. Sirohi*

MATHEMATICAL MORPHOLOGY IN IMAGE PROCESSING

EDITED BY
EDWARD R. DOUGHERTY
Rochester Institute of Technology
Rochester, New York

Marcel Dekker, Inc. New York • Basel • Hong Kong

Library of Congress Cataloging-in-Publication Data

Mathematical morphology in image processing / edited by Edward R.
Dougherty.
p. cm. -- (Optical engineering)
Includes bibliographical references and index.
ISBN 0-8247-8724-2 (alk. paper)
1. Image processing--Mathematics. 2. Image processing-
-Statistical methods. I. Dougherty, Edward R. II. Series: Optical
engineering (Marcel Dekker, Inc.)
TA1632.M363 1992
621.36'7--dc20 92-25560
CIP

This book is printed on acid-free paper.

Marcel Dekker, Inc.
270 Madison Avenue, New York, New York 10016

Current printing (last digit):
10 9 8 7 6 5 4 3 2 1

PRINTED IN THE UNITED STATES OF AMERICA

This collection is dedicated to Georges Matheron, whose seminal volume Random Sets and Integral Geometry *set forth the epistemological, topological, algebraic, and probabilistic principles for morphological image analysis.*

About the Series

The series came of age with the publication of our twenty-first volume in 1989. The twenty-first volume was entitled *Laser-Induced Plasmas and Applications* and was a multi-authored work involving some twenty contributors and two editors: as such it represents one end of the spectrum of books that range from single-authored texts to multi-authored volumes. However, the philosophy of the series has remained the same: to discuss topics in optical engineering at the level that will be useful to those working in the field or attempting to design subsystems that are based on optical techniques or that have significant optical subsystems. The concept is not to provide detailed monographs on narrow subject areas but to deal with the material at a level that makes it immediately useful to the practicing scientist and engineer. These are not research monographs, although we expect that workers in optical research will find them extremely valuable.

There is no doubt that optical engineering is now established as an important discipline in its own right. The range of topics that can and should be included continues to grow. In the About the Series that I wrote for earlier volumes, I noted that the series covers the "topics that have been part of the rapid expansion of optical engineering." I then followed this with a list of topics we have already outgrown. I will not repeat that mistake this time! Since the series now exists, the topics that are appropriate are best exemplified by the titles of the volumes listed in the front of this book. More topics and volumes are forthcoming.

Brian J. Thompson
University of Rochester
Rochester, New York

Preface

As a nonlinear branch of image and signal processing, mathematical morphology represents a dramatic break with classical linear processing and facilitates the application of various mathematical disciplines to the processing and analysis of images. These areas include nonlinear statistics, logic, geometry, geometrical probability, topology, and various algebraic systems such as the theories of lattices and groups. It is not that these disciplines have not played imaging roles outside mathematical morphology, but rather that they appear naturally within the context of mathematical morphology and are central to its development in both theory and application.

The present volume is testimony to the fertility of the original Matheron-Serra conception and to the great expansion of interest in mathematical morphology that has taken place over the past few years. Many researchers—mathematicians, statisticians, engineers, computer scientists, and natural scientists—have taken interest in the field. As a result, a number of productive research veins are currently being explored. These directions are well represented in this volume. Although it is not possible to characterize the contributions crisply, they can be roughly placed into five active areas within morphological image processing. The first three chapters are statistical, dealing with approaches to finding well-performing structuring elements and the statistical analysis of morphological operations as noise filters. Chapters 4, 5, and 11 concern morphological feature generation for classification. Chapters 6, 7, and 13 concern extension of the morphological paradigm, and they illustrate the degree to which mathematical morphology provides promising ground for algebraists. Chapters 8, 9, and 10 treat topics having to do with efficient morphological algorithms, the latter containing theoretical results whose importance extends well beyond algorithmic efficiency. Finally, Chapter 12 describes a key paradigm for morphological image segmentation. Each chapter is self-contained and can be read in any order, depending on preference. Perhaps brief descriptions of the chapters' places within mathematical morphology will prove useful.

Successful application of mathematical morphology often depends on the selection of one or more appropriate structuring elements. Except in simple problems, human structuring-element selection can prove daunting. This is especially true when one wishes to perform image restoration or shape recognition in the presence of noise. Currently, much interest is centered on automatic design of structuring elements. In the first chapter, Stephen Wilson introduces a training method based on Hebbian learning to find appropriate hit-and-miss structuring elements for character recognition in noise. Generally, overly dense structuring elements lower detection rates, whereas too sparse structuring elements raise false-positive rates. The network approach attempts to limit both types of errors by judiciously placing structuring-element pixels.

Image restoration by morphological filtering can be placed into the classical mean-square-error paradigm; however, here too human structuring-element design is not practical. Using the Matheron erosion representation as the general restoration operator, one needs to find a basis of structuring elements providing an optimal filter. Because of the nonlinear and combinatoric nature of the problem, even when computers are used design-time efficiency is problematic. In the second chapter, Robert P. Loce and I discuss design strategies based on search-space constraint and efficient estimation of image statistics. The chapter provides a practical multiple-level approach to circumventing computation limitations in the automatic design of restoration structuring elements.

If a filter is to be treated as a statistical estimator, the input image must be treated as a random process. From a statistical perspective, the input and output processes possess distributions and one would like to express the output distribution in terms of the input distribution, or, with less completeness, some output moments in terms of input moments. In the third chapter, Jaakko Astola, Lasse Koskinen, and Yrjö Neuvo derive output statistics resulting from application of certain basic flat morphological filters to independent random variables. They utilize the equivalence between stack and flat-morphological filters in their derivations. Their treatment includes analysis of flat openings and flat closings.

One of the most practically useful theories developed by Matheron concerns granulometric size distributions. Successively sieving an image modeled as a collection of grains and quantifying the sieving rate yields features that are excellent for characterizing granular and textural images. In Chapter 4, Chakravarthy Bhagvati, Dimitri Grivas, and Michael M. Skolnick apply this technique to the analysis of pavement surface condition.

A common approach to shape classification is to associate with each shape a set of features (measurements) and to distinguish shapes according to the feature set. In some instances the feature set will uniquely characterize the shapes in the pattern class; in others it will not. Just as importantly, a feature set may characterize shapes up to certain invariances, such as up to translation, rotation, or homogeneity. The basic inverse question is this: Given a collection of features, to what extent does it characterize the shapes within a given class? In Chapter 5, Michel Schmitt attacks this inverse problem for two morphologically important transformations, the geometrical covariogram and volumes of dilations by compact sets. For the covariogram, characterization applies to certain classes of polygons, and for these classes reconstruction from the covariogram is possible up to a translation and a symmetry about the origin. For dilation volumes, characterization applies to random compact sets, is stated in terms of mean dilation volumes, and is up to a random translation.

As originally conceived by Georges Matheron, mathematical morphology concerns the analysis of binary images by means of probing with structuring elements. Since then the underlying principles have been extended to various mathematical settings, with the principal vehicle for extension being lattice theory. In Chapter 6, Henk Heijmans and Luc Vincent discuss mathematical morphology on graphs and how it relates to image analysis. The extension is facilitated by the utilization of graph-theoretic neighborhoods to define dilation and erosion. Classical set probing can be replaced by probing structuring graphs, which in effect means looking for matches of the structuring graph within the large graph. Many basic morphological algorithms can be defined on graphs, including skeletons, granulometries, and watersheds.

The fundamental operations of binary set-theoretic mathematical morphology are compatible with translation in the Euclidean plane. Stated differently, they are invariant under the Euclidean translation group. For instance, translating a set and then eroding is equivalent to eroding and then translating. It has been shown that morphology can be generalized by substituting any commutative group for Euclidean translations. Such generalization is important if one wants to discuss when two objects are of the same shape, since translational congruence is not relevant in many situations. For instance, in some situations we must concern ourselves with rotational or perspective shape similarities. In Chapter 7, Jos B. T. M. Roerdink extends morphology by not requiring the invariance group to be commutative. The introduction to the chapter provides an excellent synopsis of both the motivations and directions of algebraic generalization. It is valuable reading for both mathematicians and nonmathematicians, since it provides important insight into the general nature of algebraic modeling.

A basic problem in image processing is algorithm speed. Many of the underlying morphological operations are computationally complex and require efficient algorithms. In Chapter 8, Luc Vincent discusses sequential methods for morphological computation. The basic approach is to consider only pixels that will be transformed, and the algorithms are based on image boundaries. Two classes of algorithms result: (1) loop and chain algorithms and (2) queue algorithms. The first class depends on a contour coding and dilations are computed by a set of rewriting rules in conjunction with some "readjustment." The second class is based on forming a FIFO queue of pixels and implementing morphological algorithms by performing operations on the queue.

In Chapter 9, Tapas Kanungo and Robert Haralick discuss a computational method incorporating two representations, the B-code boundary representation and the normalized half-plane representation. The representations are applicable to a restricted set of digital binary domains, but on the class of domains to which they apply, they can be employed to perform erosions and dilations in constant time.

As stated in the introduction to Chapter 8, the distance function serves as a leitmotiv for reducing binary algorithm complexity; in Chapter 10, Françoise Preteux introduces two distance functions for gray-scale mathematical morphology, the topographical and differential distance functions. These distance functions are defined, relative to a given gray-scale image, for pairs of points in the spatial domain. The topographical distance is constructed by viewing the graph of the image as a topography and considering the minimal height one must ascend to traverse between two points. The differential distance between two points is more complicated and depends on measuring the deviation cost from a topographical path of greatest slope. An important part of the chapter is the investigation of the relationships between the new distance functions, the skeleton by influence zones, and watersheds.

Chapter 11 returns to the problem of pattern recognition, with Divyendu Sinha and Hanjin Lee presenting three algorithms for the recognition of discrete multidimensional black-and-white objects. For each algorithm they solve the appropriate inverse problem, namely, the partition of the pattern class engendered by the recognition algorithm. For one algorithm, partitions contain only geometrically similar objects. The algorithms are based on different generalizations of binary covariance functions, and the authors show a link between object partitioning and solutions of certain Diophantine equations.

A prime method of morphological image segmentation is via the watershed algorithm. This algorithm (actually a class of algorithms) is discussed in depth in Chapter 12 by S. Beucher and F. Meyer. In its most basic conception, the watershed algorithm accomplishes gray-scale image segmentation by finding divide lines resulting from the morphological gradient of the original gray-scale image. Rather than finding segmentation lines from the gradient by first thresholding the gradient and then thinning, the watershed algorithm finds the crest lines of the gradient, these being the ridge lines of the gray-scale gradient image viewed topographically. The authors discuss the problem of oversegmentation and algorithm efficiency. They also provide practical examples employing the watershed.

The gray-scale theory of mathematical morphology is expressed in the context of function lattices; in the final chapter, Jean Serra extends morphological lattice theory so that it applies to multivalued images (for instance, color images) as well as to single-valued images. A basic issue is that single-valued images take values in a totally ordered lattice, whereas the range of multivalued images is not necessarily totally ordered. After reviewing the roles of flat operators and strictly increasing mappings between lattices in the single-valued-function morphology, Jean Serra extends the underlying concepts to the more general setting. A key role is played by anamorphoses, these being bijections between lattices for which they and their inverses are increasing. When the lattices involved are totally ordered, a strictly increasing mapping between them is, ipso facto, an anamorphosis, but when they are not totally ordered, strictly increasing mappings between them are not necessarily anamorphoses. Various implications of the new multivalued theory, as well as examples, are presented.

Before closing, let me offer the hope of all contributors that this collection will prove beneficial to both researchers and practitioners in image processing. Let me also offer my own appreciation to all contributors for their efforts in bringing this volume to fruition.

Edward R. Dougherty

Contents

Contributors

Jaakko Astola *Tampere University, Tampere, Finland*

Chakravarthy Bhagvati *Department of Computer Science, Rensselaer Polytechnic Institute, Troy, New York*

S. Beucher *Centre de Morphologie Mathématique, Ecole des Mines de Paris, Fontainebleau, France*

Edward R. Dougherty *Center for Imaging Science, Rochester Institute of Technology, Rochester, New York*

Dimitri A. Grivas *Department of Civil Engineering, Rensselaer Polytechnic Institute, Troy, New York*

Robert M. Haralick *Department of Electrical Engineering, University of Washington, Seattle, Washington*

Henk Heijmans *Centre for Mathematics and Computer Science, Amsterdam, The Netherlands*

Tapas Kanungo *Department of Electrical Engineering, University of Washington, Seattle, Washington*

Lasse Koskinen *Tampere University, Tampere, Finland*

Hanjin Lee *O.A. Labs, Daewoo Telecom Co., Ltd., Siheung City, Kyungki Do, South Korea*

Robert P. Loce *Joseph C. Wilson Center for Technology, Xerox Corporation, Webster, New York*

F. Meyer *Centre de Morphologie Mathématique, Ecole des Mines de Paris, Fontainebleau, France*

Yrjö Neuvo *Tampere University, Tampere, Finland*

Françoise Preteux *Département Images, Télécom Paris, Paris, France*

Jos B. T. M. Roerdink *Centre for Mathematics and Computer Science, Amsterdam, The Netherlands*

Michel Schmitt *Thomson-CSF, Laboratoire Central de Recherches, Domaine de Corbeville, Orsay, France*

Jean Serra *Centre de Morphologie Mathématique, Ecole des Mines de Paris, Fontainebleau, France*

Divyendu Sinha *Department of Computer Science, College of Staten Island, City University of New York, Staten Island, New York*

Michael M. Skolnick *Department of Computer Science, Rensselaer Polytechnic Institute, Troy, New York*

Luc Vincent *Xerox Imaging Systems, Peabody, Massachusetts*

Stephen S. Wilson *Applied Intelligent Systems, Inc., Ann Arbor, Michigan*

MATHEMATICAL MORPHOLOGY IN IMAGE PROCESSING

Chapter 1

Training Structuring Elements in Morphological Networks

Stephen S. Wilson

Applied Intelligent Systems, Inc., Ann Arbor, Michigan

I. INTRODUCTION

This chapter presents a method for automatically training structuring elements for locating specified patterns in an image. The training method is based upon Hebbian learning [1], which has been successfully used in training neural networks. In network learning, a connection strength is enhanced if it is useful in detecting a given pattern. In morphology, there is no concept of connection strength. In this case, a point in a structuring element is added if it is useful in detecting the pattern and has a minimal effect in generating false recognitions in the background. To further prevent false recognitions, hit-and-miss transforms are used, where the "miss" structuring element is trained by adding points that reduce false background detections without reducing the reliability in detecting the desired pattern. A correspondence between morphology, rank-order operators, and neural networks will become obvious.

A sequence of morphological operations can be very complex. In the most robust analysis sequences, there may be large number of feature images, each generated by a unique structuring element. The set of feature images is further processed by morphology operators. The collection of structuring elements and intermediate images that are generated is very similar to a neural network and is called a morphological network, which will be defined more formally later. The images form a vector space of images, and the structuring elements form a pattern that has a matrix structure. The morphological network can be conceptually simplified by defining a formalism in which the images and structuring elements are matrices. This formalism, called matrix morphology, is further developed in section II. One method of training structuring elements involves simulated annealing and is briefly covered in Section III. Some of the concepts defined in simulated annealing are used in the discussion of Hebbian learning.

Morphological networks are useful for determining the existence or location(s) of reasonably defined patterns, but the Hebbian learning technique that will be presented is not as useful for training ill-defined patterns, as are found, for example, in applications involving outdoor scene understanding. Examples that can be trained include classification applications such as character recognition, alignment applications such as finding pattern areas on an integrated circuit wafer, and applications in which objects are located to subpixel accuracy. It is possible to handle a wide range of distortions in the desired pattern, but false recognitions then become more probable. A trade-off mechanism is needed that will allow control over recognition when there is confusion between an object and a similar-looking one.

In Section IV it will become apparent how this mechanism can be embodied in Hebbian learning so that an automatic method of training structuring elements will allow anyone to train new patterns in a manner that is faster and more optimum than even a highly skilled user is willing or able to handle. In some cases morphological networks trained by Hebbian learning can handle patterns with occlusions, rotations of $\pm 15°$, or scale changes of around 15%. An example illustrating step-by-step training is covered in detail in Section V.

A. Pattern Detection with Morphology

The basic pattern detection mechanism to be discussed is the erosion. A particular shape in a binary image can be detected with morphology by eroding the image with a structuring element that is slightly smaller than the desired shape, as shown, for example, in Figure 1. Small shapes will completely erode away. Only a small blob at the reference point of the structuring element remains at the shape of interest. The centroid of the blob in most cases gives the location of the object to subpixel accuracy. The eroded set is easiest to view in the form given by Matheron [2]

$$A \ominus B = \{ x \in E^N \mid B_{\mathbf{x}} \subseteq A \}$$

where $B_{\mathbf{x}}$ is the translation of the structuring element B by spatial vector $\mathbf{x}$. In this form, B is a probe that is translated throughout image A. The reference point at $\mathbf{x}$ is marked in the output image wherever the probe can be included in a shape in the image. Due to the increasing property of erosions ($X \subseteq Y \Rightarrow X \ominus B \subseteq Y \ominus B$, where X and Y are images and B is any structuring element), larger shapes also result in a blob that is sometimes larger. In Figure 1, a structuring element is shown that is used to detect the desired pattern in the lower left of an image.

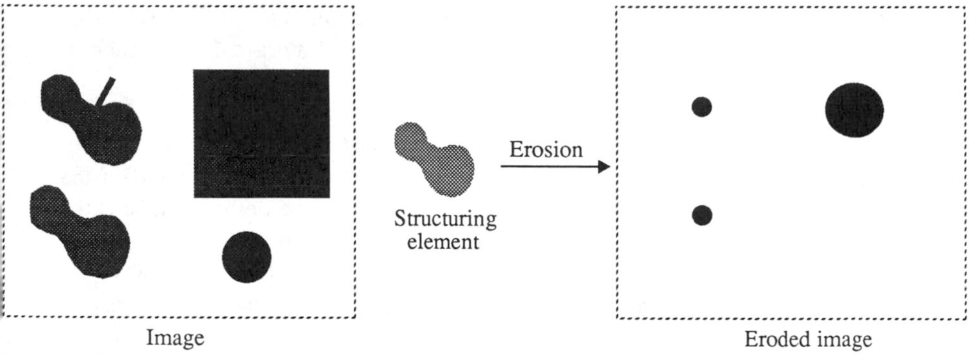

Figure 1. The erosion as a shape detection mechanism.

After an erosion of the image the smallest shape is eroded away. The larger shapes are all eroded to smaller blobs.

A simple erosion as a detection mechanism is seldom satisfactory for recognizing the desired shape and is modified to the hit-and-miss transformation defined by Serra [3] as

$$O = (I \ominus H) \cap (I^c \ominus M)$$

As a result, output image O contains only those points where structuring element H fits in a shape in image I and structuring element M fits in the background I^c; that is, H "hits" I, and M "misses" I. Figure 2 shows an example of a hit-and-miss structuring element that can detect the desired shape shown in the image. The hit part of the structuring element is the same as that shown in Figure 1. Depending on the application, the shape in the upper left corner of Figure 2 may

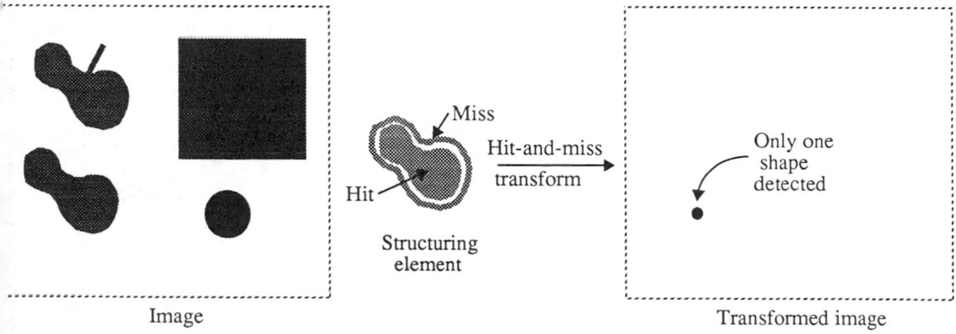

Figure 2. Operation with a hit-and-miss transform.

or may not be considered to be the desired shape. The miss part of the structuring element is designed to provide appropriate detection in marginal cases such as this.

In the simplest cases, an erosion structuring element can be taught directly from a training image by first defining the shape in the image that is to be detected. A reference point is defined in the image and is often centered in the shape. A set of the points in a slightly eroded version of the defined shape is the hit structuring element that will enable detection. This mechanism works for unoccluded objects with a fairly rigid size and little rotation. The miss part of the structuring element is more difficult to define and depends on the performance of the system for marginal shapes.

B. Extensions of Morphology to Rank-Order Operators

If an image is corrupted by salt-and-pepper noise as shown in Figure 3, then erosion by the structuring element causes the image to be completely eroded away. Dilations may cause the entire background of the image to fill in. In many cases the salt-and-pepper noise can be filtered using dilations and erosions with small structuring elements before larger probes are applied, but this is not always possible to do and still achieve the low failure rates demanded by industrial applications. For objects with a confusing background, simple erosions should be replaced by binary rank-order operators [4] (also called order statistic filters [5]) and must be trained in the context of the background.

For a binary image, the operation of ordering and ranking is not necessary and will not be used here, although the "rank-order" terminology will be used to be consistent with the literature. Very simply, a binary rank-order operator can be replaced by the operation of counting the number of image points that contact the points of the structuring element probe and marking the reference point in the output image if at least a given percentage of points contact the probe. For ex-

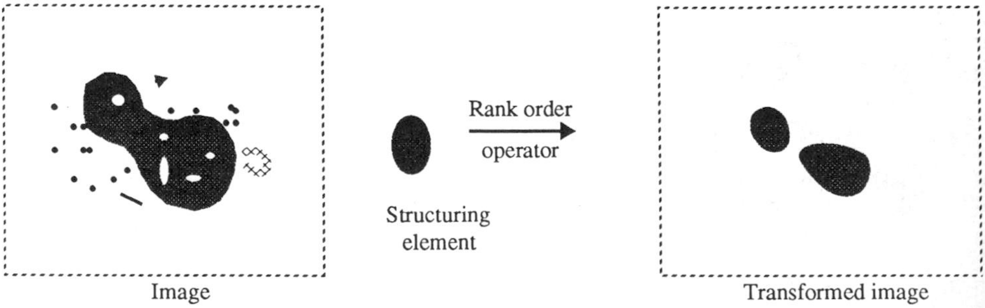

Figure 3. Operation with a rank-order operator.

ample, if only 90% of the pixels in a shape are required to contact the probe before the reference point of the probe marks a point in the output image, the operation is still very close to an erosion, but up to 10% pixel dropouts are allowed as shown, for example, in Figure 3. The percentage vote, or threshold, is a variable to be adjusted to give good performance for a particular application.

The binary rank-order operator is also called an *order statistic filter* [5], a Ξ *filter* by Preston [6], "*r-out-of-n*" *decision rule* by Justusson [7], or just *voting logic*. Counting within a moving structuring element can be accomplished by using a convolution. The convolution kernel is defined to be 1 for points corresponding to the members of a structuring element B and zero elsewhere. The binary rank-order operator is provided by thresholding the convolution [8]. A convolution of a discrete image $A(\mathbf{x})$ by kernel $B(\mathbf{u})$ is

$$C(\mathbf{p}) = \sum_{\mathbf{u}} A(\mathbf{p} - \mathbf{u})\, B(\mathbf{u}) = A * B \tag{1.1}$$

where $\mathbf{u}$ and $\mathbf{p}$ are two-dimensional spatial vectors, $\mathbf{u}, \mathbf{p} \in Z^2$, and the summation is over the domain of the image. The threshold operation X_t is applied to Eq. (1.1), at threshold t.

$$\mathrm{RO}_t(A;B)(\mathbf{p}) = X_t\left(\sum_{\mathbf{u}} A(\mathbf{p} - \mathbf{u})B(\mathbf{u})\right) = X_t(A * B) \tag{1.2}$$

For a rank-order operator expressed in this form, threshold t is the rank.

It is obvious that the rank-order operator is a generalization that includes erosions and dilations as suboperators. Erosions and dilations are thresholded convolutions with, respectively, a maximum and minimum threshold value. The hit-and-miss transform can be extended to rank-order operators. Both the hit-and-miss transform and rank-order operators are applied to morphological networks in a later section.

Even if an algorithm is to use erosion structuring elements, rank-order operators should still be used during training because the evaluation of an erosion output alone does not provide the system with enough visibility to see whether a pattern is close to being falsely recognized by a trial erosion structuring element. During the training phase, successful elimination of false recognitions with a rank threshold that is slightly lower than maximum will ensure that there is a margin for noise during the running phase, when the threshold is set to the maximum (erosion). The erosion sees only the tip of the iceberg, while the rank-order filter can see the convolution beneath the surface.

C. Segmentation

There are many methods for segmenting a gray-level image into a binary image so that binary morphology can be applied. Three techniques will be compared here.

1. Threshold Segmentation

The simplest way to create a binary segmented image is to apply a threshold. If the background is too nonuniform, the image can be normalized through three-dimensional gray-level techniques [9]. The erosion structuring element for detecting a shape in the thresholded binary image is compact and can sometimes be constructed by a concatenated series of smaller shapes that are contained in a 3×3 neighborhood window. To prevent larger shapes from being detected, a hit-and-miss structuring element is defined, as discussed earlier. The basic concept of using hit-and-miss transforms on threshold segmented images is shown in Figure 2. An image considered as a threshold compact shape is often not the best approach to applications, especially when the illumination is nonuniform to the extent that a gray-level image cannot be segmented by a thresholding operation, even with a background-normalizing algorithm.

2. Edge Segmentation

Edge detection is a local operation that can withstand uneven illumination over the scene or a variable contrast of a shape in the scene. Therefore, background normalization is not needed. With edge segmentation, there is a lower density of points in the transformed image, and the most important information is concentrated at those points.

With threshold segmentation, a large object in a scene that includes the desired object as a subset will leave a blob in the output image after the erosion, due to the increasing property of erosion. In edge segmentation, the edges of a larger object are not included in the edges of a smaller object and will leave nothing in the resulting output image after an erosion. Ordinarily, a hit-and-miss transform is not needed to differentiate between a larger and a smaller object due to the unique representation of an object with edges. However, in some cases the difference is too small to be dependable. For example, the shape in the upper left corner of the image in Figure 4 includes all the edges of the desired shape except for a small segment indicated by the arrow. The structuring element is not strictly included in that shape. Thus, an erosion of the image by the structuring element leaves only one small blob corresponding to the desired shape in the lower left corner of the eroded image in Figure 4. The edges of the image in Figure 4 are too thin to be reliably detected by the structuring element that is shown. A dilation of the edge image is necessary before using the erosion as a detection mechanism.

3. Edge Gradient Segmentation

For very low contrast or for distorted images, simple edge magnitude segmentation will often fail to give images suitable for further processing. In these cases there are advantages to segmenting the image into several images, each containing a separate edge direction computed by the difference of offset Gaussians

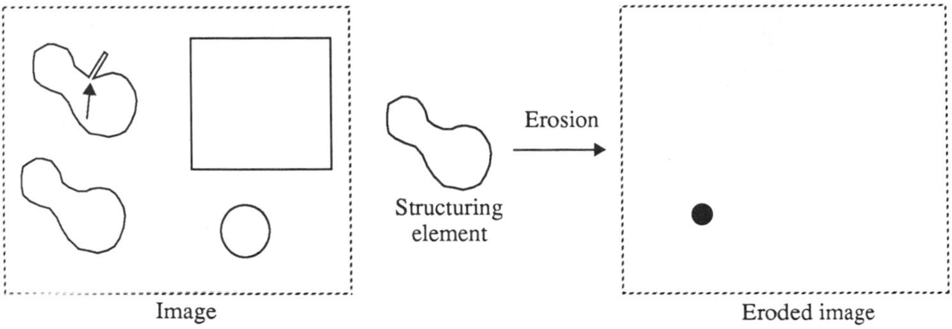

Figure 4. Erosion on image edges.

(DOOG). The DOOG is computed by convolving the image with a Gaussian function and then computing finite differences in the horizontal and vertical directions:

$$\mathrm{DOOG}(x,y) = \frac{\exp(-(x - x_0)^2 - (y - y_0)^2}{\sigma^2} - \frac{\exp(-x^2 - y^2)}{\sigma^2} \tag{1.3}$$

The offsets in equation (1.3) are $(x_0,y_0) = (0,1)$ for the vertical direction. Regions where equation (1.3) is positive indicate a north edge, and negative values indicate a south edge. Horizontal edges are given by offsets $(x_0,y_0) = (1,0)$ where positive and negative values of the DOOG, respectively, indicate east and west directions. Positive and negative thresholds applied to the DOOG with the two offset vectors give the four binary edge direction images.

Figure 5 shows the three types of segmentation discussed so far, where the edge gradient is a segmentation in four separate planes labeled by four directions. The noise density and the edge information in each of the image components, segmented as four edge directions, are much more sparse than for the other two segmentation schemes. Broken edges can more readily be repaired by closings or dilations without confusing noisy images. This technique allows a higher degree of discrimination and works especially well for complex shapes such as those in character recognition.

Because edges are generally thin and irregular, it is difficult to define structuring elements as a series of smaller elements defined in a 3×3 window. The most efficient structuring elements consist of a sparse sampling of points along the edges of the shape. The representation of an image as a set of edges that are processed with sparsely sampled structuring elements gives the best set of features to be used by a morphological network.

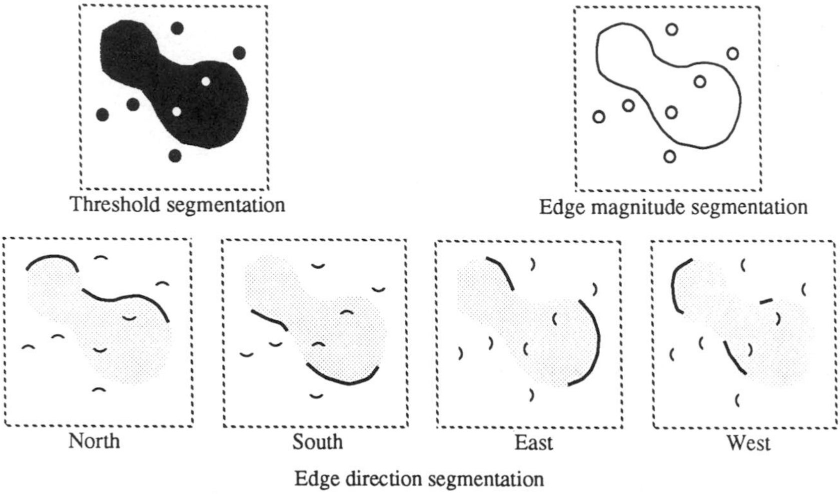

Figure 5. Three types of segmentation.

II. THE STRUCTURE OF MORPHOLOGICAL NETWORKS

In this section a coherent mathematical structure is briefly developed as a framework for morphological networks used in the rest of the chapter. The theory, called matrix morphology [10,11], involves matrices of images and structuring elements. Mathematical morphology, as currently defined in the literature [3,12], will be called scalar morphology in this chapter to avoid confusion.

A. Matrices of Images and Structuring Elements

In applications to be described later, edge gradient segmentation will be used. The set of binary edge images is most conveniently represented as a four-component row vector and is represented with a boldface symbol and indexed by the four compass directions, $\mathbf{E} = [E_N, E_S, E_E, E_W]$. Generally, morphology operations on an image row vector will require different structuring elements for each image component. Often a binary image vector is transformed by a set of operations into another binary image vector that represents a different set of features. Applications involving multiple image vectors can become confusing. The methodology called matrix morphology makes the complex processing of vector images much more comprehensible. Structuring elements that operate on image vectors are most conveniently represented symbolically with a matrix indexing. A formalism involving matrices of images and structuring elements is summarized here. Although the formalism is able to handle matrices of images, in the applications in this chapter images are always vectors while structuring elements in general are rectangular matrices.

Let $\mathbf{B}$ be an $I \times K$ matrix of images or structuring elements B_{ik}, that is

$$\mathbf{B} = \{ B_{ik} \subseteq E^N, i = 1, \ldots, I, k = 1, \ldots, K \}$$

Each matrix element B_{ik} will be a set of points in N-space, where b_{ik} are coordinate vectors to those points and $b_{n,ik}$ are the separate components:

$$B_{ik} = \{ b_{ik} \in E_N \mid b_{ik} = (b_{1,ik}, b_{2,ik}, \ldots b_{N,ik}),$$
$$i = 1, \ldots, I, k = 1, \ldots, K \}$$

The following relations emphasize the nature of the sets and members:

$$\mathbf{B} \subseteq E^N \times Z^2, \qquad B_{ij} \subseteq E^N, \qquad b_{ij} \in E^N, \qquad b_{n,ij} \in E$$

Figure 6 shows an example of a matrix in symbolic form and in a pictorial form to emphasize the image nature of the matrix elements.

The following matrices are defined and used in the remaining definitions. Let $\mathbf{A}$ be an $I \times K$ matrix, $\mathbf{B}$ be a $K \times J$ matrix, and $\mathbf{C}$ be a $I \times J$ matrix defined as follows:

$$\mathbf{A} = \{A_{ik} = \{ a_{ik} \in E^N\}, i = 1, \ldots, I \text{ and } k = 1, \ldots, K \}$$
$$\mathbf{B} = \{ B_{kj} = \{ a_{kj} \in E^N\}, k = 1, \ldots, K \text{ and } j = 1, \ldots, J \}$$
$$\mathbf{C} = \{ C_{ij} = \{ c_{ij} \in E^N\}, i = 1, \ldots, I \text{ and } j = 1, \ldots, J \}$$

Definition. A matrix transpose $\mathbf{A}^T$, is defined as a $K \times I$ matrix of images, where $\mathbf{A}^T = \{ A_{ki} \in E^n, k = 1, \ldots, K; i = 1, \ldots, I \}$.

Definition. The matrix complement is defined as $\mathbf{A}^C = \{ A_{ij}^C \}$.

Definition. The threshold or cross section is defined as $X_t(\mathbf{A}) = \{X_t(A_{ik})\}$, where $X_t(A) = \{a \in E^N : A(a) \geq t\}$.

$$\boldsymbol{B} = \begin{bmatrix} B_{11} & B_{12} \\ B_{21} & B_{22} \end{bmatrix} =$$

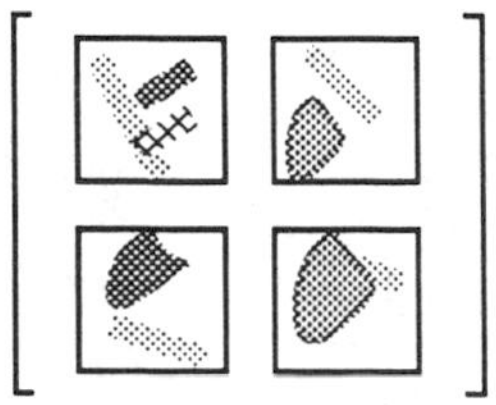

Figure 6. A symbolic and pictorial example of an image matrix.

Definition of Matrix Dilation.

$$\mathbf{A} \oplus \mathbf{B} = \{ C_{ij} \subseteq E^N \mid C_{ij} = \cup_k A_{ik} \oplus B_{kj} \quad \text{for } i = 1, \ldots, I, j = 1, \ldots, J, \text{ and } k = 1, \ldots, K \} \tag{1.4}$$

The definition will be less formally written in a form analogous to matrix multiplication: $\mathbf{A} \oplus \mathbf{B} = \{ \cup_k A_{ik} \oplus B_{kj} \}$, or $[\cup_k A_{ik} \oplus B_{kj}]$, using the square bracket notation in linear algebra.

Definition of Matrix Erosion.

$$\mathbf{A} \ominus \mathbf{B} = \{ C_{ij} \subseteq E^N \mid C_{ij} = \cap_k A_{ik} \ominus B_{kj} \quad \text{for } i = 1, \ldots, I, j = 1, \ldots, J, \text{ and } k = 1, \ldots, K \}. \tag{1.5}$$

The erosion definition will also be written in a form analogous to matrix multiplication:

$$\mathbf{A} \ominus \mathbf{B} = \{ \cap_k A_{ik} \ominus B_{kj} \} = [\cap_k A_{ik} \ominus B_{kj}]$$

Due to the nature of matrix index ranges, the opening and closing must be defined with a transpose.

Definition of Matrix Opening.

$$\mathbf{A} \bigcirc \mathbf{B} = [\mathbf{A} \ominus \mathbf{B}] \oplus \mathbf{B}^T$$

Definition of Matrix Closing.

$$\mathbf{A} \bullet \mathbf{B} = [\mathbf{A} \oplus \mathbf{B}] \ominus \mathrm{B}^T$$

If $\mathbf{A}$ is an $I \times J$ matrix then $\mathbf{A} \bigcirc \mathbf{B}$ and $\mathbf{A} \bullet \mathbf{B}$ are also $I \times J$ matrices.

The following is an example of a matrix dilation:

$$\mathbf{A} \oplus \mathbf{B} = \begin{bmatrix} A_{11} & A_{12} \\ A_{21} & A_{22} \end{bmatrix} \oplus \begin{bmatrix} B_{11} & B_{12} \\ B_{21} & B_{22} \end{bmatrix}$$

$$= \begin{bmatrix} ((A_{11} \oplus B_{11}) \cup (A_{12} \oplus B_{21})) & ((A_{11} \oplus B_{12})) \cup (A_{12} \oplus B_{22})) \\ ((A_{21} \oplus B_{11}) \cup (A_{22} \oplus B_{21})) & ((A_{21} \oplus B_{12})) \cup (A_{22} \oplus B_{22})) \end{bmatrix}$$

Further details of morphology involving matrices of structuring elements operating on matrices of images are given in refs 10 and 11, where a number of theorems and examples are given.

B. Rank-Order Operators on Matrices

Rank-order operators can be defined that also operate on matrices of images. Because a rank-order operator can be expressed as a thresholded convolution, a complete discussion of binary rank-order operators will first require a definition of convolutions on matrices of images with matrix kernels.

1. Convolutions on Matrices of Images

Let A and $B \in E^N \times R$ represent images. A scalar convolution is defined as

$$C(x,y) = \iint A(x - u, y - v)\, B\,(u, v)\, du\, dv = A * B$$

A convolution of a two-dimensional matrix of images is defined as

$$C_{ij}(x,y) = \mathbf{C}(x,y) = \iint \sum_k A_{ik}(x - u, y - v) B_{kj}(u, v)\, du\, dv = \mathbf{A} * \mathbf{B}$$

For digital image analysis, the convolution is expressed in the discrete form. Let **p** and **u** be vector integers that range over the image. Define a set of images $A_{ij} \subseteq Z^2$ and kernels $B_{kj} \subseteq Z^2$. A discrete scalar convolution is given in Eq. (1.1) and is generalized to a discrete matrix convolution given by

$$\begin{aligned} C_{ij}(\mathbf{p}) &= \sum_{\mathbf{u}} \sum_k A_{ik}\,(\mathbf{p} - \mathbf{u}) B_{kj}(\mathbf{u}) = \mathbf{A} * \mathbf{B} \\ &= \sum_k \sum_{\mathbf{u}} A_{ik}(\mathbf{p} - \mathbf{u}) B_{kj}(\mathbf{u}) = \sum_k A_{ik} * B_{kj} \end{aligned} \tag{1.6}$$

2. Matrix Rank-Order Operators

Erosions, dilations, and convolutions are easily extended to the matrix formalism. It would seem that rank-order operators would also be easy to define on matrices because they are a generalization of binary morphology and can be defined in terms of a thresholded convolution. However, there are two different ways to define rank-order operators that are both consistent with erosions and dilations of matrices.

As in the scalar definition of rank-order operators in Eq. (1.2), binary matrix rank-order operators are also represented as thresholded convolutions. One form for a matrix definition becomes

$$\mathrm{RO}(\mathbf{A};\mathbf{B}) = X_a\left(\sum_k (X_b(A_{ik} * B_{kj}))\right) \tag{1.7}$$

In this form, the two sums in Eq. (1.6) are separated into two parts with a separate threshold operation applied to each part. The threshold on the matrix convolution is analogous to the matrix erosion in definition (1.5). The first threshold X_a of the sum is analogous to the intersection over the erosions. The second threshold X_b is analogous to a set of scalar erosions expressed as thresholded convolutions.

A second definition for binary images can be written in terms of a thresholded matrix convolution:

$$\mathrm{RO}_t(\mathbf{A};\mathbf{B}) = X_t(\mathbf{A} * \mathbf{B}) \tag{1.8}$$

This form defines a homomorphism between the matrix and the scalar definition (1.2). Definition (1.8) will be used in the remainder of this chapter. One major reason that this definition is useful is that, in some applications, a matrix can have a large number of structuring elements and very few—sometimes only one—member in each structuring element. A rank-order operation is of marginal use on a very small set.

C. Rank-Order Hit-and-Miss Transforms

1. The Matrix Notation of Hit-and-Miss Transforms

Definition. Given matrix image $\mathbf{A} = [A_{ik}]$, define the posterior star operator by replacing each element of A with a submatrix that is a two-element row vector consisting of the image element and the complement of the image:

$$\mathbf{A}^* = [[A_{ik}A_{ik}^c]]$$

Example. Let $\mathbf{A} = [A_1\ A_2\ A_3\ A_4]$; then $\mathbf{A}^* = [A_1A_1^c\ A_2\ A_2^c\ A_3A_3^c\ A_4A_4^c]$.

Define a structuring element column vector, $\mathbf{B} = (H\ M)^T$. Let $\mathbf{I}$ be a scalar image. The hit-and-miss operator can easily be expressed as a vector inner product erosion:

$$\mathbf{I}^* \ominus \mathbf{B} = [I\ I^c] \ominus \begin{bmatrix} H \\ M \end{bmatrix} = (I \ominus H) \cap (I^c \ominus M)$$

This form of the hit-and-miss transformation allows it to become a natural part of the matrix formalism.

Definition. Define the anterior star notation as matrix of submatrices $\mathbf{B}_{ik}$, where each submatrix is a two-component column vector $\mathbf{B}_{ik} = [H_{ik}\ M_{ik}]^T$ and the upper and lower components are respectively a "hit" and a "miss" structuring element:

$${}^*\mathbf{B} = [[H_{ik}M_{ik}]^T$$

Suppose $\mathbf{A}$ is a two-component vector image and $\mathbf{B}$ is a 2×2 structuring element. A matrix erosion operation on $\mathbf{A}$ is given by

$$\mathbf{A} \ominus \mathbf{B} = [A_1\quad A_2] \ominus \begin{bmatrix} B_{11}B_{12} \\ B_{21}\langle \mathrm{B}_{22} \end{bmatrix}.$$

A matrix hit-and-miss operator is given by

$$\mathbf{A}^* \ominus {}^*\mathbf{B} = [\mathbf{A}_1^*\quad \mathbf{A}_2^*] \ominus \begin{bmatrix} {}^*\mathbf{B}_{11} & {}^*\mathbf{B}_{12} \\ {}^*\mathbf{B}_{21} & {}^*\mathbf{B}_{22} \end{bmatrix}$$

$$= [A_1\quad A_1^c\quad A_2\quad A_2^c] \ominus \begin{bmatrix} H_{11} & H_{12} \\ M_{11} & M_{12} \\ H_{21} & H_{22} \\ M_{21} & M_{22} \end{bmatrix}$$

2. The Scalar Hit-and-Miss Transform as a Thresholded Convolution

The scalar hit-and-miss transform defined by Serra can be written in terms of the thresholded convolution form of the binary rank-order filter according to definition (1.8):

$$\mathbf{I}^{*} \ominus {}^{*}\mathbf{B} = X_{|H|+|M|} (\mathbf{I}^{*} * {}^{*}\mathbf{B}), \tag{1.9}$$

where image $\mathbf{I}$ is a scalar image; the structuring element is $^{*}\mathbf{B} = [H\ M]^{T}$, and the threshold is the sum of the cardinalities of sets H and M. The threshold is maximum and refers to an erosion. For the scalar hit-and-miss transform expressed as a thresholded convolution, I, H, and M are not sets, but binary functions. H and M are used to represent either structuring elements or convolution kernels with weights of 0 or 1. Thus $I^{c}(\mathbf{x}) = 1 - I(\mathbf{x})$, and the hit-and-miss transformation can be written as $\mathbf{I}^{*} \ominus {}^{*}\mathbf{B} = X_{|H|+|M|} (I * H + (1 - I) * M) = X_{|H|+|M|} (I * H - I \ast M + |M|) = X_{|H|}(I * (H - M))$. The threshold is different from that in equation (1.9). This is because the star operator effectively forms a negated copy of the image and operates with a positive kernel (structuring element), whereas in the convolution definition the kernel values of M are negated rather than the image.

3. The Matrix Rank-Order Hit-and-Miss Transform

It is now easy to define a generalization of the scalar hit-and-miss transform to include both arbitrary matrices and a rank-order operation by using the matrix convolution $\mathbf{A} * \mathbf{B}$ in Eq. (1.6):

Definition. The hit-and-miss matrix rank-order filter is

$$\mathrm{RO}_{t}(\mathbf{I};\mathbf{W}) = X_{t}(\mathbf{I} * \mathbf{W}) \tag{1.10}$$

where matrix $\mathbf{W} = \mathbf{H} - \mathbf{M}$, and t is a threshold in the range $1 \leq t \leq |H|$. $\mathbf{H}$ and $\mathbf{M}$ are used to represent either structuring elements or convolution kernels interchangeably, but $\mathbf{W}$ cannot represent a structuring clcmcnt because it may contain weights of -1.

D. Relation to Neural Networks

A neural network that is useful in processing images can be defined where there is a neural cell at every pixel site. An artificial neural cell j in Figure 7 has a number of inputs indexed by i with input signals s_i and an output signal O_j. Using terminology from Rumelhart, Hinton, and McClelland [13], the various components of neural cell j include a *net*, *activation function* a_j and *output function* O_j. The net is the sum of the signals s_i multiplied by weights, where w_{ij} denotes the weight associated with input i to cell j.

$$\mathrm{net}_j = \sum_i w_{ij} s_i$$

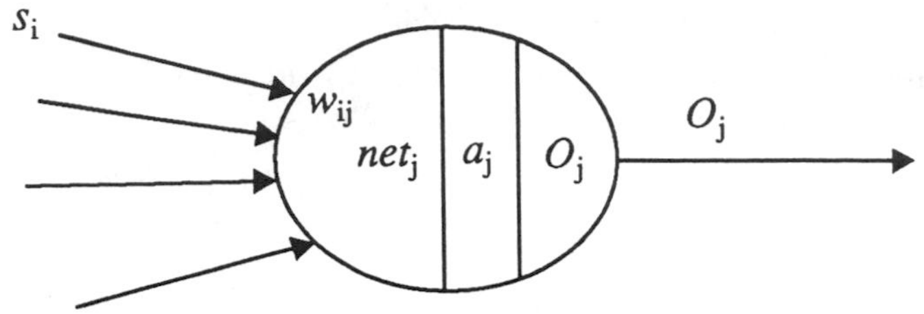

Figure 7. A single artificial neural cell.

Positive weights are called excitatory, and negative weights are called inhibitory. In this chapter the activation function is assumed to be linear and the output function is a simple threshold. Thus the output can be expressed as a threshold of the net:

$$O_j = X_t\,(\mathrm{net}_j) = X_t\Big(\sum_i w_{ij}s_j\Big) \tag{1.11}$$

Definition. An *iconic neural network* layer consists of neurons mapped in Z^2, so that one or more neural cells occupy each coordinate point of the image, and the layer is translation invariant [14].

A feedforward network consists of a number of cells in layers, where the inputs of a layer come from the previous layer and the outputs fan out to the next layer, as in Figure 8. Because of translation invariance, the weights and relative connection pattern are the same for all cells.

In ref. 15 it is shown that if a translation-invariant neural network has one neural cell at each pixel site and a simple threshold output function, then the inputs and outputs can be expressed as vector functions of position, where the vectors **u** and **p** span the domain of the input and output images, respectively:

$$O(\mathbf{p}) = X_t\Big(\sum_{\mathbf{u}} w(\mathbf{u})\ s\ (\mathbf{p} - \mathbf{u})\Big)$$

Comparing with Eq. (1.2), an iconic neural layer with binary signals and weights is isomorphic to a thresholded discrete convolution, or a rank-order operator on binary images.

$$O(\mathbf{p}) = X_t(w * s) \tag{1.12}$$

Translation invariance applied to a general network given by Eq. (1.11) has led to a layer of neurons that are all identical, although located at different coordinate positions, as shown in Figure 8.

The richness derived from a multiplicity of different types of neural cells has been lost, but it can be regained by putting a number of cells with different connectivity at each pixel site, as shown, for example, in Figure 9. Each signal input

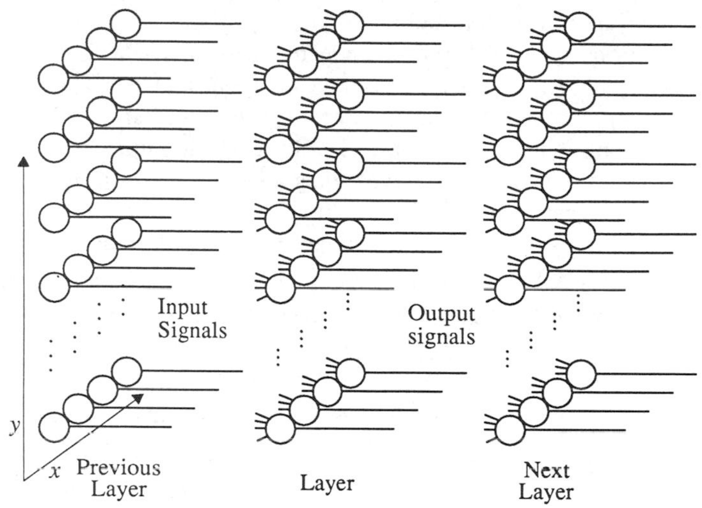

Figure 8. Iconic network.

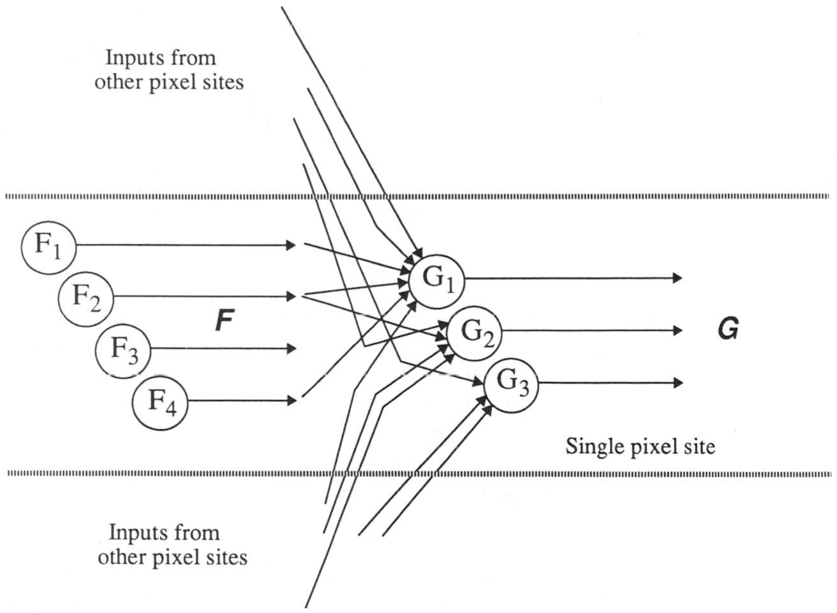

Figure 9. Neural cells operating on matrix images.

and output between layers is a vector, and the network diagram is more complicated. In the example, the input is a four-component vector image with feature components labeled $F_1, \ldots, F_4$. The output has three components labeled G_1, G_2, G_3. The weights are represented as a set of kernels that correspond to separate convolutions of each input with each output. The net for each output cell G_i is derived from a summation of different convolutions and has the form given in Eq. (1.6). In Figure 9, the set of kernels form a 4 by 3 weight matrix **W**, so that the operation of computing the net value in this example is written as $\mathbf{W} * \mathbf{F}$. Thus, iconic networks can be extended to operate with image vectors simply by substituting a matrix convolution in place of the scalar convolution in Eq. (1.12),

$$O(\mathbf{p}) = X_t\,(\mathbf{w} * \mathbf{s}) \tag{1.13}$$

The example in Figure 9 is written as $\mathbf{G} = X_t(\mathbf{W} * \mathbf{F})$. The signals and weights in Eqs. (1.11) and (1.13) are real numbers, whereas the ranges of images and structuring elements in Eqs. (1.2), (1.8), and (1.10) are binary. Thus, a rank-order operator on a binary image is a special case of an iconic neural network layer. A formal definition of a morphological network on binary integers can now be given as a special case of an iconic network:

Definition. A *binary morphological network* is an iconic network where weights are $\pm$ 1 or 0; signals are 1 or 0, and the output function is a threshold operation.

It is evident from Eq. (1.10) that the hit-and-miss matrix rank-order filter is equivalent to a neural network, where the positive weights refer to hit elements and negative weights refer to miss elements. A concatenated sequence of hit-and-miss rank-order operations is equivalent to a multiple-layer feedforward network. The input and output signals in a neural net are related to the input and output images in a morphological network. The weights and connectivity of a translation-invariant layer of neurons are related to the hit-and-miss structuring element. For both binary and gray-level images, a rank-order operator is equivalent to a matrix dilation if the threshold is 1 and equivalent to a matrix erosion if the threshold is maximum. Thus, scalar and matrix dilations and erosions are also special cases of iconic neural networks.

III. TRAINING BY SIMULATED ANNEALING

Training points that are members of structuring elements can be provided using a method called simulated annealing [16,17]. A relation between a network and a physical thermodynamic model is developed by explicitly defining temperature and energy in a manner such that the methods of thermodynamics can be applied. Details and examples of the use of simulated annealing in teaching network connections are covered in refs. 18 and 19.

A. Definition of Molecules and Quantum States

One object in Figure 1 is to be recognized, where other similar patterns are in the background. Figure 10 shows the corresponding edge gradient segmentation in four directions. Any of these edge pixels could be used as features for locating the object. That is, a structuring element made up of any of the points in these edges could be used to identify that pattern. It is not desirable to use all the edge points in defining a structuring element because as a detection mechanism the system would be too rigid and the computer time would be unnecessarily high. If a few points are used but are not well chosen, false recognitions become more probable. The goal of training is to chose optimum structuring element points according to some criteria.

The edges that define the object of interest are called spatial *quantum states*. During training, a number M of candidate points, called *molecules*, are chosen from these quantum sates to ensure that valid features are always used. The M molecular positions are the set of vectors of points that define the structuring element and become a single point in an M-dimensional *phase space*. The number of molecules will generally be much lower than the number of quantum states. The typical molecules are illustrated as asterisks in Figure 10.

B. Definition of Temperature and Energy

During training, various trial candidate structuring element points, or molecules, are chosen. A particular set of molecules are related to those in the previous trial in the sense that each molecule has moved somewhere in the neighborhood of the corresponding molecule in the previous trial. The size of that neighborhood of movement from which new trial molecules are chosen is called the *temperature*. The relation of temperature to this definition of molecular movement from one time increment to the next results in a simple physical picture: the degree of movement is analogous to the average kinetic energy of the system, which is related to the physical temperature. Thus, a very high temperature means that molecules are chosen almost at random from the quantum states and corresponds to a global scan over the molecular phase space. A low temperature means that

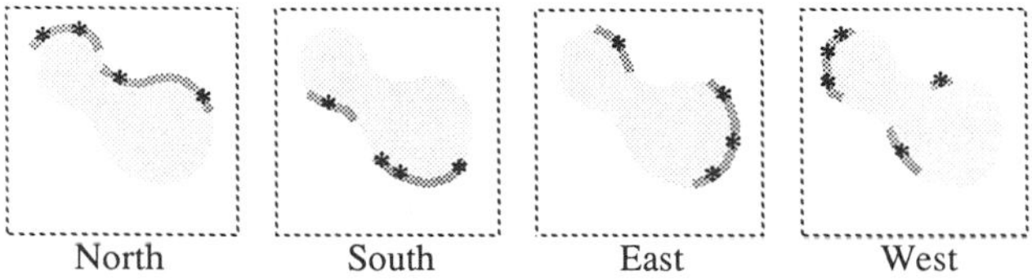

Figure 10. Edges: quantum states (lines) and molecules (asterisks).

there is only a slight difference from the previous configuration and corresponds to a scan about a local area in phase space to find a local minimum.

The potential *energy* of the system involving rank-order operators is defined to be the maximum value in the convolved image, except for the value that corresponds to the object being trained. These large values correspond to potential false recognitions. Figure 11 shows an example of a cross section of the convolution image. In this example there are 18 points in the structuring element. The cross section includes a peak at the object being trained (value, 18) along with three false near-recognitions that have the value 14. Thus the system energy has the value 14. For optimum convergence during annealing, the definition of energy is modified according to a method covered in refs. 18 and 19.

Simulated annealing has been used by Patarnello and Carnevali [20] to train Boolean networks. Their method is capable of finding good solutions to problems involving logic decisions. The temperature is used in determining the probability of energy transitions during annealing, but the temperature does not play a role in "molecular" movement. They use an energy defined as the number of wrong output bits during training, averaged over a number of training samples.

The definition of energy as an average does not eliminate false recognitions as effectively as the definition involving the maximum value used in simulated annealing. A single pattern very similar to the pattern under training can cause a very large narrow spike at the convolution output, where the vast remainder of the output image can be small. Thus, as shown in Figure 11, the average output may be very small even though there is a very serious potential for a false recognition. The concept of an energy as a maximum threshold will be carried over to Hebbian learning.

When inhibition is not important for the correct identification of a pattern, the training is close to ideal, although there is no real guarantee of convergence. The

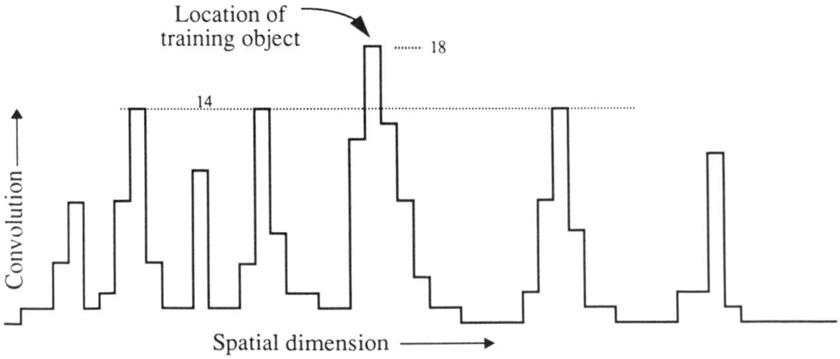

Figure 11. Convolution output with three near false recognitions.

idea of quantum states limits the choice of points for excitation structuring elements to those that are meaningful. Inhibition quantum states are defined as the points in a training image where there are no edges. The number of candidate points for an inhibition structuring element can be quite large, and many of the inhibition quantum states are not useful. The probability of randomly finding reasonable candidates for a number of inhibition points becomes vanishingly small. Training time would decrease if those inhibit candidates that are not useful in reducing false recognitions were eliminated. An effort to eliminate irrelevant inhibition candidate points leads to Hebbian learning.

IV. MORPHOLOGICAL HEBBIAN LEARNING

In the traditional development of morphology, images and structuring elements are defined as sets, and operations are logical. The operations performed during Hebbian learning are essentially numeric. Thus, the logical structure of morphology must be replaced by numeric functions in order to develop a theory of learning. The obvious numerical paradigm to pursue in morphological training is the development of iconic neural networks or rank-order operators as thresholded convolutions.

In supervised learning, the connection weights are updated according to a simple model. The concept of supervision is provided to the system in the form of a training template T, which is an image that resembles the desired output image. Let I be a training image that includes a sample of the pattern to be detected. The actual output image O results from a rank-order operation of I with structuring element B. B is incrementally changed according to a rule that attempts to find a B' that transforms input training image I to a training template T. That is, find a B' such that $\mathrm{RO}(I;B') = T$. As B changes during the training, the output change is $\mathrm{RO}(I;B) = O$. Thus the goal during training is to minimize the error between T and O. There are many possible B' that will detect the pattern of interest. Hebbian learning will attempt to find one of the most efficient B'.

A. Neural Learning Model

In neural networks, the error is defined as the mean square error of the differences between some training set t indexed by s and the network output o,

$$E_s = \frac{1}{2} \sum (t_s - o_s)^2$$

where the sum is over the image space [21]. However, in training morphological structuring elements, the error will have the same definition as energy in simulated annealing.

The delta rule [21] for training weights in neural networks is generally given by

$$\Delta W_{ij} = \eta(t_{sj} - o_{sj})i_{si} \tag{1.14}$$

where ΔW_{ij} is the connection strength from input neuron i to output neuron j, η is a learning rate constant, t_{sj} is a desired training output target indexed by s, i_{si} are the input signals for a particular training input indexed by s, and o_{sj} are the outputs computed according to the current set of weights. In practice, a training set i_{si} is input to the network; the outputs o_{sj} are computed using the existing weights; small adjustments ΔW_{ij} are computed and added to the weights W_{ij}; and the procedure is repeated until the error reaches some predetermined minimum.

The delta rule in Eq. (1.14) indexes neurons by labels i and j. For morphology or iconic networks, it is more appropriate to represent the cells as spatial functions, where $\mathbf{x}$ and $\mathbf{y} \in Z^2$ are vectors representing input and output coordinates, respectively. The input, output, and training images become $I(\mathbf{x})$, $O(\mathbf{y})$, and $T(\mathbf{y})$. The image-based delta rule becomes

$$\Delta W(\mathbf{y} - \mathbf{x}) = \eta(T(\mathbf{y}) - O(\mathbf{y}))I(\mathbf{x})$$

It is easier to compute new weights if they are referenced relative to the reference point of the structuring element. Let $\mathbf{r} = \mathbf{y} - \mathbf{x}$, where the domain of $\mathbf{r}$ is limited to the size of the structuring element

$$\Delta W(\mathbf{r}) = \eta(T(\mathbf{y}) - O(\mathbf{y}))I(\mathbf{y} - \mathbf{r}) \tag{1.15}$$

B. The Morphological Training Rule

The allowable weights in morphological networks are $+1$, 0, and -1. Because $\Delta W(\mathbf{r})$ as computed by the delta rule can be outside the allowed values of morphology, the delta training rule cannot be directly applied to morphology but can be taken as a guide to how the weights are to be handled during training. Another important difference in morphology training is that the incremental weights $\Delta W(\mathbf{r})$ are not added to the current weights $W(\mathbf{r})$ for every new input training image. A large number of incremental weights are accumulated to get a large sample of the training error before the old weights are adjusted.

There are special properties of the translation-invariant operators that allow a unique training procedure to accommodate the limited set of weights that the networks require. A large number of training images that are indexed by s in Eq. (1.14) are not used. Because each neuron (pixel) is associated with weights and connection patterns (structuring elements) that are identical throughout the image space, each pixel can serve as a separate training site for the same set of weights, and only one training image needs to be used. The delta rule in Eq. (1.15) is modified so that $\Delta W(\mathbf{r})$ is accumulated over all pixel coordinates $\mathbf{y}$ before the weights are updated:

$$\Delta W(\mathbf{r}) = \eta \sum_{\mathbf{y}} \left(\frac{T(\mathbf{y})}{\sum_{\mathbf{y}} T(\mathbf{y})} - \frac{O(\mathbf{y})}{\sum_{\mathbf{y}} O(\mathbf{y})} \right) I(\mathbf{y} - \mathbf{r})$$

where the denominators provide a normalization that is required because the number of nonzero training locations $T(\mathbf{y})$ is much smaller than the number of output locations where $O(\mathbf{y})$ is one.

In a neural network the new weights are the sum $W'(\mathbf{r}) = W(\mathbf{r}) + \Delta W(\mathbf{r})$ for the domain of $\mathbf{r}$. In a morphological network, only one weight is changed at a single position $\mathbf{r}'$, corresponding to the maximum value of $|\Delta W(\mathbf{r})|$ for all $\mathbf{r}$. If $\Delta W(\mathbf{r}')$ is positive, the new weight is $+1$ and corresponds to an excitation or "hit" structuring element. If $\Delta W(\mathbf{r}')$ is negative, the new weight is -1 and corresponds to an inhibition or "miss" structuring element. The constant η is immaterial for determining the maximum value of $|\Delta W(\mathbf{r})|$ and is therefore set to one. For the same reason, the normalization $\Sigma\, O(\mathbf{y})$ is factored out and set to one. The delta rule becomes

$$\Delta W(\mathbf{r}) = \sum_{\mathbf{y}} (N\, T(\mathbf{y}) - O(\mathbf{y}))I(\mathbf{y} - \mathbf{r}) \tag{1.16}$$

where N is the resulting normalization:

$$N = \frac{\Sigma\, O(\mathbf{y})}{\Sigma\, T(\mathbf{y})} \tag{1.17}$$

If $\Delta W(\mathbf{r})$ is maximum for more than one value of $\mathbf{r}$, then only one value of $\mathbf{r}$ is chosen at random, usually a value closest to $|\mathbf{r}| = 0$. Thus, the training rule for computing a new set of weights $W'(\mathbf{r})$ can be written as

$$W'(\mathbf{r}) = W(\mathbf{r}) + \delta\,(\mathbf{r} - \mathbf{r}')\left(\frac{\Delta W(\mathbf{r})}{|\Delta W(\mathbf{r})|}\right) \tag{1.18}$$

where $\delta(0) = 1$ is the delta function and $\mathbf{r}'$ is a location of the maximum $|\Delta W(\mathbf{r})|$. The output is defined by Eq. (1.2) as $O(\mathbf{y}) = \mathrm{RO}_t(I;W)$, where t is the highest threshold that gives at least one false recognition output. Thus, threshold t has the same definition as that for the energy in simulated annealing. Equations (1.16), (1.17), and (1.18) form what will be called the morphological training rule. Every time the training rule is applied, another point is added to the structuring element. The training rule is cycled a predetermined number of times to build up structuring elements with a specified number of points. As more structuring element points $\mathbf{r}'$ are added to $W(\mathbf{r})$, false recognitions will be eliminated and threshold t will lower. However, the training cycles can be terminated if the false recognition energy t drops below a specified level. Adding connections in neural networks during training has also been studied by Honavar and Uhr [22].

In morphological networks, the structuring element weights all start out at zero. Input image I has an area where a pattern is to be trained, centered at vector $\mathbf{t}$. The remainder of I corresponds to an area where false detections are to be minimized. The training template T is an image where $T(\mathrm{t}) = 1$ at the point $\mathbf{t}$ where the constructed structuring element is to be referenced. $T(\mathbf{t}) = 0$ elsewhere. The domain of the vector **r** relative to **t** is such that it covers only the pattern to be trained. The image O is computed according to the current $W(\mathbf{r})$.

C. Example of the Morphological Training Rule on a Scalar Image

Figure 12 shows an example of the relation of the various images in the training of weights. It is evident in the figure that the reference point of the structuring element in this case is at the lower left corner of the characters. In this example, the character F is being trained, as shown by the single point in the $T(\mathbf{y})$ image. Suppose that after a few cycles of training, the current structuring element weights cause a confusion of the character F with the characters E, H, P, and A, as evident by the output image $O(\mathbf{y})$.

1. Training an Inhibit Point

Figure 13 is a simple example of training using the morphological training rule. Example images are shown to illustrate various parts of the training algorithm. The characters to be trained are shown in Figure 12, where the F is to be trained in the context of the five other characters. In the example there are just five terms in the sum, $\Sigma_y NT(\mathbf{y}) - O(\mathbf{y})$), which are nonvanishing at only five coordinate values of y indicated by the five points in Figure 13. (Note: $N = 5$.) The five terms in the computation of $\Delta W(\mathbf{r})$ in Eq. (1.16) consist of shifts of the five associated characters in training image I to the reference point of an image plane $\Delta W(\mathbf{r})$. For clarity, the reflection about the origin $\Delta W(-\mathbf{r})$ is shown so that the $\Delta W(\mathbf{r})$ image registers with the imaged characters. The scale is enlarged and the sum is broken into two parts—the positive term, involving the training character, and the negative term, which is a sum that includes the four falsely recognized characters.

According to the morphological training rule given by Eq. (1.18), the weight to be added to the list of values that constitute the structuring element is within the region labeled -3, because it is the largest absolute value. Because it is

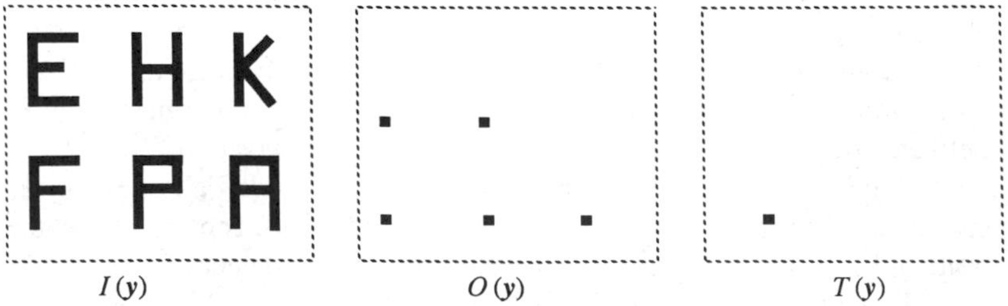

Figure 12. Training morphological weights.

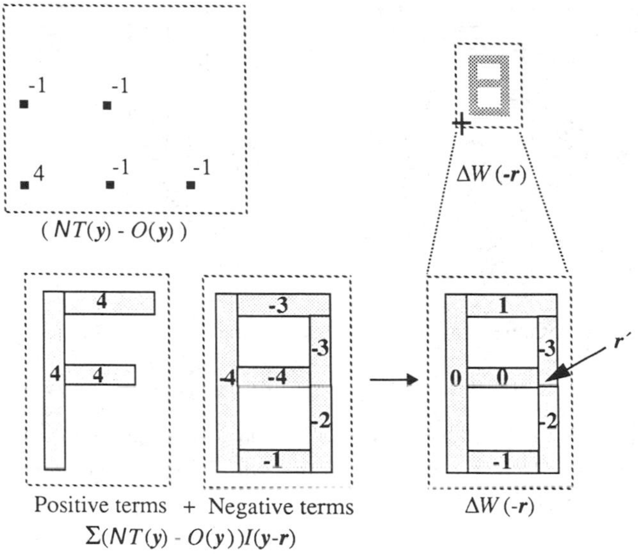

Figure 13. Example of the morphological training rule.

negative, an inhibitory point will be added to the structuring element. The value $\mathbf{r}'$ is arbitrarily chosen, usually closest to the origin $\Delta W(0)$, and is indicated by an arrow in Figure 13. Clearly, the chosen vertical stroke in this training rule is best because three false characters in Figure 12 have this stroke in common, and a inhibition in that area of the structuring element will eliminate the false recognition of those characters.

2. Training an Excitation Point

As a second example, suppose the character E were to be recognized. The sums in the training rule are those shown in Figure 14. In this case, the largest absolute value of $\Delta W(\mathbf{r})$ is a positive 4. An excitation point $\mathbf{r}'$, indicated by the arrow, is added to the structuring element. This is an optimal point because the E is the only character that has a horizontal stroke at that position, and an excitation point in that area will eliminate all false recognitions.

This method for finding best candidates for excitation and inhibition points is related to a method discovered by G. M. Berkin in 1974 [23]. The image $\Delta W(-\mathbf{r})$, for example, shown in Figure 14 is called a "*C* map," and the values in ΔW directly give the weights in a single pass. These weights were found to give excellent results for numeric networks. Multiple passes of the morphological training rule with a nonlinear energy definition are necessary to train morphological networks.

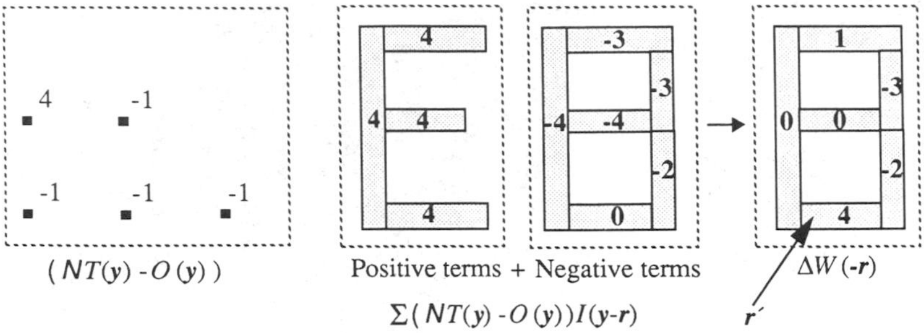

Figure 14. Training an E with the morphological training rule.

A number of heuristic modifications to the morphological training rule are needed for the procedure to result in a robust recognition. Other simplifications cut down the computation cost.

3. Thinning the Output

The recognition points shown as $O(\mathbf{y})$ in Figure 12 are generally blobs with a diameter of a few pixels. There is little advantage to using all points in a blob for the computation of ΔW. To cut down on computational cost, these blobs are thinned to single points before computing the final sum. The training image in Figure 12 consists of a single point in the image. Ordinarily this would imply that any other detections in the neighborhood of that training point would be false detections. The thinning operation on output $O(\mathbf{y})$ effectively avoids this problem. It is assumed that the thinned $O(\mathbf{y})$ overlies the $T(\mathbf{y})$ exactly at that point.

At the beginning of training, when there are very few points in the structuring elements, a large number of points in output image $O(\mathbf{y})$ will respond, and thereby result in a large number of terms in the sum of Eq. (1.16). The terms near the reference points of the structuring elements of each character are most important. At first, the domain of the sum is limited so that points at a distance greater than **d** from the reference points of characters are ignored. The sum in Eq. (1.16) is rewritten as

$$\Delta W(\mathbf{r}) = \sum_{|\mathbf{y}-\mathbf{R}_i|<\mathbf{d}} (NT(\mathbf{y}) - O(\mathbf{y}))I(\mathbf{y} - \mathbf{r}) \tag{1.19}$$

where $\mathbf{R}_i$ indicates a set of valid reference points for the character set and **d** is a small radius. As training progresses, the value of **d** increases so that a larger area of potential false recognitions can be handled.

4. Limiting the Location of Structuring Element Points

Excitation and inhibition points should not be chosen too close to the edges of the desired shape so that slight rotations or scale changes will not cause recognition to be lost. The range of excitation and inhibition points can be limited by dividing the computation into two steps as shown in Figure 15. Before computing the maximum value of ΔW, the character under training is first thinned so that the excitation points will be chosen from the center of the pattern. Also, the character is first thickened before the minimum ΔW value is found so that inhibition points will be chosen away from the character under training and thus prevent self-inhibition.

The amount of thickening and thinning creates an area of "dead zone" around edges so that structuring element points are not chosen from that zone. The size of the zone determines how much change in scale or rotation a pattern is allowed. This mechanism provides some control over a trade-off between the degree of confusion of the trained pattern with a similar one and the degree to which a pattern is allowed to deviate from the training image.

There is nothing so far in the morphological training procedure to prevent the same point from being chosen more than once. If the points are to be used in an erosion structuring element, choosing the same point more than once will have no effect except for extra unnecessary computation. In a neural net or rank-order operator, the extra point will act as a single point with a higher weight. Viewing the training operation thermodynamically as in simulated annealing, ΔW is equivalent to ΔE, where E represents energy as defined earlier. One way to re-

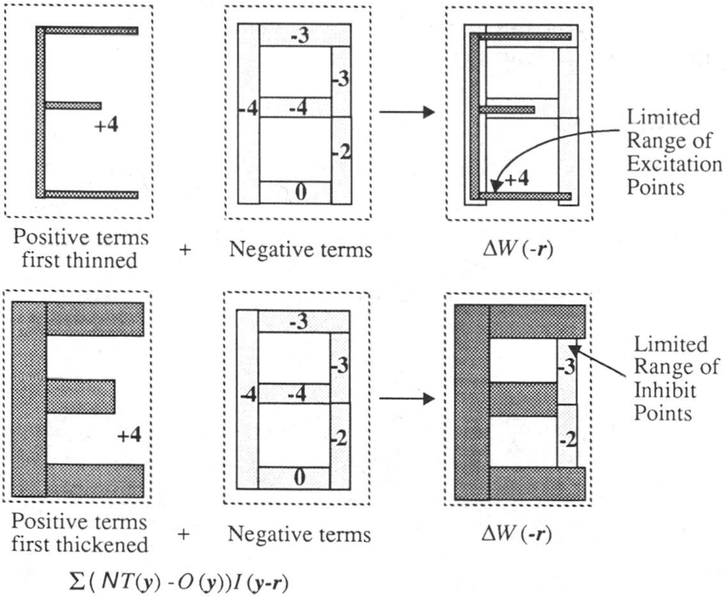

Figure 15. Avoiding points too close to character edges during training.

duce the overlap of structuring element points (molecules) at any location is to define a repulsive force. Then ΔE will exhibit a higher energy for points that are too close. Experiments show that a very light repulsive force is all that is necessary. A light repulsive force potential at $\mathbf{r}'$ with an octagonal shape of unit height is shown in Figure 16 for one point and is a constant force up to a radius of $\mathbf{r}_f$ pixels:

$$\mathbf{F}(\mathbf{r}) = 1 \text{ if } |\mathbf{r} - \mathbf{r}'| < \mathbf{r}_f, \text{ 0 otherwise}$$

where $\mathbf{r}'$ is an element of the current structuring element set.

The total repulsive energy for all points is computed by creating an image composed of hit-and-miss structuring element points $H \cup M$, dilating that image by the octagonal structuring element R in Figure 16, and then subtracting the dilated image from the ΔW image. The ΔW computation in Eq. (1.16) has the subtracted repulsive term

$$\Delta W_f = (H \cup M) \oplus R \tag{1.20}$$

The repulsion has the property of diminishing the probability that structuring element points are in a close proximity up to radius $\mathbf{r}_f$. No further influence is exhibited outside that range. The light repulsive force is used only for breaking a tie. In Figure 15 there are a number of points on the lower horizontal stroke that have the same maximum weight. Multiple structuring element points that might be chosen on that stroke by successive training cycles will be spaced $\mathbf{r}_f$ pixels apart. The only reason for reducing overlap is to reduce the sensitivity of recognition to a missing feature in any one area of the character.

D. Training Matrix Structuring Elements

The morphological training rule discussed so far is for scalar images and structuring elements. The extension to matrices is rather straightforward. The matrix structuring element is taught one column at a time. The input training image is a row vector $\mathbf{I}$. The column of the structuring element matrix that is to be taught is $\mathbf{W}_i$. The output image component O_i is the inner product of $\mathbf{I}$ with $\mathbf{W}_i$; O_i =

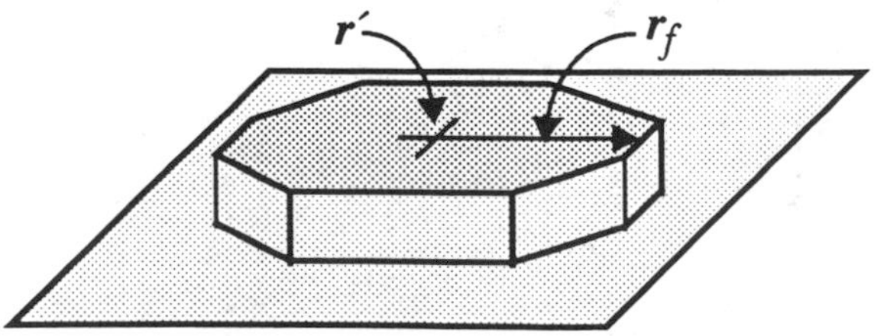

Figure 16. Repulsive force between structuring elements.

$RO(\mathbf{I};\mathbf{W}_i)$ and is a scalar. For training purposes, the matrix rank-order operator output is written as a row vector of inner products, $\mathbf{O} = X_t(\mathbf{I} * \mathbf{W}) = [X_t(\mathbf{I} \cdot \mathbf{W}_1)\ X_t(\mathbf{I} \cdot \mathbf{W}_2) \cdots X_t(\mathbf{I} \cdot \mathbf{W}_M)]$. The training image is the target image for the particular row being trained: T_i. Each weight increment becomes $\Delta\mathbf{W}_i(\mathbf{r}) = \Sigma\ (NT_i(\mathbf{y}) - O_i(\mathbf{y}))\mathbf{I}(\mathbf{y} - \mathbf{r})$, where the normalization is $N = \Sigma\ O_i(\mathbf{y})/T_i(\mathbf{y})$. The weight column vector $\Delta\mathbf{W}_i = [\Delta W_{ij}]$ consists of a number of single-element rows that must be trained. Let $\mathbf{r}'$ and j' be the spatial location and row index, respectively, of the weight component $\Delta W_{ij'}(\mathbf{r}')$ that has the maximum absolute value over all locations $\mathbf{r}$ and rows j. The weight component with the largest absolute value is changed in a manner similar to scalar morphological training:

$$W'_{ij}(\mathbf{r}) = W_{ij}(\mathbf{r}) + \delta\,(\mathbf{r} - \mathbf{r}')\,\delta\,(j - j')\left(\frac{\Delta W_{ij}(\mathbf{r})}{|\Delta W_{ij}(\mathbf{r})|}\right) \tag{1.21}$$

The matrix rule for updating the weights includes a repulsive force similar to that in Eq. (1.20) and is given by the union of all hit-and-miss structuring elements in the column $\mathbf{W}_i$, so that

$$\Delta W_f = (\cup_i(H_i \cup M_i)) \oplus R \tag{1.22}$$

The repulsive force in Eq. (1.22) applies to any weight in column vector $\Delta\mathbf{W}_i$, so that a point $\mathbf{r}$ in one component of a structuring element has a repulsive force on all other components of the structuring element in the same column, at the same point. However, the repulsive force should be made stronger if a new candidate point is not only at the same point as a current structuring element point but also in the same component. A multiplicative term is defined to enhance the repulsive force. The revised delta weight $\Delta W'(\mathbf{r})$ that includes repulsive forces used in training matrix structuring elements is

$$\Delta W'_{ij}(\mathbf{r}) = (\Delta W_{ij}(\mathbf{r}) - \Delta W_f(\mathbf{r}))\,(1 - 0.1\,\delta(\mathbf{r} - \mathbf{r}')\,\delta(j - j')) \tag{1.23}$$

where $\Delta W_f(\mathbf{r})$ is given in Eq. (1.22). Equation (1.23) is applied to equation (1.21) to define the matrix morphological training rule with repulsive forces. The repulsive force prevents too many points of a matrix structuring elements from accumulating near one spatial location, so that recognition of an object will not be too severely affected if a part of the object is occluded.

E. The Computation

1. Hardware The most efficient type of computer for training and running morphological networks is a bit-serial fine-grained array architecture. Fountain [24] has compiled a survey of various processors with this type of architecture. Systems based on single processors or digital signal processing (DSP) chips are most suitable for word-size arithmetic and not single-bit incrementing and decrementing needed for hit-and-miss operators. Pipeline processors can effectively

handle morphology in 3×3 neighborhoods but not extended sparse neighborhoods. Examples of fine-grained array architectures are the two-dimensional mesh-connected systems such as the Goodyear Massively Parallel Processor [25] and the Distributed Array Processor (DAP)[26] from Active Memory Technology. A simpler but lower-cost system is the linear array, such as the Applied Intelligent Systems AIS-5000 [27]. In this case there is one processor for each column in the image.

A linear array with only 64 fine-grained processors (AIS-3000) [28] can compute four edge directions and network computations on complex images in less than 100 milliseconds. Training characters according to the matrix training rule in equation (1.21) with the same machine requires about 25 seconds per character. Training time by simulated annealing is about 45 seconds. Larger fine-grained computer systems would train and run much faster.

2. The Algorithm

Rank-order operators with segmented edge direction structuring elements can be programmed in these systems by computing the edges and directions and then storing them in four separate image bit planes. An accumulator is defined as a byte plane to count the points in a structuring element for a rank-order operator. To pick up various points in the structuring element, the edge planes are shifted by vectors $\mathbf{v}$ that define the points of the structuring element and then summed into the stationary accumulator, which is the convolution output. Figure 17 shows the memory organization and the process of accumulating a structuring element point for a feature plane. In this algorithm all coordinate positions accumulate a new structuring element point during each loop, so that the convolution is built up gradually over the whole image. Training according to the morphological training rule uses a very similar method. The images $I(\mathbf{y} - \mathbf{r})$ in Eq. (1.16) are translated by amount $\mathbf{y}$ given by the location of nonzero points in the factor $NT(\mathbf{y}) - O(\mathbf{y})$ and summed to a small image $\Delta W(\mathbf{r})$.

V. AN EXAMPLE OF THE MORPHOLOGICAL TRAINING RULE

For the simple examples in Section IV, the morphological training rule on scalar images chooses structuring element points that are most reasonable and effective. It is much more difficult intuitively to evaluate the operation of morphological training on vector images with matrix structuring elements using the rule in Eqs. (1.21) and (1.23). In this section a complex example of training a set of characters will be evaluated. The example to be given in this section has been given in detail in terms of its morphological structure in ref. 15. In this section the structure of the morphological network will be briefly outlined and emphasis will be on the effectiveness of training matrix structuring elements.

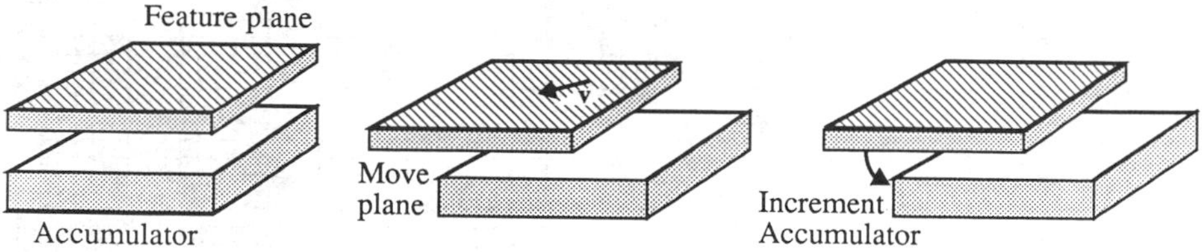

Figure 17. Convolution operations on a massively parallel processor.

A. The Morphological Network Structure

A complex example of character recognition will be presented. All images are row vectors and the structuring elements are matrices. It is assumed that the input image has undergone edge gradient segmentation and is a vector labeled $\mathbf{E} = [E_N \, E_S \, E_E \, E_W]$. The component indexes are labeled by the compass directions of the edges. There is first a matrix erosion operation on **E** that is not trained by an automatic procedure. The purpose is to transform the four component vector image **E** to an eight component vector image $\mathbf{S} = [S_1 \, S_2 \ldots S_8]$ by an erosion of an 8×4 matrix structuring element **F**: $\mathbf{S} = \mathbf{E} \ominus \mathbf{F}$. The purpose of **F** is to transform the image segmented by edges to a new image segmented into line and corner segments. It is easiest to visualize the components of **F** in a pictorial form as shown in Figure 18a

Each column of **F** is a separate segmentation operator that operates on all four edge directions and outputs a separate component of the output image. Because there are eight columns in **F**, there are eight components in the new output image. The points in the structuring elements in **F** contact edges of an image at the various points shown in Figure 18b. The first four columns detect thin lines in four directions and the last four detect corner and tee connections. A typical character is shown lightly shaded in Figure 18c with an overlay of each of the eight outputs after edge detection and an erosion by the **F** matrix.

Next, each component of image **S** is dilated by an 8×8 diagonal matrix structuring element **B**, shown in Figure 19. $\mathbf{S}' = \mathbf{S} \oplus \mathbf{B}$. The result of the dilation of the example in Figure 18c is shown in Figure 20. The purpose of dilating by **B** is to repair image noise dropouts and relax the rigidity of the training image to allow detection of characters that are slightly rotated or scaled. The dilation is a diagonal matrix and does not change the interpretation of the features of the image components. Dilations between layers are also used in the Neocognitron [29]. A dilation should not be used on threshold-segmented scalar images because the characters could become hopelessly confused.

The last step is to train the final structuring elements that will recognize characters. The training proceeds in a manner that is very similar to the examples in

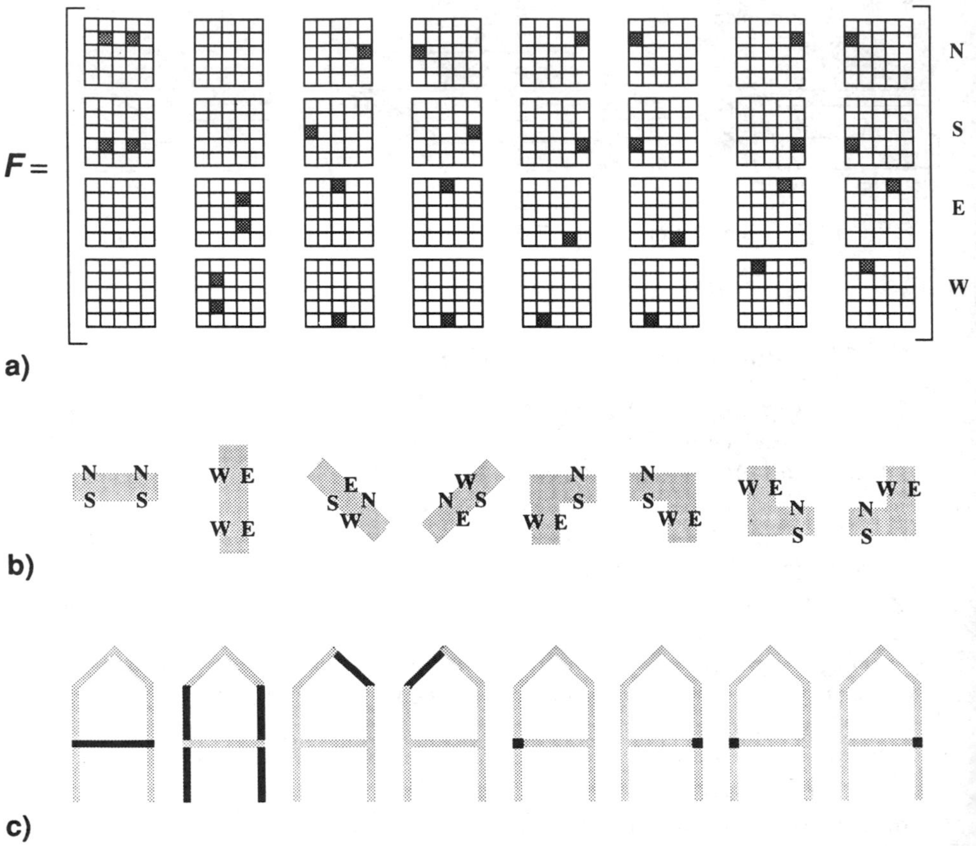

Figure 18. The F matrix erosion: (a) 8×4 structuring element; (b) contact points of F on image edges; (c) eight-component output image.

Figures 13 and 14, except that there are eight hit-and-miss components in a column to be trained rather than one, there are 36 characters in the alphanumeric set to be individually trained, and the matrix training rule in Eqs. (1.21) and (1.23) is used. The hit-and-miss structuring elements to recognize the character A is denoted ${}^*\mathbf{R}_A = [\,{}^*\mathbf{R}_{A1}\ {}^*\mathbf{R}_{A2} \ldots {}^*\mathbf{R}_{A8}]^T$. The set ${}^*\mathbf{R}_A$ is a structuring element column vector and is called the recognition vector for character A. A nonnull blob in the scalar image

$$O_A = \mathrm{RO}(\mathbf{S}'^*;{}^*\mathbf{R}_A) = X_t(\mathbf{S}'^* * {}^*\mathbf{R}_A) \tag{1.24}$$

indicates the presence of an A. O_A is called the classification plane for character A.

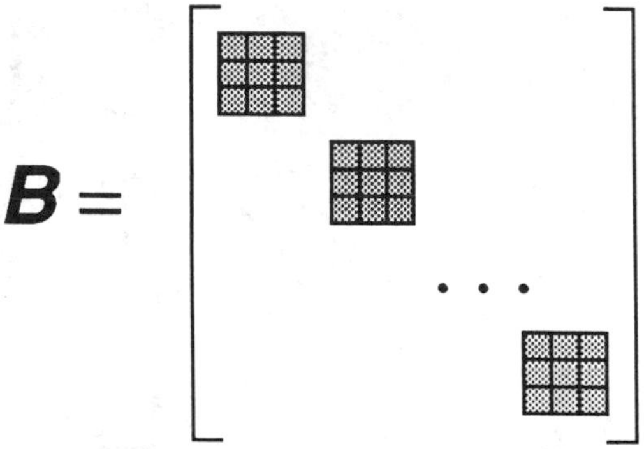

Figure 19. Matrix dilation structuring element $\boldsymbol{B}$.

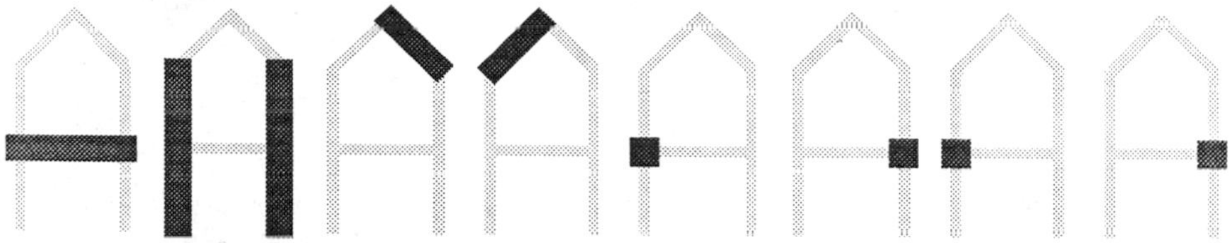

Figure 20. The result of dilating an eight-component vector image by matrix $\boldsymbol{B}$.

There are 35 more structuring elements to recognize the other alphanumeric characters and they are similarly denoted as $^*\mathbf{R}_B\ ^*\mathbf{R}_C \ldots$. There are 16 components in each recognition vector $^*\mathbf{R}_X$. All of the column vectors $^*\mathbf{R}_X$ can be grouped into one large 36×16 matrix $^*\mathbf{R} = [^*\mathbf{R}_A\ ^*\mathbf{R}_B\ ^*\mathbf{R}_C \ldots]$. Classification for all the characters can be written as $\mathbf{O} = \mathrm{RO}(\mathbf{S}'^*;^*\mathbf{R})$, where $\mathbf{O}$ is a classification vector. A nonnull blob in image component x in $\mathbf{O}$ indicates the presence of character x. In summary, the above structure for character classification can be written as $\mathbf{O} = \mathrm{RO}(\mathbf{S}'^*;^*\mathbf{R})$, where $\mathbf{S}' = \mathbf{S} \oplus \mathbf{B}$, and $\mathbf{S} = \mathbf{E} \ominus \mathbf{F}$. The matrix $^*\mathbf{R}$ is trained one column (i.e., one character) at a time according to the matrix training rule in Eq. (1.21).

B. Character Recognition Training

The character set to be trained is an OCR font used on integrated circuit wafers and is shown in Figure 21. The character set is an excellent test of the morphological training rule because there is a wide variety of problems. Some characters are very easily distinguishable from all other characters and require only a small number of points in the structuring element. Other characters need a large num-

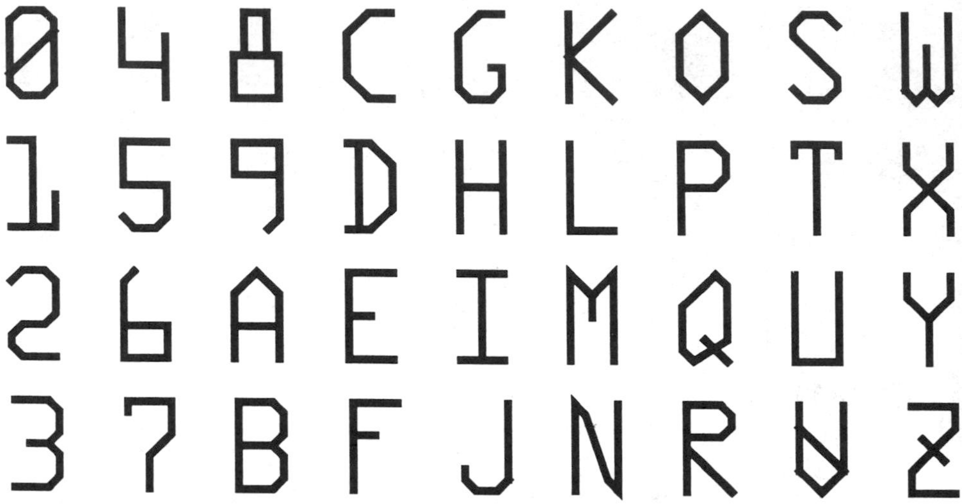

Figure 21. Alphanumeric OCR character font to be trained.

ber of points and a large percentage of inhibition points in order to distinguish them from a set of troublesome confusing characters. In this application there is only one computer-generated training example of each character. The purpose of training is to learn how to robustly distinguish one character from another, and not to train an individual character under different noise and degraded conditions. The technique of using one "golden" training image relies on the robustness of the training procedure and the several layers in the morphological network to provide good recognition under degraded conditions. Figure 22 shows a photograph of characters to be recognized as they occur in practice. These characters are in a highly noisy background but can be correctly classified after training, even though the training takes place with a clean set of characters.

All $^*\mathbf{R}_x$ recognition vector structuring elements operate on vector image $\mathbf{S}'$ using a hit-and-miss rank-order operator provided by a thresholded convolution. Because the convolution kernel consists of weights that are $+1$ and -1, corresponding to the hit-and-miss transform, the convolution can be computed by increment and decrement operations. An accumulator image is initially set to zero and is incremented or decremented according to the kernel as it is scanned across the image. A threshold is then applied to the accumulator image. Because the operator is an erosion, or "nearly" an erosion, the threshold will be close to maximum, which is the number of points constituting the structuring element.

Two examples of character training will be shown in detail. Progress of the training is tabulated in a manner that shows the efficiency of new structuring element points as they are added. For a given structuring element, the character

Figure 22. TV camera image of characters that occur in practice.

undergoing training will always respond with the maximum convolution output. Other characters will respond with less and less output as training progresses. A typical histogram of the convolution output value in Figure 23 shows the maximum false recognition value and other lower values. The scale of the histogram is such that the farthest point on the right refers to a convolution output equal to the maximum value and is the cardinality of the structuring element. The single histogram value at the far right refers to the trained character and has a full response. The next plotted value on the right refers to the largest false recognition convolution output and therefore represents the energy.

Table 1 shows a tabulated form of the convolution histogram as it changes for each additional point in the structuring element during the training cycles for the character 8. The last four columns in the table show the best structuring element point added according to the matrix morphological training rule. E and I indicate whether the added point is an excitation or inhibition. f_1 through f_8 indicate the feature component in the matrix structuring element ${}^*\mathbf{R}_8$ that is added. The last two columns in Table 1 give the x and y coordinate positions of the added point. In the first row of Table 1, the histogram shows that all 36 character reference locations respond maximally because there are initially no points in the structuring element. The entries in the histograms always add up to 36 because the histogram entry for each character is the maximum convolution value within distance $\mathbf{d}$ from training point $\mathbf{R}_i$ for each character indexed by i, as given in Eq.

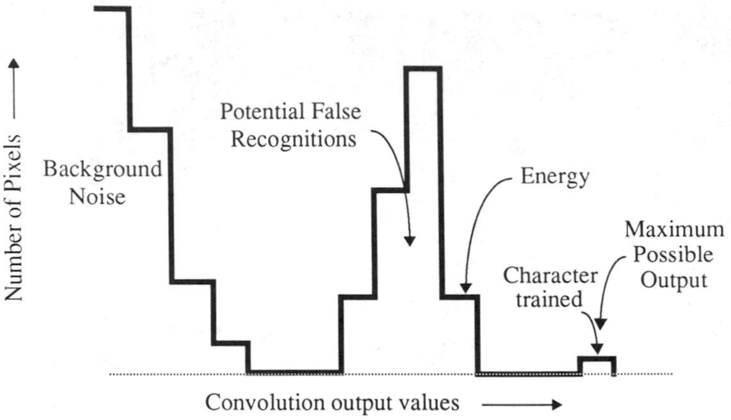

Figure 23. Typical histogram of the convolution values accumulated over an entire training image.

Table 1. Convolution Histograms and Structuring Elements for Training Character 8

Convolution histogram								Feature	Coordinate
0	0	0	0	0	0	0	36	E f_8	11, 18
0	0	0	0	0	0	21	15	E f_1	13, 16
0	0	0	0	0	13	18	5	E f_2	6, 23
0	0	0	0	1	19	15	1	E f_4	5, 17
0	0	0	1	12	22	0	1	E f_6	11, 29
0	0	1	10	22	2	0	1	E f_7	8, 18
0	1	10	11	13	0	0	1	E f_8	15, 4
0	7	14	12	2	0	0	1	E f_2	12, 23
4	10	17	4	0	0	0	1	E f_3	15, 17
11	18	5	1	0	0	0	1	E f_2	12, 26
17	14	4	0	0	0	0	1	E f_7	4, 4

(1.19). If there is more than one coordinate position with the same maximum value within distance **d**, the coordinates nearest the training point center for that character are taken.

The training rule determines that the best point to add is an excitation for feature 8 at $(x,y) = (11,18)$. It is seen on the second row of Table 1 that on the basis of the structuring element with that one point, many characters still give false recognitions. On the fourth row, after three structuring element points are added, only one character responds with maximum output, namely the character

8 being trained. The other characters in the training set will not respond if the threshold is set to maximum (an erosion). Thus, for an erosion operator, the training could be considered to be completed. However, this is too risky, and a few more training cycles result in no false recognitions even when the threshold is dropped to four below maximum. Now there is a considerable margin for missing features of a character. There are 11 excitation points in the structuring element. The coordinate positions of the structuring elements in Table 1 are plotted in Figure 24, where the plotted symbols are mnemonics for the feature components shown in Figure 18b.

The next example shows training for $^*\mathbf{R}_F$, for the recognition of the character F in Table 2. Eight inhibition and 10 excitation structuring element points are needed to give a performance similar to that for $^*\mathbf{R}_8$. There are more points in the structuring element for $^*\mathbf{R}_F$ than there are for $^*\mathbf{R}_8$, and a large percentage of inhibit points, because the F is confused with many other characters. Table 2 shows that three or four cycles are needed for each unit reduction of energy. A symbolic plot of the vector structuring element $^*\mathbf{R}_F$ is shown in Figure 25. Symbols that are circled are inhibition points.

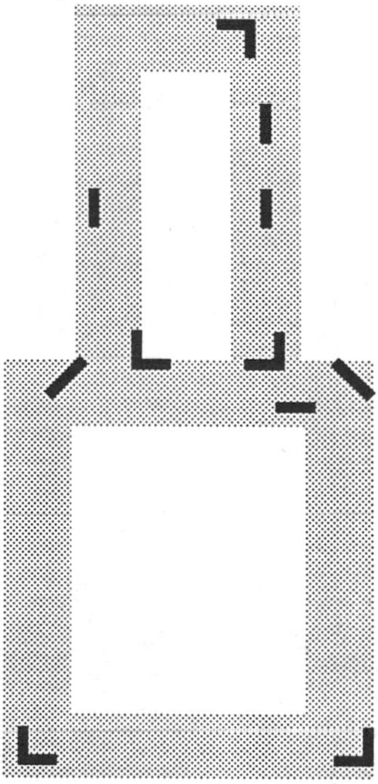

Figure 24. Locations of structuring elements points for $^*\boldsymbol{R}_8$.

Table 2. Training Cycles for the Recognition of the Character F

Convolution histogram								Feature	Coordinate
0	0	0	0	0	0	0	36	E f_7	5, 22
0	0	0	0	0	0	19	17	E f_1	18, 28
0	0	0	0	0	8	25	3	I f_2	17, 3
0	0	0	0	0	8	26	2	I f_3	10, 11
0	0	0	0	0	8	27	1	E f_5	5, 27
0	0	0	0	8	19	8	1	E f_5	6, 18
0	0	0	7	13	12	3	1	E f_2	3, 5
0	0	0	12	17	6	0	1	I f_1	16, 19
0	0	0	12	17	6	0	1	I f_2	19, 24
0	0	1	15	15	4	0	1	E f_1	10, 19
0	0	14	11	9	1	0	1	I f_1	9, 4
0	0	15	12	8	0	0	1	E f_1	17, 28
1	9	13	10	2	0	0	1	I f_1	18, 4
1	9	13	11	1	0	0	1	I f_5	5, 12
1	9	13	12	0	0	0	1	E f_1	19, 28
10	0	20	5	0	0	0	1	E f_5	6, 18
10	14	10	1	0	0	0	1	I f_7	5, 7
10	14	11	0	0	0	0	1	E f_1	9,28

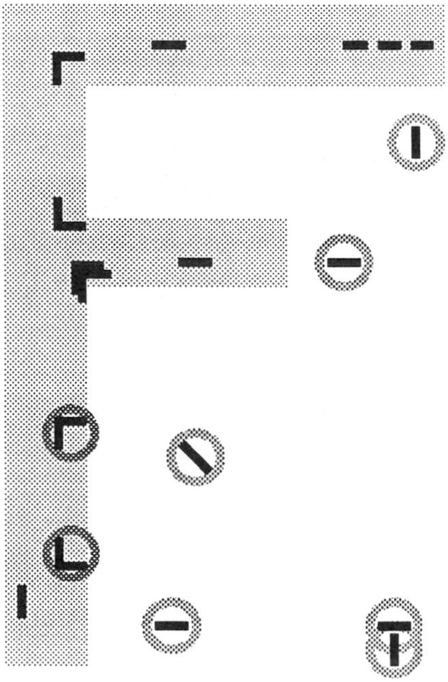

Figure 25. Locations of structuring elements points for $*\boldsymbol{R}_F$.

The photograph in Figure 26 is the negative of the $\Delta W(\mathbf{r})$ image for the training of vertical strokes in the character 8. The vertical strokes of the characters providing negative terms in $\Delta W(\mathbf{r})$ are the two wide bright vertical lines. The two dark vertical lines in the lower half of the image are the thinned version of the positive terms, which are also shown in a different example in Figure 15. Excitation points at the thinned locations at the lower half of the character are very poor candidates because many characters share vertical strokes in those positions. The best excitation point for a vertical line in the character 8 is in the upper right half of the figure, because, as seen in Figure 24, the vertical strokes are closer together at the top than at the bottom of the character and no other character shares a vertical stroke in the upper right at that position. The ΔW corresponding to seven more features as shown in Figure 18b must be computed before the matrix training rule in Eq. (1.21) can be applied. The eight features are indexed by j in Eq. (1.21).

An early step in the training of the character A is shown in the photograph in Figure 27, where the training alphabet is shown as light strokes throughout the image. The convolution part of the output image, $\mathbf{S}'^{*} * {}^{*}\mathbf{R}_{\mathrm{A}}$ in Eq. (1.24), is shown overlaid on the training alphabet and, in practice, is much more cluttered than the equivalent image $O(\mathbf{y})$ shown in Figure 12. At this early stage in training, the maximum convolution value, or "energy," is fairly high, and there are many characters with a large convolution output. Only convolution values greater than seven below maximum are shown.

At a later stage of training the character A, the corresponding $O(\mathbf{y})$ image is shown in Figure 28. There are very few areas where the convolution output is within seven units below maximum.

With a low-cost Applied Intelligent Systems AIS-3000 array processor, the $\Delta W(\mathbf{r})$ computation in Eq. (1.21) for each feature, for example as shown in Figure 26, requires about 70 milliseconds. There are eight excitation and eight inhibition ΔW computations for each of the eight features that are shown in Figure

Figure 26. Photograph of $\Delta W(r)$ for vertical strokes in the character 8.

Figure 27. Photograph of the training of the character A.

18b. Thus the total time to compute all the features (indexed by j) needed for one cycle of the training rule in Eq. (1.21) is a little over 1 second. About 20 points are needed for each structuring element for each character. The training time for each character is about 25 seconds.

VI. CONCLUSIONS

Although much of the discussion in this chapter involves the language and character of neural networks, the tools of morphology are predominant. Morphology operations, such as thickening and thinning, aid in training by avoiding marginal connections. Octagonal repulsive forces are generated with dilations. Morphological networks are useful for real-time industrial applications and can be succinctly and eloquently formalized in terms of matrix morphology, where each layer in a network is defined with a matrix structuring element. A concatenated sequence of matrix hit-and-miss rank-order operations is equivalent to a multiple-layer feedforward network. Many structuring element components must be defined. Some components are elementary and can easily be constructed by hand, but recognition vectors for character classification are too complex to be determined without an automated training procedure.

Simulated annealing and Hebbian learning have been discussed. In simulated annealing, a constant number of structuring element "molecules" undergo a re-

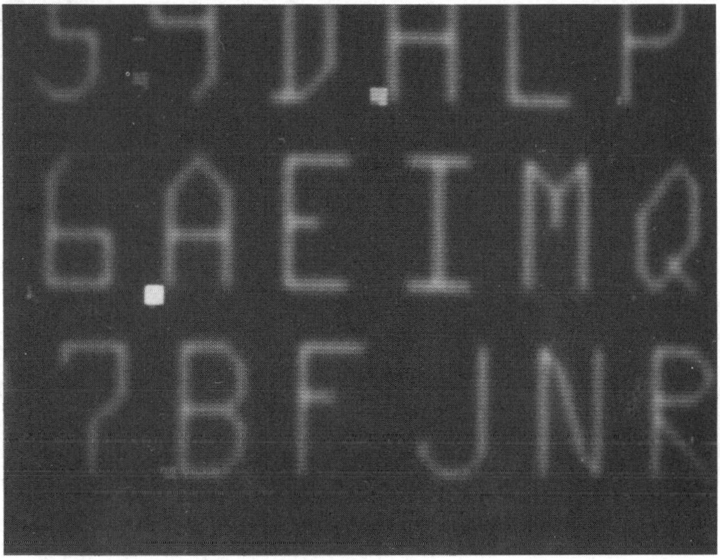

Figure 28. Training the character A near completion.

configuration to minimize the energy, whereas in morphological Hebbian learning, points are added one at a time and are chosen to minimize the energy at each step. A modification of the delta training rule of neural networks allows the positive and negative unit weights used in morphological training to be handled. The energy, defined as a maximum, is very nonlinear and reflects that fact that morphology itself is nonlinear. The energy definition directly emphasizes the reduction of false recognitions.

Translation-invariant neural networks with binary signals, and weights that are $+1$, 0, or -1, are the same as hit-and-miss rank-order operators, so that "hit" and "miss" are equivalent to "excitation" and "inhibition." The best features of mathematical morphology, such as the simplicity of the hit-and-miss transform and binary rank-order operators, are combined with the best features of the more arithmetic nature of neural networks, such as the concept of supervised learning. Morphology theory is generally based on set theory, but that must be abandoned in order to provide a more arithmetic basis for training structuring elements.

The morphological training rule has been found to be very effective at choosing the structuring element points that are most efficient at detecting a pattern and avoiding false recognitions. Although the computational complexity of training and running morphological networks looks formidable, it lends itself very readily to massively parallel computers. Even the simplest fine-grained array processor can train in less than half a minute and run in a small fraction of a second.

Mathematical morphology has generally been considered to be a low-level process in machine vision or image understanding. However, it is becoming more apparent, with the morphological networks discussed here and many other recent advances in this field, that morphology has the depth and flexibility required for mid- to high-level vision processing.

REFERENCES

1. Hebb, D. O., *The Organization of Behavior*, Wiley, New York, 1949.
2. Matheron, G., *Random Sets and Integral Geometry*, Wiley, New York, 1975.
3. Serra, J., *Image Analysis and Mathematical Morphology*, Academic Press, London, 1982.
4. Heygester, H., Rank filters in digital image processing, in *Proceedings of the 5th International Conference on Pattern Recognition*, Miami Beach, Florida, 1980, pp. 1165–1167.
5. David, H. A., *Order Statistics*, Wiley, New York, 1970.
6. Preston, K., Jr., Ξ Filters, *IEEE Trans. Acoust. Speech Signal Process., ASSP-31*(4):861–876 (1983).
7. Justusson, B. I., Median filtering: statistical properties, in *Two-Dimensional Signal Processing II* (T. S. Huang, ed.), Springer-Verlag, Heidelberg, 1981, pp. 161–196.
8. Gerritsin, F. A., and Verbeek, P. W., Implementation of cellular logic operators using 3×3 convolution and table lookup hardware, *Comput. Vision Graphics Image Process. 27*, 115–123 (1984).
9. Sternberg, S. R., Grayscale morphology, *Computer. Vision Graphics Image Process., 35*, 333–355 (1986).
10. Wilson, S. S., Applications of matrix morphology in *Proceedings of the SPIE International Symposium on Optical and Optoelectronic Applied Science and Engineering*, vol. 1350, San Diego, July 8–13, 1990.
11. Wilson, S. S., Theory of matrix morphology, Applied Intelligent Systems Tech. Rept. No. 32, 1989; *IEEE Trans. Pattern Anal. Machine Intell.*, in press.
12. Haralick R. M., Sternberg, S. R., and Zhuang, X., Image analysis using mathematical morphology, *IEEE Trans. Pattern Anal. Machine Intell., PAMI-9*(4), 532–550 (1988).
13. Rumelhart, D. E., Hinton, G. E., and McClelland, J. L., A general framework for parallel distributed processing, in *Parallel Distributed Processing* (D. E. Rumelhart and J. L. McClelland, eds.), MIT Press, Cambridge, 1986.
14. Wilson, S. S., Vector morphology and iconic neural networks, *IEEE Trans. Syst. Man Cybernet., SMC-19*(6):1636–1644, (1989).
15. Wilson, S. S., Matrix morphology—mathematical morphology on matrices of images, in *Mathematical Morphology: Theory and Applications* (R. M. Haralick, ed.), in press.
16. Metropolis, N., Rosenbluth, A., Rosenbluth, M., Teller, A., and Teller, E., Equations of state calculations by fast computing machines, *J. Chem. Phys., 21*, 1087–1091 (1953).

17. Kirkpatrick, S., Gelatt, C. D., Jr., and Vecchi, M. P., Optimization by simulated annealing, *Science*, *220*, 671–680 (1983).
18. Wilson, S. S., Teaching network connectivity using simulated annealing on a massively parallel processor, *Proc. IEEE*, *79*(4):559–566, (1991).
19. Wilson, S. S., Teaching network connections for real time object recognition, in *Neural and Intelligent Systems Integration* (B. Soucek, ed.), *Wiley Series in Sixth Generation Computer Technologies*, Wiley, New York, pp. 135–160, 1991.
20. Patarnello S., and Carnevali, P., Learning capabilities of Boolean networks, in *Neural Computing Architectures* (I. Aleksander, ed.), MIT Press, Cambridge, pp. 117–129, 1989.
21. Rumelhart, D. E., Hinton, G. E., and Williams, R. J. Learning internal representations by error propagation, in *Parallel Distributed Processing* (D. E. Rumelhart and J. L. McClelland, eds.), MIT Press, Cambridge, 1986.
22. Honavar V., and Uhr, L., A network of neuron-like units that learns to perceive by generation as well as reweighting of its links, in *Proceedings of the 1988 Connectionists Models Summer School* D. Touretzky, G. Hinton, and T. Sejnowski, eds.), Morgan Kaufmann, San Mateo, California, 1988.
23. Berkin, G. M., Kulicke & Soffa Industries, Inc., private communication.
24. Fountain, T. J., A survey of bit-serial array processor circuits, in *Computing Structures for Image Processing* (M. J. B. Duff, ed.) Academic Press, Orlando, Florida, pp. 1–14, 1983.
25. Batcher, K. E., Design of a massively parallel processor. *IEEE Tran. Comput. C-29*, 836–840 (1980)
26. Flanders, P. M., Hunt, D. J., and Reddaway, S. F., and Parkinson, D., Efficient high speed computing with the Distributed Array Processor, in *High-Speed Computer and Algorithm Organization* (D. J. Kuck, D. H. Lawrie, and A. H. Sameh, eds.), Academic Press, New York, 1979, pp. 113–127.
27. L. A. Schmitt, and Wilson, S. S., The AIS-5000 parallel processor, *IEEE Trans. Pattern Anal. Machine. Intell.*, *10*(3), 320–330 (1988).
28. Wilson, S. S., A single board computer with 64 parallel processors, in *Electronic Imaging '87*, Boston, Nov. 2–5, 1987.
29. Fukushima, K., Neocognitron: a hierarchical neural network capable of visual pattern recognition, *Neural Networks*, *1*, 119–130 (1988).

Chapter 2

Efficient Design Strategies for the Optimal Binary Digital Morphological Filter: Probabilities, Constraints, and Structuring-Element Libraries

Edward R. Dougherty

Center for Imaging Science,
Rochester Institute of Technology,
Rochester, New York

Robert P. Loce

Joseph C. Wilson Center for Technology,
Xerox Corporation,
Webster, New York

I. INTRODUCTION

Optimal mean-square morphological filters, both binary and gray scale, have been characterized by Dougherty [1–4] in terms of the Matheron representation for morphological filters, the characterization applying equally well to mean-absolute-error minimization. According to the Matheron representation [5], any increasing translation-invariant binary (or gray-scale) operator can be expressed as a union (supremum) of binary (or gray-scale) erosions, and, conversely, any union of erosions is an increasing translation-invariant operator. The paradigm of refs. 1 and 2 is to find structuring elements that yield statistically optimal representations. As developed, the theory applies to the general problem of statistical estimation and does not focus on specific problems such as restoration of degraded images or morphological prediction for decompression.

Unfortunately, even in the binary setting [1,3], design entails an excessive computational burden for even moderate-sized observation windows; indeed, the increase in complexity for increasing window size is combinatoric. To date, several strategies have been employed to facilitate design tractability for morphological filter optimization based on the Matheron representation. Dougherty, Mathew, and Swarnakar [6] have given an algorithm that derives the optimal mean-square morphological filter from the conditional expectation, and Dougherty and Haralick [7] employ a shape-based spectral approach to restore images degraded by minimum (min) noise. Moreover, we have presented some aspects of the constraint and library techniques in conference proceedings [8, 9], and the present chapter, excepting the probabilistic analysis of Section VI, appears for the most part in ref. 10.

In the present chapter we investigate a multifaceted suboptimal design methodology for binary filters based on the imposition of various constraints and small (but well-chosen) libraries, the goal being to facilitate computationally tractable design. Other approaches to morphological-filter optimization have also been taken. In the work of Schonfeld and Goutsias [11], there is a restricted optimization applying in particular to alternating sequential filters (see Serra [12] or Lougheed [13]). Because these filters are increasing and translation invariant, they too can be placed into the context of refs. 1 and 2. Finally, we mention the model-based approach taken by Haralick, Dougherty, and Katz [14] that is based on the opening spectrum. Here, the resulting optimal filter is likely not to be increasing, so it is not morphological in the sense of ref. 1; nevertheless, it can be written in terms of primitive morphological operations in conjunction with set operations.

In the following section, we review briefly some of the relevant material from refs. 1 and 3, as well as some basic morphological filter theory. In Section III we discuss general aspects of the three constraints: limiting the number of terms in the Matheron expansion, constraining the size of the observation window, and employing structuring-element libraries. Sections IV and V describe the effect of window constraint and library constraint in detail, discussing the effect on filter properties and design tractability. In Section VI we present a probabilistic analysis of the binary morphological filter that leads to a mean-square error (MSE) theorem that can be employed to search efficiently for a minimum MSE filter. In the final section, examples are given to illustrate the design strategy and performance of the optimal (suboptimal) morphological filter.

II. OPTIMAL MORPHOLOGICAL FILTERS

Given N observation random variables $X[1], X[2], \ldots, X[N]$ and a random variable Y to be estimated, the optimal mean-square (MS) estimator is the function $g(X[1]), X[2], \ldots, X[N])$ that minimizes the expected value (mean-square error),

$$\text{MSE} = E\,[\,|\,Y - g(X[1], X[2], \ldots, X[N])|^2] \tag{2.1}$$

It is well known that the optimal MS estimator is given by the conditional expectation; however, rather than find the best estimator, it is common practice to look for the best estimator among a class of estimators, thereby restricting the nature of the estimation rule g. Here we require g to be morphological.

We employ four elementary morphological operations in the present chapter: erosion, dilation, opening, and closing. In the binary setting, these are respectively defined by

$$S \ominus B = \{z : B + z \subset S\} \tag{2.2}$$

$$S \oplus B = \cup \{B + x : x \in S\} \tag{2.3}$$

$$S \bigcirc B = (S \ominus B) \oplus B \tag{2.4}$$

$$S \bullet B = [S \oplus (-B)] \ominus (-B) \tag{2.5}$$

where $B + z = \{b + z : b \in B\}$, $-B = \{-b : b \in B\}$, and B is called a *structuring element*. A key property of erosion employed throughout this chapter is the following: if $A \subset B$, then $S \ominus A \supset S \ominus B$. Both erosion/dilation and opening/closing satisfy duality relations relative to complementation:

$$S \oplus B = [S^c \ominus (-B)]^c \tag{2.6}$$

$$S \bullet B = (S^c \bigcirc B)^c \tag{2.7}$$

More generally, we consider a binary *morphological filter* to be a set mapping Ψ that is *increasing* [$S \subset T$ implies $\Psi(S) \subset \Psi(T)$] and *translation invariant* [$\Psi(S + z) = \Psi(S) + z$]. (In ref. 12, Serra requires idempotence for a filter to be called morphological. The *kernel* of an increasing, translation-invariant mapping Ψ, Ker[Ψ], is the class of all images S such that $\Psi(S)$ contains the origin. The dual is defined by $\Psi^*(S) = \Psi(S^c)^c$, it is also a morphological filter, the dual of the dual is the original filter ($\Psi^{**} = \Psi$), and

$$\text{Ker}[\Psi^*] = \{B : B^c \notin \text{Ker}[\Psi]\} \tag{2.8}$$

From Eqs. (2.6) and (2.7) we see that dilation by B is the dual of erosion by $-B$ and closing by B is the dual of opening by B. For the design methodology developed in the present chapter, it is important to recognize that Matheron [5] gave expressions for the kernels of the morphological mappings of Eqs. (2.2) through (2.5):

$$\text{Ker}[\cdot \ominus B] = \{A : A \supset B\} \tag{2.9}$$

$$\text{Ker}[\cdot \oplus B] = \{A : A \cap (-B) \neq \emptyset\} \tag{2.10}$$

$$\text{Ker}[\cdot \bigcirc B] = \{A : B - z \subset A \text{ for some } z \in B\} \tag{2.11}$$

$$\text{Ker}[\cdot \bullet B] = \{A : A \cap (B - z) \neq \emptyset \text{ for all } z \in B\} \tag{2.12}$$

(see ref. 5 or Giardina and Dougherty [15] for details). From these kernels one can easily derive corresponding bases in the digital setting.

A fundamental proposition of mathematical morphology is the Matheron representation [5]: a filter is translation invariant and increasing if and only if it can be expressed as a union of erosions by its kernel elements. As noticed by Dougherty and Giardina [16] and Maragos and Schafer [17,18], excluding certain pathological cases, a morphological filter Ψ has a *basis*, Bas[Ψ], of structuring elements such that the Matheron expansion can be taken over Bas[Ψ] instead of Ker[Ψ]:

$$\Psi(S) = \cup \{S \ominus B : B \in \text{Bas}[\Psi]\} \tag{2.13}$$

The basis is a minimal (nonredundant) class of structuring elements within the kernel: for any $B \in \text{Ker}[\Psi]$, there exists $B' = \text{Bas}[\Psi]$ such that $B' \subset B$; and there does not exist a pair of structuring elements in Bas[Ψ] properly related by set inclusion. If **B** is the basis for a filter, we will sometimes denote the filter by $\Psi_{\mathbf{B}}$. By duality, there exists a morphological filter representation in terms of dilation. (For gray-scale and lattice extensions of the Matheron representation, see refs. 12,15, and 17–20. A generalization to nonmonotonic mappings is given in ref. 21.)

If an increasing, translation-invariant mapping Ψ is also *antiextensive* [$\Psi(S) \subset S$] and *idempotent* [$\Psi\Psi = \Psi$], it is called a τ-*opening*. The *invariant class*, Inv[Ψ], of a τ-opening (or any mapping) is the class of images S such that $\Psi(S) = S$. A collection of images **B** is a *base* for Ψ if every image in Inv[Ψ] can be expressed as a union of translations of elements in **B**. There is a Matheron representation concerning τ-openings: Ψ is a τ-opening if and only if there exists a base **B** such that

$$\Psi(S) = \cup \{S \bigcirc B : B \in \mathbf{B}\} \tag{2.14}$$

(For gray-scale and lattice extensions, see refs. 12,15, and 22.)

Relative to mean-square-error optimization, erosion plays the key role. To adapt its definition to statistical estimation, consider N binary observation random variables $X[1], X[2], \ldots, X[N]$. Each realization of the random vector $X = (X[1], X[2], \ldots, X[N])$ is a 0–1 N-tuple. If we let 1 and 0 denote points that lie within or without the point set $\{1, 2, \ldots, N\}$, then each realization x of X constitutes a subset of $\{1, 2, \ldots, N\}$, and we can erode x by a deterministic structuring element $B = (b[1], b[2], \ldots, b[N])$, $b[j]$ being 0 or 1. The erosion $x \ominus B$ is a binary functional, its value being either 0 or 1.

For a random vector X and fixed structuring element B, erosion defines an estimator

$$X \ominus B = \min \{X[j] : b[j] = 1\} \tag{2.15}$$

that can be used to estimate another random variable Y. The optimal MS *erosion filter* is the one defined by the structuring element B minimizing

$$\begin{aligned} \text{MSE}\langle B\rangle &= E\left[\,\left| Y - \left((X[1], X[2], \ldots, X[N]) \ominus B\right) \right|^2\right] \\ &= E\left[\,\left| Y - \min \{X[j] : b[j] = 1\} \right|^2\right] \end{aligned} \tag{2.16}$$

Using the Matheron representation [Eq. (2.13)] as a guide, in refs. 1 and 3 an *N-observation digital morphological filter* is defined to be a functional of the form

$$\Psi(x) = \max_i \{x \ominus B(i)\} \tag{2.17}$$

where x and $B(i)$ are deterministic binary N-vectors. Assuming nonredundancy, $\{B(i)\}$ is called the *basis* of Ψ. Extension of optimality to N-observation morphological filters involves minimizing MSE $\langle\Psi\rangle$, which is defined by Eq. (2.1) with Ψ in place of g, over all possible choices of N-observation morphological filters Ψ:

$$\begin{aligned} \mathrm{MSE}\langle\Psi\rangle &= E\left[\,|\,Y - \Psi\,(X[1], X[2], \ldots, X[N])\,|^2\right] \\ &= E\left[\,|\,Y - \max_i \{x \ominus B(i)\}\,|^2\right] \end{aligned} \tag{2.18}$$

Since Ψ is fully determined by its basis, finding the optimal N-observation filter reduces to selecting the subset of the 2^N structuring elements that yields minimum MSE$\langle\Psi\rangle$. Owing to basis minimality, many of these subsets can be eliminated from consideration, and we will make much use of this fact. Nonetheless, exploitation of this redundancy still leaves us with a problematic computation.

A striking feature of the optimal MS erosion filter is that, relative to the entire image, it is not a morphological filter. Because filtering is considered as pointwise estimation, each pixel possesses its own optimal basis. Thus, it is not spatially invariant and therefore not translation invariant. As in the case of linear optimization, stationarity yields spatial invariance; however, in the morphological case it is strong stationarity that is required for spatial invariance. Even though such an assumption is often not fully warranted, it is typical of the kind of modeling assumption one must often apply in practice.

III. CONSTRAINED OPTIMALITY

Rather than select the optimal filter Ψ over an N-pixel window W, we may wish to constrain Ψ by requiring that some extra properties be satisfied. We might wish to impose some algebraic constraints on Ψ. For instance, we might desire that Ψ be antiextensive or idempotent, or perhaps both, so that it is a τ-opening. Each constraint translates into some constraint on the filter basis, and conversely. In the case of antiextensivity, the requirement is that each basis element must contain the origin. When applying algebraic constraints we gain some desirable filter property in exchange for a possible increase in MSE, because requiring the filter to possess certain properties is tantamount to restricting the set of possible filter bases to some subcollection of all potential filter bases. This type of algebraic constraint was discussed at length in ref. 1 with respect to binary τ-

openings and in ref. 2 with respect to gray-scale linear operators. Also see ref. 8, where the τ-opening constraint, as well as some material covered herein, was briefly described.

Our concern in the present chapter is design tractability, and to that end we are concerned mainly with three other constraint paradigms, each of which involves more efficient filter design in return for suboptimality (an increase in MSE). Although we will analyze each of the three individually, tractable design may very well involve using all three constraints in conjunction.

First, we might wish to constrain the number of basis elements to some prefixed limit m to obtain the *optimal m-erosion filter*. For instance, if $m = 1$, we obtain the optimal single-erosion filter. By fixing a limit m we do not require that there be exactly m erosions in the filter, only that the number be bounded by m. This convention is crucial because if $\mathbf{B}$ and $\mathbf{E}$ are two structuring-element classes such that $\mathbf{B} \subset \mathbf{E}$, then $\Psi_{\mathbf{B}} \subset \Psi_{\mathbf{E}}$. Thus, forcing too large a basis can cause overestimation: there will be too many erosions in the Matheron expansion. We call this type of constraint *size constraint*.

Rather than fix the maximum number of basis elements at the outset, we often choose to limit the number dynamically by plotting MSE against the number of structuring elements. When this *size-MSE* curve begins to flatten, we recognize that it is highly unlikely that larger bases will have a significant effect. We could, of course, wait for a definitive increase in the size-MSE curve; however, computation time increases combinatorically with increasing basis size and therefore it is pragmatic to choose some heuristic decision mechanism to judge when the basis is sufficiently large to preclude further significant reduction in MSE.

A second constraint arising from our desire to reduce computation in filter design involves observation restriction. Rather than consider all possible structuring elements in a window W, we might restrict our attention to structuring elements in some subwindow W' in W. Strictly speaking, optimization relative to W' without constraint on the basis is actually unconstrained optimization over W'; however, assuming that we would actually like to optimize over W, we can view restriction to W' as a constraint that will allow us to consider only a subclass of all W-bases, thereby likely yielding a filter with increased MSE. In effect, this type of constraint is a special case of the general constraint problem in which basis restriction is relative to window geometry. We term it *window constraint* and discussed it initially in ref. 8.

A third type of constraint is *library constraint*. Here, rather than optimize over all possible structuring elements, we restrict ourselves to some predetermined subset, or subsets, of structuring elements in the window. Specifically, we might postulate m collections of structuring elements, $\mathbf{L}_1, \mathbf{L}_2, \ldots, \mathbf{L}_m$, each of which is suited for accomplishing a certain type of filtering task. Letting $\mathbf{L} = \cup \mathbf{L}_i$, we constrain our basis selection process to $\mathbf{L}$, or perhaps to some subfamily $\mathbf{L}_{i_1}, \mathbf{L}_{i_2}, \ldots, \mathbf{L}_{i_r}$. The manner in which such a restriction is developed is key to the goodness of the filter, where (as in common statistical usage) *goodness* refers to the accuracy and precision of estimation.

We now proceed to investigate the latter two kinds of constraint in detail.

IV. WINDOW CONSTRAINT

Suppose we wish to reduce the computational burden by employing structuring elements only in a subwindow W' of the window W. Heuristically, we will delete pixels that we believe to form structuring elements that do not play a major role in reducing MSE. If our assumptions are reasonable, we will obtain a suboptimal filter that is only marginally less efficient than the optimal filter over W. For instance, we might desire a 5×5 square window; however, such a window possesses $2^{25} = 33{,}554{,}432$ structuring elements, and therefore there are $33{,}554{,}432!/m!(33{,}554{,}432 - m)!$ m-element structuring element combinations. Of course, a multitude of these are redundant due to basis minimality; nevertheless, even if we limit ourselves to small m, a search through all possible structuring-element combinations is still computationally burdensome. Rather than use the 5×5 window, we can greatly decrease the search space by eliminating 8 pixels to form the subwindow W' shown in Figure 1. The cost is a constraint that impinges ultimately on filter properties, as well as on MSE.

Window constraint can, in general, be formulated in terms of the Matheron representation, Eq. (2.13). Suppose Ψ is a filter with Bas[Ψ] over window W, which means that all basis elements lie in W. If W' is a subwindow of W, the *window-constrained filter* is given by

$$\Psi'(S) = \cup \{S \ominus (B \cap W') : B \in \text{Bas}[\Psi]\} \tag{2.18}$$

where the convention is adopted that $B \cap W'$ is deleted from the expansion if it is null. If it happens that $B \cap W' \neq \emptyset$ for all B in Bas[Ψ], then, since $B \cap W' \subset B$, each image in the constrained expansion is a superset of the original image, so that $\Psi'(S) \supset \Psi(S)$. Here one must be prudent: if the constraint eliminates some of the basis elements altogether, then there are fewer images forming the

a) b)

Figure 1. (a) Unconstrained 5×5 window W. (b) Constrained 5×5 window W'.

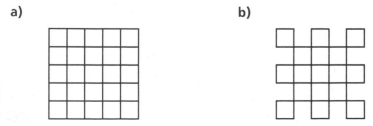

union and the constrained filter might not dominate the original filter. Moreover, note that Bas[Ψ'] need not equal $\{B \cap W' : B \in \text{Bas}[\Psi]\}$, since intersection with W' might create redundancy; however, Bas[Ψ'] can be derived by simply eliminating redundancy.

A. WINDOW-CONSTRAINED OPENINGS

To illustrate the algebraic and filtering effect of window constraints, we consider in detail their effect on τ-openings. Suppose Ψ is a τ-opening with base **B**. According to Eq. (2.14), Ψ is expressed by a union of openings by elements in **B**. But if $S \bigcirc B$ is an opening, then, as a morphological filter, it possesses a Matheron representation in terms of its own erosion basis. From the expression for the opening kernel given in Eq. (2.11), it can be seen that the basis for opening by B consists of all translates of B that contain the origin. As an example, the opening by a 2×2 square possesses four basis elements, these being depicted in Figure 2. Moreover, if statistics are known only for a given observation window W, then the τ-opening Ψ must be a union of openings (by elements in **B**) whose corresponding Matheron bases lie in W. For instance, if W is the 3×3 square, there are only 10 possible structuring elements for **B**, these being depicted in Figure 3. Assuming $\mathbf{B} = \{B_1, B_2, \ldots, B_r\}$, and assuming the opening $S \bigcirc B_i$ has Matheron basis $\{B_{i1}, B_{i2}, \ldots, B_{i,m(i)}\}$, Eq. (2.13) takes the form

$$\Psi(S) = \cup \{S \ominus B_{ij}\} \tag{2.19}$$

The collection of B_{ij} being finite, Bas[Ψ] is easily constructed by eliminating redundancy. If we then constrain Ψ to a subwindow W' in W, the resulting filter Ψ' takes the form of Eq. (2.18). It is called a *window-constrained τ-opening*, and it will very likely not be a true τ-opening.

A fundamental concern is the relationship between the invariant class of a τ-opening and a corresponding constrained τ-opening. Let us assume Ψ is a τ-opening with basis elements in W and W' is a subwindow containing the origin. For any element B in Bas[Ψ] let $B' = B \cap W'$ and assume redundancy is eliminated in the representation, Eq. (2.18). Because all basis elements of Ψ contain the origin, the basis elements of Ψ' also contain the origin, so that Ψ' is monotonic, translation invariant, and antiextensive, but it need not be idempotent.

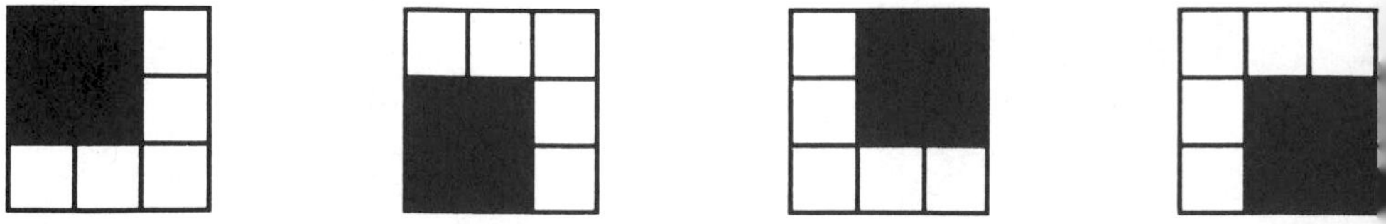

Figure 2. Basis elements of 2×2 opening.

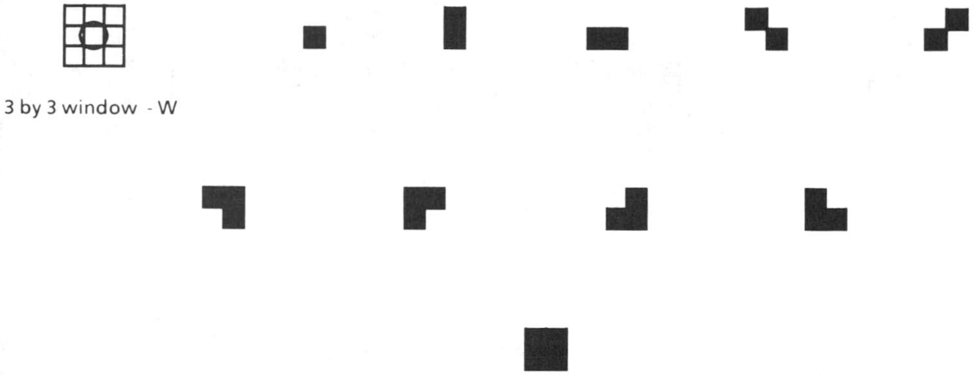

Figure 3. Opening structuring elements of a 3×3 square window.

Moreover, because $B \cap W'$ is never empty, for any image S, $\Psi(S) \subset \Psi'(S) \subset S$. Thus, if $S \in \text{Inv}[\Psi]$, $\Psi(S) = \Psi'(S) = S$, so that $S \in \text{Inv}[\Psi']$. Consequently, $\text{Inv}[\Psi] \subset \text{Inv}[\Psi']$.

Because Ψ is a τ-opening, it is idempotent: $\Psi\Psi = \Psi$. Continuing to assume that W' contains the origin, we can extend this composition relation. Since $\Psi(S) \in \text{Inv}[\Psi] \subset \text{Inv}[\Psi']$, $\Psi'\Psi(S) = \Psi(S)$. Because Ψ is a τ-opening and $\Psi' \supset \Psi$,

$$\Psi = \Psi\Psi \subset \Psi\Psi' \subset \Psi' \tag{2.20}$$

By the antiextensivity of Ψ' and monotonicity of Ψ, $\Psi \supset \Psi\Psi'$. Thus, $\Psi'\Psi = \Psi\Psi' = \Psi$.

For any image S, antiextensivity yields the decomposition

$$S = \Psi(S) \cup (S - \Psi(S)) \tag{2.21}$$

where $\Psi(S) \in \text{Inv}[\Psi]$ and $S - \Psi(S)$ is the portion of S filtered by Ψ. A similar decomposition results from Ψ'. Yet $\Psi'(S)$ may be further filtered by Ψ to obtain $\Psi(S)$. Consequently, we have the further decomposition

$$S = \Psi(S) \cup (\Psi'(S) - \Psi(S)) \cup (S - \Psi'(S)) \tag{2.22}$$

where $\Psi(S) \cup (\Psi'(S) - \Psi(S)) \in \text{Inv}[\Psi']$. $\Psi'(S) - \Psi(S)$ is the portion of S not filtered out by Ψ' that would have been filtered out by Ψ. If $\Psi(S)$ is a restoration containing the uncorrupted image, then $\Psi'(S) - \Psi(S)$ is noise that would have been filtered by Ψ but is not filtered by the constrained filter Ψ'.

As an illustration of the preceding decompositions, consider the 3×3 square window W and the opening Ψ by the 2×2 square. Let the constrained window W' consist of the origin together with its strong neighbors. Whereas the basis for Ψ is depicted in Figure 2, the basis for the constrained filter Ψ' is depicted in Figure 4. The decompositions of Eqs. (2.21) and (2.22) are illustrated in Figure 5.

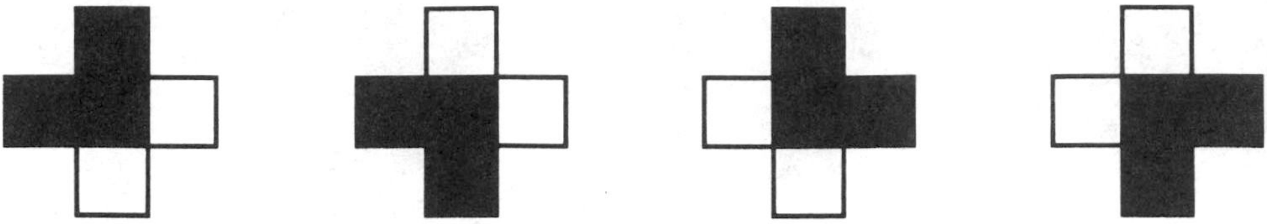

Figure 4. Basis elements for window-constrained 2×2 opening.

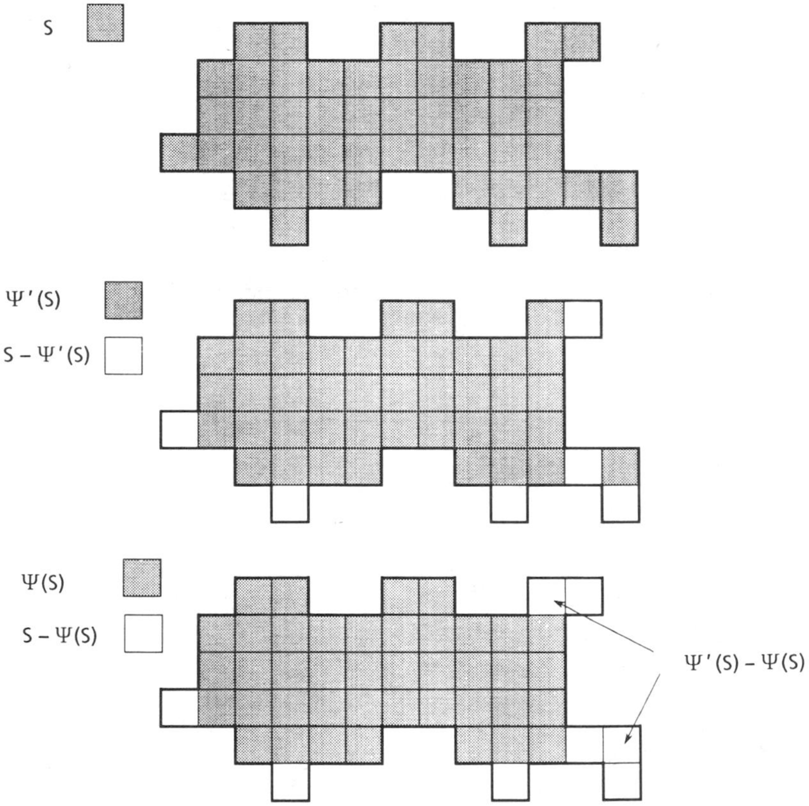

Figure 5. Example of decompositions of Eqs. (2.21) and (2.22), where Ψ is a 2×2 opening and Ψ' is the window-constrained opening with basis shown in Figure 4.

To see the effect of the window constraint on noise reduction, consider the image process consisting of rectangles uniformly randomly distributed in the grid, where each rectangle is Z_1 by Z_2, Z_1 and Z_2 being uniformly distributed over {8, 9, 10, 11, 12} and {2, 3, 4}, respectively, and where area coverage is approximately 10%. A 256 by 256 pixel^2 realization of the process is depicted in Figure 6. Consider also a noise process where each component is ragged in shape, consists of 1 to 4 pixels, and possesses approximately 30% area coverage. The cor-

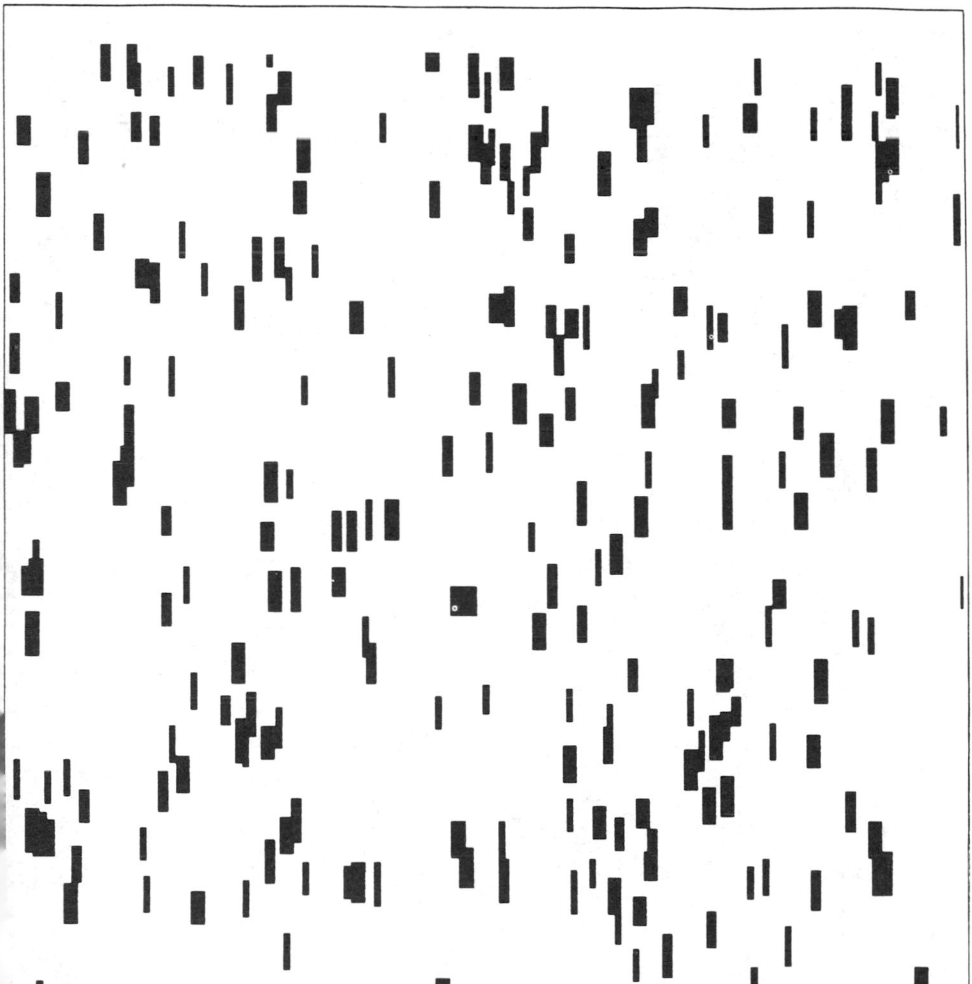

Figure 6. Realization of the image process.

rupted image is the union of the uncorrupted image and the noise. A realization of the image-noise process is given in Figure 7, where MSE = 0.2711. For each of the 10 possible openings for the 3 × 3 window (Figure 3), MSE relative to image restoration is given in Table 1. Owing to symmetry of image and noise components, the two diagonal structuring elements and the four 3-point structuring elements result in similar MSE values, respectively, and are shown as aver-

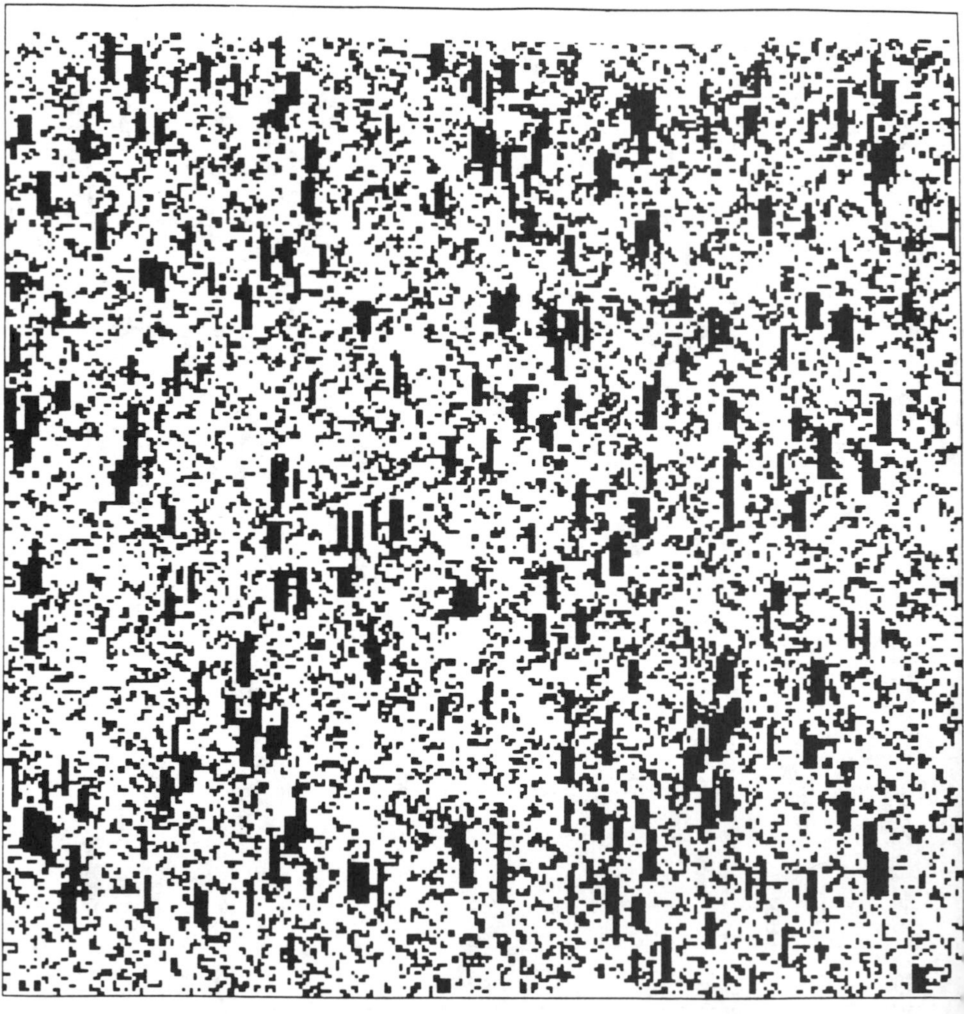

Figure 7. Realization of the image-noise process, MSE = 0.2711.

Table 1. MSE of Openings Allowed by a 3 × 3 Window

Filter type	MSE	Number of error pixels
No filter (identity opening)	0.2711	16,540
1 × 2 vertical opening	0.2010	12,263
2 × 1 horizontal opening	0.2250	13,726
2-point diagonal opening	0.1748	10,668
3-point opening	0.1226	7,479
2 × 2 opening (optimal opening)	0.1218	7,430
2 × 2 constrained opening	0.1640	10,004
General optimal filter (2-erosion)	0.0525	3,210

ages in the table. The optimal opening is given by the 2 × 2 square element, and the filtered image, where MSE = 0.1217, is depicted in Figure 8.

Letting Ψ denote the 2 × 2 opening, consider the window-constrained opening Ψ' (which is not an opening) resulting from restriction to W', the origin together with its strong neighbors. The basis for Ψ' is illustrated in Figure 4 and the result of applying Ψ' to the noisy realization of Figure 6 is shown in Figure 9. As expected, there is an increase in MSE relative to application of Ψ, the MSE corresponding to Ψ' being 0.1639.

For the subwindow of the preceding paragraph we can easily design the overall optimal morphological filter for the image-noise model. The optimal filter relative to W' possesses the two-element basis depicted in Figure 10, MSE for the optimal filter is 0.0525, and the filtered version of the noisy realization of Figure 6 is shown in Figure 11. It is interesting to compare this MSE, which is suboptimal relative to the 3 × 3 window W, to MSE = 0.1217 for the 2 × 2 opening. The signal is in the invariant class of the opening, the majority of the noise components are proper subsets of the opening structuring element, and the degradation is union noise. Hence, opening represents a classical approach to filtering the noisy image. By giving up the opening constraint and turning to erosions, we obtain a filter whose output possesses an MSE that is only 43% as large as the open-filtered image. Of course, if one looks closely at the image-noise model, it becomes apparent that the opening would have performed far better had we employed a 3 × 2 rectangle; nonetheless, relative to statistical analysis, such a rectangle requires a 5 × 3 observation window, whereas we were restricted to a 3 × 3 window. Moreover, and this point is crucial, if we were

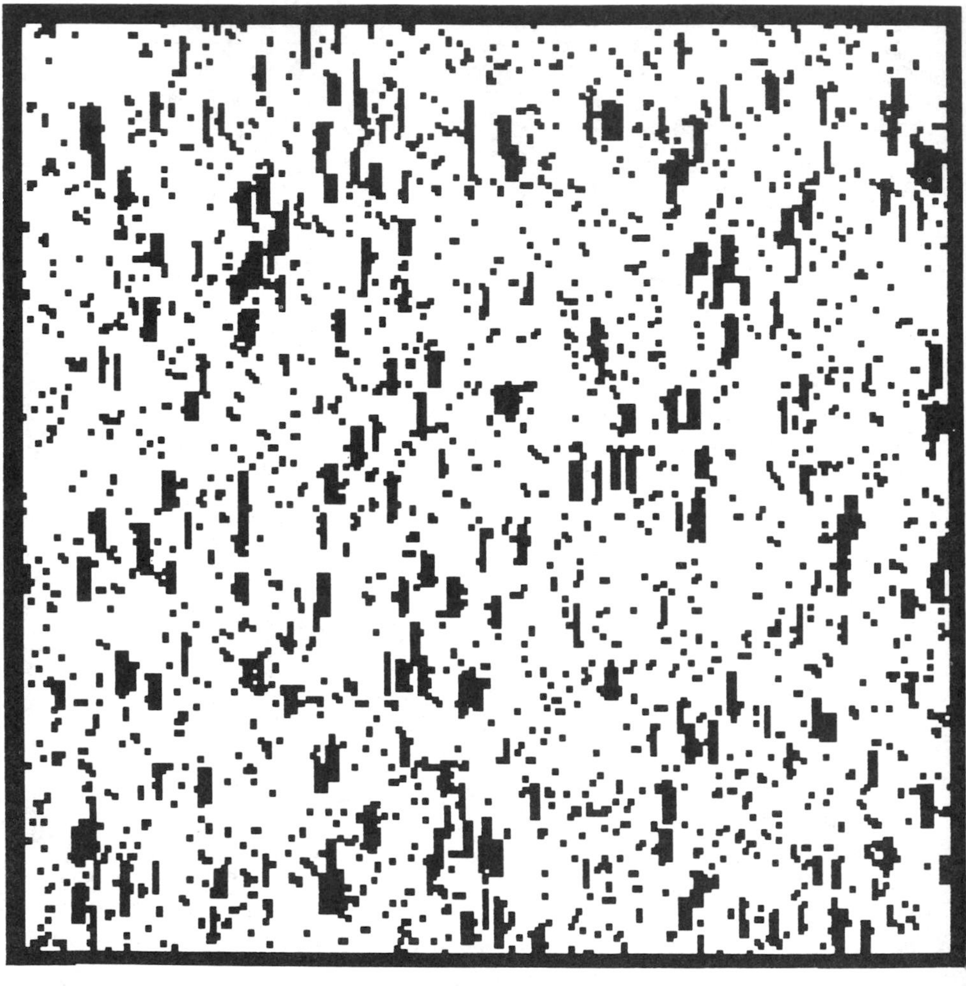

Figure 8. Image resulting from optimal opening (2 × 2), MSE = 0.1217.

to optimize over a 5 × 3 window, we would expect to obtain a different erosion basis, and we are guaranteed that the resulting Matheron expansion will provide a better filter than that resulting from the opening, the latter representing a strong algebraic constraint on optimization.

V. LIBRARY-OPTIMAL FILTERS

A key method of achieving design tractability over a given window W is to limit the set of potential basis elements: rather than optimize over all nonredundant

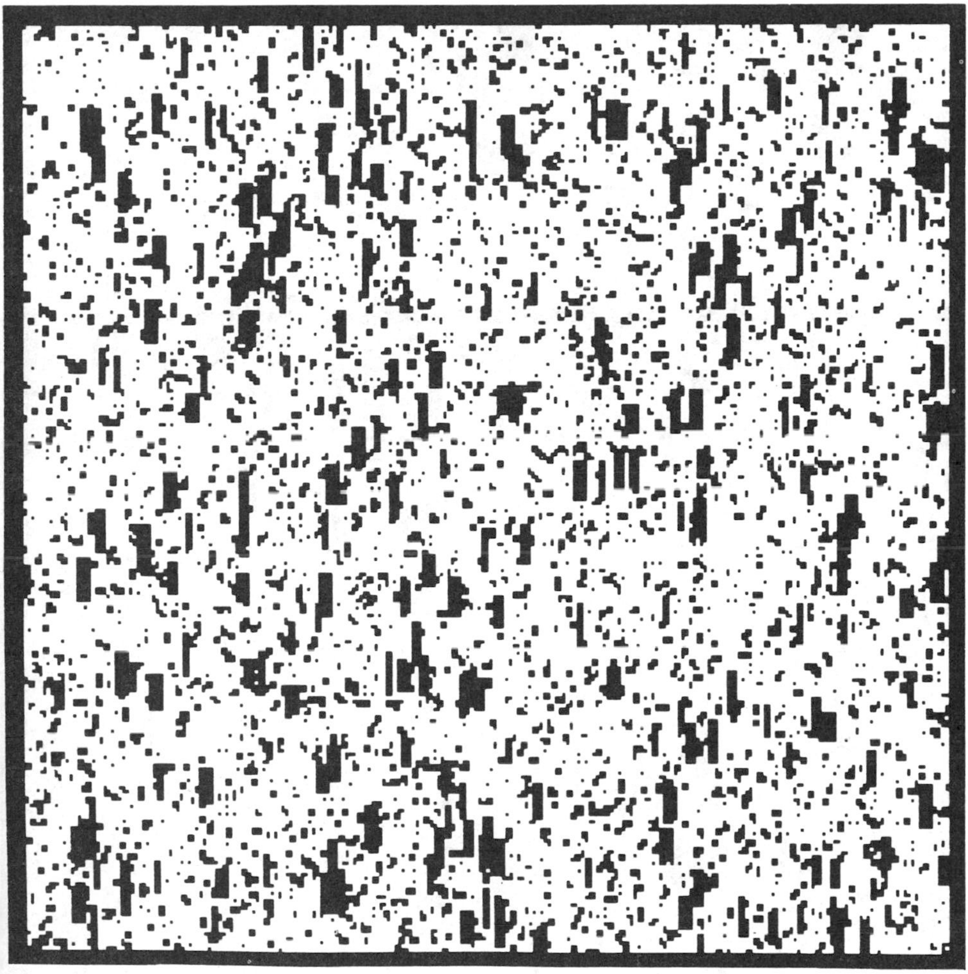

Figure 9. Image resulting from window-constrained 2×2 opening, MSE = 0.1639.

collections of structuring elements within W, we preselect a *library* **L** from which to form bases, and we say that the optimal filter from among those whose bases are formed from **L** is **L**-*optimal*. Library optimization constraint is characterized by the limitations of **L**. In selecting **L**, two conditions must be met: (1) bases formed from **L** must produce a class of filters that provides good suboptimality over the image range of interest, and (2) the size of **L** must be sufficiently small to yield design tractability. Aside from these global constraints, there are other properties that will prove useful, and these will become apparent as we proceed.

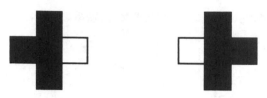

Figure 10. Basis elements of window-constrained optimal morphological filter.

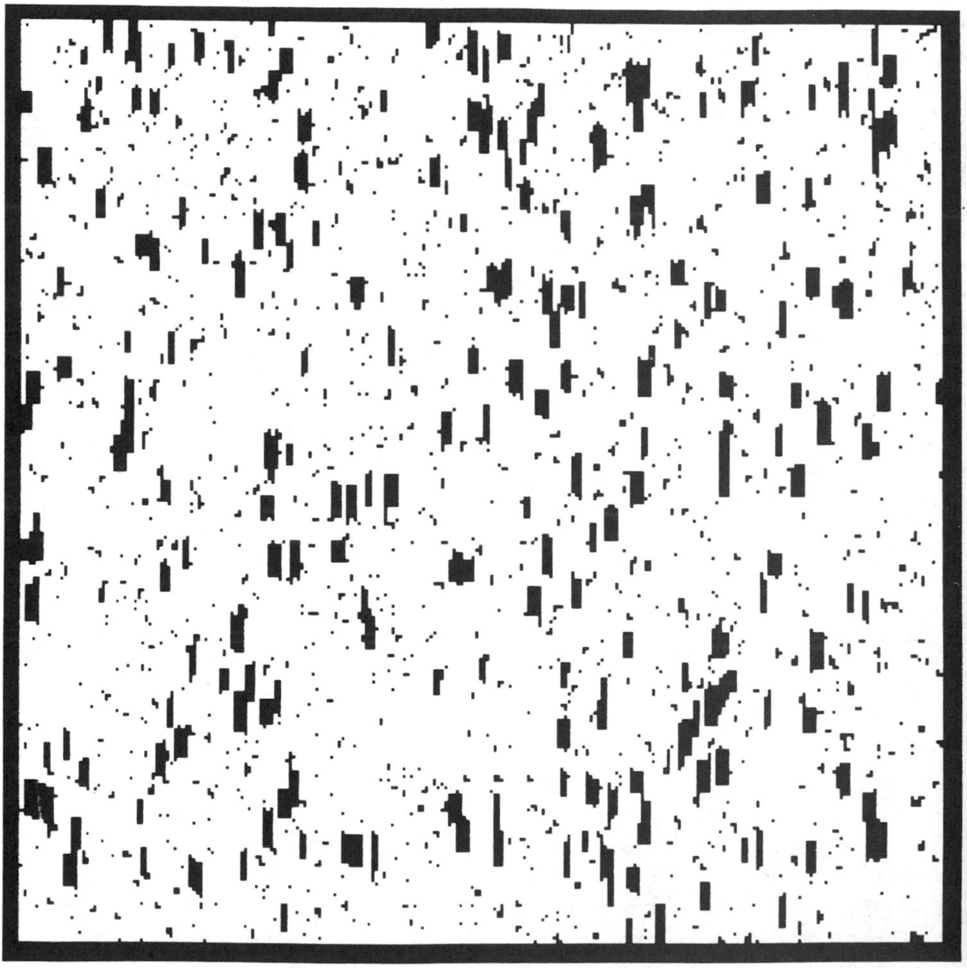

Figure 11. Image resulting from optimal window-constrained morphological filter, MSE = 0.0525.

While one might approach the construction of a library in a number of ways, here we will describe two methods. One will be based on knowledge of important filter bases. Because knowledge of filter behavior will be contained in the final library, we call it an *expert library*. We will also discuss a method of library construction based on first-order statistical properties of structuring elements. This we call a *first-order library*.

Concerning properties of libraries constructed by either method or a combination of methods, consider a class of filters

$$\mathbf{G} = \{\Psi_1, \Psi_2, \ldots, \Psi_n\} \tag{2.23}$$

with bases Bas[Ψ_i], $i = 1, 2, \ldots, n$. If **C** is a class of structuring elements such that $\mathbf{C} \supset \cup \text{Bas}[\Psi_i]$, and Ψ is **C**-optimal, then MSE$\langle\Psi\rangle \leq$ MSE $\langle\Psi_i\rangle$ for all i, so that Ψ is better than Ψ_i for all i. In particular, if we let $\mathbf{L} = \cup \text{Bas}[\Psi_i]$, then the **L**-optimal filter is better than Ψ_i for all i.

Taking a slightly more general approach, let $\mathbf{G}_1, \mathbf{G}_2, \ldots, \mathbf{G}_m$ be collections of filters, let

$$\mathbf{L}_i = \cup \{\text{Bas}[\Psi_{ij}] : \Psi_{ij} \in \mathbf{G}_i\} \tag{2.24}$$

and let $\mathbf{L} = \cup \mathbf{L}_i$. If Ψ is **L**-optimal, we say that it is *generated* by $\mathbf{G} = \cup \mathbf{G}_i$, and for each i we say the $\mathbf{L}_i$-optimal filter Ψ_i is *generated* by $\mathbf{G}_i$. The filter Ψ is better than Ψ_i, for each i, and each Ψ_i is better than any of the filters constituting $\mathbf{G}_i$. If we let

$$\mathbf{L}\langle i_1\, i_2 \cdots i_r\rangle = \mathbf{L}_{i_1} \cup \mathbf{L}_{i_2} \cup \cdots \cup \mathbf{L}_{i_r} \tag{2.25}$$

$$\mathbf{L}\langle j_1\, j_2 \cdots j_s\rangle = \mathbf{L}_{j_1} \cup \mathbf{L}_{j_2} \cup \cdots \cup \mathbf{L}_{j_s} \tag{2.26}$$

where $\{i_1, i_2, \cdots, i_r\} \subset \{j_1, j_2, \cdots, j_s\}$, then the $\mathbf{L}\langle j_1 j_2 \cdots j_s\rangle$-optimal filter is better than the $\mathbf{L}\langle i_1 i_2 \cdots i_r\rangle$-optimal filter. In particular, the **L**-optimal filter is better than the $\mathbf{L}\langle i_1, i_2 \cdots i_r\rangle$-optimal filter, or, in words, the *library-optimal* filter is better than any *sublibrary-optimal* filter. Generally, if Ψ is any filter whose basis lies in $\mathbf{L}\langle ij \cdots\rangle$, then Ψ is said to be *generated* by $\mathbf{L}\langle ij \cdots\rangle$, or by $\mathbf{G}\langle ij \cdots\rangle = \mathbf{G}_i \cup \mathbf{G}_j \cup \cdots$. Note that in our notation, $\mathbf{L}\langle i\rangle = \mathbf{L}_i$ and $\mathbf{G}\langle i\rangle = \mathbf{G}_i$.

Given the preceding methodology, our task is to form an appropriate set of $\mathbf{G}_i$ or, equivalently, an appropriate set of sublibraries $\mathbf{L}_i$.

A. A Specific Expert Library

For an expert library, our approach is to select important filter classes $\mathbf{G}_i$. Keep in mind that a particular sublibrary labeling will no longer be general, but instead will be relative to the specific library that we develop. Throughout the present study, we will be focusing our attention on a centered 5 × 5 window, but the actual library will involve the constrained 5 × 5 window of Figure 1. Thus, we will actually be employing two constraint criteria in conjunction. Henceforth, we will denote the constrained 5×5 window by W_5 and the full centered 3×3 window by W_3.

We will now describe the particulars of a specific expert library. To begin with, there are nine singleton elements in W_3 and eight in $W_5 - W_3$ (Figure 12a). Each of these represents a translation: if $B = \{z\}$ is a singleton, then $S \ominus B = S - z$. Consequently, if $\mathbf{B} = \{B_1, B_2, \ldots, B_q\}$ is a collection of q singletons, $B_i = \{z_i\}$, then the filter $\Psi_{\mathbf{B}}$ defined by Matheron expansion is a dilation,

$$\Psi_{\mathbf{B}}(S) = \cup \{S \ominus B_i\} = \cup \{S - z_i\} = S \oplus (- \cup \{B_i\}) \tag{2.27}$$

Let $\mathbf{L}_1$ and $\mathbf{L}_2$ denote the singletons in W_3 and $W_5 - W_3$, respectively. $\mathbf{G}_1$ and $\mathbf{G}_2$ denote the collections of translations by pixels in W_3 and $W_5 - W_3$, respectively. The filters generated by $\mathbf{G}_1$ are dilations by structuring elements in W_3, and the filters generated by $\mathbf{G}_1 \cup \mathbf{G}_2$ are dilations by structuring elements in W_5.

There are four length-two linear openings: vertical, horizontal, and two diagonal. These constitute $\mathbf{G}_3$. Each possesses a basis consisting of two elements, the two translates of the opening structuring element that contain the origin. These eight structuring elements make up $\mathbf{L}_3$ and are depicted in the first two rows of Figure 12b. Because any τ-opening formed from the length-two linear-opening structuring elements is a union of the openings, and therefore a union of erosions formed from $\mathbf{L}_3$, these τ-openings are generated by $\mathbf{G}_3$. The four length-three linear openings constitute $\mathbf{G}_4$. Each has a basis with three elements, and the 12 elements of $\mathbf{L}_4$ are shown in the last four rows of Figure 12b. Taken together, $\mathbf{L}_3$ and $\mathbf{L}_4$ generate all τ-openings with linear structuring elements of lengths two and three.

In addition to linear openings, we desire square openings. Letting $\mathbf{G}_5$ consist of the single 2×2 opening, $\mathbf{L}_5$ consists of the four structuring elements in Figure 12c, these forming the basis for the 2×2 opening. A 5×5 observation window can hold the nine basis elements for the 3×3 opening; however, here we must contend with the constrained window W_5. Consequently, $\mathbf{L}_6$ consists of the nine elements depicted in Figure 12d, these constituting the basis of the window-constrained opening.

An erosion structuring element comprises its own single-element basis. Letting $\mathbf{G}_7$ consist of the four centered length-three linear erosions, $\mathbf{L}_7$ is composed of the four structuring elements shown in the first row of Figure 12e. Each of these is a subset of W_3. $\mathbf{L}_8$ consists of the eight length-four, noncentered linear structuring elements shown in the second and third rows of Figure 12e. $\mathbf{L}_9$ consists of the four length-five, centered linear structuring elements in the last row of Figure 12e. $\mathbf{G}_8$ and $\mathbf{G}_9$ are composed of the erosions corresponding to $\mathbf{L}_8$ and $\mathbf{L}_9$, respectively. For both $\mathbf{L}_8$ and $\mathbf{L}_9$, window constraint plays no role.

As for nonlinear antiextensive erosions, we consider two classes, the elements of which are to some extent "ball-like." $\mathbf{L}_{10}$ has three elements, each lying in W_3, and these are shown in the first row of Figure 12f. $\mathbf{L}_{11}$ has two elements, each requiring W_5. These are shown in the second row of Figure 12f.

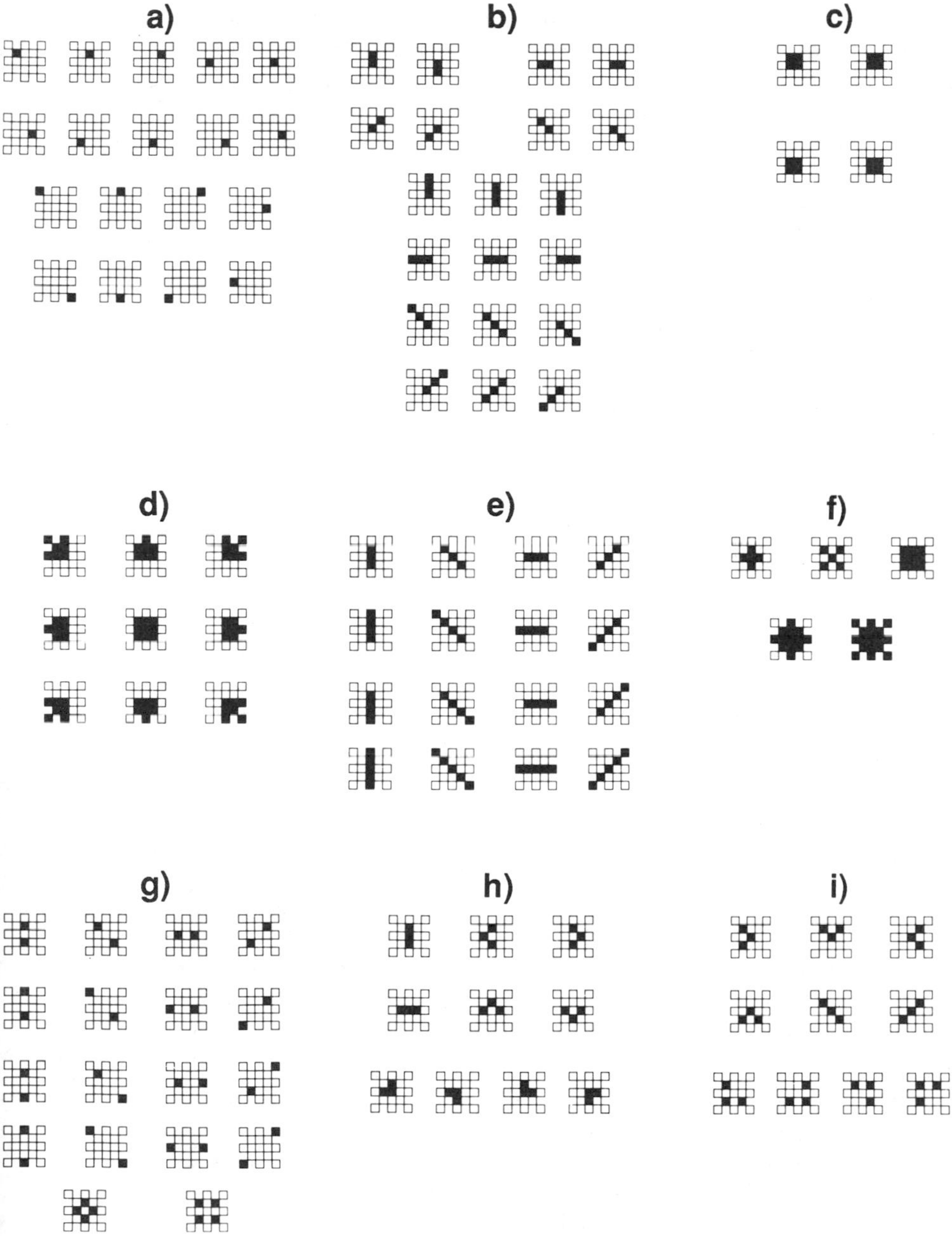

Figure 12. A specific expert library.

The next three sublibraries correspond to hole-filling erosions and are of particular importance for restoring images degraded by salt noise. We call them *hole-fillers*, and their genesis can be seen in the closing kernel, given in Eq. (2.12). To analyze these we introduce the following notation: $(b_{-p}, \ldots, b_{-1}, \underline{b}_0, b_1, \ldots, b_q)$ represents the structuring element with value b_i at $(i, 0)$ in the grid and having all other pixel values zero.

Consider closing by $B = (\underline{1}\ 1)$. There are two translates $B - z$, $z \in B$, these being given by B and $(1\ \underline{1})$. The singleton set $(\underline{1})$ and the two-point set $(1\ \underline{0}\ 1)$ both lie in Ker[Ψ], and it is readily seen that any kernel element must contain one of these. Thus,

$$\mathrm{Bas}[\Psi] = \{(\underline{1}), (1\ \underline{0}\ 1]\} \tag{2.28}$$

and, since erosion by the origin is the identity map,

$$S \bullet B = S \cup [S \ominus (1\ \underline{0}\ 1)] \tag{2.29}$$

In effect, for a horizontal two-point linear element, the closing maintains S while filling any inactivated pixel with activated pixels both to the left and to the right. In our terminology, the element $(1\ \underline{0}\ 1)$ is a hole-filler.

As a second illustration relevant to our study, suppose we consider closing by a horizontal linear three-point element B. This closing fills inactivated pixels lying horizontally between two activated pixels and fills horizontal pixel pairs lying horizontally between activated pixels. There are three translates $B - z$, $z \in B$: $(\underline{1}\ 1\ 1)$, $(1\ \underline{1}\ 1)$, and $(1\ 1\ \underline{1})$. The following nonredundant elements lie in the kernel: $(\underline{1})$, $(1\ \underline{0}\ 1)$,$(1\ 0\ \underline{0}\ 1)$, and $(1\ \underline{0}\ 0\ 1)$. Moreover, every element in the kernel must contain at least one of them, so that they form the basis for closing by B. Thus, $S \bullet B$ operates by unioning the results of three hole-fillers and the input image S. Note that $(1\ \underline{0}\ 1)$ fills single-pixel holes, and, working in conjunction, $(1\ \underline{0}\ 0\ 1)$ and $(1\ 0\ \underline{0}\ 1)$ fill horizontal two-pixel holes.

Turning to the hole-filler sublibraries, the elements in the first row of Figure 12g compose $\mathbf{L}_{12}$. Each requires only W_3 as an observation window. The elements in the second, third, and fourth rows of Figure 12g constitute $\mathbf{L}_{13}$. They require W_5 and are not affected by the window constraint. The power of the hole-fillers can be seen in the following example. Suppose an image contains a 2×2 square hole, whose external boundary is contained in the image. The hole will be filled by eroding by the second and fourth elements of the first row, that is, by using only hole-fillers requiring W_3 as an observation window.

Whereas inclusion of the elements in the second two rows of Figure 12g is explained by the closing illustrations discussed previously, the elements of the last row are not so explained. While $(1\ \underline{0}\ 0\ 1)$ and $(1\ 0\ \underline{0}\ 1)$ will work in conjunction to fill two-pixel horizontal holes, $(1\ 0\ \underline{0}\ 0\ 1)$ cannot, by itself, fill a three-pixel horizontal hole. Nevertheless, it can be of use. For instance, if the image contains a 3×3 square hole, whose external boundary is contained in the image, then the interior boundary of the square hole can be filled by the elements in the first three rows of Figure 12g, while the center is filled by any of the four in the fourth row.

In order to protect against overestimation when filling pixels that have been removed from the uncorrupted image by the noise process, we include the hole-fillers in $\mathbf{L}_{14}$, these being depicted in the fifth row of Figure 12g. Like the hole-fillers of $\mathbf{L}_{12}$, these fill single pixels, but the use of four pixels around the center protects against overfilling. As a case in point, consider the image *S* and its noisy salted version *T* of Figure 13. Erosion by any structuring element in $\mathbf{L}_{12}$ will fill all noise holes; however, each will erroneously fill certain pixels that are properly outside *S*. On the contrary, erosion by an element of $\mathbf{L}_{14}$, the strong-neighbor hole-filler, will fill all noise holes without any overfilling.

Our final sublibrary concerns the median. Specifically, we let $\mathbf{G}_{15}$ consist of 3×3 strong- and weak-neighbor medians, both being useful for restoring images degraded by certain types of salt-and-pepper noise. As discussed by Maragos and Schafer [17, 18] and Giardina and Dougherty [15], both strong- and weak-neighbor medians possess 10-element bases, these being depicted in Figure 12h and i, respectively. $\mathbf{L}_{15}$ consists of the 20 elements in the two figures. Note that we have restricted our attention to strong- and weak-neighbor medians because the full W_3 median possesses 126 basis elements. Using W_5 would have presented an even worse computational problem.

If we have knowledge of the degradation process, then we will use a subset of sublibraries that applies to the particular degradation. For instance, if the noise is maximum noise, we would not apply the hole-fillers of $\mathbf{L}_{13}$ and $\mathbf{L}_{14}$. A particularly important sublibrary collection consists of only those involving W_3. The examples will illustrate use of the sublibraries. Table 2 lists the sublibraries. Because there is some redundancy between sublibraries, the total number of struc-

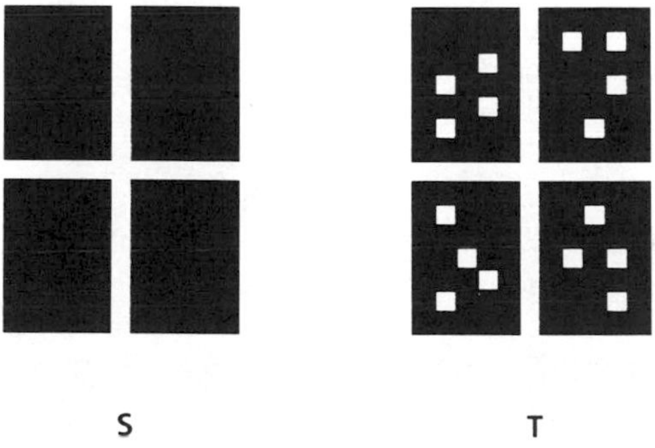

Figure 13. Ideal image *S* and salt-noise-corrupted image *T*. Certain hole-fillers of the expert library would have overfilled upon filtering.

Table 2. Elements of a Specific Expert Library

Sublibrary	Description	Number of elements
$\mathbf{L}_1$	3×3 singletons	9
$\mathbf{L}_2$	5×5–3×3 singletons	8
$\mathbf{L}_3$	Length-2 τ-openings	8
$\mathbf{L}_4$	Length-3 τ-openings	12
$\mathbf{L}_5$	2×2 opening	4
$\mathbf{L}_6$	Constrained 3×3 opening	9
$\mathbf{L}_7$	Length-3 erosions	4
$\mathbf{L}_8$	Length-4 erosions	8
$\mathbf{L}_9$	Length-5 erosions	4
$\mathbf{L}_{10}$	Nonlinear 3×3 erosions	3
$\mathbf{L}_{11}$	Nonlinear 5×5 erosions	2
$\mathbf{L}_{12}$	1-pixel hole-fillers	4
$\mathbf{L}_{13}$	2,3-pixel hole-fillers	12
$\mathbf{L}_{14}$	Strong- and weak-neighbor hole-fillers	2
$\mathbf{L}_{15}$	Strong- and weak-neighbor medians	20

turing elements in the full library $\mathbf{L}$ is less than the sum of the sublibrary counts, although the difference is not great.

It is important to recognize that our specific full library, and specific partition into sublibraries, is based on our view that our suboptimal filters must perform better than certain recognized filters. In practice, one may wish to (1) include more sublibraries, (2) employ a different partition, or (3) condense some of the sublibraries to achieve even greater computational savings.

Regarding the first point, it might be desirable to include the element collection depicted in Figure 14. Like the elements of $\mathbf{L}_{14}$, these are hole-fillers that protect against overestimation. The trade-off between larger libraries (and a hoped-for reduction in MSE) and small libraries (and computational tractability) is ever present in our approach. As noted above, our library choice has been made to accomplish certain ends, and different ends will require different libraries.

A second choice is partitioning. For instance, one might choose to partition the hole-fillers into more classes, thereby allowing certain irrelevant sublibraries to be eliminated from the design procedure. Although this certainly provides more flexibility, it also introduces more complexity into the optimization procedure. On the other hand, a coarser partition might include the hole-fillers in a single sublibrary. Here, however, there would be an immediate cost: the W_3 elements would be grouped with the W_5 elements.

Regarding sublibrary condensation, suppose one only wishes to employ the elements of $\mathbf{L}_{13}$ to help fill 3×3 holes in conjunction with elements of $\mathbf{L}_{12}$. Then

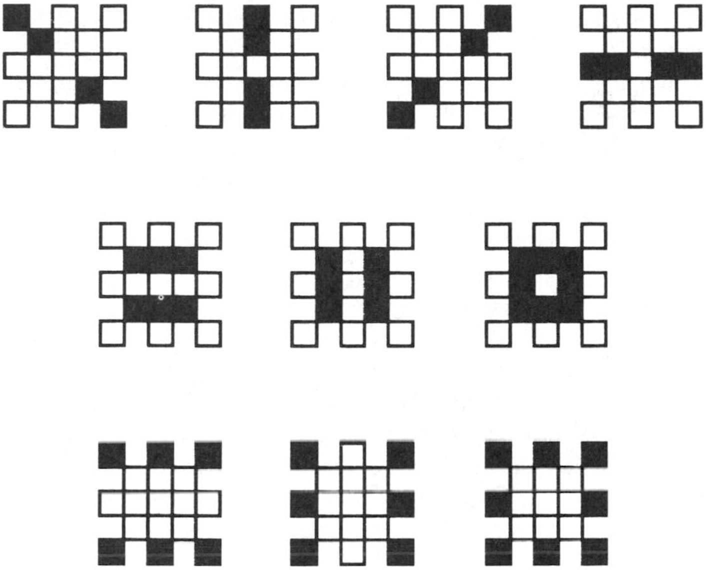

Figure 14. Hole-fillers that prevent overfilling.

one need only employ a single element from $\mathbf{L}_{13}$ and discard the rest; of course, some other filling capability would be sacrificed. For instance, should we only retain the first element of the fourth row of Figure 12g, then we would lose some capacity to fill pixels in a 4×4 hole.

B. First-Order Library

A library or sublibrary may be constructed on the basis of statistics of the individual structuring elements. One key statistic of a structuring element, relative to forming filter bases, is the MSE incurred by single-erosion filtering with that element. Limiting a library to structuring elements individually possessing low MSE is a *first-order-statistics* approach. Various sublibraries can be constructed using this method. These *first-order* sublibraries may be used in conjunction with one another, or with sublibraries formed by other methods, such as the expert approach. In designing a first-order library, the number of elements employed will typically be limited so that the design procedure is tractable and computationally feasible using the available computing tools.

The simplest methodology for obtaining a first-order library is to begin with a class $\mathbf{C}$ of structuring elements (perhaps all possible structuring elements in the window), find MSE for each single-erosion filter arising from a structuring element in $\mathbf{C}$, and let the library $\mathbf{L}$ be composed of the q structuring elements possessing the least MSE. The **L**-optimal or m-erosion **L**-optimal filter can then be found in the usual manner.

A more general approach is to suppose we have n classes of structuring, $\mathbf{C}_1$, $\mathbf{C}_2, \ldots, \mathbf{C}_n$, and, for $i = 1, 2, \ldots, n$, the q_i structuring elements possessing least first-order MSE are chosen to form the sublibrary $\mathbf{L}_i$. In fact, the expert- and first-order-library methodologies can be combined by supposing that $\{\mathbf{C}_1, \mathbf{C}_2, \ldots, \mathbf{C}_n\}$ is itself an expert library partition and the first-order statistics are being used to provide a derived sublibrary collection having fewer structuring elements.

For instance, suppose we have two classes $\mathbf{C}_1$ and $\mathbf{C}_2$ of structuring elements, where no element of $\mathbf{C}_1$, except the identity element $(\underline{1})$, which is assumed to be in $\mathbf{C}_1$, contains the origin, and all the elements of $\mathbf{C}_2$ contain the origin. Employed alone, $\mathbf{C}_1$ will yield filters that fill holes created by min noise but do not eliminate max noise; on the other hand, filters arising from $\mathbf{C}_2$ alone can eliminate max noise but cannot fill holes. It may happen that both $\mathbf{C}_1$ and $\mathbf{C}_2$ provide tractable filter design if used alone. Therefore, in the presence of either min or max noise alone we choose the appropriate class for filter design. However, in the presence of both min and max noise, we may very likely wish to select some structuring elements from each.

Suppose $\mathbf{C}_1$ and $\mathbf{C}_2$ contain n_1 and n_2 elements, respectively. Used alone (and without considering nonredundancy), $\mathbf{C}_i$ yields $n_i!/m!(n_i - m)!$ combinations of m basis elements. Used in conjunction, and supposing we wish m_1 elements from $\mathbf{C}_1$ and $m_2 = m - m_1$ elements from $\mathbf{C}_2$, there are

$$\frac{n_1!n_2!}{m_1!m_2!\,(n_1 - m_1)!\,(n_2 - m_2)!} \tag{2.30}$$

possible basis-element combinations. Thus, the computational burden increases dramatically, even though the number of structuring elements generating the filter remains the same. To reduce the computational burden, we can apply first-order statistical information to find smaller subclasses $\mathbf{L}_1$ and $\mathbf{L}_2$ in $\mathbf{C}_1$ and $\mathbf{C}_2$, respectively, thereby forming the sublibrary partition $\{\mathbf{L}_1, \mathbf{L}_2\}$.

Use of first-order libraries has implications for algebraic properties of the library-optimal filters. The Matheron expansion allows a morphological filter to be described by the interaction of a set of basis elements. Commonly employed filters, such as an opening filter, possess bases where each of the individual structuring elements may or may not contribute greatly to the reduction of estimation MSE. Basis elements contributing little to the estimation accuracy may not be members of the first-order library. Yet each basis element, regardless of its contribution to the goodness of estimation, is required for certain algebraic properties, such as idempotence. However, quite often in estimation applications (e.g., image restoration) these algebraic properties are not required. This form of basis-element interaction, to obtain certain algebraic properties, is a constraint on filter design and is not considered in construction of the first-order library. Consequently, it may not be possible to obtain certain traditional morphological filters (e.g., open, close) when using a first-order library, but those filters may not be relevant to the given estimation problem. Note that design of the traditional filters can be allowed through the expert approach.

VI. PROBABILISTIC ANALYSIS OF MEAN-SQUARE ERROR

The present section states a morphological filter MSE theorem that can be employed in a computer algorithm to search for an optimal basis. In effect, the theorem states that the MSE of a morphological filter can be expressed as a linear combination of the MSEs of its individual basis elements and their unions. An alternative approach would be to utilize Eqs. (2.16) and (2.18), which give the numerical MSE expressions for single-erosion and multiple-erosion filters, respectively. Under the assumption of stationarity, these equations can be employed to estimate MSE from image realizations by comparing filtered noise-corrupted realizations to corresponding uncorrupted realizations; however, rather than actually filtering images to determine the structuring elements constituting an optimal-filter basis, it is computationally more efficient to employ the general theorem regarding MSE for morphological filters. The theorem is proved in the Appendix.

In deriving the theorem we will employ certain structuring element "fit," or "subset," statistics based on an image model and image degradation model (or, more generally, image transformation model), or realizations of such models. First, we will describe the fit statistics, how they pertain to two types of estimation errors that can occur upon single-erosion filtering, and how they lead to an expression for MSE of the single-erosion filter. To provide understanding of how the general theorem for m erosions will lead to a basis search algorithm, we will next derive the simpler two-erosion MSE formula. Strong stationarity of the image process and degradation process is assumed throughout the derivation. A form of the m-erosion MSE expression is the basis of our optimal filter design algorithm. The algorithm, exercised on a computer, provides estimates of MSE for filter bases of a given size (number of structuring elements). Exercised for a range of basis sizes, to an allowed limit, the algorithm provides the basis that yields minimum MSE.

To proceed, let S and S' respectively denote the uncorrupted and corrupted image, let B_z denote the structuring element B translated by the vector z, $B_z = B + z$, and let K_z denote the set of observed pixels about the pixel z, $K_z = S' \cap W_z$, where W_z is an observation window W translated by z. Under stationarity, we can speak of the MSE for a filter as being an image error, since it is pixel independent. For a single erosion by B,

$$\begin{aligned}\mathrm{MSE}\langle B\rangle &= E\left[\,|\, S(z) - (S' \ominus B)(z)|^2\right] \\ &= P\left[S(z) \neq (S' \ominus B)(z)\right] \\ &= P\left[\left(S(z) = 1\right) \cap \left((S' \ominus B)(z) = 0\right)\right] \\ &\quad + P\left[\left(S\,(z) = 0\right) \cap \left((S' \ominus B)(z) = 1\right)\right]\end{aligned} \tag{2.31}$$

As written, MSE$\langle B\rangle$ is evaluated relative to sets being treated as {0,1}-valued functions. However, each of the events in Eq. (2.31) possesses a random-set formulation. For instance, the random-function event $[(S' \ominus B)(z) = 0]$ possesses the equivalent random-set formulation $[B_z \not\subset K_z]$. Similar considerations apply to more general filters, Eq. (2.31) holding for MSE $\langle\Psi\rangle$ with $\Psi(S')$ in place of $S' \ominus B$. If $\Psi(S') = \max\{S' \ominus B_i\}$, then the following random-function and random-set events are equivalent:

$$\left[\Psi(S')(z) = 0\right] = \left[\bigvee (S' \ominus B_i)(z) = 0\right] \tag{2.32a}$$

$$\left[z \notin \Psi(S')\right] = \bigcap \left[(B_i)_z \not\subset K_z\right] \tag{2.32b}$$

In Eq. (2.32b), it must be kept in mind that the intersection is an intersection of events. To avoid needless notational pedantry, in the sequel we will frequently mix random-function and random-set formulations.

When estimating $S(z)$ by $(S' \ominus B)(z)$, one of two types of estimation error can occur. We define *type-0 error* as that which occurs when $S(z) = 1$ but $(S' \ominus B)(z) = 0$; eroding the observed image results in a zero at a location where the ideal value is one. *Type-1 error* occurs when $S(z) = 0$ but $(S' \ominus B)(z) = 1$; the erosion estimate is one where the ideal state of the pixel is zero. The probabilities of type-0 and type-1 errors occurring can be written as probabilities of event intersections:

$$\begin{aligned}p_0[B] &= P\left[\left((S' \ominus B)(z) = 0\right) \cap \left(S(z) = 1\right)\right] \\ &= P\left[\left(B_z \not\subset K_z\right) \cap \left(z \in S\right)\right]\end{aligned} \tag{2.33}$$

and

$$\begin{aligned}p_1[B] &= P\left[\left((S' \ominus B)(z) = 1\right) \cap \left(S(z) = 0\right)\right] \\ &= P\left[\left(B_z \subset K_z\right) \cap (z \notin S)\right]\end{aligned} \tag{2.34}$$

where $p_0[B]$ and $p_1[B]$ are the probabilities of type-0 and type-1 estimation errors, respectively, when erosion by B is the estimation rule. Note the equivalent events $[(S' \ominus B)(z) = 0] = [B_z \not\subset K_z]$ and $[(S' \ominus B)(z) = 1] = [B_z \subset K_z]$, where erosion is treated as set exclusion or inclusion for derivation purposes. The mean-square error of estimation at z is the sum of these two mutually exclusive error probabilities:

$$\begin{aligned} \text{MSE}\,\langle B\rangle &= p_0[B] + p_1[B] \\ &= P\left[\left(B_z \not\subset K_z\right) \cap \left(z \in S\right)\right] + P\left[\left(B_z \subset K_z\right) \cap \left(z \notin S\right)\right] \end{aligned} \tag{2.35}$$

It is advantageous to view Eq. (2.35) from a slightly different perspective:

$$\text{MSE}\,\langle B\rangle = P\left[\left(B_z \subset K_z\right) \Delta \left(z \in S\right)\right], \tag{2.36}$$

where Δ is the symmetric difference of the events.

The assumption of strong stationarity results in the MSE relative to the entire image being equal to the pointwise MSE of Eq. (2.36). Provided with a suitable image model and image transformation model, we may extract the probabilities of type-0 and type-1 errors and thus calculate MSE for a given single-erosion filter. Depending on the nature of the models, it may be more convenient to state p_0 and p_1 in a particular form, such as conditional probabilities, set inclusions, or erosions.

Toward our goal of deriving and understanding a theorem relating single-erosion MSE and m-erosion MSE, we next examine the two-erosion case. The probability of type-0 error can be written

$$\begin{aligned} p_0[B_1, B_2] &= P\left[\left((S' \ominus B_1)(z) \vee (S' \ominus B_2)(z) = 0\right) \cap \left(S(z) = 1\right)\right] \\ &= P\left[\left[\left((B_1)_z \not\subset K_z\right) \cap \left((B_2)_z \not\subset K_z\right)\right] \cap \left(z \in S\right)\right] \end{aligned} \tag{2.37}$$

Equation (2.37) can be simplified to a sum of single-erosion type-0 probabilities by using $P[C \cap D] = P[C] + P[D] - P[C \cup D]$, with $C = [(B_1)_z \not\subset K_z] \cap [z \in S]$ and $D = [(B_2)_z \not\subset K_z] \cap [z \in S]$:

$$\begin{aligned} p_0[B_1, B_2] &= P\left[\left((B_1)_z \not\subset K_z\right) \cap \left(z \in S\right)\right] \\ &\quad + P\left[\left((B_2)_z \not\subset K_z\right) \cap \left(z \in S\right)\right] \\ &\quad - P\left[\left[\left((B_1)_z \not\subset K_z\right) \cup \left((B_2)_z \not\subset K_z\right)\right] \cap \left(z \in S\right)\right] \end{aligned} \tag{2.38}$$

The event that at least one structuring element is not a subset of the observation set is equivalent to the union of structuring elements not fitting in K_z:

$$\left((B_1)_z \not\subset K_z\right) \cup \left((B_2)_z \not\subset K_z\right) = \left((B_1 \cup B_2)_z \not\subset K_z\right) \tag{2.39}$$

Substituting into Eq. (2.38), we obtain

$$\begin{aligned} p_0[B_1, B_2] &= P\left[\left((B_1)_z \not\subset K_z\right) \cap \left(z \in S\right)\right] \\ &\quad + P\left[\left((B_2)_z \not\subset K_z\right) \cap \left(z \in S\right)\right] \\ &\quad - P\left[\left((B_1 \cup B_2)_z \not\subset K_z\right) \cap \left(z \in S\right)\right] \end{aligned} \tag{2.40}$$

and finally, by recognizing the form of Eq. (2.33),

$$p_0[B_1,B_2] = p_0[B_1] + p_0[B_2] - p_0[B_1 \cup B_2] \tag{2.41}$$

From Eq. (2.41) we see that knowing the probability of type-0 estimation error, individually, of any two structuring elements and their union, we may simply calculate the probability of type-0 error when the two elements are used as a morphological filter basis.

Probability of type-1 error can be calculated similarly:

$$\begin{aligned} p_1[B_1,B_2] &= P\left[\left((S' \ominus B_1)(z) \vee (S' \ominus B_2)(z) = 1\right) \cap \left(S(z) = 0\right)\right] \\ &= P\left[\left[\left((B_1)_z \subset K_z\right) \cup \left((B_2)_z \subset K_z\right)\right] \cap \left(z \notin S\right)\right] \end{aligned} \tag{2.42}$$

Apply $P[C \cup D] = P[C] + P[D] - P[C \cap D]$, where $C = [(B_1)_z \subset K_z] \cap [z \notin S]$ and $D = [(B_2)_z \subset K_z] \cap [z \notin S]$:

$$\begin{aligned} p_1[B_1,B_2] &= P\left[\left((B_1)_z \subset K_z\right) \cap \left(z \notin S\right)\right] \\ &\quad + P\left[\left((B_2)_z \subset K_z\right) \cap \left(z \notin S\right)\right] \\ &\quad - P\left[\left[\left((B_1)_z \subset K_z\right) \cap \left((B_2)_z \subset K_z\right)\right] \cap \left(z \notin S\right)\right] \end{aligned} \tag{2.43}$$

The event where both structuring elements are subsets of the observation set is equivalent to the union of the structuring elements being a subset of K_z:

$$\left((B_1)_z \subset K_z\right) \cap \left((B_2)_z \subset K_z\right) = \left((B_1 \cup B_2)_z \subset K_z\right) \tag{2.44}$$

Substituting into Eq. (2.43), we obtain an expression that can be written in the same form as Eq. (2.41):

$$p_1[B_1,B_2] = p_1[B_1] + p_1[B_2] - p_1[B_1 \cup B_2] \tag{2.45}$$

Mean-square error of the 2-erosion filter is the sum of the two mutually exclusive error probabilities:

$$\begin{aligned}\text{MSE}\langle B_1,B_2\rangle &= p_0[B_1,B_2] + p_2[B_1,B_2] \\ &= p_0[B_1] + p_0[B_2] - p_0[B_1 \cup B_2] + p_1[B_1] \\ &\quad + p_1[B_2] - p_1[B_1 \cup B_2]\end{aligned} \tag{2.46}$$

which itself is a sum of mean-square errors,

$$\text{MSE}\langle B_1,B_2\rangle = \text{MSE}\langle B_1\rangle + \text{MSE}\langle B_2\rangle - \text{MSE}\,\langle B_1 \cup B_2\rangle \tag{2.47}$$

The key result of this probabilistic analysis is that the MSE of a 2-erosion filter is equal to a linear combination of the MSEs of the individual structuring elements and their union. We found a similar result with the individual error probabilities [Eqs. (2.41) and (2.45)]. Equation (2.47) leads us to a minimum MSE filter basis search strategy. Knowing the MSE values for individual structuring elements, which may have been chosen for, say, their filter properties, and knowing the corresponding unions, one may calculate all 2-erosion MSE values, and the optimal erosion basis is the pair that minimizes Eq. (2.47).

It is also worth interpreting Eq. (2.47) from a morphological filtering perspective. Equation (2.47), in essence, counts estimation errors incurred upon filtering with a 2-erosion filter. The terms MSE$\langle B_1\rangle$ and MSE$\langle B_2\rangle$ count the errors incurred upon eroding individually with B_1 and B_2, respectively. When overall estimation error (MSE$\langle B_1, B_2\rangle$) is written in terms of individual errors, a subtraction of the mean-square error of the structuring-element union (MSE$\langle B_1 \cup B_2\rangle$) must be performed so that errors are not counted twice.

To provide a key intermediate step to the understanding and derivation of the general m-erosion mean-square error, we end the 2-erosion analysis by stating the MSE in terms of the symmetric difference. De Morgan's law indicates that Eqs. (2.37) and (2.42) contain the complementary events

$$\left[\left((B_1)_z \not\subset K_z\right) \cap \left((B_2)_z \not\subset K_z\right)\right] = \left[\left((B_1)_z \subset K_z\right) \cup \left((B_2)_z \subset K_z\right)\right]^c \tag{2.48}$$

Using this relationship, we can express 2-erosion mean-square error as the probability of the symmetric difference of events:

$$\text{MSE}\langle B_1,B_2\rangle = P\left[\left[\left((B_1)_z \subset K_z\right) \cup \left((B_2)_z \subset K_z\right)\right] \Delta \left[z \in S\right]\right] \tag{2.49}$$

We now turn to the general case, stating the morphological filter MSE theorem. The theorem states that the MSE of a morphological filter can be expressed as a linear combination of the MSEs of its individual basis elements and their unions. A proof is given in the Appendix.

Theorem. An m-erosion morphological filter Ψ possessing basis Bas[Ψ] = $\{B_1, B_2, \ldots, B_m\}$ will provide a point estimate with mean-square error given by

$$\mathrm{MSE}\langle\Psi\rangle = \sum_{j=1}^{m} (-1)^{j+1} \sum_{1\le i_1<i_2<\cdots<i_j\le m} \mathrm{MSE}\langle \cup_{k=1}^{j} B_{i_k} \rangle \tag{2.50}$$

Besides providing insight into MSE incurred upon morphological filtering, the morphological filter MSE theorem provides a mechanism for searching for an optimal filter basis. To see this, we write the theorem in expanded form for a specific basis size. The MSE of a 3-erosion filter is given by

$$\begin{aligned}\mathrm{MSE}\langle B_1,B_2,B_3\rangle = {} & \mathrm{MSE}\langle B_1\rangle + \mathrm{MSE}\,\langle B_2\rangle + \mathrm{MSE}\,\langle B_3\rangle \\ & -\mathrm{MSE}\,\langle B_1\cup B_2\rangle - \mathrm{MSE}\,\langle B_1\cup B_3\rangle \\ & - \mathrm{MSE}\,\langle B_2\cup B_3\rangle \\ & + \mathrm{MSE}\,\langle B_1\cup B_2\cup B_3\rangle\end{aligned} \tag{2.51}$$

Single-erosion-filter MSE for structuring elements B_1, B_2, B_3 and their unions may be obtained through image models and image degradation models. MSE of the corresponding 3-erosion filter may be calculated through the linear operations indicated in Eq. (2.51). An optimal 3-erosion filter may be designed by examining libraries of structuring-element candidates, where, knowing single-erosion statistics, all 3-tuples are evaluated by Eq. (2.51) and the combination yielding minimum MSE is optimal. In general, the morphological filter MSE theorem may be exercised for a range of basis size, to an allowed limit m, thereby providing the m-optimal filter.

Note that the generality of the MSE theorem does allow for evaluation of redundant structuring-element combinations but reduces to nonredundant forms. For example, consider the structuring elements B_1, B_2, and B_3 as a potential basis, where B_1 and B_2 are redundant ($B_1 \subset B_2$). By examining Eq. (2.51) we see that $\mathrm{MSE}\langle B_1,B_2,B_3\rangle = \mathrm{MSE}\langle B_1,B_3\rangle$.

VII. EXAMPLES

To illustrate constrained optimal morphological filter performance we will examine several image restoration examples involving different image types and image degradation types. Images in their ideal state, degraded, and filtered will be shown with the corresponding filter bases and mean-square error.

A. Rectangular Image Primitives, Ragged Min Noise

Consider an image process and noise process, similar to the example in Figure 6, where the image process consists of rectangles uniformly randomly distributed in the grid, and each rectangle is Z_1 by Z_2, Z_1 and Z_2 being uniformly distributed over {8, 9, 10, 11, 12} and {5, 6, 7}, respectively, and where area coverage is approximately 10%. A 256 by 256 pixel^2 realization of the process is depicted in Figure 15a. For the noise process, each component is ragged in shape and con-

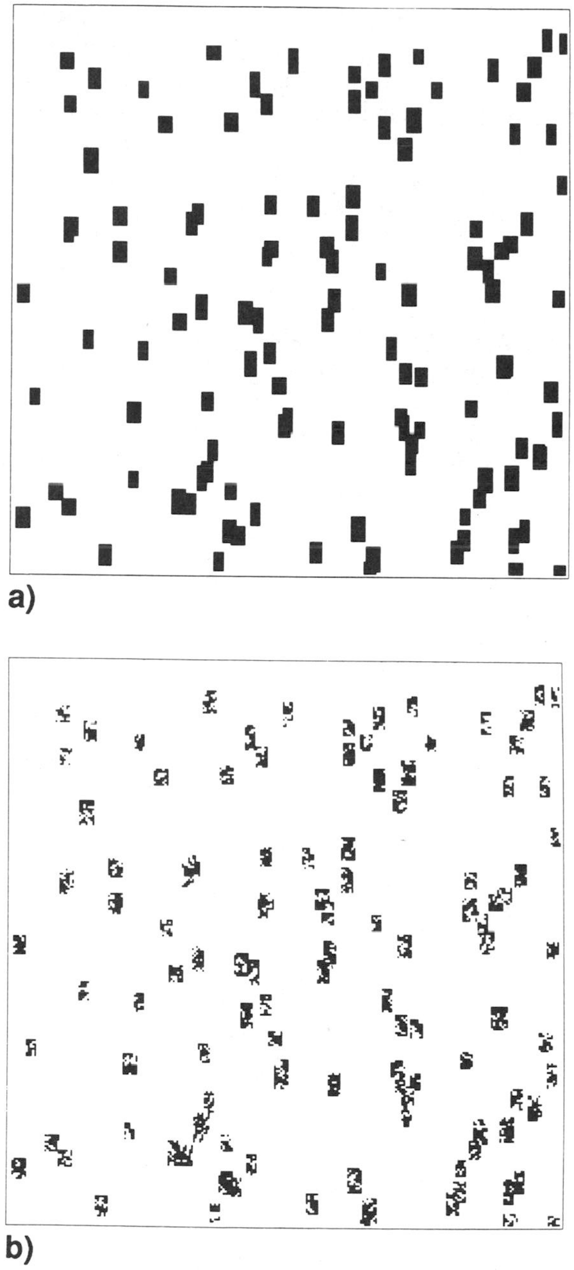

Figure 15. For example A: (a) realization of image process; (b) realization of image-noise process.

sists of one to four pixels. These are uniformly randomly distributed in the grid and together possess approximately 30% area coverage. The corrupted image is the set subtraction of the noise from the uncorrupted image (min noise). A realization of the image-noise process is shown in Figure 15b with MSE = 0.0310.

Filters have been designed using the expert library and first-order-library methods. A graph showing MSE versus basis size is given in Figure 16. We see that the curve is leveling off in the region of six through eight basis elements, and beyond this point there is only small improvement for increasing basis size. Filtered versions of the noisy image are shown in Figure 17a and b for 8-erosion filters designed using an expert library and a first-order library, respectively. Because of their dilation, hole filling (closing), and median properties, the expert-library method employed sublibraries $\mathbf{L}_1$,$\mathbf{L}_2$,$\mathbf{L}_{12}$,$\mathbf{L}_{13}$,$\mathbf{L}_{14}$, and $\mathbf{L}_{15}$, resulting in a restored image with MSE=0.0101. The first-order method employed a 60-element *min type* library, which consisted of the single-pixel origin element and the 59 structuring elements possessing lowest MSE and having an unactivated pixel at the origin. It is termed a "min library" because the elements have a re-

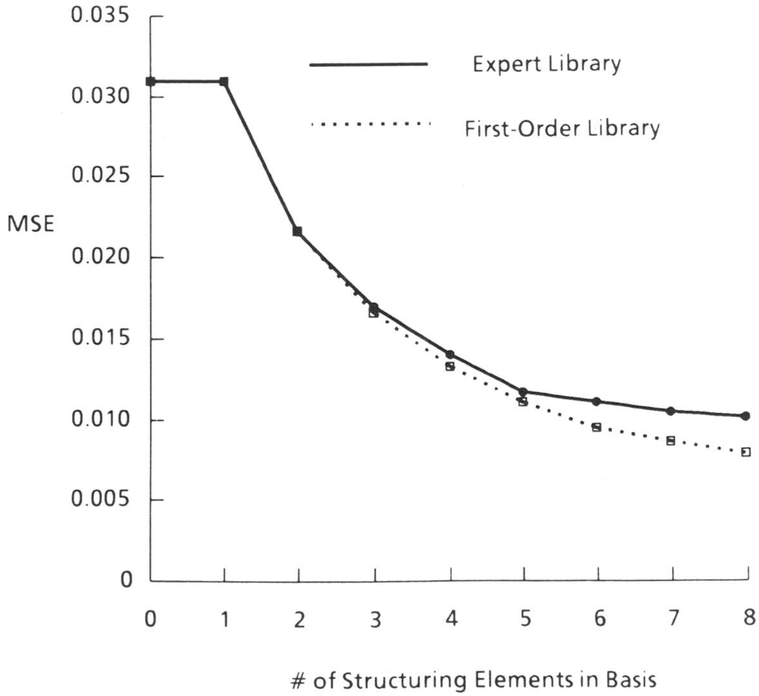

Figure 16. For example A: MSE versus basis size.

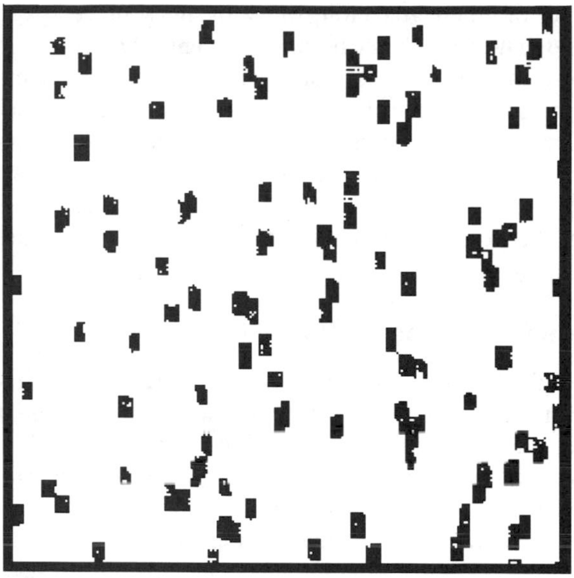

a)

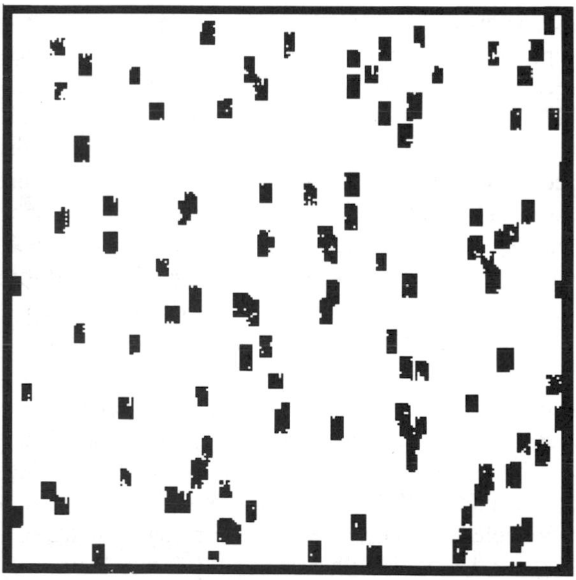

b)

Figure 17. For example A: image filtered using (a) expert method and (b) first-order method.

storative effect in the presence of min degradation (where the noise process consists of set subtractions). The first-order library is shown in Figure 18, where the MSE increases from left to right and top to bottom. Notice that there is some grouping of mirror-symmetric structuring elements, which is due to the symmetry of the image primitives and noise components. Given a larger realization from which to extract the statistics, the grouping would likely have been more ideal. A restored image with MSE = 0.0079 resulted from the first-order-library filter. The optimal filter bases for the expert approach and the first-order method are shown in Figure 19a and b, respectively.

B. Rectangular Image Primitives, Salt-and-Pepper Noise

Conventional morphological filtering can often be applied with reasonable success when either max or min noise alone is present and the image consists of relatively simple primitives, the appropriate filters being τ-openings and τ-closings, respectively. If the noise is mixed, so that pixels are both adjoined and deleted, alternating sequential filters (iterated openings and closings) work well if the noise components are small relative to the image components and the noise components are not overly dense. Now consider the image process of the previous example with a uniformly randomly distributed point noise degradation process consisting of 15% area coverage max noise and then 15% min noise, to yield a salt-and-pepper-like effect as shown in the 256×256 pixel^2 realization of Figure 20 (MSE = 0.1300). Opening or closing will not have a restorative effect. The expert library employed was the union of the 15 expert sublibraries described above, $\mathbf{L} = \mathbf{L}_1 \cup \mathbf{L}_2 \cdots \cup \mathbf{L}_{15}$. A 2-partition first-order library possessing 80 structuring elements was also employed. The partitioning was such that 71 min elements (as described in the previous example) and 8 max elements were employed. Max elements have an activated origin pixel and have a restorative effect when union noise is the degradation process. The numbers of elements in the sublibraries were chosen so that the total was sufficiently small to optimize over in a timely manner, and 90% and 10% of the elements were max noise correcting and min noise correcting, respectively, to correspond to the observed area coverage of these two noise types.

A graph showing basis size versus MSE is given in Figure 21, where we see a flattening curve beyond six structuring elements. Employing the expert-library 6-erosion optimal filter, we obtain the restored image shown in Figure 22a. Mean-square error has been reduced to 0.0214. The basis elements are shown in Figure 22(b): two 2×2 opening elements ($\mathbf{L}_5$), two linear-erosion elements ($\mathbf{L}_7$, $\mathbf{L}_8$), a protected hole-filler ($\mathbf{L}_{14}$), and a median element ($\mathbf{L}_{15}$). It is highly unlikely that a complex basis, such as shown here, could be designed by common heuristic guidelines and methods. In comparing the expert-library method to the first-order method, we note that a very simple form of first-order library was used

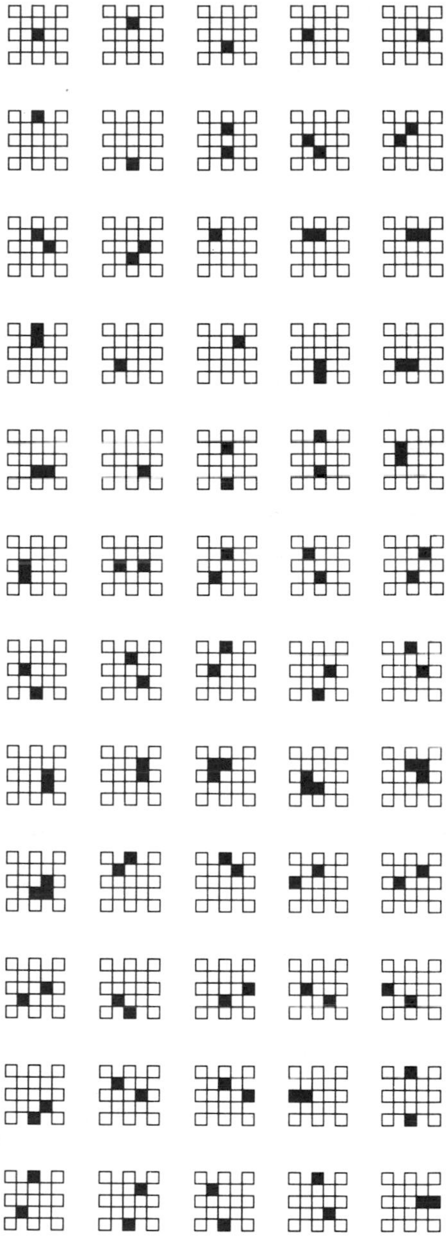

Figure 18. For example A: first-order library.

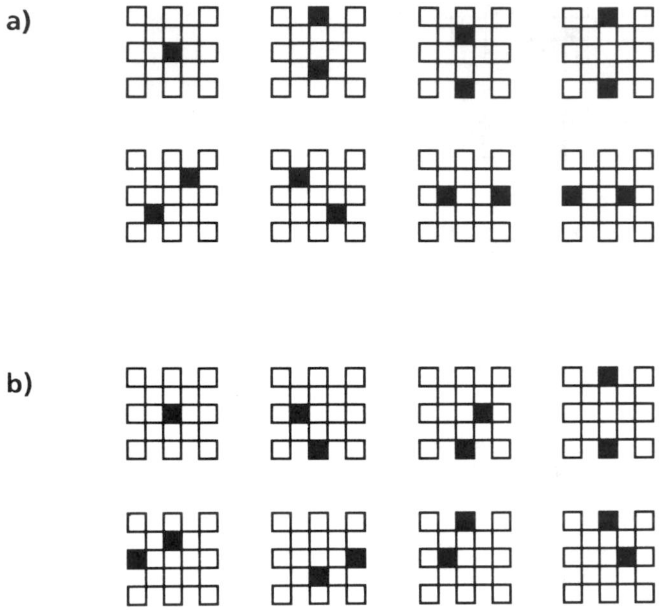

Figure 19. For example A: filter bases for (a) expert method and (b) first-order method.

here, and a combination of well-chosen first-order sublibraries may have performed much better. Then again, the expert library used here is only an example of the expert-type method of library construction.

In designing a filter, one must consider robustness with respect to variations in the degradation process. It is desirable for a filter to provide good estimation over a range of degradation conditions. To illustrate optimal morphological filter robustness, the noise process shown in Figure 20 was varied to produce 5%, 10%, and 15% area coverage. Optimal filters were designed for each noise level. The basis size-MSE curves and the 6-erosion bases are shown in Figures 23 and 24, respectively. We see a similarity among the bases and might assume that they could perform with similar restoration efficiency over the variation in noise level. To test this, each filter was applied to restoring the image at the three noise levels. The resulting MSEs and filter efficiencies are shown in Table 3, where we see that a filter designed for a given noise level performed with <2% loss in efficiency when applied at other noise levels. Note that we refer to filter efficiency as the fractional decrease in MSE upon filtering.

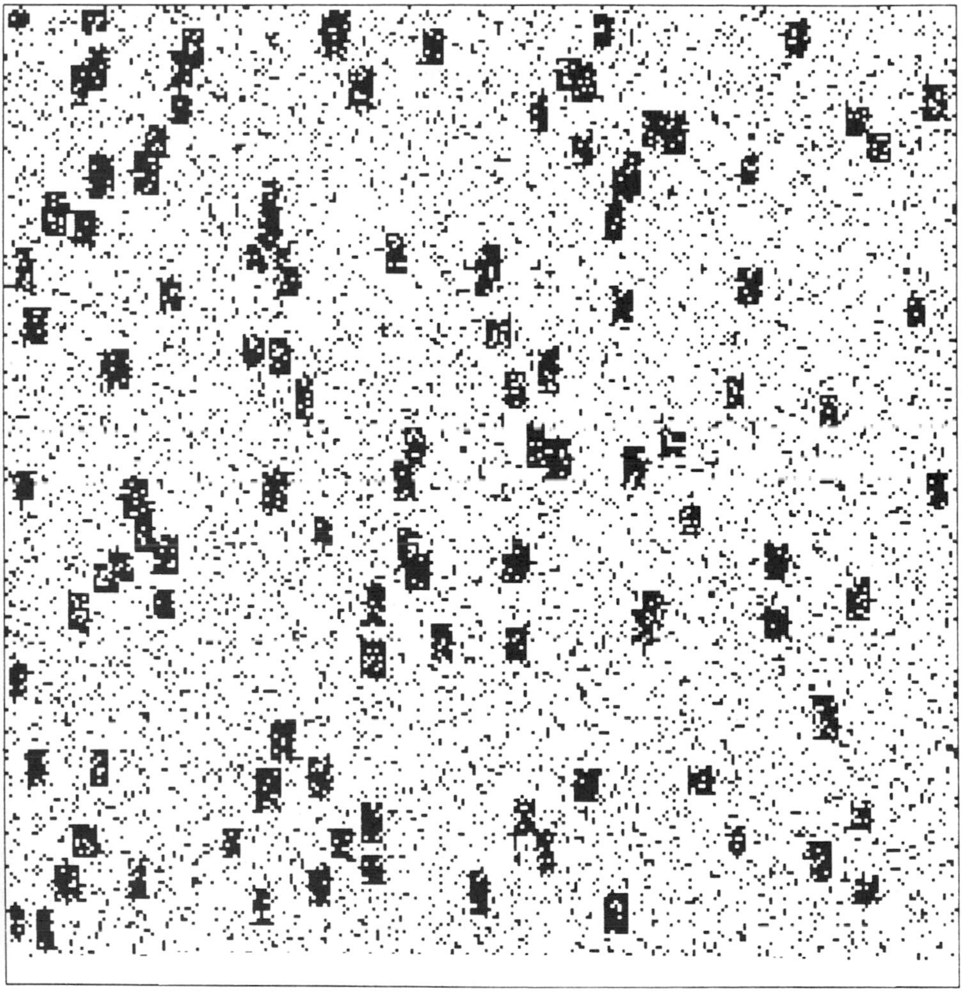

Figure 20. For example B: realization of image-noise process.

C. Disk-Shaped Image Primitives, Ragged Max and Min Noise

We now examine an image process with disk-shaped image primitives. The disks are uniformly randomly distributed on the grid, and prior to digitization each disk possesses a radius R uniformly distributed over $\{5, 6, 7, 8, 9, 10, 11, 12\}$. Also consider a noise process where max noise and then min noise is applied, each at 10% area coverage, the components are randomly chosen subsets of a

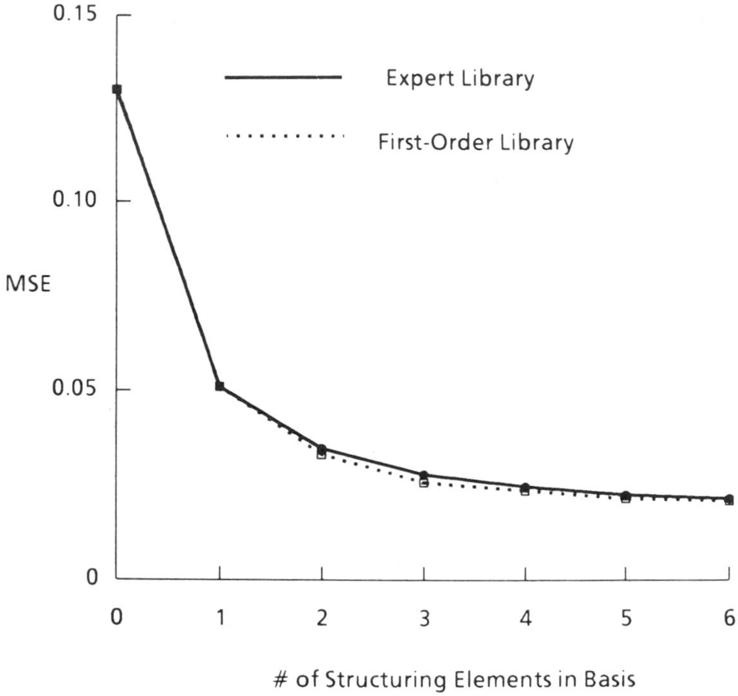

Figure 21. For example B: MSE versus basis size.

2×2 pixel block, and the components are uniformly randomly distributed in the grid (MSE $= 0.0926$). Realizations of the image process and the image-noise process are shown in Figure 25a and b, respectively. Concerning filter design, this image type is an increase in complexity over previous examples because sampling effects cause the primitives to be represented in a nonideal manner. The noise process is also more complex, making design by heuristic methods very difficult. MSE versus basis size is plotted in Figure 25c. A 6-erosion filter designed using the complete expert library produced the restoration shown in Figure 25d (MSE $= 0.0263$). The filter basis is shown in Figure 25e, where we see it consists of two linear-horizontal opening elements ($\mathbf{L}_4$) and four structuring elements from the weak-neighbor median basis ($\mathbf{L}_{15}$).

D. Text Image, Ragged Max and Min Noise

Thus far the images of the examples have been synthetic; the process was the union of simple primitives and the image-noise process was stationary. Next let us consider a real text image digitized at 400 spots per inch by a Xerox 7650

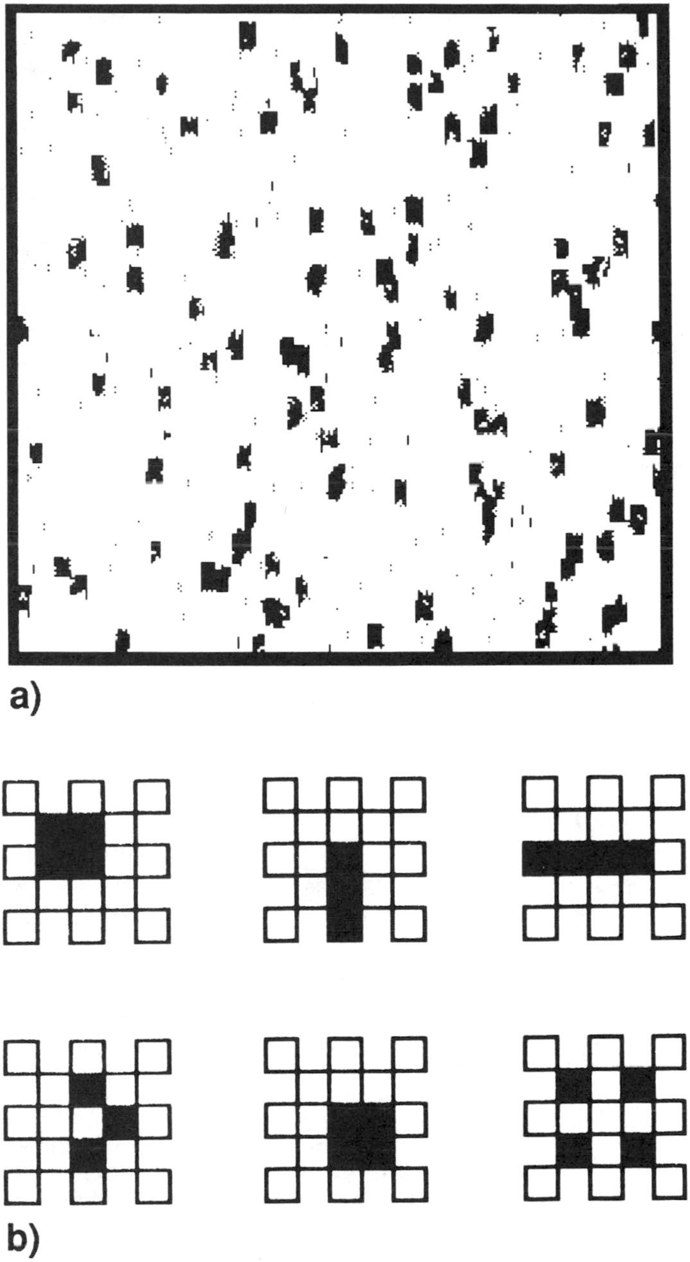

Figure 22. For example B: (a) image restored using expert library method; (b) filter basis.

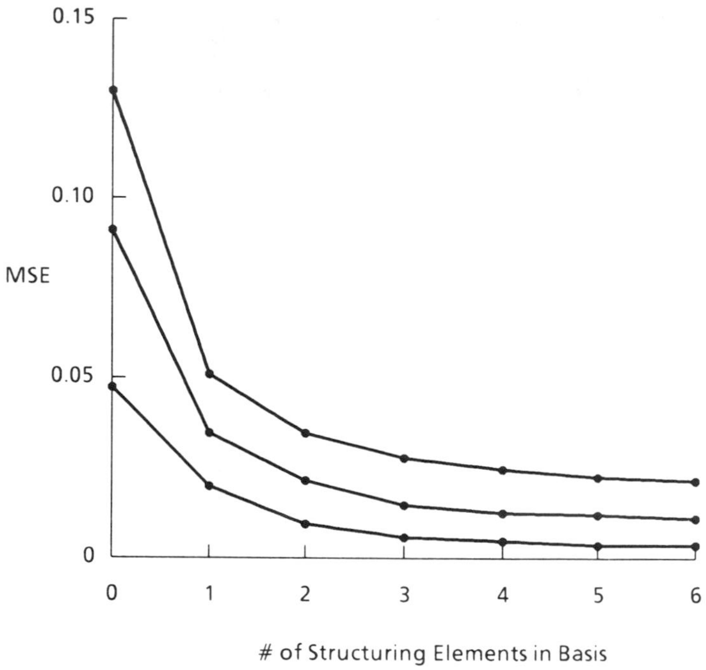

Figure 23. MSE versus basis size for the image-noise process of example B, with area coverage of 5%, 10%, and 15%.

document scanner. It is roughly 18% area coverage and has been thresholded to 1 bit per pixel. The image shown in Figure 26a is a 10-point classic font in a 512×512 pixel^2 window. It may be considered a typically good scan for the sampling resolution and size of the serifed font that is present. Image degradation was again performed artificially in a controlled manner so that we may be able to quantify the restoration ability of the filter. In this case the 10% min/max noise process of the previous example was used. A realization of the image degraded by this process, and having MSE = 0.0925, is shown in Figure 26b. The complete expert library was employed, and the result of optimal 6-erosion filtering is shown in Figure 26c. Mean-square error has been reduced to 0.0341. The basis elements are shown in Figure 26d, where we see they are a mixture of horizontal, vertical, and diagonal linear-erosion structuring elements, as well as strong- and weak-neighbor medians.

VIII. CONCLUSION

Several methods have been developed for reducing computation in the design of optimal binary digital morphological filters. These include basis-size, observa-

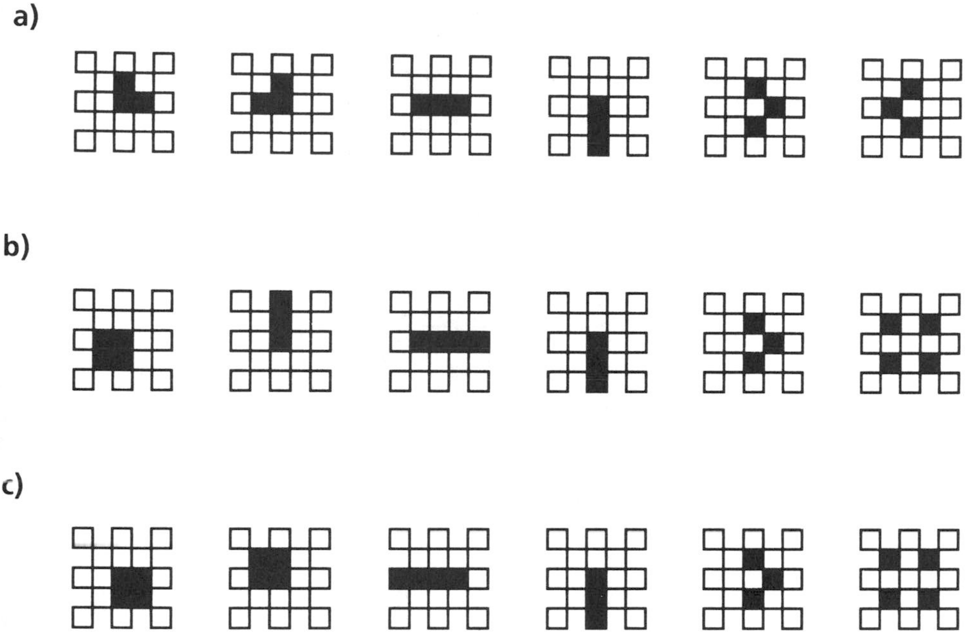

Figure 24. Optimal 6-erosion bases for (a) 5% noise, (b) 10% noise, and (c) 15% noise.

tion-window, and structuring-element-library constraints. Particular attention has been paid to the latter two approaches. Window constraints eliminate certain pixels from consideration to reduce search complexity, and library constraints restrict filter bases to either prechosen structuring elements or those providing good single-erosion filtering (expert libraries and first-order libraries, respectively). All constraint methodologies achieve design tractability at the cost of higher MSE (and hence suboptimality), the key point being to proceed in a manner that provides efficiency from both perspectives, design time and image restoration. A side effect of constraint can be a loss of desired algebraic properties; however, if restoration is the key issue, the loss of desirable properties may simply have to be accepted. A natural problem that deserves further investigation, and was only briefly touched upon in the current study, is the effect of constraint on filter properties.

An MSE theorem for morphological filters has been given that can be used to search efficiently for an optimal filter basis. The theorem states that the MSE of a morphological filter is equal to a linear combination of the MSEs of its individual basis elements and their unions.

Table 3. MSE and Filter Efficiency for the Three Optimal 6-Erosion Bases of Figure 24

Area coverage of noise	Unfiltered MSE	Filter Optimized for 5% Noise		Filter Optimized for 10% Noise		Filter Optimized for 15% Noise	
		MSE	Efficiency	MSE	Efficiency	MSE	Efficiency
5%	4.740×10^{-2}	3.507×10^{-3}	0.926	4.278×10^{-3}	0.910	4.262×10^{-3}	0.910
10%	9.910×10^{-2}	1.231×10^{-2}	0.876	1.111×10^{-2}	0.888	1.169×10^{-2}	0.882
15%	1.300×10^{-1}	2.478×10^{-2}	0.809	2.219×10^{-2}	0.829	2.136×10^{-2}	0.836

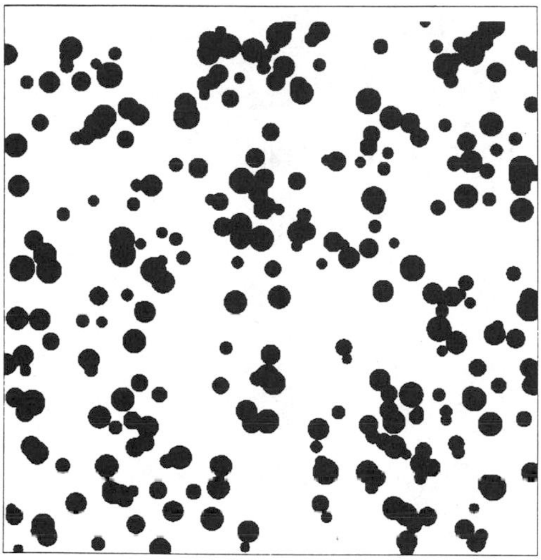

a)

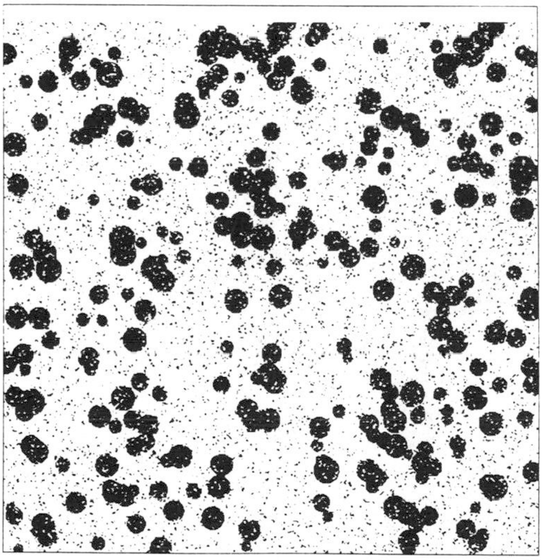

b)

Figure 25. For example C: (a) realization of image process; (b) realization of image-noise process; (c) MSE versus basis size; (d) filtered image; (e) filter basis.

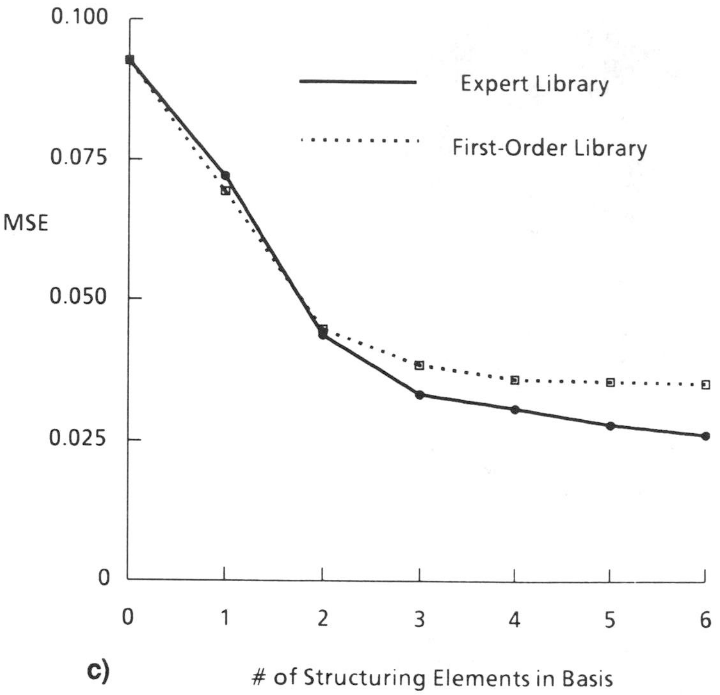

c)

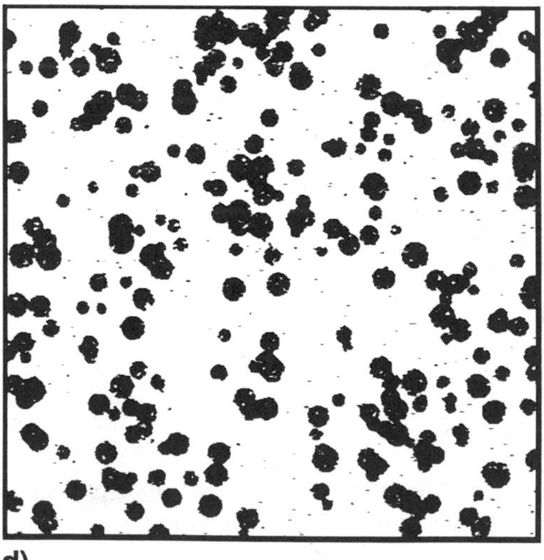

d)

Figure 25. (Cont'd)

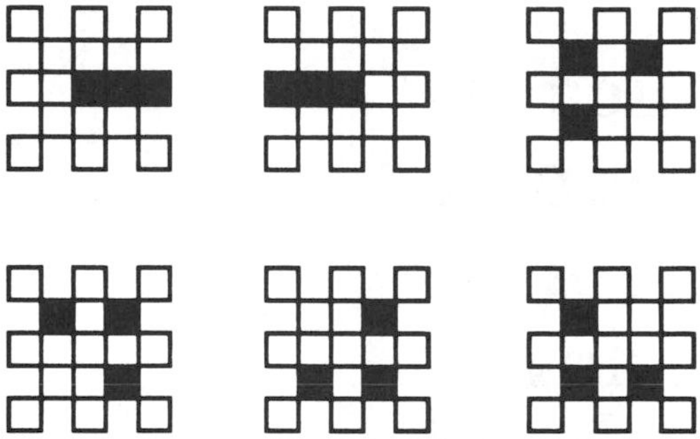

e)

REFERENCES

1. Dougherty, E. R., Optimal mean-square *N*-observation digital morphological filters. Part I: Optimal binary filters, *J. Comput. Vision Graphics Image Process, 55* (1), 36–54, (1992).
2. Dougherty, E. R., Optimal mean-square *N*-observation digital morphological filters. Part II: Optimal gray scale filters, *J. Comput. Vision Graphics Image Process, 55(1)*, 55–72, (1992).
3. Dougherty, E. R., The optimal mean-square *N*-observation digital binary morphological filter, in *Proceedings SPSE 43rd Annual Conference*, Rochester, New York, June 1990.
4. Dougherty, E. R., Minimal search for the optimal mean-square digital gray-scale morphological filter, *Proc. SPIE, 1360*, 214–226 (1990).
5. Matheron, G., *Random Sets and Integral Geometry*, Wiley, New York, 1975.
6. Dougherty, E. R., Mathew, A., and Swarnakar, V., A conditional-expectation-based implementation of the optimal mean-square binary morphological filter, *Proc. SPIE, 1451*, 137–147 (1991)
7. Dougherty, E. R., and Haralick, R. M., The hole spectrum—model-based optimization of morphological filters, *Proc. SPIE, 1568*, 224–232 (1991).
8. Dougherty, E. R., and Loce, R. P. Constrained optimal digital morphological filters, in *Proceedings of the 25th Annual Conference on Information Sciences and Systems*, Baltimore, March 1991.
9. Loce, R. P., and Dougherty, E. R., Using structuring element libraries to design suboptimal morphological filters, *Proc. SPIE, 1568*, 233–246 (1991).
10. Dougherty, E. R., and Loce, R. P., Facilitation of optimal morphological filter design via structuring element libraries and design constraints, *SPIE Opt. Eng., 31* (5), (1992).
11. Schonfeld, D., and Goutsias, J., Optimal morphological pattern restoration from noisy binary images, *IEEE PAMI, 13*(1), 14–29 (1991).
12. Serra, J., ed., *Image Analysis and Mathematical Morphology*, vol. 2: *Theoretical Advances*, Academic Press, New York, 1988.

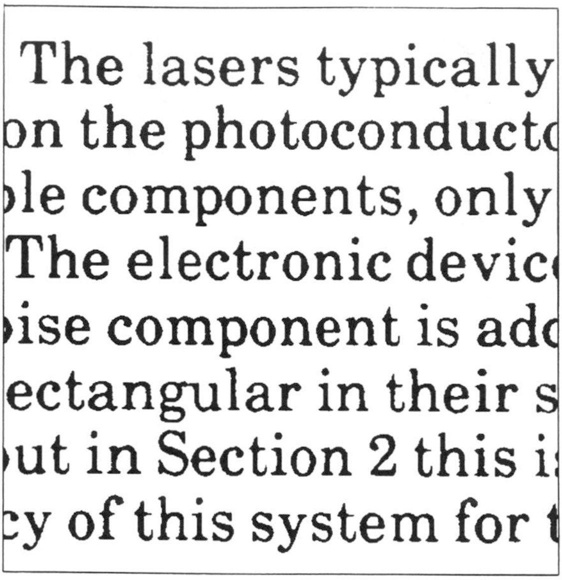

a)

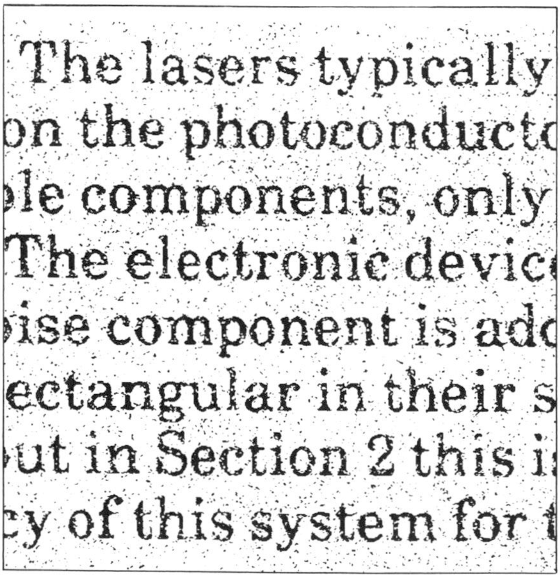

b)

Figure 26. For example D: (a) realization of image process; (b) realization of image-noise process; (c) filtered image; (d) filter basis.

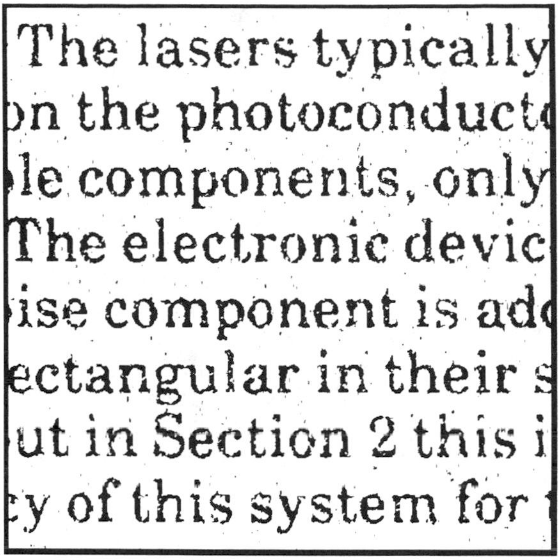
The lasers typically
on the photoconduct
le components, only
The electronic devic
ise component is ad
ectangular in their s
ut in Section 2 this i
y of this system for

c)

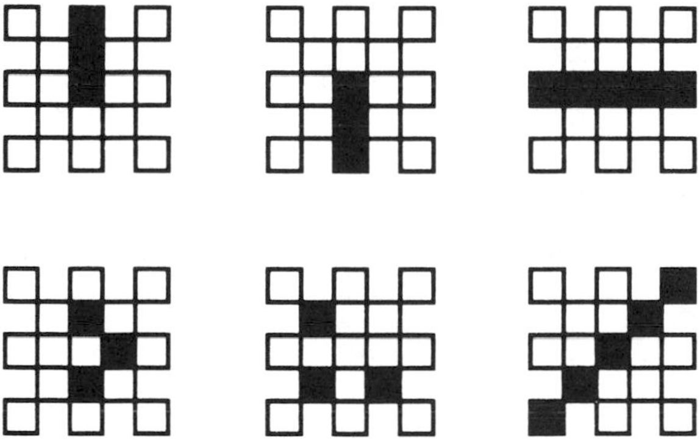

d)

13. Lougheed, R., Lecture notes for summer course in mathematical morphology, University of Michigan, 1983.
14. Haralick, R. M., Dougherty, E. R., and Katz, P., Model-Based Morphology, *Proc. SPIE*, *1472*, (1991).
15. Giardina, C. R., and Dougherty, E. R., *Morphological Methods in Image and Signal Processing*, Prentice-Hall, Englewood Cliffs, New Jersey, 1988.
16. Dougherty, E. R., and Giardina, C. R., A digital version of Matheron's theorem for increasing τ-mappings in terms of a basis for the kernel, *IEEE Comput. Vision Pattern Recog.*, 534–536 (1986).
17. Maragos, P., and Schafer, R. W., Morphological filters—Part I: Their set-theoretic analysis and relations to linear shift-invariant filters, *IEEE Trans. Acoust. Speech Signal Process. ASSP-35*, 1153–1169 (1987).
18. Maragos, P., and Schafer, R. W., Morphological filters—Part II: Their relations to median, order statistic, and stack-filters, *IEEE Trans. Acoust. Speech Signal Process. ASSP-35*, 1170–1184 (1987) (corrections, *ibid.*, *ASSP-37*, 597 (1989).
19. Maragos P., A representation theory for morphological image and signal processing, *IEEE PAMI*, *11*(6), (1989).
20. Heijmans, H. J., and Ronse, C., The algebraic basis of mathematical morphology—Part I: Dilations and erosions, *J. Comput. Vision Graphics Image Process.*, *50*, (1990).
21. Banon, G. J. F., and Barrera, J., Minimal representation of translation invariant set mappings by mathematical morphology, *SIAM J. Appl. Math.*, 51(6), (1991).
22. Ronse, C., and Heijmans, H. J., The algebraic basis of mathematical morphology. II. Openings and closings, *J. Comput. Vision Graphics Image Process.*, 54, (1991).

APPENDIX

In this appendix, we prove the morphological filter MSE theorem. The proof employs the probability addition theorem, which we first state here for reference.

Probability Addition Theorem. The probability of the union of m events is given by

$$
\begin{aligned}
P\left[\cup_{k=1}^{m} E_k\right] &= \sum_{k=1}^{m} P\left[E_k\right] - \sum_{k=1}^{m-1} \sum_{j=k+1}^{m} P\left[E_k \cap E_j\right] \\
&\quad + \sum_{k=1}^{m-2} \sum_{j=k+1}^{m-1} \sum_{i=j+1}^{m} P\left[E_k \cap E_j \cap E_i\right] \\
&\quad - \cdots + (-1)^{m+1} P\left[\cap_{k=1}^{m} E_k\right] \\
&= \sum_{j=1}^{m} (-1)^{j+1} \sum_{1 \le i_1 < i_2 < \cdots < i_j \le m} P\left[\cap_{k=1}^{j} E_{i_k}\right]
\end{aligned}
\tag{2.A1}
$$

Morphological Filter MSE Theorem. An m-erosion morphological filter Ψ possessing basis $\text{Bas}[\Psi] = \{B_1, B_2, \ldots, B_m\}$ will provide a point estimate with mean-square error given by

$$\text{MSE}\langle\Psi\rangle = \sum_{j=1}^{m} (-1)^{j+1} \sum_{1 \le i_1 < i_2 < \cdots < i_j \le m} \text{MSE}\langle \cup_{k=1}^{j} B_{i_k} \rangle \tag{2.A2}$$

Proof. The probabilities of type-0 and type-1 errors are given by

$$p_0[B_1, B_2, \ldots, B_m] = P\left[\left(\vee_{k=1}^{m} (S' \ominus B_k)(z) = 0\right) \cap \left(S(z) = 1\right)\right] \tag{2.A3}$$

$$= P\left[\left(\cap_{k=1}^{m} [(B_k)_z \not\subset K_z]\right) \cap \left(z \in S\right)\right] \tag{2.A4}$$

and

$$p_1[B_1, B_2, \ldots, B_m] = P\left[\left(\vee_{k=1}^{m} (S' \ominus B_k)(z) = 1\right) \cap \left(S(z) = 0\right)\right] \tag{2.A5}$$

$$= P\left[\left(\cup_{k=1}^{m} [(B_k)_z \subset K_z]\right) \cap \left(z \notin S\right)\right] \tag{2.A6}$$

respectively. Analogous to the 2-erosion case, De Morgan's law indicates that the probability of type-0 error and the probability of type-1 error are related by complementary events:

$$\left[\cap_{k=1}^{m} [(B_k)_z \not\subset K_z]\right] = \left[\cup_{k=1}^{m} [(B_k)_z \subset K_z]\right]^c \tag{2.A7}$$

Combining Eqs. (2.A4) and (2.A6) in a manner analogous to the 2-erosion case, we can write a general expression for the MSE in terms of the symmetric difference:

$$\text{MSE}\langle B_1, B_2, \ldots, B_m\rangle = P\left[\left(\cup_{k=1}^{m} [(B_k)_z \subset K_z]\right) \Delta \left(z \in S\right)\right] \tag{2.A8}$$

The probability of the symmetric difference may be written

$$P\left[\left(\cup_{k=1}^{m} C_k\right) \Delta\, D\right] = P\left[\cup_{k=1}^{m} C_k\right] + P\left[D\right] - 2\,P\left[\left(\cup_{k=1}^{m} C_k\right) \cap D\right] \tag{2.A9}$$

where $C_k = [(B_k)_z \subset K_z]$ and $D = [z \in S]$. Next, we distribute the intersection to obtain

$$P\left[\left(\cup_{k=1}^{m} C_k\right) \Delta\, D\right] = P\left[\cup_{k=1}^{m} C_k\right] + P\left[D\right] - 2\,P\left[\cup_{k=1}^{m} \left(C_k \cap D\right)\right] \tag{2.A10}$$

Employing the probability addition theorem on the first and third items of the right-hand side of Eq. (2.A10) yields

$$P\left[\left(\cup_{k=1}^{m} C_k\right) \Delta D\right] = P\left[D\right] + \sum_{j=1}^{m} (-1)^{j+1} \sum_{1 \le i_1 < i_2 < \cdots < i_j \le m} \left(P\left[\cap_{k=1}^{j} C_{i_k}\right] - 2\,P\left[\cap_{k=1}^{j} \left(C_{i_k} \cap D\right)\right]\right) \quad (2.A11)$$

To move $P[D]$ into the summation, use the relationship

$$\sum_{k=1}^{m} (-1)^{k+1} \binom{m}{k} = 1 \quad (2.A12)$$

so that

$$\begin{aligned} P\left[\left(\cup_{k=1}^{m} C_k\right) \Delta D\right] &= \sum_{j=1}^{m} (-1)^{j+1} \sum_{1 \le i_1 < i_2 < \cdots < i_j \le m} \left(P\left[\cap_{k=1}^{j} C_{i_k}\right] + P\left[D\right] - 2\,P\left[\cap_{k=1}^{j} \left(C_{i_k} \cap D\right)\right]\right) \\ &= \sum_{j=1}^{m} (-1)^{j+1} \sum_{1 \le i_1 < i_2 < \cdots < i_j \le m} P\left[\left(\cap_{k=1}^{j} C_{i_k}\right) \Delta D\right] \end{aligned} \quad (2.A13)$$

Equations (2.A11)–(2.A13) state that the probability addition theorem applies in a symmetric difference setting. Next, we note the event equivalence

$$\cap_{k=1}^{j} \left[(B_{i_k})_z \subset K_z\right] = \left[(\cup_{k=1}^{j} B_{i_k})_z \subset K_z\right] \quad (2.A14)$$

Combining Eqs. (2.A8) and (2.A13) yields

$$\text{MSE}\langle\Psi\rangle = \sum_{j=1}^{m} (-1)^{j+1} \sum_{1 \le i_1 < i_2 < \cdots < i_j \le m} P\left[\left((\cup_{k=1}^{j} B_{i_k})_z \subset K_z\right) \Delta \left(z \in S\right)\right] \quad (2.A15)$$

which, upon application of Eq. (2.36), proves the theorem.

Chapter 3

Statistical Properties of Discrete Morphological Filters

Jaakko Astola, Lasse Koskinen, and Yrjö Neuvo

Tampere University,
Tampere, Finland

I. INTRODUCTION

The basic goal in the development of mathematical morphology has been to create an algebraic theory for the investigation of the geometrical structure of sets. It has shown to be a powerful tool in automated image analysis, and its application in image and signal processing is now known as morphological filtering. The way that morphological filtering has evolved makes it only natural that the understanding of the effects of noise in signals and images is sporadic, at least compared to the highly developed deterministic theory.

Recently, morphological filtering has become more popular in applications in which not only image analysis but also some type of image restoration plays a central role. If we are filtering an image that is thought to be corrupted by some kind of a noise process, it is essential also to know the statistical behavior of the filter.

As the statistical analysis of linear filters is much simpler than that of nonlinear filters, linear restoration methods have been extensively studied. The importance of edge preservation and the nonstationarity of real images present serious problems for linear filtering and have resulted in a search for more effective methods among nonlinear filters. One of the most widely used methods is median filtering [4]. Several authors have presented methods that outperform the standard median filter under specific conditions. These filters include ranked-order filters [15], linear median hybrid filters [6] [2], in place growing filter structures [23], L-filters [16], weighted median filters [7], [24], stack filters [22], and many others. As these filters were designed to restore noisy signals, their statistical behavior is rather well understood (usually by knowledge and experience accumulated in statistics) even though an exact analysis is sometimes impossible.

Although it has been known that morphological filters are very sensitive to noise, quantitative analysis of the effects of noise has received attention only recently. Certain consequences of noise in the signals to be morphologically processed have been studied by Stevenson et al. [21]. The connection between stack and morphological filters [11] has been utilized in Koskinen et al. [8], [9] to derive formulas for the output distributions of morphological filters. Restoration and representation properties of noisy binary images have been studied by Schonfeld et al. [17], [18], [19].

Median-type filters have their roots in statistical estimation theory, and their analysis can be satisfactorily carried out using standard methods in statistics. On the other hand, morphological filters are based on geometrical concepts and cannot be directly analyzed by standard statistical methods. However, by utilizing the connection between stack and morphological filters it is possible to develop methods for the analysis of the statistical input/output relations of morphological filters. The purpose of this chapter is to develop these methods and show that by applying them we can gain a deeper understanding of the behavior of morphological filtering.

This chapter is organized as follows. In Section II continuous stack filters are defined, and in Section III the relation of morphological filters to stack filters is discussed and explicit formulas for stack filter expressions of morphological filters are derived using Boolean functions indexed by sets. To understand sufficiently the filtering of noisy signals, it is desirable to determine the output distribution of the filter in terms of the input distribution. To this end, general formulas for output distributions of morphological filters are derived in Section IV in the case of independent and identically distributed inputs. Certain symmetry properties stemming from the duality properties of morphological filters are presented in Section V, and in Section VI numerical examples are presented. In Section VII closed-form expressions for output distributions are derived for one-dimensional filters and their second-order statistical properties are studied. Finally, comparisons between morphological and median filters are made.

In the following we recall the definitions of morphological filters which process discrete signals by sets; see, for example, Serra [20] and Maragos et al. [10], [11], [12].

The *structuring set* B is a finite subset of $\mathbf{Z}^m$. The *symmetric set* B^s of B is defined by $B^s = \{-x : x \in B\}$; and the *translated set* B_x, where the set B is translated by $x \in \mathbf{Z}^m$, is defined by $B_x = \{x + y : y \in B\}$; and the *Minkowski sum* of structuring sets, $A, B \subset \mathbf{Z}^m$, is defined by $A + B = \{x_1 + x_2 : x_1 \in A, x_2 \in B\}$.

The operations dilation, erosion, closing, and opening by B, $B \subset \mathbf{Z}^m$, transform a signal f, $f : \mathbf{Z}^m \to \mathbf{R}$, to another signal by the following rules.

The *dilation* of f by B is denoted by $f \oplus B^s$ and is defined by

$$(f \oplus B^s)(x) = \max_{y \in B_x}\{f(y)\}, \qquad x \in \mathbf{Z}^m$$

The *erosion* of f by B is denoted by $f \ominus B^s$, and is defined by

$$(f \ominus B^s)(x) = \min_{y \in B_x}\{f(y)\}, \qquad x \in \mathbf{Z}^m$$

The *closing* of f is denoted by $(f \bullet B)$ and is defined by

$$(f \bullet B)(x) = [(f \oplus B^s) \ominus B](x), \qquad x \in \mathbf{Z}^m$$

The *opening* of f by B is denoted by $f \bigcirc B$ and is defined by

$$(f \bigcirc B)(x) = [(f \ominus B^s) \oplus B](x), \qquad x \in \mathbf{Z}^m$$

In the same way as closing and opening were defined as dilation followed by erosion and erosion followed by dilation, the clos-opening by the structuring set B is defined as the closing by B followed by the opening by B, and the open-closing by B is defined as the opening by B followed by the closing by B. These two operations have many properties similar to those of median type filtering operations.

The *clos-opening* of f by B is denoted by $(f \bullet B) \bigcirc B$ and is defined by

$$((f \bullet B) \bigcirc B)(x) = [(((f \oplus B^s) \ominus B) \ominus B^s) \oplus B](x), \qquad x \in \mathbf{Z}^m$$

The *open-closing* of f by B is denoted by $(f \bigcirc B) \bullet B$ and is defined by

$$((f \bigcirc B) \bullet B)(x) = [(((f \ominus B^s) \oplus B) \oplus B^s) \ominus B](x), \qquad x \in \mathbf{Z}^m$$

II. CONTINUOUS STACK FILTERS

Stack filters are nonlinear filters that are defined via threshold decomposition and positive Boolean functions. The class of stack filters is relatively large and includes all ranked-order operators. The key to the analysis of stack filters comes from their definition by threshold decomposition, which basically says that stack filters are completely characterized by their operation on binary signals. The basic properties of stack filters are discussed below. More extensive treatments of stack filters and the methods for analyzing them can be found in Wendt et al. [22] and Yli-Harja et al. [24].

In Boolean expressions we use $x \wedge y$ for x AND y, $x \vee y$ for x OR y, and $\bar{x}$ for NOT x, where x and y are Boolean variables. In some formulas, binary values are to be understood as real 1's and 0's. The number of elements in a finite set A is denoted by $|A|$.

A Boolean function is called *positive* if it can be written as a Boolean expression that contains no complemented input variables. The concept of positive Boolean functions is closely related to partial ordering. The relation $\geq$ of binary vectors $\underline{x} = (x_1, x_2, \cdots, x_n)$ and $\underline{y} = (y_1, y_2, \cdots, y_n)$ is defined as $\underline{x} \geq \underline{y}$ if and only if $x_i \geq y_i$ for all $i \in \{1,2, \cdots, n\}$. As this relation is reflexive, antisymmetric, and transitive, it defines a partial ordering on the set of binary vectors. This order property is known as the *stacking property* and it is said that $\underline{x}$ and $\underline{y}$ "stack" if $\underline{x} \geq \underline{y}$. A Boolean function f is said to *possess the stacking property* if the relation $\underline{x} \geq \underline{y}$ implies the relation $f(\underline{x}) \geq f(\underline{y})$. Filters that are defined by positive Boolean functions possess the stacking property and are called *stack filters*.

Filters whose definitions are based on positive Boolean functions have many theoretical advantages, but filters whose definitions are based on nonpositive Boolean functions also have some attractive properties. For example Astola et al. [1] have applied them to construct a noise-insensitive edge detector.

In the following we define the concept of a Boolean function indexed by a set [8], [9]. By this definition we link the geometrical concept of a set to the Boolean function.

Definition 1. Let A be a finite subset of $\mathbf{Z}^m$. Then the *Boolean function g indexed by the set* A is a Boolean expression of variables z_a, $a \in A$, denoted by

$$g(\underline{z}) = g(z_a : a \in A)$$

Example 1. Let $A \subset \mathbf{Z}^2$, $A = \{(0,0), (0,1), (1,0)\}$. An example of the positive Boolean functions that have their variables indexed by A is

$$g(z_{(0,0)}, z_{(0,1)}, z_{(1,0)}) = (z_{(0,0)} \wedge z_{(0,1)} \wedge z_{(1,0)}) \vee (z_{(0,0)} \wedge z_{(0,1)}) \vee z_{(0,0)}$$

Consider a signal $f : \mathbf{Z}^m \rightarrow \mathbf{R}$. Then the output $y(b)$ of a *median filter with filter window* A ($A \in \mathbf{Z}^m$) at point b ($b \in \mathbf{Z}^m$, $|A|$ odd) is defined by

$$y(b) = \text{median of set}\{f(a + b) : a \in A\}$$

Example 2. Consider the median filter with the "cross" $A = \{(-1,0), (0,1), (0,0), (1,0), (0,-1)\}$ as the moving window. Then the equivalent stack filter is defined by the Boolean function

$$\begin{aligned} g(\underline{z}) = &(z_{(0,0)} \wedge z_{(-1,0)} \wedge z_{(0,1)}) \vee (z_{(0,0)} \wedge z_{(-1,0)} \wedge z_{(1,0)}) \\ &\vee (z_{(0,0)} \wedge z_{(-1,0)} \wedge z_{(0,-1)}) \vee (z_{(0,0)} \wedge z_{(0,1)} \wedge z_{(1,0)}) \\ &\vee (z_{(0,0)} \wedge z_{(0,1)} \wedge z_{(0,-1)}) \vee (z_{(0,0)} \wedge z_{(1,0)} \wedge z_{(0,-1)}) \\ &\vee (z_{(0,-1)} \wedge z_{(-1,0)} \wedge z_{(0,1)}) \vee (z_{(1,0)} \wedge z_{(-1,0)} \wedge z_{(0,1)}) \\ &\vee (z_{(1,0)} \wedge z_{(-1,0)} \wedge z_{(0,-1)}) \vee (z_{(1,0)} \wedge z_{(0,-1)} \wedge z_{(0,1)}) \end{aligned}$$

Let $u(t)$ denote the real *unit step function*

$$u(t) = \begin{cases} 1, & \text{if } t \geq 0 \\ 0, & \text{otherwise} \end{cases}$$

We use a Boolean function with its variables indexed by a set to define the corresponding continuous stack filter in the following way, Koskinen et al. [8], [9].

Definition 2. Let A be a finite subset of $\mathbf{Z}^m$, $g(\underline{z})$ a positive Boolean function indexed by A, and $f : \mathbf{Z}^m \rightarrow \mathbf{R}$ a real signal. Then the *continuous stack filter S* corresponding to $g(\underline{z})$ is defined by

$$S(f)(x) = \max\{t \in \mathbf{R} : g(u(f(x + a) - t) : a \in A) = 1\}$$

The moving window corresponds to the index set A of the defining positive Boolean function. In the definition of the continuous stack filter, the positivity of the Boolean function is related to the stacking property and the unit step function is related to the threshold decomposition. In fact, $u(f(x) - t)$ is obtained by thresholding $f(x)$ at level t, that is, $u(f(x) - t) = T_t(f(x))$, where

$$T_t(\alpha) = \begin{cases} 1, & \text{if } \alpha \geq t \\ 0, & \text{otherwise} \end{cases}$$

Example 3. Let f be an image, that is, function $f : \mathbf{Z}^2 \rightarrow \mathbf{R}$. Suppose that f is filtered by a two-dimensional median filter having the filter window $A = \{(-1,0), (0,0), (0,1)\}$. Then the equivalent stack filter is

$$\begin{aligned} S(f)(a,b) = \max\{t \in \mathbf{R} : &(u(f((a,b) + (-1,0)) - t) \\ &\wedge u(f((a,b) + (0,0)) - t)) \vee (u(f((a,b) + (-1,0)) - t) \\ &\wedge u(f((a,b) + (0,1)) - t)) \vee (u(f((a,b) + (0,0)) - t) \\ &\wedge u(f((a,b) + (0,1)) - t)) = 1\} \end{aligned}$$

An attractive property of stack filters is that it is possible to derive analytical results for their statistical properties. For example, the output distribution of a continuous stack filter can be expressed using the following proposition [24].

Proposition 1. Let the input values X_b, $b \in B$, in the window B of a stack filter be independent random variables having the distribution functions $F_b(t)$, respectively. Then the output distribution function $G(t)$ of a stack filter S defined by a positive Boolean function $g(\underline{z})$ is

$$G(t) = \sum_{\underline{z} \in g^{-1}(0)} \prod_{b \in B} (1 - F_b(t))^{z_b} F_b(t)^{1 - z_b}$$

where $g^{-1}(0)$ is the preimage of 0, that is, $g^{-1}(0) = \{\underline{z} : g(\underline{z}) = 0\}$.

Proof. Let $g(\underline{z})$ be the positive Boolean function that defines S. Then the output distribution function is, by definition,

$$
\begin{aligned}
G(t) &= P\{S \leq t\} \\
&= P\{g(u(X_b - t) : b \in B) = 0\} \\
&= P\{(u(X_b - t) : b \in B) \in g^{-1}(0)\} \\
&= \sum_{\underline{z} \in g^{-1}(0)} P\{(u(X_b - t) : b \in B) = \underline{z}\} \\
&= \sum_{\underline{z} \in g^{-1}(0)} \prod_{b \in B} (1 - F_b(t))^{z_b} F_b(t)^{1 - z_b}
\end{aligned}
$$

In the case of independent and identically distributed input values we get the following immediate corollary.

Corollary 1.1. Let the input values $X_1, X_2, \cdots, X_n$ in the window of a stack filter be independent random variables having a common distribution function $F(t)$. Then the output distribution function $G(t)$ of a stack filter S defined by a positive Boolean function $g(\underline{z})$ is

$$
G(t) = \sum_{\underline{z} \in g^{-1}(0)} (1 - F(t))^{w(\underline{z})} F(t)^{n - w(\underline{z})}
$$

where $w(\underline{z})$ is the number of 1's in $\underline{z}$.

Example 4. Consider the filtering by a 3-point median where the input values in the moving window are independent random variables and two of the input values in the moving window have the distribution function $F_1(t)$ and one of the input values has distribution function $F_2(t)$. Then the output distribution function $G(t)$ of the median filter is, by Proposition 1,

$$
\begin{aligned}
G(t) &= F_1(t)^2 F_2(t) + (1 - F_1(t))F_1(t)F_2(t) \\
&\quad + F_1(t)(1 - F_1(t))F_2(t) + F_1(t)^2(1 - F_2(t)) \\
&= F_1(t)^2 F_2(t) + 2F_1(t)(1 - F_1(t))F_2(t) + F_1(t)^2(1 - F_2(t))
\end{aligned}
$$

Example 5. Consider the filtering by a 3-point median where the input values in the moving window are independent random variables having a common distribution function $F(t)$. Then the output distribution function $G(t)$ of the median filter is, by Corollary 1.1,

$$
G(t) = F(t)^3 + 3F(t)^2(1 - F(t))
$$

III. STACK FILTER EXPRESSIONS OF MORPHOLOGICAL FILTERS

In this section we consider the relation of morphological filters to stack filters. It has been shown by Maragos et al. [10], [11] and Wendt et al. [22] that discrete morphological filters are, in fact, stack filters. Explicit formulas for stack filter expressions of morphological filters are derived in Koskinen et al. [8], [9] using positive Boolean functions indexed by sets. This stack filter representation of morphological filters has proved useful because it allows morphological filters to be analyzed in the framework of positive Boolean functions.

From the definition of the continuous stack filter we see that the dilation and the erosion by a structuring set B correspond to the outputs of certain continuous stack filters whose defining positive Boolean functions are indexed by B. Furthermore, cascaded operations of dilations and erosions correspond to a single stack filter because all compositions of positive Boolean functions can be expressed as a single positive Boolean function. The concept of a positive Boolean function indexed by a set plays a key role when we study the connection between stack and morphological filters. The index set is the link between the structuring set of morphological filters and the moving window of stack filters.

By using simple relations between structuring sets and index sets, the following proposition gives explicit stack filter expressions for morphological filters [8], [9].

Proposition 2. Let B be a structuring set, B^s the symmetric set, and $B+B^s$ the Minkowski sum of B and B^s. Then the positive Boolean function that corresponds to stack filter expression of

(a) dilation by B is $g_d(\underline{z}) = \bigvee_{b\in B} z_b$,

(b) erosion by B is $g_e(\underline{z}) = \bigwedge_{b\in B} z_b$,

(c) closing by B is $g_c(\underline{z}) = \bigwedge_{a\in B^s} (\bigvee_{b\in Ba} z_b)$,

(d) opening by B is $g_o(\underline{z}) = \bigvee_{a\in B^s} (\bigwedge_{b\in Ba} z_b)$,

(e) clos-opening by B is $g_{co}(\underline{z}) = \bigvee_{a\in B^s} (\bigwedge_{b\in(B+B^s)a}(\bigvee_{c\in B_b} z_c))$,

(f) open-opening by B is $g_{oc}(\underline{z}) = \bigwedge_{a\in B^s}(\bigvee_{b\in(B+B^s)a}(\bigwedge_{c\in B_b} z_c))$,

Proof. We prove only case (e); the other cases can be proved in a similar way. Let f be the signal to be clos-opened. We get the equations

$$\begin{aligned}((f \bullet B) \circ B)(x) &= [(((f \oplus B^s) \ominus B) \ominus B^s) \oplus B](x) \\ &= \max_{a\in B^s}(\min_{b\in B_a}(\min_{c\in B^s_b}(\max_{d\in B_c} f(x+d)))) = \\ &\quad \max_{a\in B^s}(\min_{b\in(B+B^s)_a}(\max_{c\in B_b} f(x+c))) \\ &= \max\{t\in \mathbf{R} : \textstyle\bigvee_{a\in B^s}(\bigwedge_{b\in(B+B^s)}a \\ &\quad (\textstyle\bigvee_{c\in B_b} u(f(a+c)-t))) = 1\}\end{aligned}$$

So, by the definition of the continuous stack filter, the Boolean function that corresponds to the stack filter expression of clos-opening by B is

$$g_{co}(\underline{z}) = \textstyle\bigvee_{a\in B^s}(\bigwedge_{c\in(B+B^s)}a\,(\bigvee_{c\in B_b} z_c))$$

From Proposition 2 it follows that the Boolean function g_d corresponding to the stack filter expression of dilation is the *dual* of the Boolean function g_e corresponding to erosion, and the Boolean function g_c corresponding to closing is the dual of the Boolean function g_o corresponding to opening. Similarly, the Boolean function g_{co} corresponding to clos-opening is the dual of the Boolean function g_{oc} corresponding to open-closing.

For practical filter implementations these expressions are not useful because of their high computational complexity. An advantage of expressing cascaded operations as a single stack filter is that it is then possible to derive analytical results for the statistical properties of the filtering operations.

The following example shows how dilation and opening can be expressed as stack filters.

Example 6. Let $f(i,j)$ be a real image. Consider morphological filtering of f with the structuring set

$$B = \{(0,0), (0,1), (1,0)\}$$

Then, by Proposition 2, the stack filter expression of the dilation by B corresponds to the positive Boolean function

$$g(\underline{z}) = z_{(0,0)} \vee z_{(0,1)} \vee z_{(1,0)}$$

and the opening by B corresponds to the positive Boolean function

$$g(\underline{z}) = (z_{(0,0)} \wedge z_{(0,1)} \wedge z_{(1,0)}) \vee (z_{(-1,1)} \wedge z_{(-1,0)} \wedge z_{(0,0)}) \\ \vee (z_{(0,0)} \wedge z_{(0,-1)} \wedge z_{(1,-1)})$$

Thus, the output of the stack filter expression of the dilation by B is

$$S(f)(i,j) = \max\{t \in \mathbf{R} : u(f(i,j) - t) \vee u(f(i+1,j) - t) \\ \vee u(f(i,j+1) - t) = 1\}$$

and the output of the stack filter expression of the opening by B is

$$S(f)(i,j) = \max\{t \in \mathrm{R}: (u(f(i,j) - t) \wedge u(f(i,j+1) - t) \\ \wedge u(f(i+1,j) - t)) \vee (u(f(i-1,j+1) - t) \\ \wedge u(f(i-1,j) - t) \wedge u(f(i,j) - t)) \vee (u(f(i,j) - t) \\ \wedge u(f(i,j-1) - t) \wedge u(f(i+1,j-1) - t)) = 1\}$$

Example 7. Consider the morphological filtering with "cross" $B = \{(-1,0), (0,-1), (0,0), (0,1), (1,0)\}$ as the structuring set. Then the positive Boolean function that corresponds to dilation by B is

$$g_d(\underline{z}) = z_{(-1,0)} \vee z_{(0,-1)} \vee z_{(0,0)} \vee z_{(0,1)} \vee z_{(1,0)}$$

and the positive Boolean function that corresponds to closing by B is

$$\begin{aligned} g_c(\underline{z}) = & (z_{(-2,0)} \vee z_{(-1,-1)} \vee z_{(-1,0)} \vee z_{(-1,1)} \vee z_{(0,0)}) \\ & \wedge (z_{(-1,-1)} \vee z_{(0,-2)} \vee z_{(0,-1)} \vee z_{(0,0)} \vee z_{(1,-1)}) \\ & \wedge (z_{(-1,0)} \vee z_{(0,-1)} \vee z_{(0,0)} \vee z_{(0,1)} \vee z_{(1,0)}) \\ & \wedge (z_{(-1,1)} \vee z_{(0,0)} \vee z_{(0,1)} \vee z_{(0,2)} \vee z_{(1,1)}) \\ & \wedge (z_{(0,0)} \vee z_{(1,-1)} \vee z_{(1,0)} \vee z_{(1,1)} \vee z_{(2,0)}) \end{aligned}$$

Table 1 shows stack filter expressions for morphological filters. Both Boolean functions and moving windows (the index sets of the Boolean functions) are presented. Because $B + B^s$ and $(B + B^s) + (B + B^s)$ are always symmetric with respect to the origin, the moving windows of stack filters corresponding to closing, opening, clos-opening, and open-closing are always symmetric with respect to the origin. On the other hand, the moving windows corresponding to stack filter expressions of dilation and erosion are symmetric if and only if the structuring set B is symmetric with respect to the origin. The sizes of $B + B^s$ and $(B + B^s) + (B + B^s)$ depend on both the size and the shape of B but $B + B^s$ is always larger than B and $(B + B^s) + (B + B^s)$ is larger than $B + B^s$. So the Boolean functions that correspond to closing and opening have more variables than the Boolean functions corresponding to dilation and erosion. Similarly, the Boolean functions corresponding to clos-opening and open-closing have more variables than the Boolean functions corresponding to closing and opening.

The moving windows corresponding to morphological operations (figure 1) expand as the operations become more complicated compositions of the basic operations.

IV. GENERAL FORMULAS FOR OUTPUT DISTRIBUTIONS

The objective of this section is to derive output distributions of morphological filters. This knowledge is important when we are applying morphological filters

Table 1 Morphological Filtering Operations by Structuring Set B, Positive Boolean Functions, and Moving Windows of Corresponding Stack Filters

Filtering Operating	Moving Window	Boolean Function
Dilation	B	$g_d(\underline{z}) = \bigvee_{b\in B} z_b$
Erosion	B	$g_e(\underline{z}) = \bigwedge_{b\in B} z_b$
Closing	$B + B^s$	$g_c(\underline{z}) = \bigwedge_{a\in Ba}(\bigvee_{b\in Ba} z_b)$
Opening	$B + B^s$	$g_o(\underline{z}) = \bigvee_{a\in Bs}(\bigwedge_{b\in Ba} z_b)$
Clos-Opening	$(B + B^s) + (B + B^s)$	$g_{co}(\underline{z}) = \bigvee_{a\in B^s}(\bigwedge_{b\in(B+B^s)a}(\bigvee_{c\in B_b} z_c))$
Open-closing	$(B + B^s) + (B + B^s)$	$g_{oc}(\underline{z}) = \bigwedge_{a\in Bs}(\bigvee_{b\in(B+B^s)a}(\bigwedge_{c\in B_b} z_c))$

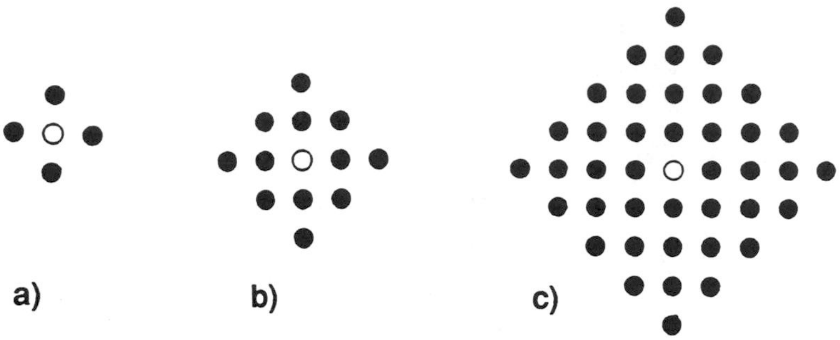

Figure 1. The moving windows corresponding to (a) erosion and dilation (that is, the structuring set itself), (b) closing and opening, and (c) clos-opening and open-closing.

to noisy signals. To understand the filtering process, it is desirable to determine the output distribution in terms of the input distributions. This analysis will give an idea of the noise suppression capability and the biasing effects of the filter.

Because discrete morphological filters that process signals by sets can be expressed as stack filters, their statistical properties can also be analyzed using the methods that were developed for stack filters. In this section, we utilize this connection to derive simple expressions for the output distributions of basic morphological filters. These expressions are studied in Koskinen et al. [8], [9].

When we are applying morphological filters to a noisy image, it is very important to know the statistical behavior of the filter not only on uniform parts of the image but also near edges. The situation is similar to filtering an image by a median filter. A median filter performs ideally when no noise is present, but if the image is noisy the edges are more or less smeared. In the following, the results that concern identically distributed inputs apply on the constant parts of images, and results that concern nonidentically distributed inputs apply on the edges of images.

In the following we analyze the effect of morphological filters on noise in the case in which the values of the input signal are independent random variables. Proposition 3 shows that statistical properties of dilation and erosion can easily be analyzed and do not depend on the shape of the structuring set.

Proposition 3. Consider a discrete signal f and the dilation or the erosion by a structuring set B of size n at point x_0. Let the values $f(x_0 + b)$, $b \in B$, be independent random variables with distribution functions $F_1(t), F_2(t), \cdots, F_n(t)$. Then the distribution function of the value $h(x_0)$ of the filtered signal h is $G_d(t) = \Pi_{i=1}^{n} F_i(t)$, for dilation, and $G_e(t) = 1 - \Pi_{i=1}^{n} (1 - F_i(t))$, for erosion.

Proof. Let the structuring set of the dilation and the erosion be B, where the size of B is n. Then by Proposition 2 the positive Boolean function that corresponds to the stack filter expression is

$$g(\underline{z}) = \vee_{b \in B} z_b \qquad \text{for dilation}$$

and

$$p(\underline{z}) = \wedge_{b \in B} z_b \qquad \text{for erosion}$$

Since $g^{-1}(0) = \{\underline{0}\}$, Proposition 1 implies that

$$G_d(t) = \sum_{\underline{z}=\underline{0}} \prod_{i=1}^{n} (1 - F_i(t))^{z_i} F_i(t)^{1 - z_i} = \prod_{i=1}^{n} F_i(t)$$

Similarly, since $p^{-1}(0) = \{0,1\}^n - \{\underline{1}\}$ (all components of $\underline{1}$'s are 1's), using Proposition 1 we get the result

$$G_e(t) = \sum_{\underline{z} \neq \underline{1}} \prod_{i=1}^{n} (1 - F_i(t))^{z_i} F_i(t)^{1 - z_i}$$

$$= 1 - \prod_{i=1}^{n} (1 \quad F_i(t))$$

Corollary 3.1. Consider a discrete signal f and the dilation or the erosion by a structuring set B ($|B| = n$) in the case where the values of the signal f are independent and identically distributed with a common distribution function $F(t)$. Then the distribution function of the values of the filtered signal is $G_d(t) = F(t)^n$, for dilation, and $G_e(t) = 1 - (1 - F(t))^n$, for erosion.

Propositions 4 and 5 are direct consequences of Propositions 1 and 2.

Proposition 4. Consider a discrete signal f and the closing by a structuring set B at point x_0. Let the values $f(x_0 + b)$, $b \in B + B^s$, be independent random variables with distribution functions $F_b(t)$, respectively. Then the distribution function $G_c(t)$ of the value $h(x_0)$ of the closed signal h is

$$G_c(t) = \sum_{\underline{z} \in g^{-1}(0)} \prod_{b \in B + B^s} (1 - F_b(t))^{z_b} F_b(t)^{1 - z_{bmu3}}$$

where $g(\underline{z}) = \wedge_{a \in B^s} (\vee_{b \in Ba} z_b)$.

Corollary 4.1. Consider a discrete signal f and the closing by a structuring set B in the case where the values of f are independent and identically distributed with a common distribution function $F(t)$. Then the distribution function $G_c(t)$ of the values of the closed signal h is

$$G_c(t) = \sum_{\underline{z} \in g^{-1}(0)} (1 - F(t))^{w(\underline{z})} F(t)^{n - w(\underline{z})}$$

where $g(\underline{z}) = \wedge_{a \in B^s} (\vee_{b \in B_a} z_b)$, $w(\underline{z})$ is the number of 1's in $\underline{z}$, and $n = |B + B^s|$.

Example 8. Consider a one-dimensional signal f and the dilation and the closing by structuring set $B = \{-1,0,1\}$ at point 0. Let the values of the signal f

be independent random variables where the values $f(-2)$, $f(-1)$, and $f(0)$ have a common distribution function $F_1(t)$ and the values $f(1)$ and $f(2)$ have distribution function $F_2(t)$. Then the distribution function $G_1(t)$ of the value $h_1(0)$ of the dilated signal h_1 is

$$G_1(t) = F_1(t)^2 F_2(t)$$

and the distribution function $G_2(t)$ of the value $h_2(0)$ of the closed signal h_2 is

$$G_2(t) = F_1(t)^3 + F_1(t)^2 F_2(t) - F_1(t)^3 F_2(t) + F_1(t)F_2(t)^2 - F_1(t)^2 F_2(t)^2$$

Proposition 5. Consider a discrete signal f and the clos-opening by a structuring set B at point x_0. Let the values $f(x_0 + b)$, $b \in (B + B^s)$, be independent random variables with distribution functions $F_b(t)$, respectively. Then the distribution function $G_{co}(t)$ of the value $h(x_0)$ of the clos-opened signal h is

$$G_{co}(t) = \sum_{\underline{z} \in g^{-1}(0)} \prod_{b \in (B+B^s)+(B+B^s)} (1 - F_b(t))^{z_b} F_b(t)^{1-z_b}$$

where $g(\underline{z}) = \bigvee_{a \in B^s} (\bigwedge_{b \in (B+B^s)_a} (\bigvee_{c \in B_b} z_c))$.

Corollary 5.1. Consider a discrete signal f and the clos-opening by a structuring set B in the case where the values of signal f are independent and identically distributed random variables with a common distribution function $F(t)$. Then the distribution function $G_{co}(t)$ of the values of the clos-opened signal h is

$$G_{co}(t) = \sum_{\underline{z} \in g^{-1}(0)} (1 - F(t))^{w(\underline{z})} F(t)^{n - w(\underline{z})}$$

where $g(\underline{z}) = \bigvee_{a \in B^s} (\bigwedge_{b \in (B+B^s)_a} (\bigvee_{c \in B_b} z_c))$, and $n = |(B + B^s) + (B + B^s)|$.

V. SYMMETRY PROPERTIES

Propositions 4 and 5 offer us a straightforward way to calculate output distributions of closing and clos-opening. On the other hand, if we know the distribution function of the closing by B or the clos-opening by B, Proposition 6 gives us an easy way to find the output distributions of the opening by B or the open-closing by B, because the Boolean functions corresponding to the stack filter expressions of opening and open-closing are duals of those that correspond to closing and clos-opening, [8].

Proposition 6. Consider the filtering of a discrete signal f by a stack filter S at point x_0 where the values in the moving window A ($|A| = n$) of the stack filter S are independent random variables with distribution functions $F_1(t), F_2(t), \ldots,$ $F_n(t)$. Let g be a positive Boolean function, g_D the dual function of g, S the stack filter defined by g, and S_D the stack filter defined by g_D. If the distribution function of the value $S(f)(x_0)$ is

$$G(F_1(t), F_2(t), \ldots, F_n(t))$$

then the distribution function of the value $S_D(f)(x_0))$ is

$$G_D\,(F_1(t), F_2(t), \cdots, F_n(t)) = 1 - G((1 - F_1(t)),\\ (1 - F_2\,(t)), \ldots, (1 - F_n\,(t)))$$

Proof. As the filtering by S corresponds to the positive Boolean function $g(\underline{z})$, then Proposition 1 implies that the distribution function of the value $S(f)(x_0)$ is

$$G(F_1(t), F_2(t), \cdots, F_n(t)) = \sum_{\underline{z}\in g^{-1}(0)} \prod_{i=1}^{n} (1 - F_i(t))^{z_i}\, F(t)^{1-z_i}$$

where n is the size of the moving window A that corresponds to stack filter S. The positive Boolean function $g_D(\underline{z})$ that corresponds to filtering by S_D is the dual of g. So we get the relation $\bar{g}(\bar{\underline{z}}) = g_D(\underline{z})$. Thus, $g_D(\underline{z}) = 0$ if and only if $g(\bar{\underline{z}}) =$ 1 and Proposition 1 implies that

$$\begin{aligned} G_D(F_1(t), F_2(t), \ldots, F_n(t)) &= \sum_{\underline{z}\in g_D^{-1}} \prod_{i=1}^{n} (1 - F_i(t))^{z_i}\, F_i(t)^{1-z_i} \\ &= 1 - \sum_{\underline{z}\in g^{-1}(0)} \prod_{i=1}^{n} (1 - F_i(t))^{1-z_i}\, F_i(t)^{z_i} \\ &= 1 - G((1 - F_1(t)), (1 - F_2(t)), \ldots, \\ &\quad (1 - F_n(t))) \end{aligned}$$

Corollary 6.1. Consider the filtering of a discrete signal f where the values $f(z)$ are independent and identically distributed random variables with a common distribution function $F(t)$. Let g be a positive Boolean function, g_D the dual function of g, S the stack filter defined by g, and S_D the stack filter defined by g_D. If the distribution function of the values $S(f)(x)$ after filtering by S is $G(F(t))$, then the distribution function of the values $S_D(f)(x)$ after filtering by S_D is

$$G_D(F(t)) = 1 - G(1 - F(t))$$

Example 9. Consider an image $f(i,j)$ where the gray-level values of the pixels are independent random variables with a common distribution function $F(t)$. Then by Corollary 4.1 the distribution function $G_c(t)$ of the pixels after closing by $B = \{(-1,0), (0,0), (0,1), (1,0)\}$ is

$$G_c(t) = -F(t)^{11} + F(t)^{10} + 2F(t)^9 + F(t)^8 \\ - 4F(t)^7 - 2F(t)^6 + 4F(t)^4$$

Similarly, by Corollary 5.1, after opening the distribution function is

$$G_o(t) = 1 + (1 - F(t))^{11} - (1 - F(t))^{10} - 2(1 - F(t))^9 \\ - (1 - F(t))^8 + 4(1 - F(t))^7 + 2(1 - F(t))^6 - 4(1 - F(t))^4$$

See figure 2.

In Section III, we saw that the basic morphological filters, which are duals (in a morphological sense) of each other, are duals also in stack filter sense. This implies the following statistical symmetry properties, which fit well our intuitive ideas.

Proposition 7. Consider the filtering of a discrete random signal f where the values $f(x)$ are independent and identically and symmetrically distributed with distribution function $F(t)$ and expectation E. Let g be a positive Boolean function, g_D the dual of g, S the stack filter defined by g, and S_D the stack filter defined by g_D. If the expectation of the values of the signal after filtering by S is $E - \mu$, then the expectation after filtering by S_D is $E + \mu$. Moreover, the output variances of S and S_D are equal.

Proof. Corollary 1.1 implies that the distribution function of the values of the signal after filtering by S is

$$G(F(t)) = \sum_{\underline{z} \in g^{-1}(0)} (1 - F(t))^{w(\underline{z})} F(t)^{n - w(\underline{z})}$$

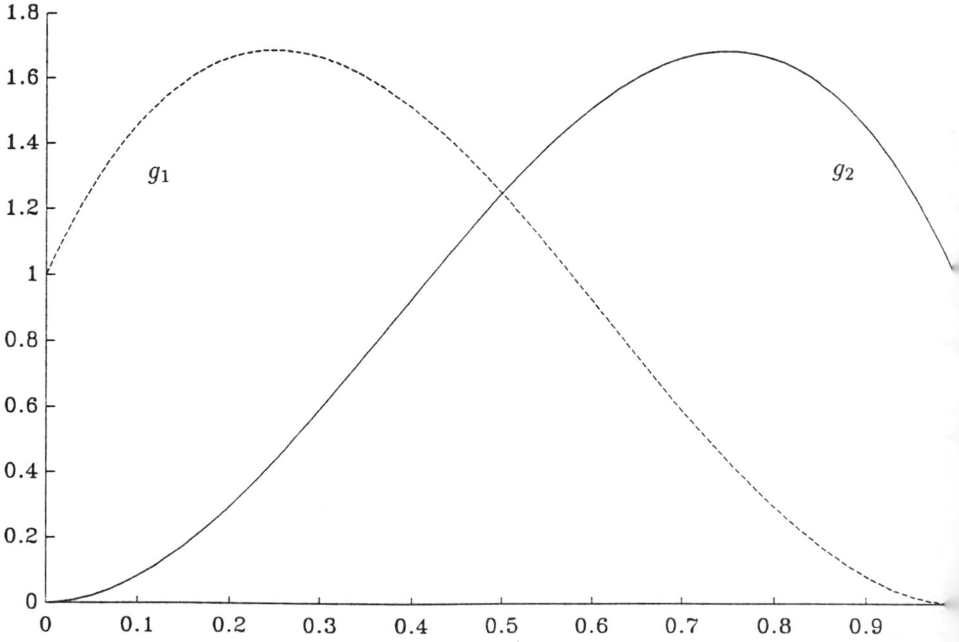

Figure 2. The probability density function of the values of the signal after opening (g_1) and after closing (g_2), where the distribution function of the values of the signal before filtering is uniform and the structuring set is $B = \{-1,0,1\}$.

and that the distribution function of the values of the signal after filtering by S_D is

$$G_D(F(t)) = \sum_{z \in g_D^{-1}(0)} (1 - F(t))^{w(z)} F(t)^{n-w(z)}$$

Now, Corollary 6.1 implies that

$$\sum_{z \in g^{-1}(0)} (1 - F(t))^{w(z)} F(t)^{n-w(z)} = 1 - \sum_{z \in g_D^{-1}(0)} (1 - F(t))^{n-w(z)} F(t)^{w(z)}$$

Since $F(t)$ is symmetric, $F(E + \mu) = 1 - F(E - \mu)$, implying

$$\begin{aligned} G(F(E + \mu)) &= \sum_{z \in g^{-1}(0)} (1 - F(E + \mu))^{w(z)} F(E + \mu)^{n-w(z)} \\ &= 1 - \sum_{z \in g_D^{-1}(0)} (1 - F(E + \mu))^{n-w(z)} F(E + \mu)^{w(z)} \\ &= 1 - \sum_{z \in g_D^{-1}(0)} (1 - F(E - \mu))^{w(z)} F(E - \mu)^{n-w(z)} \\ &= 1 - G_D(F(E - \mu)) \end{aligned}$$

Thus, the output distributions of S and S_D are mirror images with respect to E. This implies Proposition 7.

Corollary 7.1. Consider the filtering of a discrete signal f where the values $f(x)$ are independent and identically and symmetrically distributed with expectation E. Let the expectation of the output after dilation by B be $E + \mu_1$, after closing be $E + \mu_2$, and after clos-opening by B be $E + \mu_3$. Then

(a) after erosion by B the expectation of the output is $E - \mu_1$,
(b) after opening by B the expectation of the output is $E - \mu_2$,
(c) after open-closing by B the expectation of the output is $E - \mu_3$.

Corollary 7.2. Consider the filtering of a discrete signal f where the values $f(x)$ are independent and identically and symmetrically distributed. Then

(a) the variances of the outputs after dilation by B and erosion by B are equal.
(b) the variances of the outputs after closing by B and opening by B are equal.
(c) the variances of the outputs after clos-opening by B and open-closing by B are equal.

VI. NUMERICAL EXAMPLES

In this section, we utilize the theoretical results achieved in preceding sections to numerically compute output distributions and other basic statistical properties of some morphological filters. In Examples 10–13 we illustrate the filtering of constant parts of the noisy signals. This means that the values of the signals to be filtered are independent random variables with a common distribution function. These examples show that morphological filters bias the expectation and that the variance of the output is small. We also give an example of the morphological filtering of an edge of a signal. This example shows that the morphological filters smear the edges of noisy signals.

Example 10. Consider the opening by 2×2-square structuring set $B = \{(0,0), (1,0), (0,1), (1,1)\}$ where the distribution of the values of the signal before filtering is biexponential with means = 0 and variance = 1. Then the values of the opened signal have mean = -0.68 and variance = 0.83 (Figure 3).

Example 11. Consider the closing by structuring set $B = \{(-1,0), (0,1), (0,0), (1,0)\}$ where the distribution function of the values of the signal before filtering is uniform with mean = 0.5 and variance = 1/12. Then the values of the closed signal have mean = 0.67 and variance = 0.026 (Figure 4).

Example 12. Consider the clos-opening by structuring set $B = \{-1,0,1\}$ where the distribution of the values of the signal before filtering is uniform with mean = 0.5 and variance = 1/12. Then the clos-opened signal values have mean = 0.62 and variance = 0.038 (Figure 5).

Example 13. Consider a noisy one-dimensional signal f and the dilation and the closing by structuring set $B = \{-1,0,1\}$ at point 0. Let the values of the

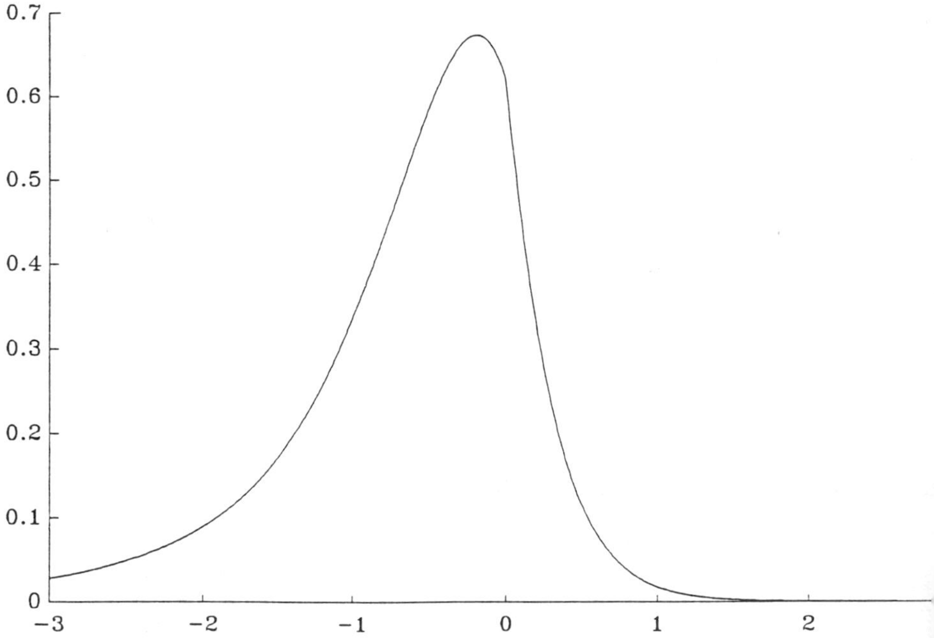

Figure 3. The probability density function of the values of the opened signal in Example 10.

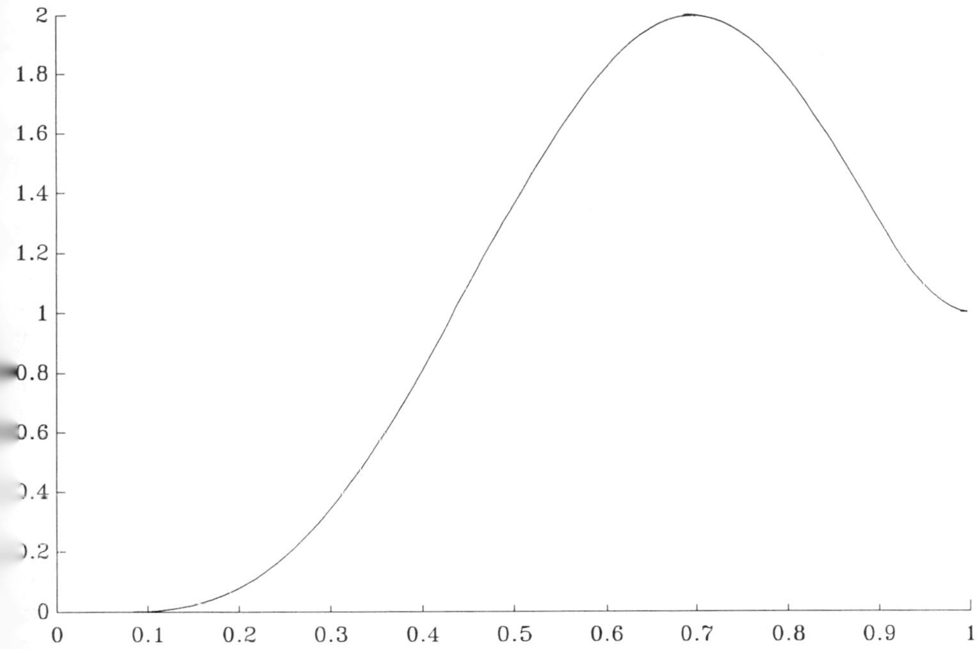

Figure 4. The probability density function of the values of the closed signal in Example 11.

signal f be independent random variables where the values $f(-2)$, $f(-1)$, and $f(0)$ have a common uniform distribution function $F_1(t)$ on [0,2] and the values $f(1)$, $f(2)$ have a common uniform distribution function $F_2(t)$ on [1,3]. Then the distribution function $G_1(t)$ of the value $h_1(0)$ of the dilated signal is

$$G_1(t) = \begin{cases} 0 & \text{if } t \leq 1 \\ -0.125t^2 + 0.125t^3 & \text{if } 1 < t \leq 2 \\ -0.5 + 0.5t & \text{if } 2 < t \leq 3 \\ 1 & \text{if } 3 < t \end{cases}$$

and the distribution function $G_2(t)$ of the value $h_2(0)$ of the closed signal (Figure 6) is

$$G_2(t) = \begin{cases} 0 & \text{if } t \leq 0 \\ 0.125t^3 & \text{if } 0 < t \leq 1 \\ 0.125t - 0.4375t^2 + 0.5625t^3 - 0.125t^4 & \text{if } 1 < t \leq 2 \\ 1 & \text{if } 2 < t \end{cases}$$

This clearly shows that both dilation and closing smear the edge of a noisy signal.

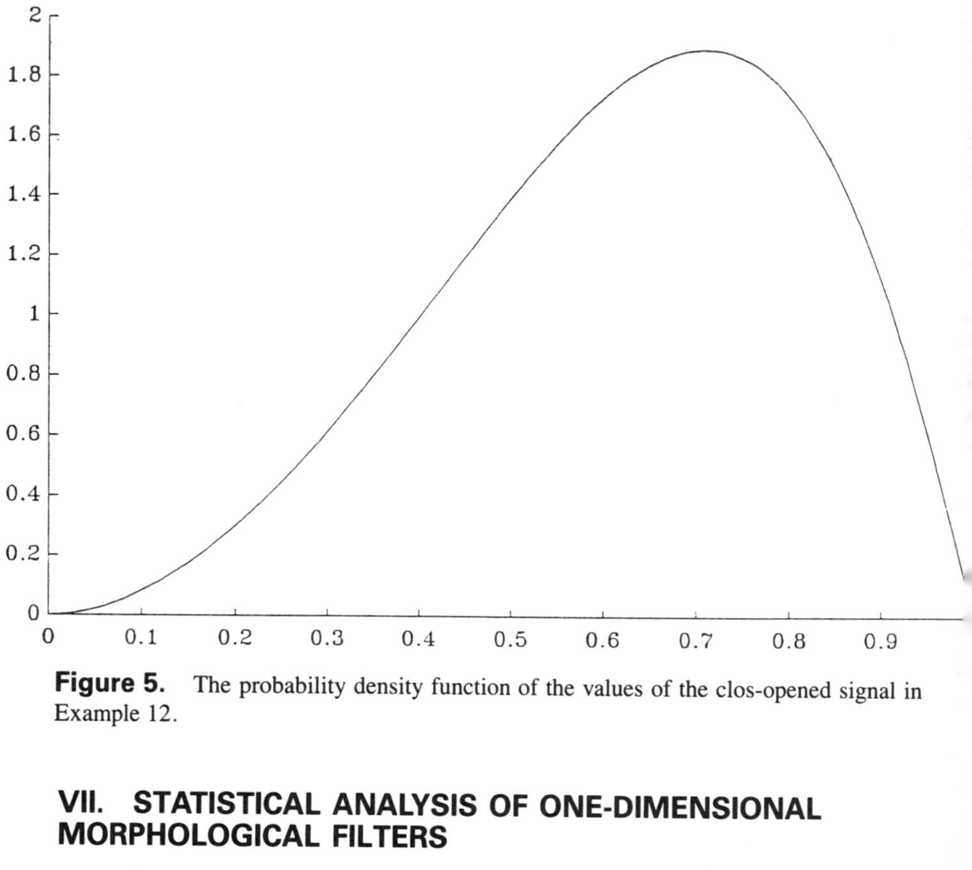

Figure 5. The probability density function of the values of the clos-opened signal in Example 12.

VII. STATISTICAL ANALYSIS OF ONE-DIMENSIONAL MORPHOLOGICAL FILTERS

In general, the statistical properties of morphological filters depend on both the shape and the size of the structuring set. As a result, it is difficult to derive closed-form expressions for these properties. Koskinen et al. [8] have studied one-dimensional morphological filters using the stack filter method, and in this section we extend that approach. In the case of a one-dimensional convex structuring set we derive analytical expressions for the output distributions and study the second-order statistical properties of closing, opening, clos-opening, and open-closing when the values of the input signal are independent and identically distributed. These results can also be used to approximate the statistical properties of two- and higher-dimensional morphological filters.

Proposition 8. Consider a discrete signal f where the values $f(x)$, $x \in \mathbf{Z}^m$, are independent and identically distributed random variables with a common distribution function $F(t)$. Let f be closed using a convex one-dimensional structuring set B of length n. Then the distribution function $G_c(t)$ of values $h(x)$, $x \in \mathbf{Z}^m$, of the closed signal h is

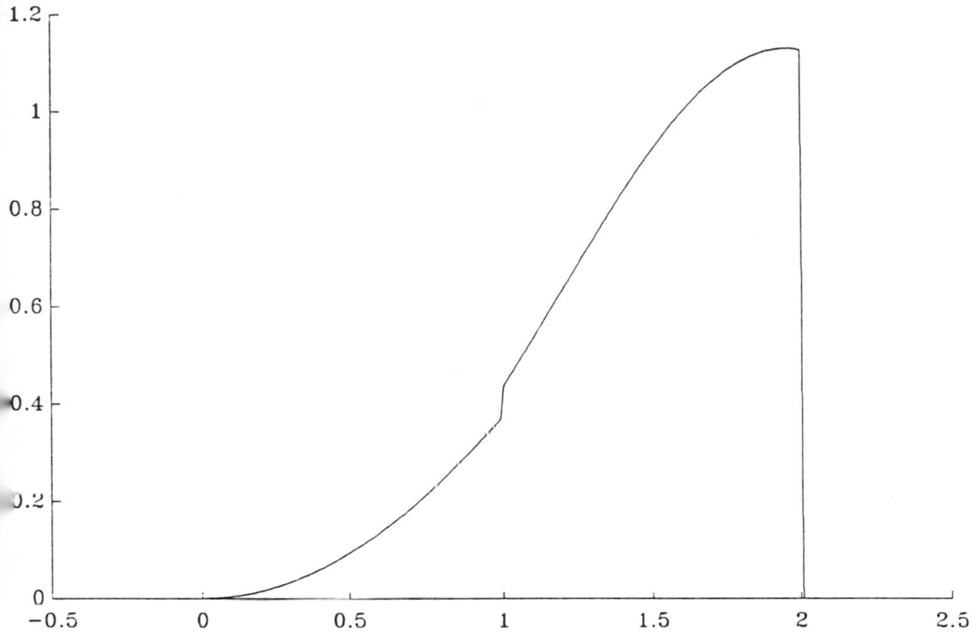

Figure 6. Probability density function of the value $h_2(0)$ of the closed signal in Example 13.

$$G_c(t) = nF(t)^n - (n - 1)F(t)^{n+1}$$

Proof. Consider the closing by a convex one-dimensional structuring set B of length n. By Corollary 4.1 we need to compute the number of vectors $\underline{z} = (z_1, z_2, \cdots, z_{2n-1})$ of each weight in preimage $g^{-1}(0)$ of the Boolean function

$$g(\underline{z}) = \bigwedge_{a\in B^s}(\bigvee_{b\in B_a} z_b)$$

that corresponds to the stack filter expression of closing.

Now, the number of vectors of weight $2n - 1 - s$ in $g^{-1}(0)$ equals the number of vectors of weight $2n - 1 - s$ containing at least n consecutive 0's. Divide these vectors into n distinct classes in the following way. The class C_0 contains all vectors

$$b^{(0)} = (\overbrace{0,0,\ldots,0}^{n\text{ times}}, *, *, \ldots, *)$$

and the class C_l contains all vectors such that

$$b^{(l)} = (\overbrace{*,*, \ldots, *}^{l \text{ times}}, 1, \overbrace{0,0, \ldots, 0}^{n \text{ times}}\ 8,7, \ldots, *)$$

where $*$ denotes 0 or 1. It is easy to see that C_0 consists of $\binom{n-1}{s-n}$ vectors of weight $2n - 1 - s$ and C_l consists of $\binom{n-2}{s-n}$ vectors of weight $2n - 1 - s$ for all $l = 1, 2, \ldots, n - 1$. So there are $\binom{n-1}{s-n} + (n - 1)\binom{n-2}{s-n}$ vectors of weight $2n - 1 - s$ in $g^{-1}(0)$. Now, Corollary 4.1 implies that the distribution function $G_c(t)$ of the closed signal is

$$\begin{aligned} G_c(t) &= \sum_{i=n}^{2n-1} [\binom{n-1}{i-n} + (n-1)\binom{n-2}{i-n}](1 - F(t))^{2n-1-i}F(t)^i \\ &= \sum_{i=0}^{n-1} [\binom{n-1}{i} + (n-1)\binom{n-2}{i}](1 - F(t))^{n-1-i}F(t)^{i+n} \\ &= F(t)^n \sum_{i=0}^{n-1} \binom{n-1}{i}(1 - F(t))^{n-1-i}F(t)^i \\ &\quad + (n-1)F(t)^n(1 - F(t)) \sum_{i=0}^{n-1} \binom{n-2}{i}(1 - F(t))^{n-2-i}F(t)^i \\ &= nF(t)^n - (n-1)F(t)^{n+1} \end{aligned}$$

Using Corollary 6.1, we get the following corollary.

Corollary 8.1. Consider a discrete signal f where the values $f(x)$, $x \in \mathbf{Z}^m$, are independent and identically distributed random variables with a common distribution function $F(t)$. Let f be opened using a convex one-dimensional structuring set B of length n. Then the distribution function $G_o(t)$ of values $h(x)$, $x \in \mathbf{Z}^m$, of the opened signal h is

$$G_o(t) = 1 - n(1 - F(t))^n + (n - 1)(1 - F(t))^{n+1}$$

Proposition 9. Consider a discrete signal f, where the values $f(x)$, $x \in \mathbf{Z}^m$, are independent and identically distributed random variables with a common distribution function $F(t)$. Let f be clos-opened using a convex one-dimensional structuring set B of length n, where $n > 2$. Then the distribution function $G_{co}(t)$ of the values $h(x)$, $x \in \mathbf{Z}^m$, of the clos-opened signal h is

$$\begin{aligned} G_{co}(t) &= \frac{n^2 - n - 2}{2}F(t)^{2n+2} + (-n^2 + n + 1)F(t)^{2n+1} \\ &\quad + \frac{n^2 - n}{2} F(t)^{2n} - (n - 1)F(t)^{n+1} + nF(t)^n \end{aligned}$$

Proof. Consider the clos-opening by a convex one-dimensional structuring set B of length $n > 2$. By Corollary 5.1 we need to compute the number of

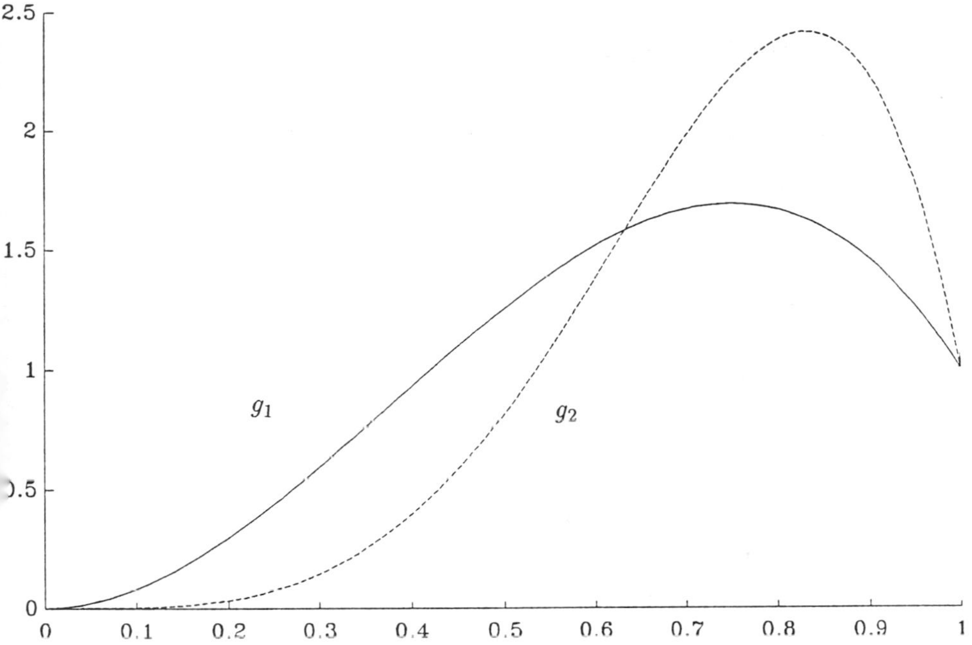

Figure 7. The probability density function of the values of signal f after closing by a one-dimensional convex structuring set of length 3 (g_1) and length 5 (g_2) where the values of f have a uniform distribution function before filtering.

vectors $\underline{z} = (z_1, z_2, \ldots, z_{4n-3})$ of each weight in the preimage $g^{-1}(0)$ of the Boolean function

$$g(\underline{z}) = \bigvee_{a \in B^s}(\bigwedge_{b \in (B+B^s)_a} (\bigvee_{c \in B_b} z_c))$$

that corresponds to the stack filter expression of clos-opening.

The number of vectors of weight $4n - 3 - s$ in $g^{-1}(0)$ equals the number of vectors $\underline{z} = (z_1, z_2, \ldots, z_{4n-3})$ of weight $4n - 3 - s$ such that each subvector $(z_i, z_{1+i}, \ldots, z_{3n-3+i})$ contains at least n consecutive 0's for all $i \in \{1, 2, \ldots, n\}$.

First we divide $g^{-1}(0)$ into three distinct classes A, B, and C:

$$A = \{\underline{z} \in \{0,1\}^{4n-3} : \text{for some } i \in \{0, \ldots, n-1\}, z_{n+i} = 0, \ldots, z_{2n-1+i} = 0\}$$

$$B = \{\underline{z} \in \{0,1\}^{4n-3} : z_{2n-1} = 0, \text{ and for some nonnegative } i, j \text{ such that } i + j \le n - 2, z_{n-1-i} = 0, \ldots, z_{2n-2-i} = 0 \text{ and}$$

$z_{2n+j} = 0, \ldots, z_{3n-l+j} = 0$, and for some nonnegative l, m such that $l + m \leq n - 4, z_{2n-2-l} = z_{2n+m} = 1\}$

$C = \{\underline{z} \in \{0,1\}^{4n-3} : z_{2n-l} = 1$, and for some nonnegative i, j such that $i + j \leq n - 2, z_{n-l-i} = 0, \ldots, z_{2n-2-i} = 0$ and $z_{2n+j} = 0, \ldots, z_{3n-l+j} = 0\}$

Then we divide C again in three distinct classes $C1$, $C2$, and $C3$:

$C1 = \{\underline{z} \in C \mid z_{2n-1} = 0, \ldots, z_{2n-2} = 0\}$

$C2 = \{\underline{z} \in C \mid z_{2n} = 0, \ldots, z_{3n-1} = 0$ and for some i such that $0 \leq i \leq n-3\ z_{2n-2-i} = 1\}$

$C3 = \{\underline{z} \in C \mid$ for some nonegative i, j such that $i, j \leq n - 4$ $z_{2n-2-i} = z_{2n+j} = 1\}$

Using the same method as in the proof of Proposition 8, we see that class A consists of vectors of weight $4n - 3 - s$. We can calculate directly that class B consists of

$$\sum_{i=1}^{n-3} i \binom{2n-6}{s-2n-1} = \frac{n^2-5n+6}{2} \binom{2n-6}{s-2n-1}$$

and class $C3$ of

$$\sum_{i=1}^{n-3} i \binom{2n-6}{s-2n} = \frac{n^2-5n+6}{2} \binom{2n-6}{s-2n}$$

vectors of weight $4n - 3 - s$.

It is easy to see that class $C1$ consists of

$$\binom{2n-4}{s-2n} + (n-2)\binom{2n-5}{s-2n}$$

and class $C2$ of

$$(n-2)\binom{2n-5}{s-2n}$$

vectors of weight $4n - 3 - s$.

So the number $t(s)$ of vectors of weight $4n - 3 - s$ in $g^{-1}(0)$ is

$$t(s) = \binom{3n-3}{s-n} + (n-1)\binom{3n-4}{s-n} + \frac{n^2-5n+6}{2}\binom{2n-6}{s-2n-1} + \frac{n^2-5n+6}{2}\binom{2n-6}{s-2n} + \binom{2n-4}{s-2n} + (2n-4)\binom{2n-5}{s-2n}$$

Now, the output distribution $G_{co}(t)$ for clos-opening is given by

$$\begin{aligned} G_{co}(t) &= \sum_{i=0}^{4n-3} t(i)(1 - F(t))^{4n-3-i} F(t)^i \\ &= (2n-4)(1-F(t))^2 F(t)^{2n} \\ &+ \frac{n^2-5n+6}{2}(1-F(t))^3 F(t)^{2n} \\ &+ \frac{n^2-5n+6}{2}(1-F(t))^2 F(t)^{2n+1} + F(t)^n \\ &+ (n-1)(1-F(t))F(t)^n \\ &= \frac{n^2-n-2}{2} F(t)^{2n+2} + (-n^2+n+1) F(t)^{2n+1} \\ &+ \frac{n^2-n}{2} F(t)^{2n} - (n-1) F(t)^{n+1} + nF(t)^n \end{aligned}$$

Using Corollary 6.1, we get the following corollary.

Corollary 9.1. Consider a discrete signal f, where the values $f(x)$, $x \in \mathbf{Z}^m$, are independent and identically distributed random variables with a common distribution function $F(t)$. Let f be open-closed using a convex structuring set B of length n, where $n > 2$. Then the distribution function $G_{oc}(t)$ of the values $h(x)$, $x \in \mathbf{Z}^m$, of the open-closed signal h is

$$\begin{aligned} G_{oc}(t) = 1 &- \frac{n^2-n-2}{2}(1-F(t))^{2n+2} - (-n^2+n+1)(1-F(t))^{2n+1} \\ &- \frac{n^2-n}{2}(1-F(t))^{2n} + (n-1)(1-F(t))^{n+1} - n(1-F(t))^n \end{aligned}$$

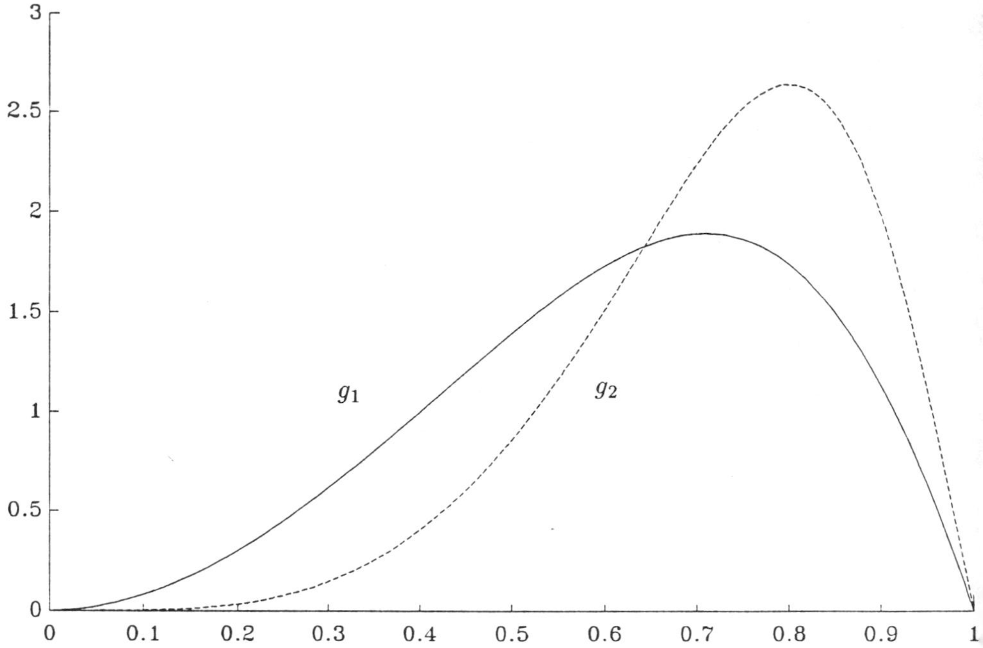

Figure 8. The probability density function of the values of signal f after clos-opening by a one-dimensional convex structuring set of length 3 (g_1) and length 5 (g_2) where the values of f are uniformly distributed before filtering.

The analytical forms of the output distributions enable us also to compute the asymptotic approximations for the expectation and the variance in the case of independent and identically uniformly distributed inputs.

Using Propositions 3, 8, and 9, we get after lengthy computations the following results. Computations were performed using a symbolic program.

Proposition 10. Consider a discrete signal f where the values $f(x)$ are independent and identically uniformly distributed on $[0,1]$, and let B be a convex one-dimensional structuring set of length n. Then

(a) after dilation or erosion by B the variance of the output is $1/n^2 + O(1/n^3)$,
(b) after closing or opening by B the variance of the output is $2/n^2 + O(1/n^3)$,
(c) after clos-opening or open-closing by B the variance of the output is $119/64n^2 + O(1/n^3)$.

Proposition 11. Consider a discrete signal f where the values $f(x)$ are independent and identically uniformly distributed on $[0,1]$, and let B be a convex one-dimensional structuring set of length n. Then

(a) after dilation by B the expectation of the output is $n/(n+1)$,
(b) after erosion by B the expectation of the output is $1 - n/(n+1)$,
(c) after closing by B the expectation of the output is $1 - 2/n + O(1/n^2)$,
(d) after opening by B the expectation of the output is $2/n + O(1/n^2)$,
(e) after clos-opening by B the expectation of the output is $1 - 17/18n + O(1/n^2)$,
(f) after open-closing by B the expectation of the output is $17/18n + O(1/n^2)$.

The results provide the theoretical basis for the smoothing capabilities and biasing effects experienced with morphological filters. Proposition 10 shows that morphological filters have very good smoothing capabilities, and Proposition 11 shows that morphological filters bias expectation heavily.

Figure 9 illustrates how the length of the structuring set affects the expectations of the outputs of dilation, closing, and clos-opening. The inputs were assumed to be independent and identically and uniformly distributed on $[0,1]$.

Median and morphological filters are nonlinear filters that have many similarities. These filters are often used in the same kinds of applications, so it is important to compare their properties. Maragos et al. [11] have also examined the

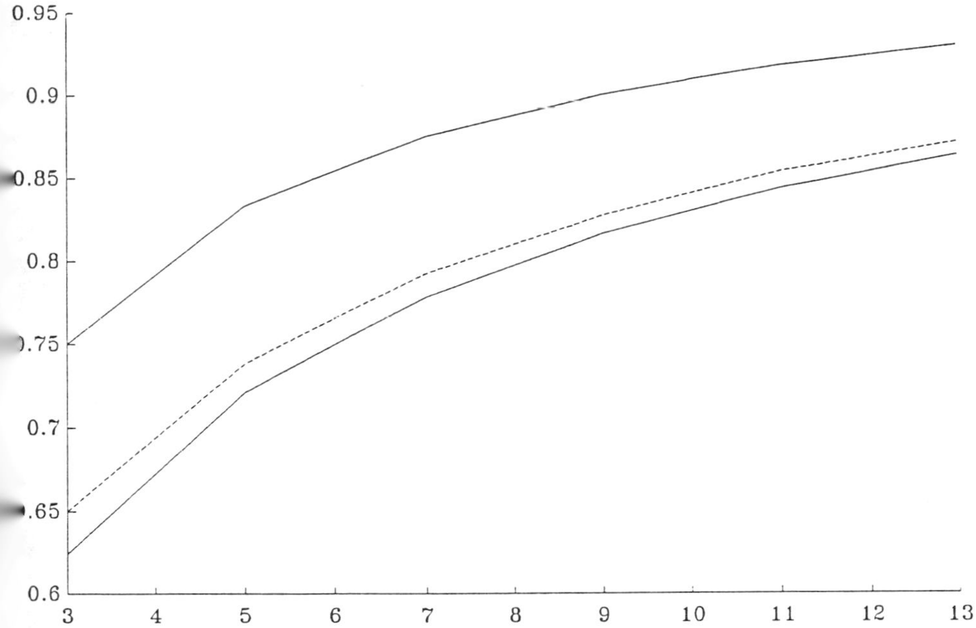

Figure 9. The expectations of outputs plotted against the length of the structuring elements, dilation (top), closing (middle), and clos-opening (bottom).

connection between morphological and median filters and shown that morphological filters have good spiky noise attenuation properties. Neejärvi et al. [14] have compared sinusoidal and pulse responses of morphological and median filters.

Results of this chapter show that morphological filters bias the expectation heavily. On the other hand, it is a well-known result in statistics that the median is an unbiased estimate of expectation for symmetric distributions. In the case of inputs that are independent and identically uniformly distributed on [0,1], the variance of the output of a median filter is $1/(4n + 8)$, where n is the size of the moving window of the median filter; see,e.g., Justusson [7].

Consider the morphological filtering by a convex one-dimensional structuring set B of length n and median filtering where the window size is n. Then Proposition 10 shows that the variance of the output of the morphological filter is an order of magnitude smaller than that of the median filter. Thus, morphological filters attenuate noise more efficiently than median filters *in the case of uniform distribution*.

In Fig. 10 the variances of median and morphological filters are presented where inputs were assumed to be independent and identically and uniformly distributed on [0,1].

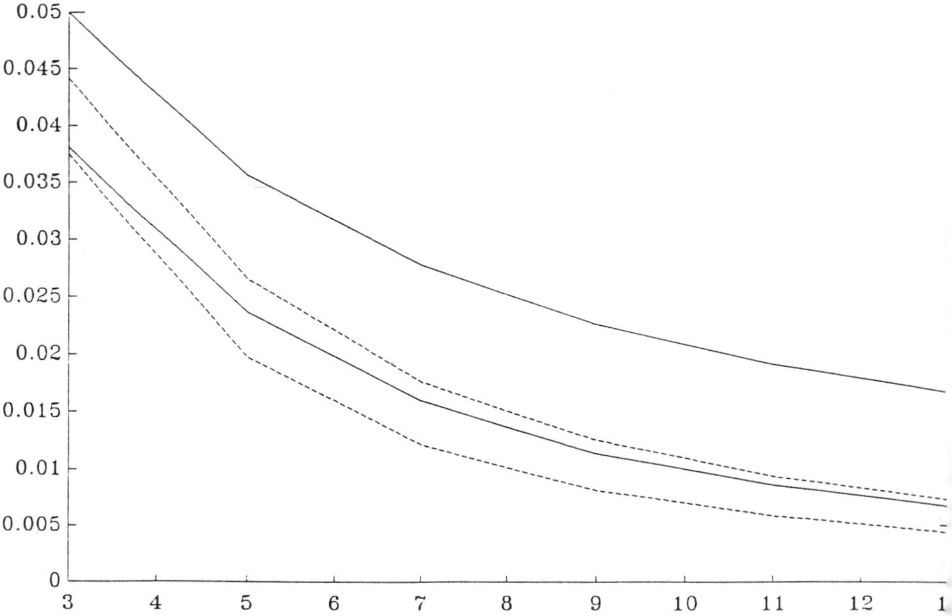

Figure 10. The output variances plotted against the length of the structuring element of median, closing, clos-opening, and dilation, respectively, from top to bottom.

REFERENCES

1. Astola, J., Koskinen, L., Yli-Harja, O., and Neuvo, Y., Digital filters based on threshold decomposition and Boolean functions, in *Proceedings of the SPIE Symposium on Visual Comm., Image Proc.*, Philadelphia, Nov. 1989, pp. 461–470.
2. Astola, J., Heinonen, P., and Neuvo, Y., Linear median hybrid filters, *IEEE Trans. Circuits Syst. CAS-36*, 1430–1438 (1989).
3. Chu, C., and Delp, E., Impulsive noise suppression and background normalization of electrocardiogram signals using morphological operators, *IEEE Trans. Biomed. Eng. 36*, 226–273 (1989).
4. Gallagher, N. C., Jr., Median filters: a tutorial, in *Proceedings of the 1988 IEEE International Conference on Circuits and Systems*, Helsinki, Finland, June 1988, pp. 1737–1744.
5. Haralick, R., Stenberg, S., and Zhuang, X., Grayscale morphology, in *Proceedings of the IEEE Conference on Computer Vision and Pattern Recognition*, Miami, June 1986, pp. 543–550.
6. Heinonen, P., and Neuvo, Y., FIR-median hybrid filters with predictive FIR substructures, *IEEE Trans. Acoust. Speech Signal Process., ASSP-36*, 892–899 (1988).
7. Justusson, B., Median filtering: statistical properties, in *Topics in Applied Physics, Two Dimensional Digital Signal Processing*, vol. II (T. S. Huang, ed.), Springer-Verlag, Berlin, 1981.
8. Koskinen, L., Astola, J., and Neuvo, Y. Analysis of noise attentuation in morphological image processing, in *Proc. SPIE Symp. on Nonlinear Image Processing II*, San Jose, February 1991, pp. 102–113.
9. Koskinen, L., Astola, J., and Neuvo, Y., Morphological filtering of noisy images, in *Proceedings of the SPIE Symposium on Visual Comm., Image Proc.*, Lausanne, Switzerland, November 1990, pp. 155–165.
10. Maragos, P., and Schafer, R., Morphological filters. Part I. Their set theoretic analysis and relations to linear shift-invariant filters, *IEEE Trans. Acoust. Speech Signal Process. ASSP-35*, 1153–1169 (1987).
11. Maragos, P., and Schafer, R., Morphological filters. Part II. Their relations to median, order-statistics, and stack filters, *IEEE Trans. Acoust. Speech Signal Process. ASSP-35*, 1170–1184 (1987).
12. Maragos, P., and Schafer, R., Applications of morphological filters to image processing and analysis, in *Proceedings of the IEEE International Conference on Acoustics, Speech, and Signal Processing, ICASSP-1986*, April 1986, pp. 2067–2070.
13. Matheron, G., *Random Sets and Integral Geometry*, Wiley, New York, 1975.
14. Neejärvi, J., and Neuvo, Y., Sinusoidal and pulse responses of morphological filters, in *Proceedings of the IEEE International Symposium on Circuits and Systems, ISCAS-1990*, May 1990, pp. 2136–2139.
15. Nodes, T. A., and Gallagher, N. C., Jr., Median filters: some modifications and their properties, *IEEE Trans. Acoust. Speech Signal Process., ASSP-30*, 739–746 (1982).

16. Palmieri, F., and Boncelet, C. G., Jr., *Ll*-filters—a new class of order statistic filters, *IEEE Trans. Acoust. Speech Signal Process. ASSP-37*, 691–701 (1989).
17. Schonfeld, D., and Coutsias, J., Optimal morphological pattern restoration from noisy binary images, *IEEE Trans. Pattern Anal. Machine Intell.*, PAMI-13, no. 1, January 1991, pp. 14–29.
18. Schonfeld, D., and Coutsias, J., On the morphological representation of binary images in noisy environment, submitted to *J. Visual Commun. Image Represent.*, submitted.
19. Schonfeld, D., and Coutsias, J., Robust morphological representation of binary images, in *Proceedings of the IEEE International Conference on Acoustics, Speech, and Signal Processing, ICASSP-1990*, April 1990, pp. 2065–2068.
20. Serra, S., *Image Analysis and Mathematical Morphology*, Academic Press, London, 1988.
21. Stevenson, R., and Arce, G., Morphological filters: statistics and further syntactic properties, *IEEE Trans. Circuits Syst., ASSP-34*, 1292–1305 (1987).
22. Wendt, P., Coley, E., and Callager, N., Stack filters, *IEEE Trans. Acoust. Speech Signal Process. ASSP-34*, 898–911 (1986).
23. Wichman, R., Astola, J., Heinonen, P., and Neuvo, Y., FIR-median hybrid filters with excellent transient response in noisy conditions, *IEEE Trans. Acoust. Speech Signal Process. ASSP-38*, 2108–2117 (1990).
24. Yli-Harja, O., Astola, J., and Neuvo, Y., Analysis of properties of median and weighted median filters using threshold logic and stack decomposition, *IEEE Trans. Signal Processing, ASSP-39*, February 1991, pp. 395–410.

Chapter 4

Morphological Analysis of Pavement Surface Condition

Chakravarthy Bhagvati

Dimitri A. Grivas

Michael M. Skolnick

Rensselaer Polytechnic Institute, Troy, New York

I. INTRODUCTION

Mathematical morphology has its roots in texture analysis and problems of determining material properties [1]. Pavement structures are made of composite materials, with particles (aggregates) bound together by asphalt or Portland cement mix. Establishment of measures for the condition of pavement surfaces from sensed images is of great practical importance and provides the rationale for exploring the applicability of methods available in mathematical morphology [2–4].

The present study addresses the development of a morphological approach based on particle size distributions for the analysis of the condition of pavement surfaces. It represents the first phase of a broader research effort that aims to develop and implement morphological measures that can characterize both the surface condition and the material properties of pavement structures.

The text describes the analysis and the results of the research effort to date. Section II provides an overview of pavement inspection techniques based on distress surveys and describes the set of images used to develop and test morphological algorithms. Section III outlines procedures used for deriving opening distributions to categorize texture and distresses appearing in the test images. An evaluation of derived distributions with respect to their ability to classify pavement surface condition is given in Section IV, and an approach is presented for normalizing "raw" distributions using models of the underlying distribution of the texture.

Two specific models considered are based on the simplifying assumptions of constant number of particles and constant amount of area, respectively, at each scale level of the structuring element used to obtain particle distributions. Despite the fact that these models represent simplifications of the "true" texture, they are useful because they can provide an a priori structure (or, "pravda" [1]) for texture, which improves upon classifications produced by raw distributions. Perhaps of greater significance is the framework created for the development of other (more realistic) models to normalize texture distributions. Modeling distribution data (rather than using different classes of structuring elements to probe images) can result in substantial savings in computation time. Of interest is the trade-off between applying a diverse (and computationally costly) set of structuring elements to the image data versus applying a diverse (and computationally less costly) set of models to the much reduced data of size distributions.

Section IV also provides preliminary results on a new set of measures based on the changes of global centroid position over the sequence of the images used to compute the opening distribution data. These new measures appear to be useful in distinguishing and identifying asymmetries in the texture field. They also have the advantage that their computational cost is negligible, as the needed data are available from the process of computing size distributions.

II. PAVEMENT SURFACE INSPECTION AND TEST IMAGES

A. Distress Surveys

Distress surveys are essential components of pavement management systems [5]. They are used to evaluate and monitor pavement condition over time. Collected data are analyzed to characterize pavement surface condition and establish causes of deterioration. Distress data are also used to evaluate the performance of individual pavement designs and determine appropriate maintenance strategies.

There are no standard procedures for conducting distress surveys at the present time. Available methods vary widely in both objectivity (data quality) and cost. They include manual field mapping, windshield surveys, visual surveys with automated data logging, and condition rating of previously collected images. Thus, distress surveys are currently conducted manually and suffer from subjectivity, nonrepeatability, high personnel costs, simplifications (so as to be reliable), and personnel danger [6].

The obvious approach to automating the survey process is via sensed images of pavement surfaces. Although systems employing laser and radar sensors have been used [7–9], the majority of systems employ digitizing cameras to generate a reflected image of the pavement surface. An advantage of using reflected light is that it corresponds to what is visible to the human eye. Thus, observed objects can be naturally related to those examined in current systems for pavement distress assessment [10]. Several image processing systems already exist—for example, Pavedex, Videocomp, and PCES [11]. These demonstrate the potential for automated processing, even though they use relatively simple algorithms for distress analysis of cracks based on edge detection and thresholding [8]. Furthermore, existing systems are not sufficiently robust, especially in the presence of texture, and they require substantial manual intervention for their proper functioning.

B. Test Images

Digitized data from a pavement condition survey are used to detect and classify distress into various categories such as cracking (e.g., transverse, longitudinal), raveling, rutting, and others. Severity and extent of each individual distress type are used to evaluate the overall pavement condition. The sample test data used in this study are taken from three classes of distress, namely G, N, and M, shown in Figures 1, 2, and 3, respectively.

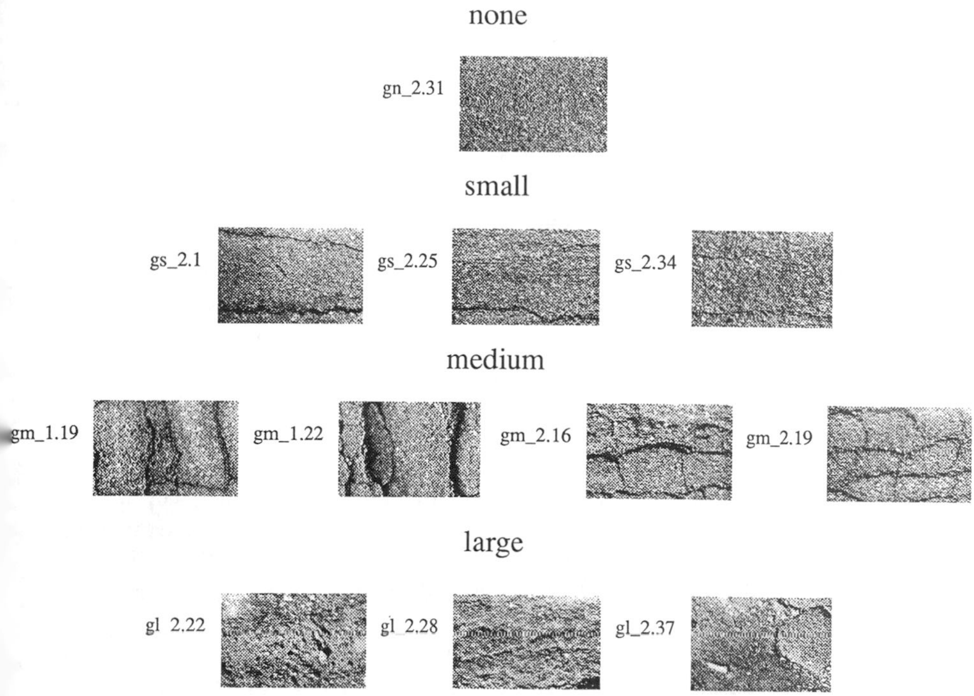

Figure 1. Pavement images belonging to class G—asphalt shoulder surfaces—categorized as to four levels of distress.

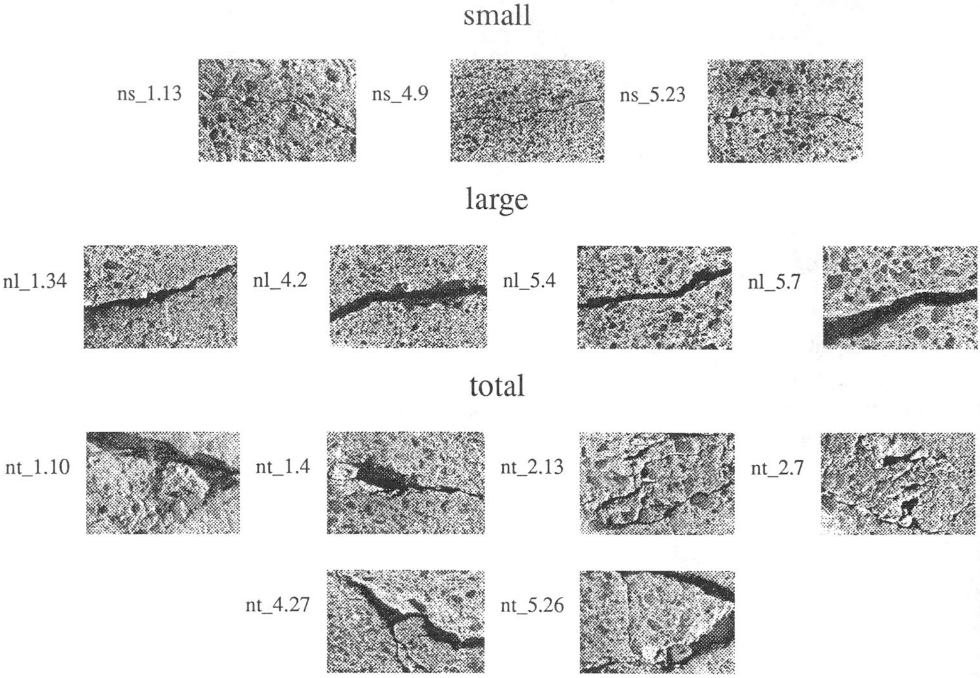

Figure 2. Pavement images belonging to class N—surface cracking of Portland cement concrete—categorized as to three levels of distress.

The images in the distress classes are categorized in terms of severity. Severity is measured using a linguistic scale having five levels: none (n), small (s), medium (m), large (l), and total (t). Although the test images do not cover the full range of distress severities for each distress class, the analyzed image data are fairly typical and diverse and are considered adequate for the purposes of this initial investigation. Furthermore, as described below, different distress categories (G, N, and M) correspond to different components of a pavement structure (shoulder, slab, and joint) as well as different types of material (asphalt cement and Portland cement concrete). Algorithms developed for this study are intended to distinguish the levels of severity within each distress category. The additional task of developing algorithms that can also distinguish between distress categories is part of future work.

Distress G characterizes shoulder surface defects, and the scale used for its measurement is based on the magnitude of cracking that is present. Thus, severity value none (n) denotes the absence of any cracking on the shoulder surface, while value small (s) indicates the presence of longitudinal or transverse cracks

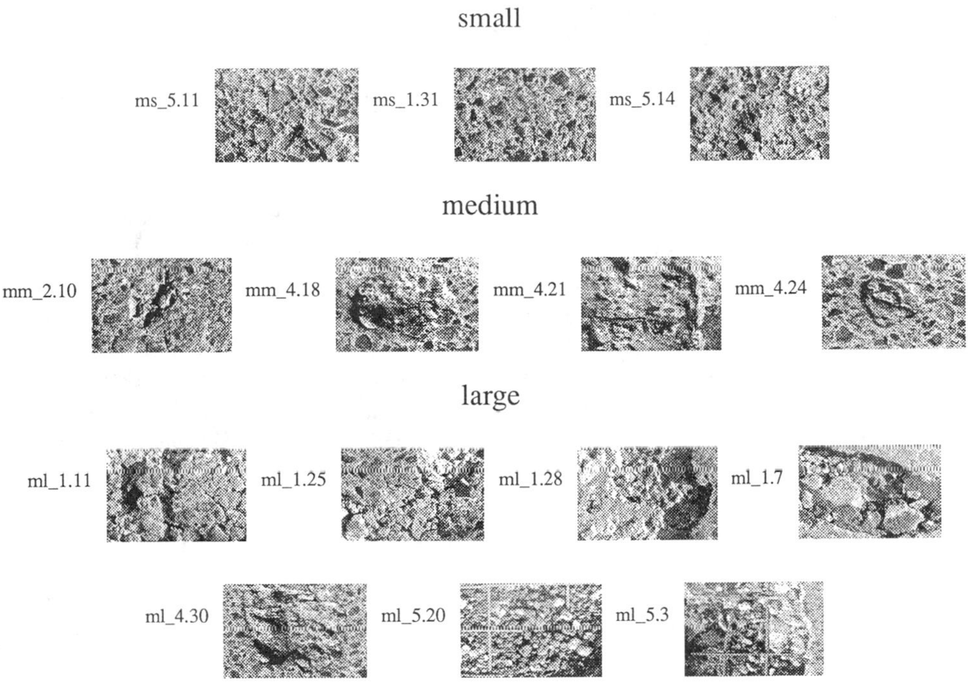

Figure 3. Pavement images belonging to class M—concrete slab surface defects (other than cracks)—categorized as to three levels of distress.

As severity values increase (from small (s) to medium (m) to large (l), the size of existing cracks and the engineering importance of surface defects also increase. Value medium (m) denotes the presence of alligator cracking or the occurrence of milling, and value large (l) denotes significant material loss (e.g., potholes).

Distress N characterizes slab surface cracking for Portland cement concrete pavements. Severity value none (n) denotes no slab cracking, and small (s) denotes cracks at the joints of the slabs that have been repaired using full-depth asphalt cement. Value medium (m) indicates the presence of tight cracks and large (l) the presence of open cracks. Both medium (m) and large (l) values correspond to cracks that are generally spall free. The existence of spalled cracks is identified through value total (t).

Distress M characterizes slab surface defects (other than cracks) for Portland cement concrete pavements. Such defects include patches, pitting of the material, and spalls that are not related to cracks and appear isolated (potholes) on the slab surface. Severity value small (s) denotes the presence of squared-off patches, medium (m) material pitting, large (l) small surface spalls (less than 6 inches), and total (t) large surface spalls (more than 6 inches).

III. OPENING DISTRIBUTIONS FOR DISTRESS SEVERITY ANALYSIS

The primary visual indicators of distress severity in the test images (Figures 1–3) are cracking and abnormal texture. The presence of cracking is easily detected visually because of the spatial extent of its features. Abnormal texture is harder to identify precisely but can reveal itself as inhomogeneities, large-scale features, excess of features at a certain scale, etc. As an example, differences in the severities for distress class G (Figure 1) are due to the transition from a normal asphalt surface (none) to various degrees of cracking (small and medium), until cracking becomes so severe that the top surface is entirely removed (large). The last stage results in an undersurface that can be distinguished from the normal surface (none) only by texture. Thus, whether due to cracking or to abnormal texture, surface defects are apparent in the distribution of particles within sensed images. Consequently, morphological operations of openings [1], which reveal information about the size distribution of particles, are valuable tools for the analysis and classification of pavement surfaces.

Before computing openings, the input gray-scale images must be thresholded to generate binary images. The threshold value chosen for each image is based on the criterion that 40% of the darkest pixels in the initial image define the binary image. Three thresholded images from the N distress class representing small (s), large (l), and total (t) distress severities are shown in the left column of Figure 4. Each binary image is opened with a horizontal line structuring element.

In general, an opening operation by a structuring element of a certain size removes all features from an image that are smaller than the size and geometry of the structuring element. The effects of opening the three binary images in the left column of Figure 4 with varying length of horizontal lines can be seen in the middle and right columns of Figure 4. For example, ns_1.13 (open) Line_{10} (shown at the top middle of Figure 4) is the result of opening binary image ns_1.13 with a horizontal line of length 10. In the notation used, the first and second letters of each image name denote distress class and severity, respectively, and are followed by a unique numeric identifier of the image.

A horizontal line structuring element is used because it preserves cracking in the images (a large percentage of the distress severities analyzed involve horizontal cracking). In addition, it offers the advantage of a single structuring element upon which to base the particle distributions, simplifying the final analysis. Another reason for the use of a single element is computational efficiency, as employment of a battery of structuring elements, each tuned to a different distress indicator, is computationally more costly. It is also intellectually intriguing to investigate the limitations of a single geometry upon which to base the analysis of derived morphological measures. Of particular interests are issues concerning the effectiveness of a horizontal linear structuring element in capturing information, not only about the cracking for which it is intended but also about changes in the underlying texture.

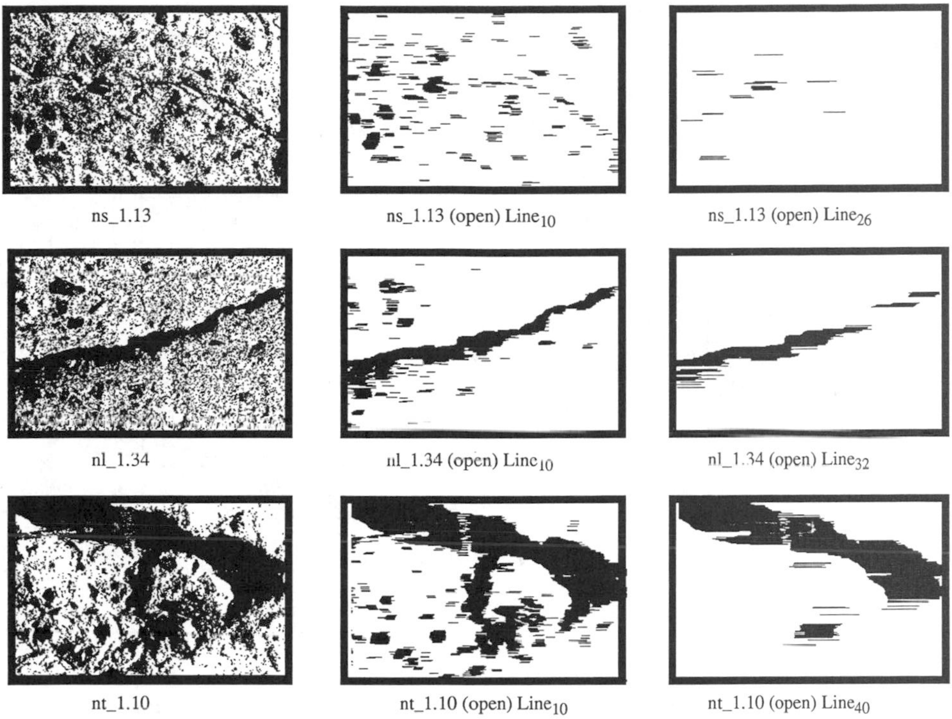

Figure 4. Three binary images—ns_1.13, nl_1.34, and nt_1.10—in the left column, with corresponding images in each row showing the results of openings with various sizes of linear structuring elements; for example, ns_1.13 (open) Line 10 is the result of opening ns_1.13 with a horizontal line of length 10.

An opening distribution [1] is generated by opening an image with a series of structuring elements that span the range of spatial scales within the image. In the present study, the distributions are generated by opening each binary image with a succession of horizontal linear structuring elements of increasing size. This is pursued in steps of 2 pixels, starting at length 0 (the original binary image), to a maximum size of 46 pixels. The area of each opened image is extracted, and the area remaining is expressed as a percentage of the area that existed prior to opening and is plotted against the size of the structuring element involved in the opening.

Figure 5 shows three opening distributions generated from the binary images ns_1.13, nl_1.34, and nt_1.10 (Figure 4). The distribution plots are coded so that the darkness of each line is inversely proportional to the severity of the distress. Thus, in Figure 5 the distributions of distress images with small severity are plotted in black, those with large severity in dark gray, and, finally, those with total severity in light gray. The intuition behind this coding is that black is used

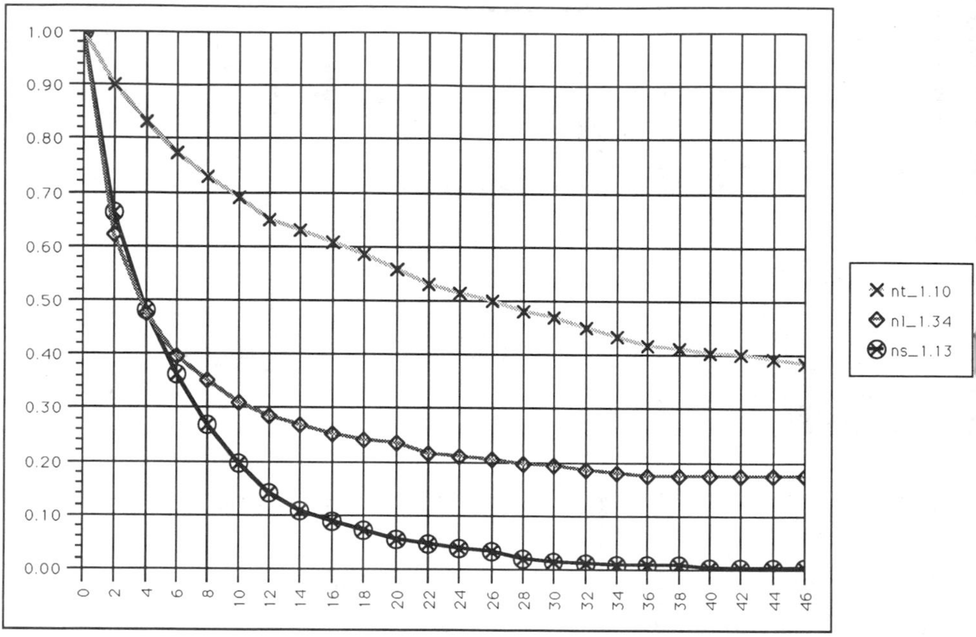

Figure 5. Opening distributions for selection of three images in distress class N. The x-axis values—for all distribution graphs—are the lengths of horizontal lines used to open the three binary images. The y axis is the percentage area remaining after each opening operation. The distress categories of the curves are displayed in gray scales such that the small (ns_1.13) distress is darkest, large (nl_1.34) distress is medium dark, and total (nt_1.10) is lightest; visualize the gray scales as representing increasing levels of pavement deterioration from darkest to lightest.

to represent solid pavement, and as the severity of each distress increases, pavement condition is coded in lighter shades of gray. In this manner, the darkness of each line decreases with increasing deterioration.

The opening distributions shown in Figure 5 are ordered according to the severity of distress. Distribution ns_1.13 has the smallest residue and is the closest to the X axis, nt_1.10 is at the top of the figure with a very large residue at large scales, and nl_1.34 lies between the two (as it should). Thus, at least in the case of these three images, the opening distribution can distinguish between the different distress levels. Texture tends to be removed at an early stage in the application of openings. With texture removed, cracking becomes apparent as long plateaus that are eventually followed by sharp drops. The long plateaus result from the fact that a size range of structuring elements fit inside cracks until a size is reached that exceeds the extent of cracking. Within such a size range of spatial scales, there is little area to remove. When the structuring element exceeds the scale of a particular crack (or part of a crack), a large amount of area is removed by the opening operation, which results in an abrupt downward shift of the distribution. (The maximum size of element used did not exceed the size required for cracks to be entirely removed.) In general, when distress becomes a less dominant feature in an image, opening distributions tend to slope gradually as texture slowly disappears with increasing scale. The opening distribution may have an occasional sharp slope or plateau to reflect the unequal presence of texture at different scales, but it is typically not as distinct as in the case of cracking.

Under ideal conditions, opening distributions should reflect the existence of cracks, normal texture, and changes in texture in the presence of cracks. In practice, such a delineation of effects on opening distributions is difficult to achieve. Figures 6, 7 and 8 show the opening distributions for all test images taken from distress classes N, G, and M, respectively. From these figures, it is clear that plots of opening distributions may become quite confusing as curves correspond-

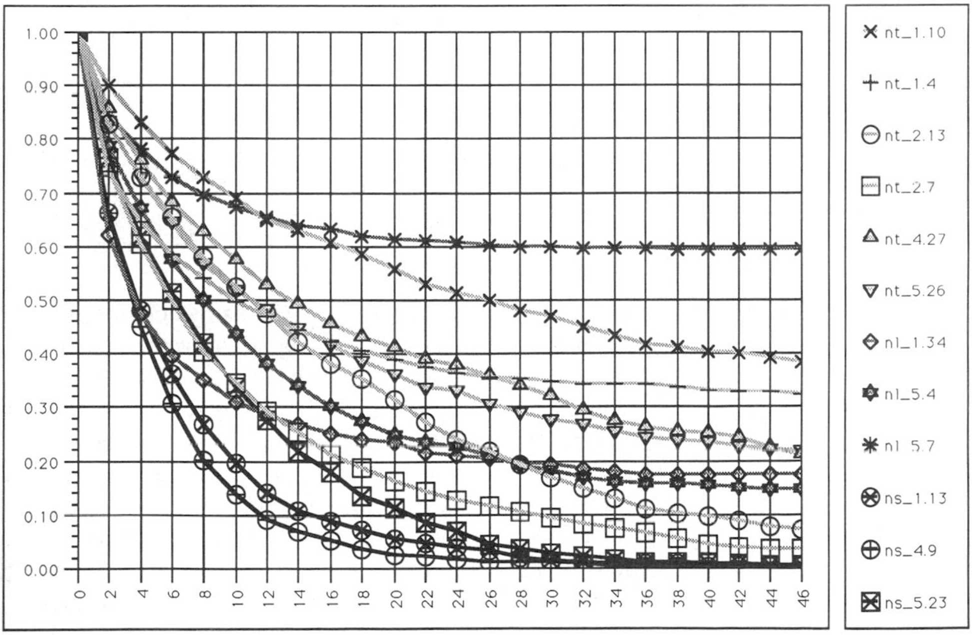

Figure 6. Opening distributions for all images in distress class N. The y axis is the percentage area remaining after each opening operation. The distress categories of the curves are displayed in gray scales such that small distress is darkest, large distress is medium dark, and total is lightest; visualize the gray scales as representing increasing levels of pavement deterioration from darkest to lightest.

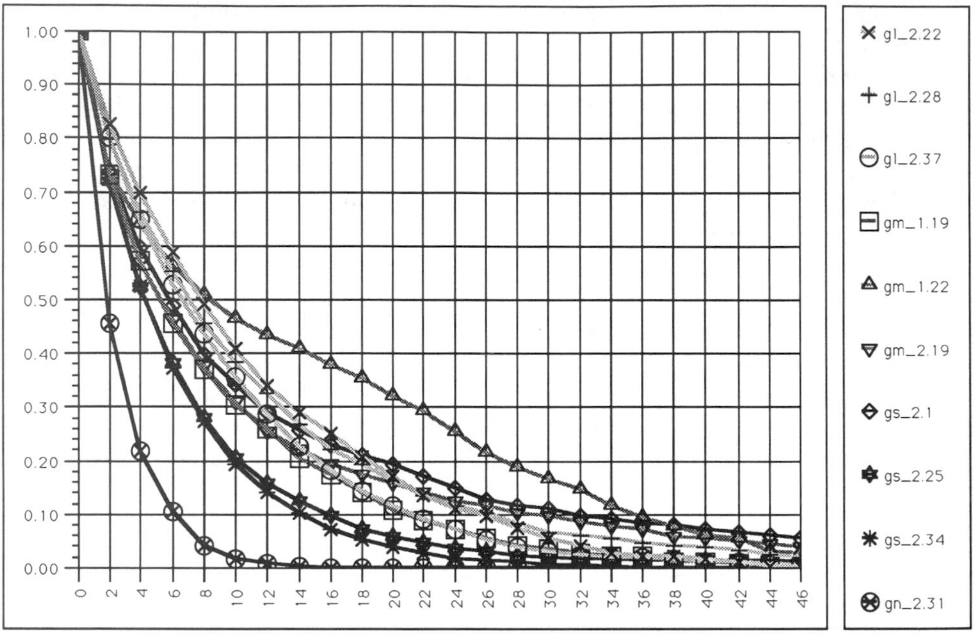

Figure 7. Opening distributions for all images in distress class G. The *y* axis is the percentage area remaining after each opening operation. The distress categories of the curves are displayed in gray scales such that no distress is black, small distress is less black, medium distress is medium dark, and large distress is lightest; visualize the gray scales as representing increasing levels of pavement deterioration from darkest to lightest.

ing to different severity classes cross one another. The clear distinction of distribution plots seen in Figure 5 disappears. Consequently, classifying images on the basis of opening distributions alone becomes a difficult endeavor. Limitations associated with the use of simple opening distributions for distress classification are discussed in the next section. This is followed by a presentation of the method used in this study to normalize opening distributions and improve distress severity classification. Finally, although the use of closing distributions is not addressed in this study, future developments will explore their role in distress classification, especially in terms of their ability to detect spatial relationships between particles. This study focuses exclusively on opening distributions and their sufficiency in achieving the task of classification.

IV. NORMALIZING OPENING DISTRIBUTIONS WITH TEXTURE MODELS

Opening distributions provide a partially accurate ordering based on the severity of distress. This is due to the tendency in the distributions (Figures 6–8) to reflect

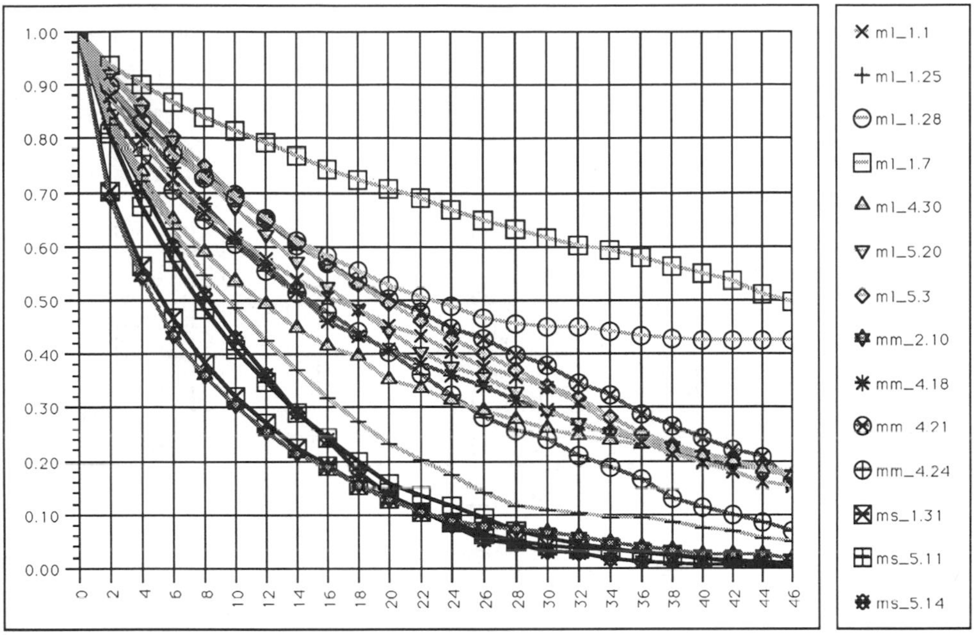

Figure 8. Opening distributions for all images in distress class M. The *y* axis is the percentage area remaining after each opening operation. The distress categories of the curves are displayed in gray scales such that small distress is darkest, medium distress is medium dark, and large distress is lightest; visualize the gray scales as representing increasing levels of pavement deterioration from lightest to darkest.

the severity of the distress at the top—encoded in lighter gray-scale plot lines—corresponding to higher levels of distress. It is also clear that the distinction between different severity levels can become blurred.

This limitation may be illustrated by considering Figure 9, which shows the opening distributions of the images gl_2.22, gm_2.19, gs_2.25, and gn_2.31. There is a reversal of the large- and medium-scale distresses midway through the range of structuring element sizes. Reasons for this are not hard to identify and they highlight some of the weaknesses of using raw opening distributions. Basically, opening distributions generated by horizontal structuring elements favor extended horizontal features in the images; that is, there is greater severity to features that appear as cracking versus features due to differences in texture. Thus, gm_2.19 is ranked distressed because of well-defined cracking, while gl_2.22 is associated with texture.

Changing the shape of a structuring element results only in a change of the sensitivity of the opening distributions between cracking and texture, with some shapes favoring cracking and others texture. For example, in experiments with disk structuring elements, it was possible to distinguish different textures but at

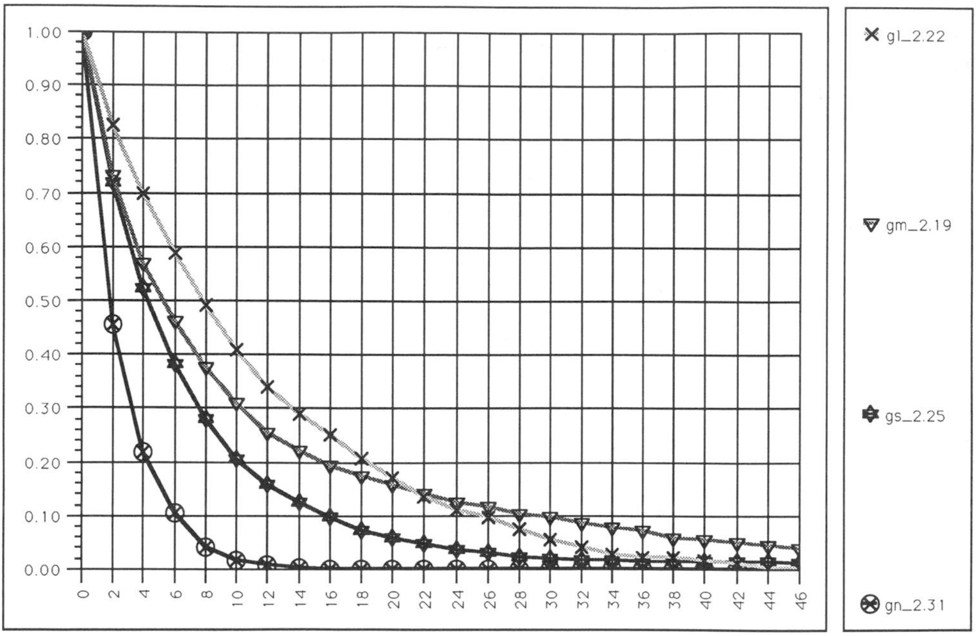

Figure 9. Four selected opening distributions from distress class G that illustrate the difficulties in using the raw opening distributions to realize the distress types. Note the crossing over of medium and large distress categories at linear structuring element size 22.

the cost of reduced sensitivity to the extent of cracking. Thus, opening distributions tend to be sensitive either to cracking or to texture, but not to both. This is due to the fundamentally different morphological characteristics of these two pavement surface features. Consequently, opening distributions can function as reliable indicators of surface features only if all images are graded according to either texture or cracking.

An approach to resolving the problem of unreliable classification based on opening distributions is to extract and combine information from various shapes of structuring elements, edge detectors, etc. Such an approach, while worthy of exploration, is computationally costly because it requires that openings on image data be performed for each class of structuring elements. In order to be consistent with the current goal of exploring the capabilities and limitations of using a single class of structuring elements, an alternative approach is followed in this study that incorporates in opening distributions a morphological model of the expected "normal" texture, as measured by a single class of structuring element.

The concept involved in the applied approach is that a distressed pavement surface can be viewed as one that in some way deviates from an *ideal* surface. If the visual characteristics of the ideal image can be precisely described, then distresses can be identified as departures from that description. These departures can be present visually as cracking, ravelling, rutting, abnormal texture, or any other feature associated with distress. Using an *ideal* model of the normal pavement surface, one expects to achieve greater sensitivity to all forms of distress.

As a starting point, a model is incorporated into a *normalized* opening distribution. The reason for normalization is to generate a description that is very simple, such that any deviations or anomalies can be emphasized, making them easy to identify. A model incorporates the distribution for an ideal surface, which is in turn used as a basis for identifying departures from such a surface.

Two simple models of the ideal surface are considered, based on the assumption that the ideal surface has either a constant number of particles at all scales or a constant area removed at all scales. Although, admittedly, these two specific models may be overly simple (and, thus, unrealistic), they can help illustrate how models can be incorporated into opening distributions. In addition, they produce improved performance of the pavement distress analysis. Perhaps of greater general significance is that this method is an example of pravda [1] defined on morphological measures themselves, as opposed to the more traditional approach involving the use of various classes of structuring elements to probe images.

A. Constant Number of Particles Model

1. Definition

The basic premise of this model is that the ideal surface corresponds to a texture that has an identical number of particles at each scale of observation up until a scale, t, at which all texture is removed (the "range" of the distribution [1], or in our parlance the texture scale). It is assumed that particles at each scale do not overlap. With a constant number of particles at each scale, most of the area of the ideal image is present at scales relatively close to the texture range of the distribution, t. Therefore, the opening distribution shows a relatively flat slope at small scales with increasing slope as the size of the structuring element approaches the texture scale. The increase in the slope is proportional to the number of particles and the size of the structuring element at that scale.

Let the initial area in the image be A_0 and the number of particles at any scale be κ. As there are κ particles at each scale, the area at scale i in the ideal image is $S_i\kappa$, where S_i is the size of the structuring element at scale i. The area removed from the initial image by an opening at scale i is

$$A_0^i = \left(\sum_{j=1}^{i} S_j\right)\kappa = (A_S^i)\kappa \qquad (4.1)$$

The initial area can be expressed as

$$A_0 = \left(\sum_{j=1}^{t} S_j\right)\kappa \tag{4.2}$$

because all the area is removed at scale t. Thus, the area remaining at scale i—the raw opening distribution—is given by

$$A_i = A_0 - \left(\left(\sum_{j=1}^{i} S_j\right)\kappa\right) = A_0 - (A_S^i)\kappa \tag{4.3}$$

The distribution varies as $\Sigma_{j=1}^{i}$, S_j as both A_0 and κ are constants in the current model. Thus, if S_j increases linearly with scale (as is the case for line-shaped structuring elements), the opening distribution exhibits a *quadratic* slope. Similarly, if S_j increases quadratically (as in the case of a box- or disk-shaped structuring element), the opening distribution displays a *cubic* curve.

The opening distribution does not exploit directly the basic assumption of the model, that is, that the number of particles at each scale is constant. As an opening distribution is dependent on both the particle distribution and the structuring element, the latter dependence serves to hide the former. Therefore, the distribution should be normalized to eliminate the effects of the structuring element so that it reflects the particle distribution alone. Such a normalization scheme is now developed for the case of a horizontal line-shaped structuring element.

The contribution of the structuring element to the opening distribution at scale i is given by the term A_S^i, appearing in Eq. (4.1). As the structuring element is assumed to be a line, the size of the structuring element is $S_i = i$, at scale i. It is also noted that S_t, which is the structuring element size at which all texture is removed in the ideal case, has a value of t (for linear structuring elements). Therefore, $A_S^i = S_i(S_i + 1)/2 = i(i + 1)/2$. Elimination of A_S^i from the opening distribution results in the following expression for the normalized distribution:

$$\eta_\kappa(i) = \begin{cases} \dfrac{2A_0}{t(t + 1)} & \text{for } i = 0 \\ \dfrac{2(A_0 - A_i)}{i(i + 1)} & \text{for } 0 < i \le t \\ \dfrac{2(A_0 - A_i)}{t(t + 1)} & \text{for } i > t \end{cases} \tag{4.4}$$

This expression reflects solely the effects of the particle distribution, independent of the effects of structuring element size.

The value of the function at $i = 0$ is $\eta_\kappa(0) = \kappa$, from Eq. (4.2). As the area removed in the ideal surface, A_0^i, follows Eq. (4.1) for all scales until the texture scale t is reached, the function is again equal to κ for all scales below t. At all scales i above the texture scale t, all the area has already been removed from the image and $A_i = 0$, with the result that $\eta_\kappa(i) = \kappa$ at all scales above the texture scale. Thus, the normalized distribution has a constant value κ reflecting the basic premise of the "constant number of particles" model.

It is clear that this approach to normalization is applicable even if the structuring element is nonlinear. As long as an expression can be found for A_S^i appearing in Eq. (4.1), it can be used to eliminate the effects of the structuring element. If the structuring element is irregular, A_S^i can be computed by "accumulating" the area at different scales. Therefore, a normalizing function based on the constant number of particles model is obtainable over a wide range of structuring elements.

Consider a nonideal image from which the normalized distribution is extracted. If the image has an excess of particles at any scale below the texture scale, more area is removed at that scale than predicted by the ideal surface and the residue is smaller. As a result, the value $(A_0 - A_i)$ is larger and the distribution $\eta_\kappa(i)$ is larger than κ. Therefore, the discrepancy is visible as an overshoot from the ideal distribution. A similar argument indicates that a particle deficiency at any scale is visible as an undershoot at that scale. Thus, any anomaly in the texture below the texture scale appears as a deviation from the "expected" flat distribution. If any texture is present in the image above the texture scale, A_i has a nonzero value and $A_0 - A_i$ is smaller than the ideal value, which results in an undershoot until all the texture is removed. Thus, the severity of large-scale distress is visible in the magnitude of the undershoot just beyond the texture scale. Moreover, information about the visual nature of the distress, that is, whether it is cracking or abnormal texture, is obtained from the shape of the distribution beyond the texture scale. Cracking, by definition, is an extended feature and therefore is not removed by anything but an extremely large horizontal structuring element. Consequently, the shape of the distribution will be flat. Abnormal texture exhibits a more gradual sloping because the structuring element slowly removes it.

2. Experimental Results

Application of the normalizing function to the images shown in Figures 1, 2, and 3 indicates that the model-based scheme is powerful enough to classify correctly a larger number of images than were classified in the raw opening distributions—in spite of the obvious inadequacies of the model. The raw distribution data described in Section III are used as input to the normalizing function and plotted with κ normalized to 1.0.

G-Rated Images. Figure 10 shows the results of applying the constant number of particles model embodied in Eq. (4.4) to the images of the G-rated distress category. The texture scale is empirically fixed at $t = 8$, as the texture from the "no distress" image (gn_2.31) tends to be eliminated at 8. There is a distinct peak (not shown) in the normalized distribution of all images at $i = 2$, indicating that a larger number of particles than the number predicted by the model have a size

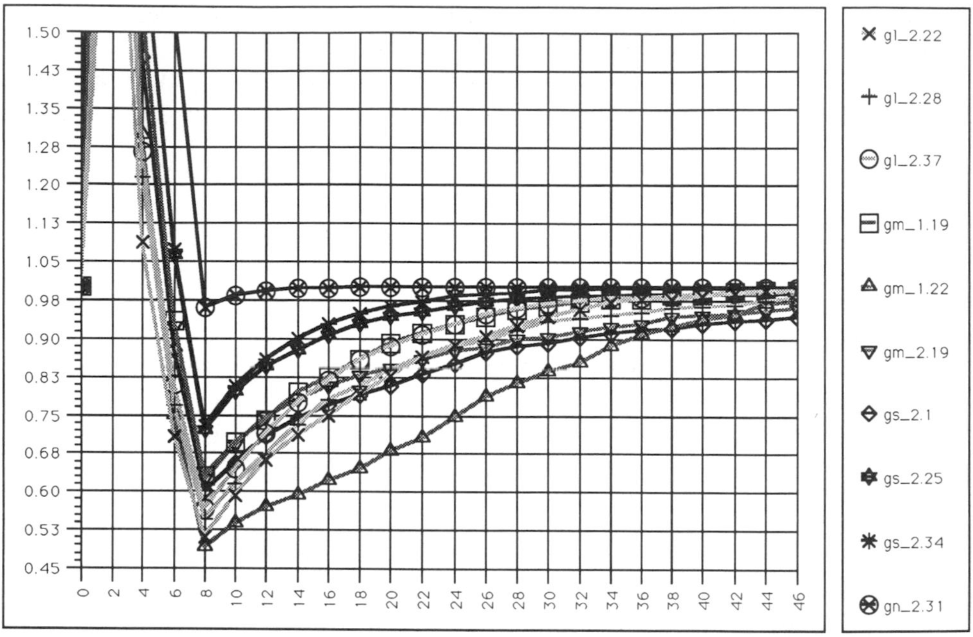

Figure 10. Equal particle model normalized opening distribution of images in class G. Note that the full y range has been cut off in the low x values in order to provide more resolution over the entire range.

2. There is clear deficiency of particles at scales 6 to 8. A rough measure of the severity of distress is then obtained by the magnitude of the undershoot at the texture scale of 8. The "small" category images show the smallest undershoot with the no distress image showing an almost insignificant undershoot. The "medium" distress images show medium undershoots, while the "large" distress images fall to the bottom. At the same time, two images are not classified correctly: gs_2.1 and gm_1.22. Unequal illumination causes features other than cracking to be detected in gs_2.1. Consequently, it is classified as "large." The effects of unequal illumination can be reduced to insignificant amounts either by providing artificial lighting when taking pictures of the pavement surfaces or by preprocessing. Vertical cracking in gm_1.22 is the reason for incorrect classification. The problem is that a horizontal structuring element does not preserve vertical cracking.

In general, an improved method would further normalize the distribution with respect to the entire ideal distribution produced by gn_2.31, as opposed to simply using the texture range value of 8. This would result in a completely flat normalized curve for the ideal image, gn_2.31. It is now possible to measure all normalized distributions in terms of how much they deviate from the entire normalized distribution of gn_2.31. For example, Figure 11 shows the magnitude of deviations of the various images from the ideal distribution defined by gn_2.31. The magnitudes are obtained by taking the absolute sum of the deviations at each scale from the values defined by the gn_2.31 distribution. All images, with the exception of the two discussed in the previous paragraph, are ordered by their magnitudes according to the severity of the distress, an improvement over the raw distributions.

N-Rated Distress. Figure 12 shows the results obtained from images belonging to the N-rated distress category. It is noted that, in this case, there are no images belonging to the "no distress" category. As texture in images with "small" severity tends to be eliminated around a scale of 14 to 18 (Figure 6), the texture scale is empirically fixed at $S_t = t = 12$. Most of the images in this category show distinct cracking and, consequently, the selection of a potentially unrepresentative texture scale does not affect the results to a large extent. However, this is not the case with images in the M-rated distress category discussed next.

The N-Rated results support some of the conclusions obtained from the G-rated images, for example, particle excess at $i = 2$ and particle deficiency at scales of 6 to 12. Once again, images tend to be classified according to severity

Image	Magnitude of Deviation from gn_2.31
gl_2.22	9.70
gl_2.28	9.24
gl_2.37	8.49
gm_1.19	7.09
gm_1.22	10.18
gm_2.19	7.84
gs_2.10	8.54
gs_2.25	5.90
gs_2.34	5.69
gn_2.31	0.00

Figure 11. Magnitude of deviation of the equal particle model curves (of Figure 10) from the no distress curve (gn_2.31) of the G class. See text for discussion of the two images, gs_2.10 and gm_1.22, ranked incorrectly.

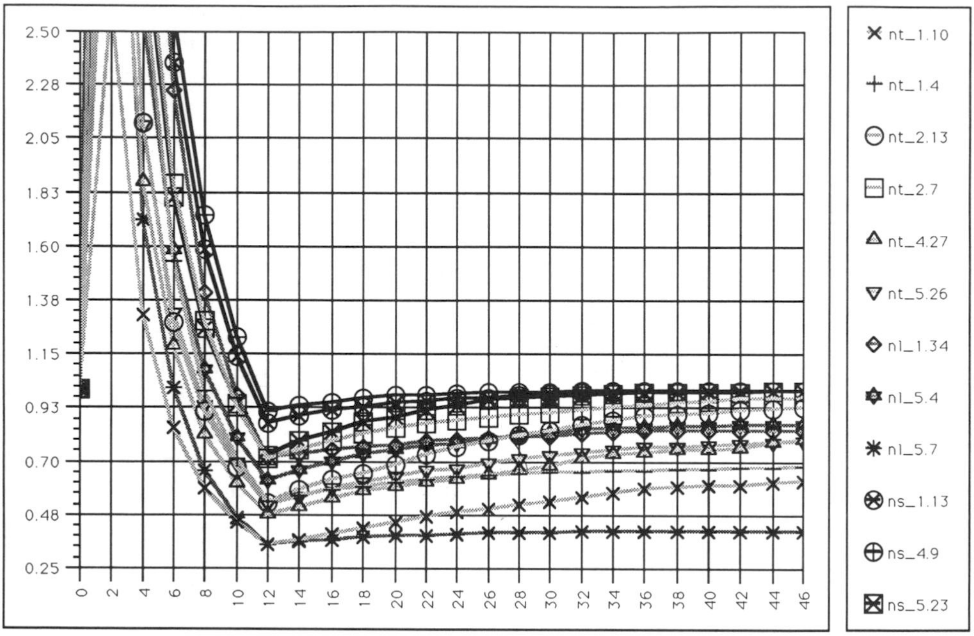

Figure 12. Equal particle model normalized opening distribution of images in class N. Note that the full y range has been cut off in the low x values in order to provide more resolution over the entire range.

at scales just beyond the texture scale, $t = 12$, where the images are classified correctly except for the two images nt_2.7 and nl_5.7. With the exception of these two cases, the three "small" images are on top, followed by the "large" and "total" images in that order. Further information about the surface condition is obtained from the distribution plots beyond the texture scale. All distributions are flat except for nt_2.13, indicating that all images (except nt_2.13) show distinct cracking. The exceptional flatness of nl_5.7 results from large and well-defined cracking. In addition, similarities in the large-scale cracks in nt_5.26 and nt_1.10 are reflected in the close tracking of their normalized distributions. This suggests possibly ability to "predict" similarities in large-scale features from normalized distributions. For example, distributions nl_5.4 and nl_1.34 are nearly coincident, suggesting that cracks associated with these two distributions are similar.

Distress severities associated with distributions nl_5.7 and nt_2.7 are classified incorrectly. Distribution nl_5.7 displays a large, well-defined crack running across the image, which places the image on the boundary between large and total. Distribution nt_2.7 has the form of abnormal texture rather than distinct cracking. Detection of abnormal texture in this case requires a better model of the ideal texture, with capabilities not captured in the current normalization scheme, which is based only on the value of the texture range, $t = 12$. Also, the absence of a "no distress" image in the current data set (corresponding to gn_2.31 in the G-rated distress category) prevents attempts at classification using such an image as a reference and at computing deviations in a manner identical to the one that produced the results shown in Figure 11.

M-Rated Images. Figure 13 shows the results from images belonging to the M-rated distress category. The texture scale is empirically fixed at $t = 22$. This is based on the observation that texture from "small distress" images tends to be eliminated around 26 to 28, and, presumably, at a smaller-scale value for the "no distress" case. Images in this category differ significantly from those in the other two categories. All images show abnormal texture and very little cracking. As a result, modeling of the ideal surface has greater significance here than it did in the other categories.

The normalized distributions reveal that all images have an excess of particles at almost all scales below the texture scale, indicating that the current model does

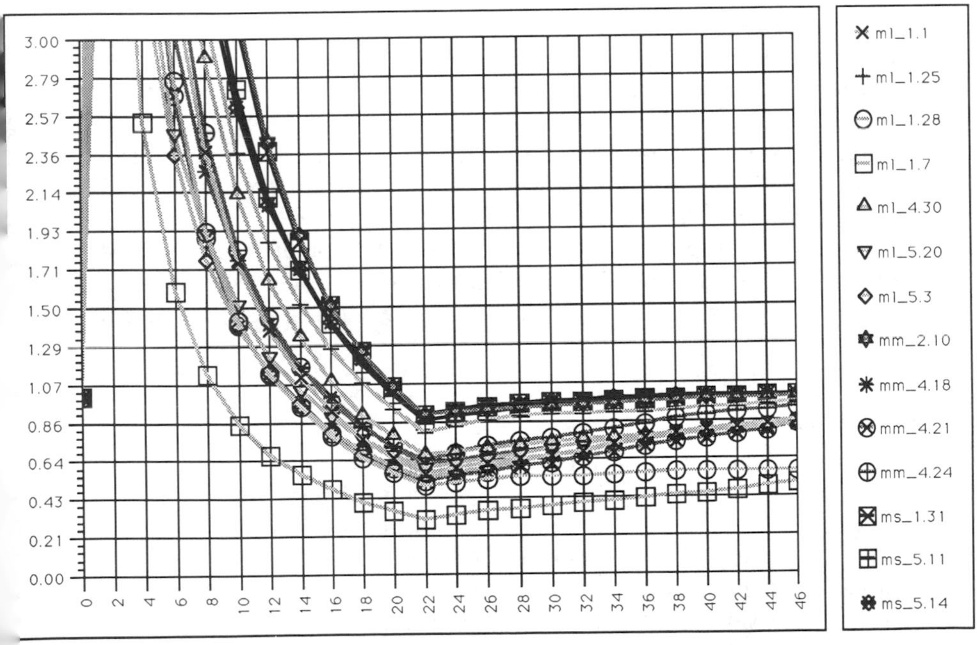

Figure 13. Equal particle model normalized opening distribution of images in class M. Note that the full y range has been cut off in the low x values in order to provide more resolution over the entire range.

not accurately reflect the ideal surface. There is again a peak in the number of particles at scale $i = 2$. The distributions exhibit three distinct clusters just beyond the texture scale (at $t = 22$). All "small distress" images and one "medium distress" image (mm_2.10) form one cluster. A second cluster comprises mm_4.18, mm_4.21, mm_4.24, ml_1.11, ml_1.25, ml_4.30, ml_5.20, and ml_5.3, while a third cluster comprises ml_1.28 and ml_1.7. Thus, as each cluster can be associated with a single scale of distress severity, such a classification results in a number of discrepancies.

It is interesting to compare the visual appearance of images in this category (Figure 3) with the classifications suggested by the normalized distributions. It is clear that the images are roughly classified according to the similarity of their textures. For example, mm_4.18 and ml_4.30, mm_4.21 and ml_5.3 show similar textures, although they are classified into different severity classes by human experts. Bars that are visible in ml_5.3 (exposed reinforcement rods) are removed at smaller scales by the opening operation, and, without their presence, the resulting texture is quite similar to that of mm_4.21. Also, mm_2.10 displays a texture similar to that of ms_1.31, except for very localized pitting.

As explained in Section II, classification of M category distress is based on diverse pavement surface defects, such as patching, spalling, and pitting. The present model, based on the assumption of a constant number of particles, is sensitive to any inhomogeneities in particle distribution and, thus, unable to distinguish such defects. The development of a more sophisticated model (Section D) may result in better estimates of "normal texture." On the other hand, the performance of models on this set of images may also indicate some of the limitations of the model-based approach using an opening distribution with a single structuring element. The study of these limitations may, in turn, lead to the establishment of an interesting trade-off between application of various models to a reduced data set, as opposed to the method of extracting various parameters through different morphological operations on the much larger image data sets.

B. Constant Area Model

1. Definition

This model defines an ideal surface as one that has a constant area at each scale of the structuring element. Thus, the opening distribution of the ideal surface (which plots the residue at each scale) is expected to have a constant slope.

As is the case with the first model, it is assumed that there exists no texture beyond a certain texture scale t. The procedure followed for normalizing the distribution is similar to the one described in Section IV.A. Let the area at each scale be α. The area removed at each scale by opening with a structuring element is α, and the area removed with a linear structuring element from the ideal image at scale i is $A_0^i = i\alpha$. As the entire area in the image is below the texture scale t, the initial area in the image is $A_0 = t\alpha$. From this expression, the normalizing function is defined as

$$\eta_\alpha(i) = \begin{cases} \dfrac{A_0}{t} & \text{for } i = 0 \\ \dfrac{A_0 - A_i}{i} & \text{for } 0 < i \leq t \\ \dfrac{A_0 - A_i}{t} & \text{for } i > t \end{cases} \tag{4.5}$$

The value of the function is initially α, based on the expression for A_0. For the ideal distribution, the area A_0^i removed at any scale i below the texture scale t is $i\alpha$, and the function evaluates to α. Above the texture scale, $A_i = 0$, and the normalizing function receives the value α. Thus, $\eta_\alpha(i)$ defines a distribution that has a constant value of α, reflecting the constant area model.

As was the case with the constant particle model, if the normalizing function is applied to any image that does not define an ideal surface, the resulting distribution exhibits deviations from the constant value α. If there is excess area at any scale below the texture scale, more area is removed from the image at that scale. Consequently, the residue is smaller than that predicted by the distribution, and the function reveals this as an overshoot above the value of α. Similarly, deficiency in area at any scale is revealed as an undershoot at that scale. Large-scale distress produces a nonzero value of the residue A_i at scales above the texture scale and results in an undershoot. The slope of the distribution at large scales provides information about the nature of the surface condition: large slope indicates texture, while small slope (or flat shape) indicates cracking.

2. Experimental Results

Much of the analysis and results of this section mirrors those obtained using the equal particle model. Both models tend to give similar results, such as large deviations from the ideal texture, because the actual data tend to violate the assumptions of both models. There are differences, however, and additional investigation is required to understand and explain all reasons for their existence.

Figures 14, 15, and 16 show the normalized distributions based on the equal area model for the images belonging to the G, N, and M distress categories, respectively. This model succeeds in classifying correctly the same images as did the previous model. Raw distribution data described in Section III are used as input to the normalizing function, and the distributions are plotted so that $\alpha = 1.0$. The texture scales are the same as in the previous section; $S_t = t = 8$ for G images, 12 for N images, and 22 for M images.

Figure 14 shows the normalized distributions of images belonging to the G category. The peak at a scale of 2—a primary feature of the constant particle model distributions—is absent for all images classified as "large" on the basis of

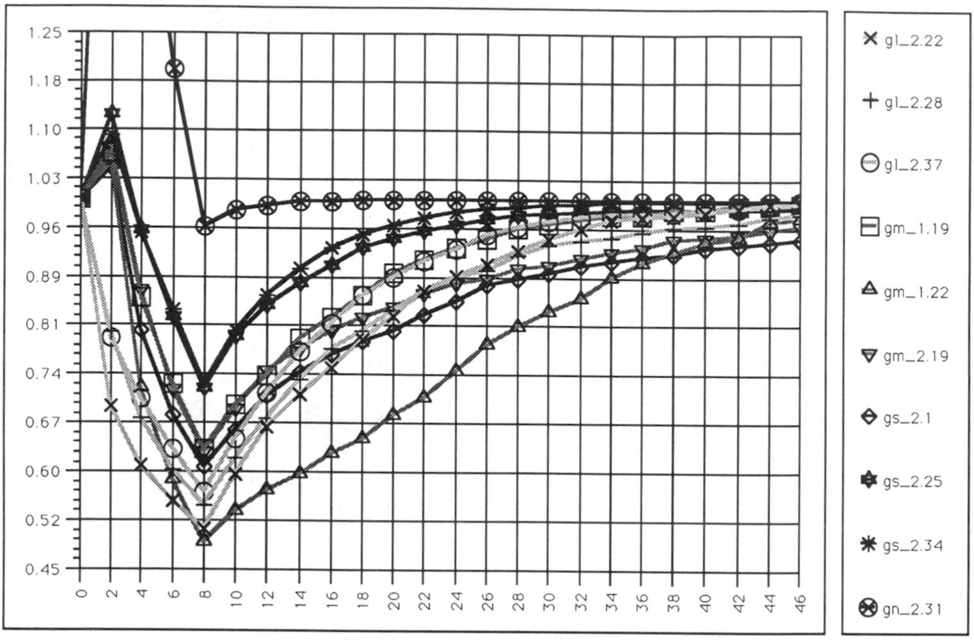

Figure 14. Equal model area normalized opening distribution of images in class G. Note that the full y range has been cut off in the low x values in order to provide more resolution over the entire range. The texture scale t is equal to 8.

distress severity. This phenomenon may reflect a difference in the way deviations from the ideal surface are weighted by the equal area versus the equal number of particles models. The images are classified according to the severity of distress at a texture scale of 8. The range of the magnitude of the undershoots is larger than that for the constant number of particles model, and this provides a higher resolution for classification. Also, the undershoots at the texture scale show a cleaner separation between various severity levels than does the previous model. However, as in the other model, the two images gs_2.1 and gm_1.22 are incorrectly classified. The reasons are the same as those mentioned in Section IV.A.2.

Figure 15 shows the results for N-rated images. Most of the discussion appearing in Section IV.A.2. is also valid for the current model. An exception is given by images nt_2.7 and nl_5.7, which are ordered by their deviations from the ideal flat curve according to their distress at large scales.

Figure 16 shows the results of applying the current model to the M-rated images. There is excess area at all scales below the texture scale compared to that predicted by the model. This is the case for all images except those with large severity of distress. The division of the distributions into three distinct clusters

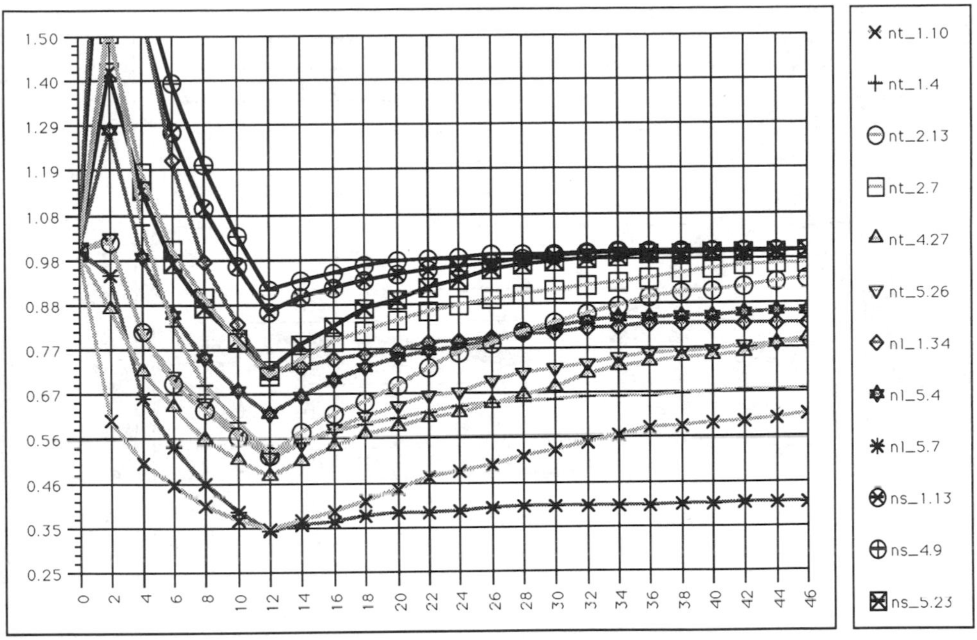

Figure 15. Equal area model normalized opening distribution of images in class N. Note that the full y range has been cut off in the low x values in order to provide more resolution over the entire range. The texture scale t is equal to 12.

is also valid in this model. In fact, the results are identical to those obtained from the previous model, an indication that neither of these models is sufficiently powerful to isolate abnormal texture from these images.

C. Centroid Movement as Indicator of Texture Inhomogeneity

1. Definition

In the process of gathering size distributions, area measurements of altered images are taken. A measure that can be obtained with insignificant additional computational cost is that of the path followed in the process by the area centroid. Thus, it is of interest to explore what additional information might be obtained from the sequence of centroids along the distribution. Previous work on measures of centroid movement over the sequence of opened and closed images has focused on shape description [12,13]. Some preliminary results on the use of centroid trajectory measures applied to texture analysis are considered below.

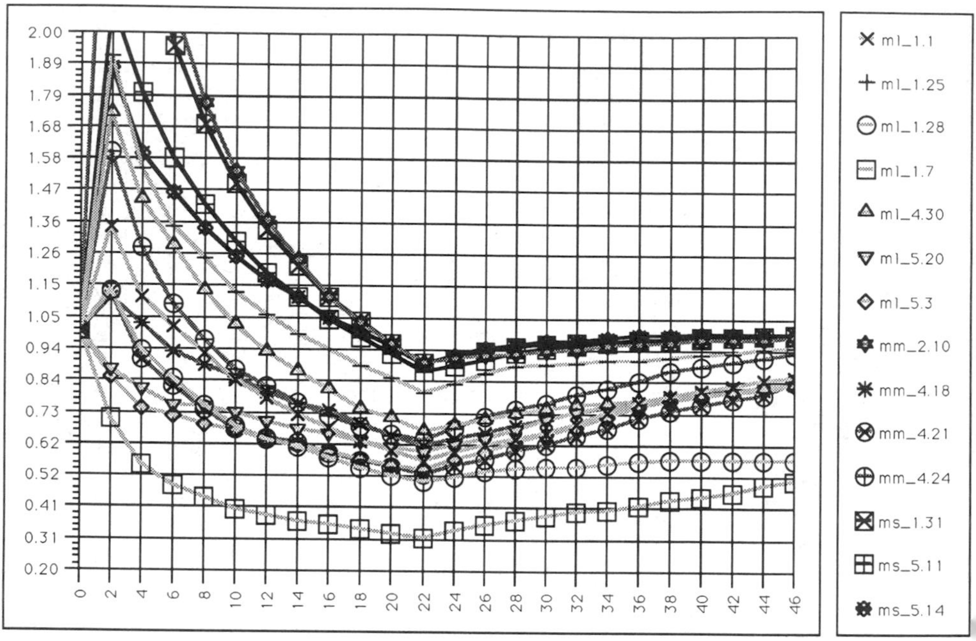

Figure 16. Equal area model normalized opening distribution of images in class M. Note that the full y range has been cut off in the low x values in order to provide more resolution over the entire range. The normalizing texture scale t is equal to 22.

Opening an image with a structuring element of a certain size isolates texture that is at least of that scale. As the centroid is a reflection of the spatial distribution of the pixels present in the image, its position in the opened image provides condensed information about the spatial arrangement of texture features at that scale. If the positions of the centroid are computed for every opened image in an opening distribution, its movement reflects the change in spatial arrangement of texture features at different scales. Thus, it is possible to establish whether certain texture features are localized or whether they are globally distributed. If the texture is randomly distributed about the center of the image, the centroid executes a random walk around the true centroid. On the other hand, if there is an uneven distribution of texture at a given scale, then the centroid executes a large jump when the texture is removed by a structuring element of the given scale. This scheme potentially provides a computationally less intensive alternative to methods based on subwindows [14], where the image is divided into several smaller subwindows and morphological distributions are extracted from each subwindow to detect inhomogeneities.

2. Experimental Results

Figures 17, 18, and 19 show the paths followed by the centroids of selected images belonging to the G, N, and M categories, respectively, when opened by the same horizontal line-shaped structuring with size varying from 0 (initial image) to a maximum of 46 pixels. The dimension of the image data is 256 rows by 384 columns.

G-Rated Images. The centroids of the images gs_2.1 and gs_2.25 show a steady downward movement, indicating that most of the large-scale features are present in the lower portions of the image. The initial position of the centroid in gs_2.1 indicates that most of the area is present toward the lower left corner of the image, while the starting point of gs_2.25 indicates that the area is distributed fairly evenly. The path of the centroid of the image gl_2.37 moves more slowly downward, indicating that the large-scale texture is at the bottom of the image.

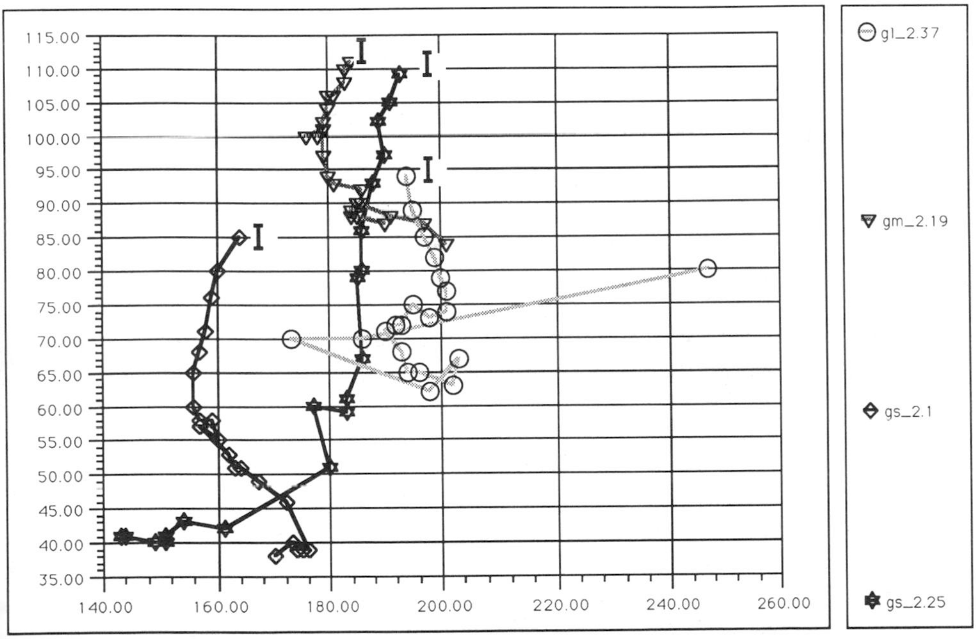

Figure 17. Centroid movements from sequence of opened images generated from selected binary images of G severity class. The letter I is placed near the centroid position generated from the initial binary image. The succeeding points are the centroids from each of the opened images used to generate the size distribution.

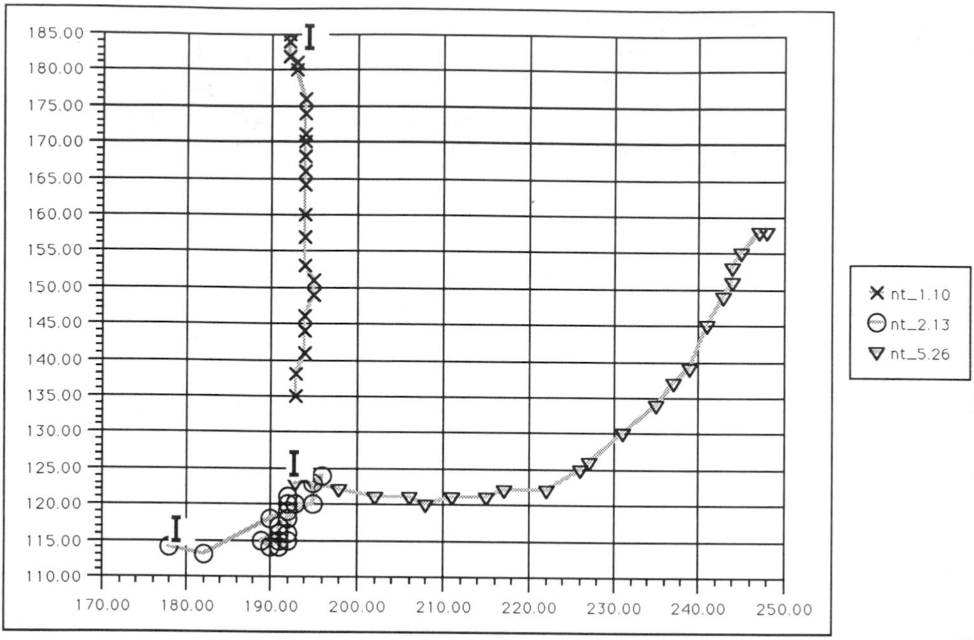

Figure 18. Centroid movements from sequence of opened images generated from selected binary images of N severity class. The letter I is placed near the centroid position generated from the initial binary image. The succeeding points are the centroids from each of the opened images used to generate the size distribution.

However, at very large scales there is an abrupt jump to the right, indicating that the largest-scale features are on the right side of the image. The image gm_2.19 does not display as much movement as the other images. This suggests that the texture is fairly evenly distributed around the center of the image at all scales. There is a slight downward movement that is accounted for by a slightly greater number of cracks toward the bottom of gm_2.19.

N-Rated Images. The centroid of the image nt_1.10 shows a steady upward drift, indicating the presence of a large-scale feature at the top. On the other hand, nt_5.26 shows a large movement toward the right. This suggests that there is a large-scale feature on the right side of the image. In contrast to these two images, image nt_2.13 shows very little movement of the centroid. This suggests that texture is distributed evenly in the image at all scales.

M-Rated Images. Four images belonging to the M category are selected to show that tracking centroids across a series of openings provides information about texture inhomogeneities. Three of them (ml_1.25, ml_5.20, and ml_5.3) exhibit directional movement indicating asymmetries in the presence of texture

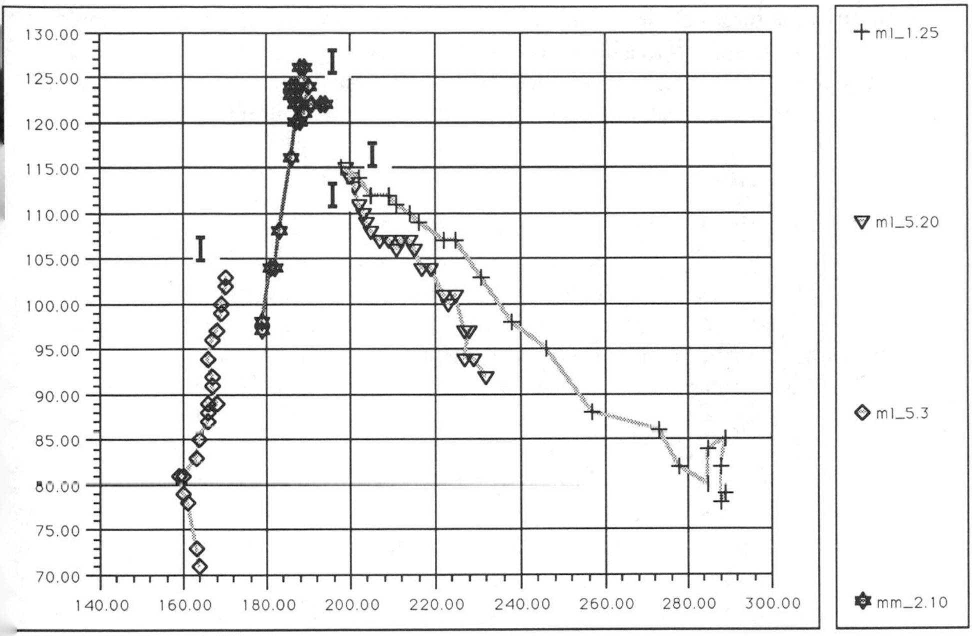

Figure 19. Centroid movements from sequence of opened images generated from selected binary images of N severity class. The letter I is placed near the centroid position generated from the initial binary image. The succeeding points are the centroids from each of the opened images used to generate the size distribution.

features, while mm_2.10 shows very little movement, except at very large scales, indicating that most of the texture is symmetrically distributed around the origin.

These results indicate that centroid movement provides useful information about symmetries in texture at various scales. However, centroid movement can result in confusing and incorrect results in certain situations. Consider the presence of two localized features on opposite sides of the centroid at a certain scale. When they are removed by opening the image, the centroid executes very little movement, and one may conclude that there are no localized features at that scale. In fact, a single localized feature with the centroid right at its center results in an identical situation. In these situations, methods of extracting parameters over various subwindows may provide the finer discriminations [14]. However, combining the results of centroid movement with clustering information obtained from closing distributions may provide a method of correctly interpreting the results. If a good model is developed for the opening distributions, the magnitude of the deviations might provide information about the particle distribution that can assist the interpretation of the measures on centroid movement. For example, if the model-based approach suggests that there are many particles at that scale, it may be reasonable to conclude that the lack of centroid movement is the result of symmetric arrangement of these particles around the centroid.

D. Discussion

The results of the three test cases indicate that classifications involving cracking were relatively easily obtained with a linear structuring element. Cracking is such a distinct large-scale feature that it stands out from any texture. All that is required is the identification of the largest scale at which all texture disappears from the image (the value t of the texture scale). Cracking persists as an opening residue above texture scales almost until the image dimensions themselves are reached, as long as cracks extend horizontally across the image. This is considered the primary reason for the good performance of both the constant particle and constant area models on all images that exhibited cracking (almost all N-category images and some G-category images).

Identification of abnormal texture is difficult when the analysis is based on raw opening distributions derived from a single class of structuring elements. The ability to identify abnormal texture is intimately linked to the selection of a "good" model of an ideal texture. If a good model is not available, then all images show significant deviations from the ideal surface below the texture scale. In these cases, it becomes difficult to isolate the deviations that occur because of errors in the model from those that result from surface defects, as reflected in texture changes. Both equal particle and equal area models displayed their inadequacy in their performance on images belonging to the M category, where there is little cracking and most of the surface features are due to abnormal texture.

The results presented in Section IV.B suggest that neither the constant number of particles nor the constant area normalizing model is adequate for modeling pavement texture, although both models resulted in improvements over the distress classifications produced by nonnormalized opening distributions. For images with no or small distress, there tend to be distinct overshoot peaks that appear at a small size scale, followed by a rapid decrease beyond that scale. This indicates the need for an improved model (e.g., exponential or Gaussian) to represent particle size distribution. Such a model would account better for the empirical notion that the particle distribution increases initially with scale, reaches a peak at a certain scale, and then decreases as particle size continues to increase.

Work is in progress toward developing normalizing functions for the above distributions and testing them on the image set. A more long-term objective, one that is more truly morphological in spirit (e.g., similar to the early morphological work relating openings to porosity measures), is to develop models based on pavement surface specifications themselves. Such models will identify dominant engineering factors (e.g., reflecting materials, methods of construction) and incorporate them in the expressions for opening distributions. Finally, the possibility also exists to extend the models to predict not only opening distributions but also other morphological measures like closing distributions.

V. SUMMARY AND CONCLUSIONS

The problem of pavement distress classification has been approached by normalizing opening size distributions of pavement texture images, where the normalization is done with respect to various models of ideal distributions formed from nondistress images. Two basic models of texture have been formulated based on the characterization of ideal texture as having either a constant number of particles or a constant area removed at all scales of the size distribution. Although in a strict sense both models are unrealistic, the resulting normalizations of the opening distributions have improved on the categorizations produced by raw distributions.

The advantage of a model-based approach was that it allowed increased sensitivity to pavement surface features involving both inhomogeneities in the ideal texture (e.g., as may result from pavement cracking) and appearances of new textures (e.g., as may result from large pits and holes in the pavement). The applied normalization method was general in that other models of particle texture distributions (e.g., Gaussian) could also be the basis for normalization.

This work also represents an investigation into the utility of a single class of horizontal linear structuring elements as the basis for the morphological measures. A pragmatic reason for this restriction is that the eventual goal of this work is a real-time system and, thus, it is desirable to minimize the computational burden of probing the images with diverse structuring element geometries. Thus, starting from a small set of raw distribution data—produced using computationally costly pixel operations of openings and area measurements—various morphological texture models can be used to probe the data. There is a clear computational advantage in probing the data at a time following the application of the more costly image processing operations. In addition, alternative measures can be obtained in this process of gathering the distribution data—for example, the sequence of opening centroid trajectories used to characterize symmetries in the texture. The central issue concerns what is lost in probing image data with only one class of structuring elements and how this loss can be compensated by applying various models to the distribution data. The preliminary results show that the model-based normalization is capable of distinguishing between inhomogeneities in texture due to certain surface features (e.g., cracks) and changes in the texture itself. This, in turn, points to the existence of an interesting trade-off between the pravda [1] of exploring a diverse structuring element geometry and the pravda of applying texture models to the size distributions derived from single geometries.

ACKNOWLEDGMENTS

Equipment supporting this work was provided by a grant from the New York State Center for Advanced Technology in Automation and Robotics at R.P.I. In addition, assistance in gathering test images was provided by personnel of the New York State Thruway Authority.

REFERENCES

1. Serra, J. *Image Analysis and Mathematical Morphology*, Academic Press, New York, 1982.
2. Grivas, D. A., and Skolnick, M. M., Morphological algorithms for the analysis of pavement structure, *Proc. SPIE, 1199*, 211–219 (1989).
3. Grivas, D. A., and Skolnick, M. M., Morphology-based image processing for pavement surface analysis, in *Powders and Grains* (Biarez and Gurves, eds.), Balkema, Rotterdam, 1989, pp. 67–74.
4. Grivas, D. A., and Skolnick, M. M. Morphology in particulate media, in *Powders and Grains* (Biarez and Gurves, eds.), Balkema, Rotterdam, 1989, pp. 63–66.
5. American Association of State Highway and Transportation Officials (AASHTO), Guidelines for Pavement Management Systems, Washington, D.C., 1990.
6. Hudson, W. R., and Uddin, W. Future pavement evaluation technologies: prospects and opportunities, *Proceedings of the Second North American Conference on Managing Pavements, 3*, Toronto, Canada, November 1987, pp. 233-258.
7. Longnecker, K. E., Pavement surface video image work in Idaho, Automated Pavement Distress Data Recognition Seminar, Ames, Iowa, June 12–15, 1990.
8. Copp, R., Field test of three video distress recognition systems, Automated Pavement Distress Data Recognition Seminar, Ames, Iowa, June 12–15, 1990.
9. Lee, H., Evaluation of Pavedex computerised pavement image processing system in Washington, Automated Pavement Distress Data Recognition Seminar, Ames, Iowa, June 12–15, 1990.
10. Grivas, D. A., Thruway distress survey manual, Report No. RPI.NYSTA-1.3, Civil Engineering Department, Rensselaer Polytechnic Institute, Troy, New York, 1990.
11. Hudson, W. R., Elkins, G. E., Uddin, W., and Reilly, K. T., Improved methods and equipment to conduct pavement distress surveys, Report FHWA-TS-87-213, FHWA, U.S. Department of Transportation, 1987.
12. Skolnick, M. M., Brown, R. H., Chakravarthy, B., and Wolf, B. R., Morphological algorithms for centroid normalization in relational matching, in *Proceedings of the IEEE International Symposium on Circuits and Systems*, Portland, Oregon, May, 1989, pp. 987–990.
13. Bhagvati, C., Jacobs, D. B., Skolnick, M. M., Morphological distributions as shape signatures, in *23rd Asilomar Conference on Signals, Systems and Computers*, Pacific Grove, California, October 1989.
14. Doughtery, E. R., and Pelz, J. B., "Morphological granulometric analysis of electrophotographic images—size distribution statistics for process control," *SPIE J. Opt. Eng.* in press.

Chapter 5

On Two Inverse Problems in Mathematical Morphology

Michel Schmitt

Thomson-CSF, Laboratoire Central de Recherches
Domaine de Corbeville, Orsay, France

I. INTRODUCTION

The usual approach to statistical shape recognition involves two successive steps:

1. The shape to be recognized is transformed (or measured) into a small number of parameters reflecting the interesting features of this shape.
2. Then a classifier does the final identification of these parameters [3,16].

This measurement step is crucial, and the overall performance of a recognition system generally depends on the quality of these measures. In order to help the classifier, the measurements usually have some invariances; for example, two shapes differing only by a translation or a rotation give rise to the same set of measurements. In this case, the classifier does not have to learn these invariances, which otherwise can be discovered only from a huge amount of data.

The price for this approach is that for each problem a new set of measurements has to be designed in order to fit the specificities of the problem. But some general morphological measurements, such as granulometry [5] or covariogram [4], have proved their generality and their quality on many classification tasks.

In this chapter we are interested in the following question: the small number of measurements is supposed to reflect the structure of the shape, but to what extent? There are other ways to ask the same question:

Having a set of measurements, what can I say about the original shape?
Is it uniquely determined by the measurements, or for instance up to a translation or a rotation?

What information has been lost during this measurement process?

We call these generic questions *inverse problems* (see Figure 1).

A. Example of a Trivial Case

In some cases the inverse problem may be trivial. Let us examine an example that illustrates this general approach to shape recognition.

Let X be a binary planar shape, that is, a compact set in the plane $\mathbf{R}^2$. We identify $\mathbf{R}^2$ with the complex plane $\mathbf{C}$. Suppose X is limited by a smooth curve $\gamma(s)$, where s is the arc length on γ, γ taking complex values. γ is periodic with period L, the length of the boundary of X. Expanding γ into Fourier series, we obtain Fourier coefficients that fully describe γ; no information has been lost and reciprocal formulas allow retrieval of γ from its expansion.

If we want to recognize shape X, a classifier is likely to work better on the first most significant coefficients than directly on curve γ. Moreover, these coefficients can be "normalized" so that two shapes X' and X'' equal up to a translation, a rotation, and a scaling factor have the same expansion. The learning phase of the classifier is simplified because it does not have to discover these transformations, which do not matter if we are interested only in shape.

B. Example of a Complex Case

On the other hand, the example of the granulometric curve [5,2,17,14] gives an illustration of unsolved inverse problems.

The proposed measurements are the following: for each $r > 0$, we compute the area of X opened by a disk of radius r, that is, the area swept by a disk of radius r when it is forced to be included in X. Many applications in image analysis are based on this function, but no one knows what information is extracted from the shape.

In the convex polygonal case, an easy case allowing analytical computations, properties can be inferred. For instance, it can be shown that the number of edges cannot be computed from the granulometric curve. More precisely:

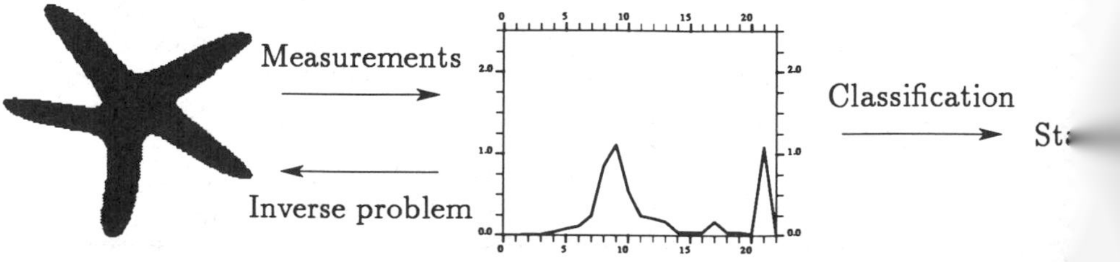

Figure 1. What is an inverse problem?

Proposition 1.1. Given a granulometric curve Γ of a convex polygon, there exists $n_0 > 0$ such that for each $n \geq n_0$, a convex polygon with n sides can be constructed which has Γ as granulometric curve.

In fact, the only information contained in the granulometric curve is

$$\sum_{i=1}^{n} \tan\left(\frac{\delta_i}{2}\right)$$

where δ_i is the angle of the polygon at vertex i. Some examples are given in Figure 2. However, the general case seems very complicated and needs more research [7].

C. Organization of the Chapter

In this chapter we examine two particular inverse problems illustrated in Figure 3. The first one, developed in Section II., concerns the geometric covariogram C_x of X [4], defined by $C_X(h) = S(X \cap X_h)$ where $h \in \mathbf{R}^2$, X_h is the translation of X by h, and S is the surface area. It is obvious that X and Y differing by a translation and a symmetry about the origin verify $C_X = C_Y$. We prove the reciprocal statement in some cases.

We state a general theorem on the structure of C_X in the polygonal case and show how it can be used in the convex polygonal case and in the generic polygonal nonconvex case. It is the first time a result concerning the covariogram of a nonconvex shape is stated. Moreover, an explicit construction is given to retrieve the shape from the covariogram, up to a translation and a symmetry about the origin.

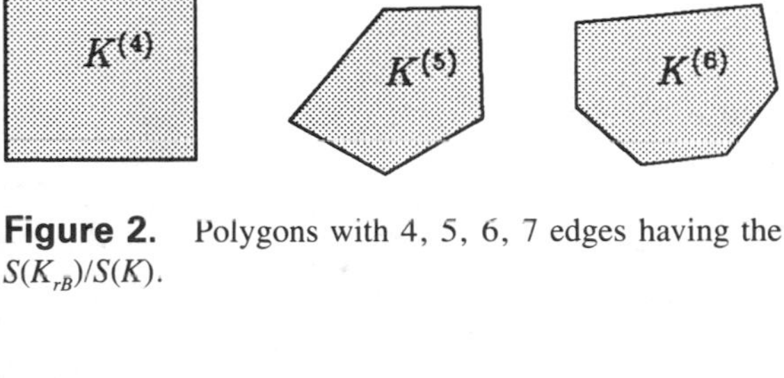

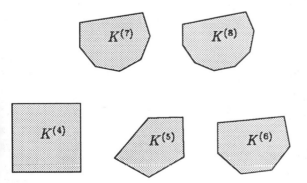

Figure 2. Polygons with 4, 5, 6, 7 edges having the same granulometric curve $r \to S(K_{rB})/S(K)$.

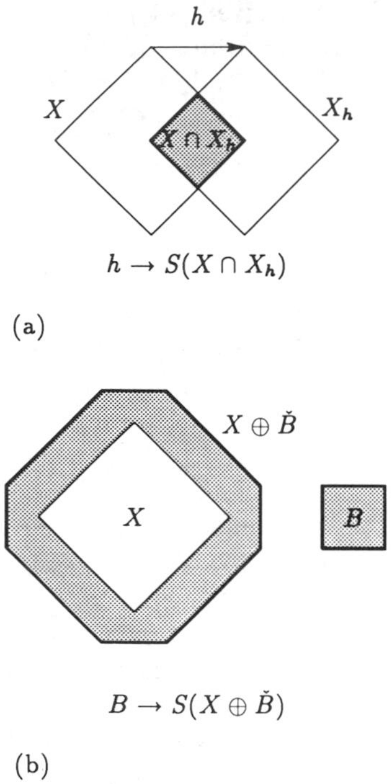

Figure 3. Two inverse problems: (a) geometric covariogram; (b) dilation by compacts.

The second one, developed in Section III, concerns the functional in $\mathbf{R}^n$ associating to a random compact set X the mean value of the volume of the dilation of X by any compact K, $D_X(K) = \bar{\mu}(X \oplus \check{K})$, where μ stands for the Lebesgue measure, $\bar{\mu}$ for its expectation, and $\check{K}$ for the symmetry of K with respect to the origin. We show that this functional characterizes X up to a random translation. If X is a deterministic set, the mean values become ordinary values and the theorem holds. But we present a stronger result that allows explicit reconstruction of X (up to translation) knowing $\mu(X \oplus \check{K})$ only for a given sequence of test compacts.

All the proofs are given here. A short version without proofs has appeared in [13]. Some notation is summarized in Section V.

II. GEOMETRIC COVARIOGRAM

Let X be a planar compact set.

Definition 2.1. The geometric covariogram of X is the function associating to each vector h in the plane the quantity

$$h \in \mathbf{R}^2 \rightarrow C_X(h) = S(X \cap X_h) \tag{5.1}$$

where X_h is the translation of X by h and S the surface area.

This geometric covariogram is of particular importance in geostatistics, for modeling second-order properties of random sets, and also in optics, because its Fourier transform is the spectrum of X and optical devices allow fast computation of this spectrum. The spectrum has been used in optical pattern recognition [1]. Our results give some answers about the knowledge that is introduced by the geometric covariogram in these two fields.

The following proposition is a trivial one:

Proposition 2.2. Two sets that are equal up to a translation and a symmetry about the origin have the same covariogram.

G. Matheron [6] conjectured the reciprocal statement in the *convex* case but did not fully prove it:

Conjecture 2.3. Two *convex sets* having the same covariogram are equal up to a translation and a symmetry about the origin.

W. Nagel [8] gives an answer in the *polygonal convex* case:

Theorem 2.4. Two *convex polygons* having the same covariogram are equal up to a translation and a symmetry about the origin.

We first prove a structure theorem concerning singularities of the covariogram, deduce that *generic polygons* are characterized by their covariograms, and derive two explicit reconstruction procedures under special conditions, one in the convex case and one in the nonconvex case.

A. The Structure Theorem

Let $P = (x_i)_{i=0}^{n-1}$ be a simple polygon in the plane. P is a compact set. We denote by ∂P its boundary (∂P is a loop). The singularities we are interested in are directional discontinuities of the second derivative.

Definition 2.5. A function $f : \mathbf{R}^2 \rightarrow \mathbf{R}$ exhibits a directional discontinuity of its second derivative at x if there exists a direction $\mathbf{u}$ (a unit vector) such that function $\lambda \rightarrow f(x + \lambda\mathbf{u})$ has a discontinuity in its second derivative at $\lambda = 0$.

Let us study C_P, the covariogram of P in a neighborhood of a given vector h in the plane, and let $\mathbf{u}$ be a unit vector.

The intersection $P \cap P_h$ is composed of some connected components. The boundary of each connected component is composed of a connected part of the boundary of P and one of P_h (see Figure 4). These two chains meet at a and b. The difference $C_P(h + \lambda\mathbf{u}) - C_P(h)$ is obtained by translating the chain of P_h by $\lambda\mathbf{u}$ and measuring the difference in surface area. According to Figure 4, this difference is

$$\lambda\Delta + A(\lambda) + B(\lambda) \tag{5.2}$$

where $A(\lambda)$ and $B(\lambda)$ may be negative quantities.

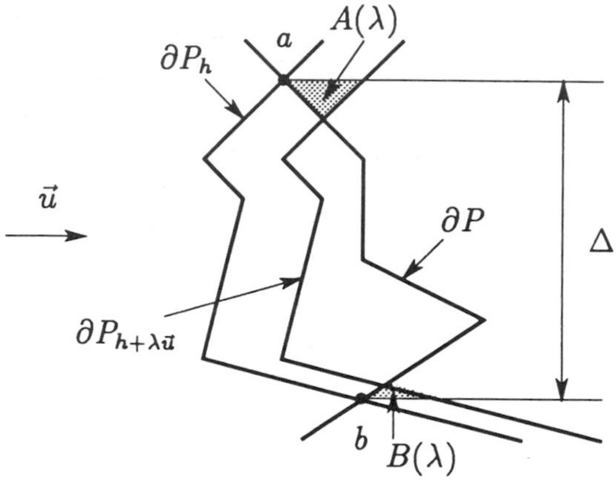

Figure 4. One connected component of $P \cap P_h$.

If a and b are neither vertices of P nor vertices of P_h, for sufficiently small λ, $A(\lambda)$ and $B(\lambda)$ are triangles whose area is proportional to λ^2. Summing up the contributions of all the connected components of $P \cap P_h$, we obtain

$$C_P(h + \lambda\mathbf{u}) = C_P(h) + \lambda D + \lambda^2 T \tag{5.3}$$

where D and T depend only on $\mathbf{u}$. C_P has a continuous directional second derivative at h in direction $\mathbf{u}$.

Suppose now that a is a vertex of P or P_h and b is not (see Figure 5). Then in Eq. (5.2), $A = \lambda^2 A_+$ for $\lambda > 0$ and $A = \lambda^2 A_-$ for $\lambda < 0$ with two different constants A_+ and A_-, λ small enough, and for a direction $\mathbf{u}$ that is different from all those of edges of P. If this appears only for one connected component of $P \cap P_h$, by summing up the contributions, Eq. (5.3) must be replaced by

$$\begin{aligned} \lambda > 0, \qquad C_P(h + \lambda\mathbf{u}) &= C_P(h) + \lambda D + \lambda^2 T_+ \\ \lambda < 0, \qquad C_P(h + \lambda\mathbf{u}) &= C_P(h) + \lambda D + \lambda^2 T_- \end{aligned} \tag{5.4}$$

with $T_+ \neq T_-$. So C_P exhibits a directional discontinuity in its second derivative at h. In this case h is characterized by:

$$\text{if } a \text{ is a vertex of } P,\ a \in \partial P_h \qquad \Leftrightarrow h \in (-\partial P)_a$$

$$\text{if } a \text{ is a vertex of } P_h,\ a + h \in \partial P \Leftrightarrow h \in (\partial P)_{-a}$$

Suppose now that more than one a or b in different connected components of $P \cap P_h$ are vertices of either P or P_h. We cannot extend the previous arguments because differences may compensate, so that possibly $T_+ = T_-$ in Eq. (5.4). But if we make the following assumption,

Assumption I. Any two edges of polygon P are not colinear.

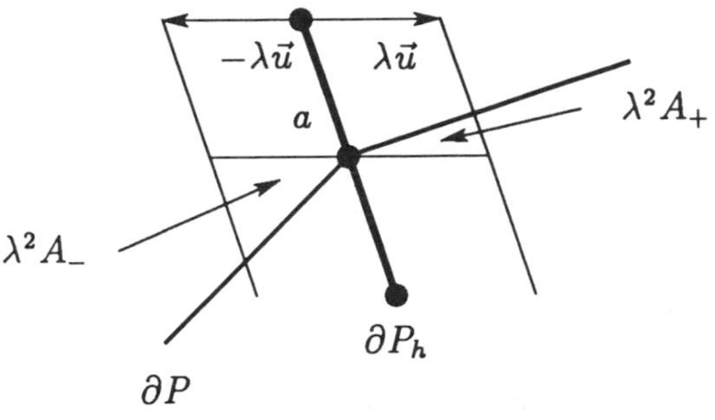

Figure 5. Case of a corner point $a \in \partial P$.

this situation occurs only for discrete h, which are always adherent to h where Eq. (5.4) holds with $T_+ \neq T_-$.

Combining these three cases, we get the desired structure theorem:

Theorem 2.6. Under assumption I, the adherence of the set of points where the covariogram of P has a directional discontinuity of its second derivative, denoted by H, is given by

$$H = \left(\bigcup_{0 \le i < n} \partial P_{-x_i} \right) \cup \left(\bigcup_{0 \le i < n} (-\partial P)_{x_i} \right) \tag{5.5}$$

where $-\partial P$ is the symmetry of ∂P about the origin. Figure 6 illustrates H.

The knowledge of C_P allows the computation of H under assumption I. Under which other assumptions is the knowledge of H sufficient to reconstruct P up to a translation and a symmetry?

B. Nonconvex Reconstruction

For the moment, H is a set of points in the plane. We transform it into a graph G (not necessarily planar) by the following procedure: the edges of G are the maximal line segments included in H and the vertices are all the end points of these line segments plus the origin. Because of assumption I, the edges of the graph are exactly edges of P.

Now consider a point at distance one (according to the graph structure) from the origin, that is, an edge of the form $[0,a]$. Under the following assumption,

Assumption II. For any pair of vertices x_i and x_j and any edge $[x_k, x_{k+1}]$, we have $x_j - x_i \neq x_{k+1} - x_k$.

a is only a triple point of graph G. Graph G is illustrated in Figure 7.

The reconstruction procedure includes the following steps illustrated in Figure 8:

1. Take an edge ending at the origin, say $[0,a]$. G has only a triple point at a.
2. The structure of this vertex is $[0,a]$, $[a,b]$, $[a,c]$. One of the two sequences,

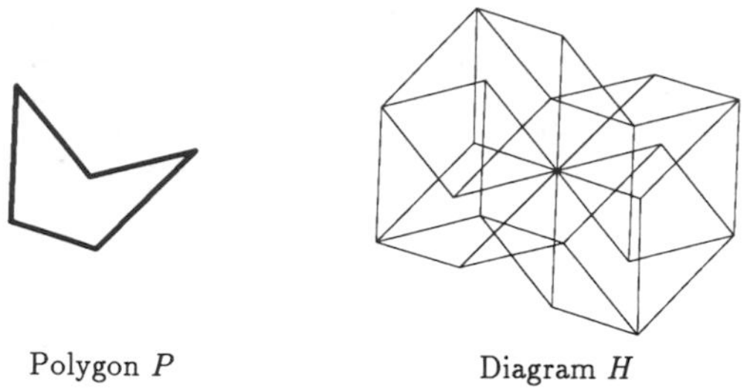

Figure 6. Directional discontinuities of the second derivative of C_P.

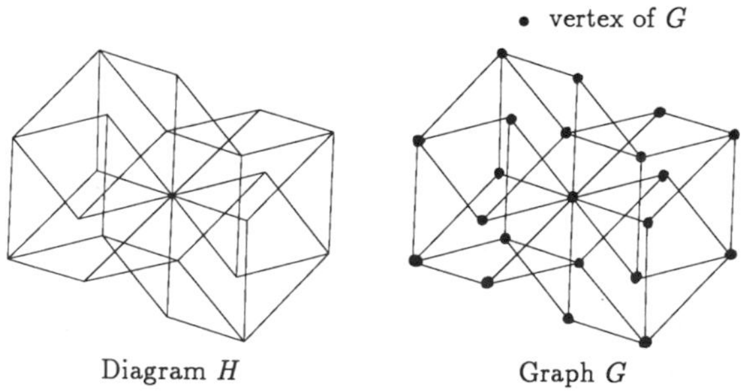

Figure 7. Graph structure G derived from diagram H.

say $(0,a,b)$, has been obtained with a translation of P, the other $(0,a,c)$ by a translation of $-P$. Because only one edge of P is equal to $[0,a]$ up to a translation and because $-P$ enumerates its vertices in reverse order, $(a - c, 0, a, b)$ are four successive vertices of a translation of P.

3. Translate now at the origin $(a - c, 0, a)$, which is part of the boundary of P. We get $(0, c - a, c)$, which is included in G. G exhibits a triple point at $c - a$. The three edges ending at $c - a$ are $[0, c - a]$, $[c - a, c]$, $[c - a, d]$. The chain $(0, c - a, d)$ corresponds to a translation of $-P$, so that $(c - a - d, 0, c - a, c, c + b - a)$ are successive vertices of P.
4. The process continues until the polygon has been closed.

It is worth noting that this procedure uses only a neighborhood of the origin, namely a neighborhood of size 2 in the sense of graph G. Therefore the information contained in the geometric covariogram is highly redundant: a neighborhood of size 2 of the directional discontinuities of the second derivative allows one to reconstruct P up to a translation and symmetry, and also the hole covariogram, under the two stated assumptions.

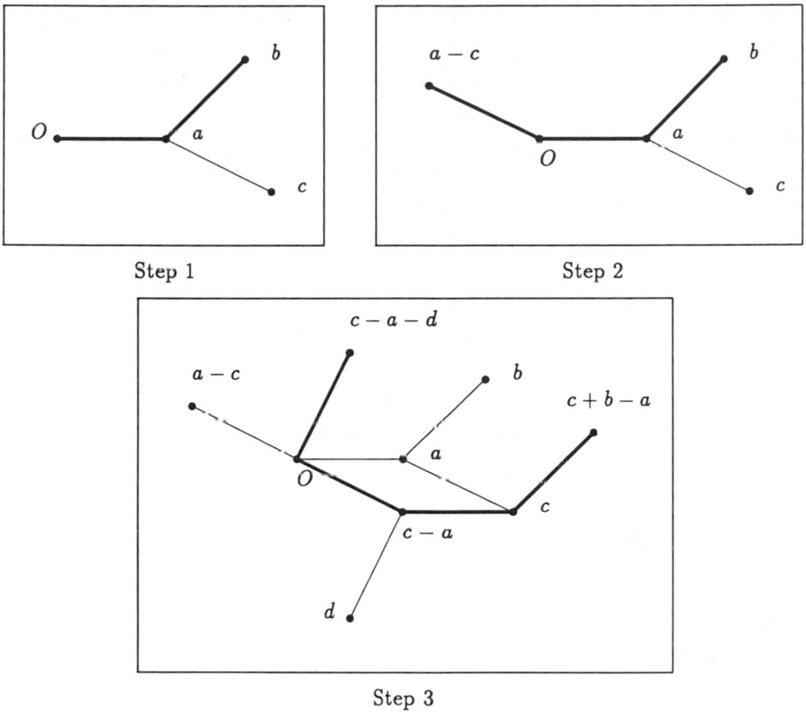

Figure 8. Reconstruction of P in the nonconvex case. The bold edges represent edges of P up to a translation.

C. Convex Reconstruction

The nonconvex reconstruction may also be used in the convex case. We now present another reconstruction technique only under assumption I.

The principle is to study the border of H. It is a convex polygon, symmetric with respect to the origin. It has twice the number of edges of P, each edge of P appearing twice.

We orient $P = (x_i)_{i=0}^{n-1}$ counterclockwise. Then we say that the line segment $[a,b]$ is an edge of P (up to a translation) if $b - a = x_{i+1} - x_i$ and an edge of $-P$ if $a - b = x_{i+1} - x_i$. Because of assumption I, $[a,b]$ cannot be an edge of P and $-P$ simultaneously. The edges of the border of H are also oriented counterclockwise.

Let us take an arbitrary (oriented) edge $[a,b]$ of the border of H as a beginning edge in our reconstruction. Because we cannot distinguish P from $-P$ using H, we assume that $[a,b]$ belongs to P. The following procedure gives the vertex of P before a:

The local structure of H near a is given in Figure 9. a is a vertex of only one translation of P, say P_1, and only one of $-P$, say P_2. So there are four segments of H ending at a, two of them being identical when P is a triangle. Let us enumerate these line segments in counterclockwise order: $[a,b]$, $[a,c]$, $[a,d]$, and $[a,e]$. When a is a triple point, $[a,c]$ and $[a,d]$ are identical. The two polygons P_1 and P_2 are convex and have the origin 0 as vertex, so that the line segment $[0,a]$ is totally included in both. We deduce that two of the four edges $[a,b]$, $[a,c]$, $[a,d]$, and $[a,e]$ are on one side of $[0,a]$, two on the other side. Moreover, on the same side, one belongs to P_1 and one to P_2. We conclude that $[a,c]$ belongs to P_2, that is, $-P$, because $[a,b]$ belongs to P_1, that is P. To assign $[a,d]$ and $[a,e]$, we look at the local structure of H near b. Only one edge $[b,f]$ is on the same side of $[0,b]$ as $[b,a]$. Then $[f,b]$ belongs to $-P$. This edge is equal to one of $[e,a]$ and $[d,a]$. The other then belongs to P, so that we have the vertex before edge $[a,b]$ of P in counterclockwise order. Doing the same reconstruction procedure again yields the hole polygon P.

D. Interpretation in the Fourier Plane

The interpretation of our theorem in the Fourier plane gives rise to a strange mathematical theorem.

If we denote by 1_X the characteristic function of X,

$$\begin{cases} 1_X(x) = 1 \Leftrightarrow x \in X \\ 1_X(x) = 0 \Leftrightarrow x \notin X \end{cases}$$

and Tf the Fourier transform, the Fourier transform of 1_X is $Tf(1_X)$ and that of C_X is $|Tf(1_X)|^2$, the square of its modulus. Under the assumptions on X in our theo-

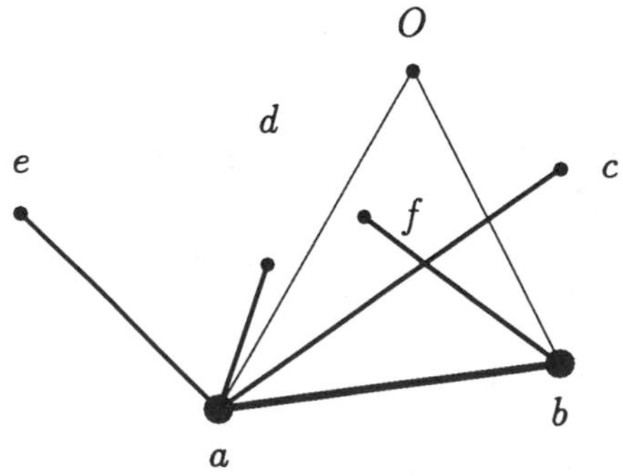

Figure 9. Reconstruction of P in the convex case.

rem, the modulus of $Tf(1_X)$ is sufficient to know the complete $Tf(1_X)$. In other words, knowing that we deal with the Fourier transform of a characteristic function of an unknown set X, there exists only one possible phase for $Tf(1_X)$.

E. Miscellaneous

1. Nagel's result [8] is stronger than our result in the convex case: he does not make assumption I. But our result can be generalized to the nonconvex case and gives an explicit reconstruction based on H.
2. The approach based on H seems very difficult to adapt in order to improve the results in the polygonal case, because H can be computed from the covariogram only under assumption I and H does not characterize P if P does not verify assumption I, as illustrated in Figure 10.
3. In the convex case, assumption I can be tested on the covariogram itself, using the following proposition:

Proposition 2.7. The convex polygon P has two parallel edges in direction $\mathbf{u}$, the smaller one having length l if and only if

$$\begin{aligned} 0 \le \lambda \le l, \quad & C_P(\lambda\mathbf{u}) = C_P(0) - D_{\mathbf{u}} \times \lambda \\ \lambda > l, \quad & C_P(\lambda\mathbf{u}) < C_P(0) - D_{\mathbf{u}} \times \lambda \end{aligned} \tag{5.6}$$

for some $D_{\mathbf{u}}$

4. The class of polygons verifying assumptions I and II is dense in the set of all polygon in the sense that, by slightly moving its vertices, any polygon

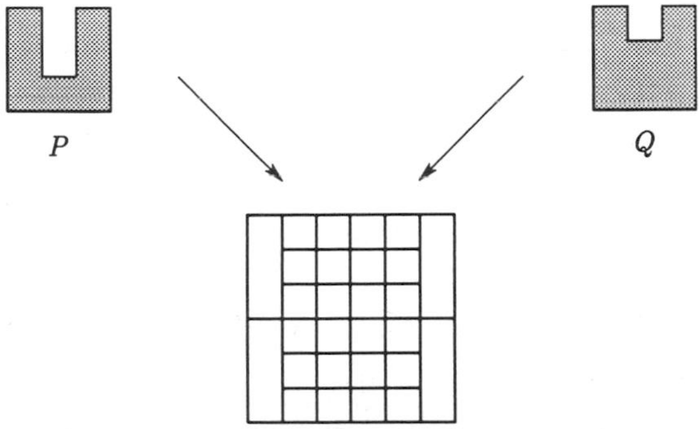

$$H = \left(\bigcup_{0 \le i < n} \partial P_{-x_i} \right) \cup \left(\bigcup_{0 \le i < n} (-\partial P)_{x_i} \right)$$

Figure 10. H cannot distinguish P from Q.

can be transformed into a polygon verifying assumptions I and II. That is why we call it the "generic" case.

5. The covariogram is highly redundant: we use only singularities of the second derivative and only a neighborhood of the origin in the nonconvex reconstruction.
6. The methodology we propose is not restricted to polygons. In fact, if the boundary of X is a finite union of C^2 arcs, a discontinuity of the second derivative appears as soon as the boundary of X has exactly one corner point falling on ∂X_h or ∂X_{-h}. Then Eq. (5.5) giving diagram H is still valid if the set of vectors h for which many corner points of X fall on ∂X_h or ∂X_{-h} is discrete in $\mathbf{R}^2$. The generalization of assumptions I and II is very tricky and does not lead to tractable results. One simple (and restrictive!) result is that a convex set with only one corner point is characterized by its covariogram.

The point to note is that our methodology works only in the presence of corner points on the boundary of X.

III. DILATIONS OF A SET

We now study the second inverse problem. Let X be a random almost sure (a.s.) compact set in $\mathbf{R}^n$. We recall that a random closed set in the sense of Matheron [5] is $(\mathcal{F}, \sigma_f, P_X)$, where $\mathcal{F}$ is the set of all closed sets, σ_f the hit-or-miss σ-algebra, and P_X a probability measure on $\mathcal{F}$. X is a.s. compact means that $P_X(\mathcal{K}) = 1$, where $\mathcal{K}$ stands for the set of all compact sets in $\mathbf{R}^n$.

In this section we examine the measurements consisting of associating to each compact K the mean value (taken according to P_X) of the n-dimensional volume of the dilation of X by K:

$$K \in \mathcal{K} \to D_X(K) = \bar{\mu}(X \oplus \check{K}) \tag{5.7}$$

where μ is the Lebesgue measure and $\bar{\mu}$ its expectation.

The following theorem shows exactly what information is contained in this functional:

Theorem 3.1. Let X_1 and X_2 be two random a.s. compact sets. We suppose that $D_{X_1}(B) < +\infty$ and $D_{X_2}(B) < +\infty$ for the unit ball B.

X_1 and X_2 are equal up to a random translation if and only if $D_{X_1} = D_{X_2}$.

The meaning of random translation will be given later.

This theorem was first proved in the context of Boolean models, where it gives a canonical description of this random process. The theorem has already been stated in [10,11] and some parts of the proofs in [9,12]. Here we give a complete proof in the random case and present a construction of X knowing D_X in the deterministic case.

A. Inverse Theorem in the Random Case

Using the notation of the theorem, we first define the notion of random translation.

1. Notion of Random Translation

The basic idea is to translate at the origin all the compact sets, that is, force the sets to have their circumscribed circle (smallest circle containing the set) centered at the origin.

Lemma 3.2. The applications:

$c(X)$, associating to a compact set X the center of its circumscribed circle
$\tau(X) = X_{-c(X)}$, translating a set at the origin

are continuous for the myope topology [5] on $\mathcal{K}'$.

Then τ can be considered as a random variable, so that the following definition makes sense:

Definition 3.3. Let $X_1 = (\mathcal{F},\sigma_f,P_1)$ and $X_2 = (\mathcal{F},\sigma_f,P_2)$ be two random a.s. compact sets. We say that X_1 and X_2 are equal up to a random translation if for all events $\mathcal{V} \in \sigma_f$, $P_1(\tau^{-1}(\mathcal{V})) = P_2(\tau^{-1}(\mathcal{V}))$.

Definition 3.4. Let $X = (\mathcal{F},\sigma_f,P)$ be a random a.s. compact set. The new random set $\tau(X) = (\mathcal{F}, \sigma_f, P \bigcirc \tau^{-1})$ is called the "centered set associated to X."

This denomination comes from the property that $P \bigcirc \tau^{-1}$ is a measure concentrated on the sets centered at the origin: $P(\tau^{-1}(c^{-1}\{0\})) = 1$. Then X_1 and X_2 are equal up to a random translation if and only if their associated centered sets have the same probability law.

2. If Part of the Theorem

We introduce a measure on σ_f, associated to an a.s. compact random set X.

Lemma 3.5. Let $X = (\mathcal{F},\sigma_f,P)$ be a random closed set. The mapping θ_X defined by

$$\forall \mathcal{V} \in \sigma_f, \qquad \theta_X(\mathcal{V}) = \int_{\mathbf{R}^n} P(\mathcal{V}_x)\, dx \tag{5.8}$$

is a measure.

In fact, this measure has been used by Matheron to define the stationary Boolean model as the union of a Poisson process on $\mathcal{F}$.

Proposition 3.6. Let $X_1 = (\mathcal{F},\sigma_f,P_1)$ and $X_2 = (\mathcal{F},\sigma_f,P_2)$ be two random a.s. compact sets. Then

$$\forall K \in \mathcal{K},\ D_{X_1}(K) = D_{X_2}(K) \Leftrightarrow \forall \mathcal{V} \in \sigma_f,\ \theta_{X_1}(\mathcal{V}) = \theta_{X_2}(\mathcal{V}) \tag{5.9}$$

Proof. Let us prove that $D_X(K) = \theta_X(\mathcal{F}_K)$. This comes from

$$\begin{aligned} D_X(K) &= \int_{\mathbf{R}^n} P(x \in X \oplus \check{K})\, dx \\ &= \int_{\mathbf{R}^n} P(K_x \cap X \neq \emptyset)\, dx \\ &= \int_{\mathbf{R}^n} P(\mathcal{F}_{K_x})\, dx \\ &= \theta_X(\mathcal{F}_K) \end{aligned}$$

Using Choquet's theorem [5], $\theta_{X_1}(\mathcal{F}_K) = \theta_{X_2}(\mathcal{F}_K) \Leftrightarrow \theta_{X_1} = \theta_{X_2}$.

Let us now examine what happens when two random sets are equal up to a random translation:

Proposition 3.7. Let $X_1 = (\mathcal{F},\sigma_f,P_1)$ and $X_2 = (\mathcal{F},\sigma_f,P_2)$ be two random a.s. compact sets. If X_1 and X_2 are equal up to a random translation, then $\theta_{X_1} = \theta_{X_2}$.

Proof. We study events of the form $\tau^{-1}(\mathcal{V}) \cap c^{-1}(H)$ where $\mathcal{V} \in \sigma_f$ and H is a Borelian set of $\mathbf{R}^n$. If K is a compact set centered at the origin in $\mathcal{V}$, the event $\tau^{-1}(\mathcal{V}) \cap c^{-1}(H)$ contains all the translates of K having their reference point in H. The family of these events generates σ_f and is stable by intersection, so that it suffices to study θ_X on them.

Let us compute

$$\begin{aligned} \theta_X(\tau^{-1}(\mathcal{V}) \cap c^{-1}(H)) &= \textstyle\int_{\mathrm{R}^n} P(\tau^{-1}(\mathcal{V})_x \cap c^{-1}(H)_x)\, dx \\ &= \textstyle\int_{\mathrm{R}^n} P(\tau^{-1}(\mathcal{V}) \cap c^{-1}(H_x))\, dx \end{aligned}$$

But $P(\tau^{-1}(\mathcal{V}) \cap c^{-1}(H))$ is a measure in H, say $\alpha(H)$. Using the equality

$$\int_{\mathrm{R}^n} \alpha(H_x)\, dx = \mu(H)\alpha(\mathbf{R}^n)$$

(which is a simple application of Fubini's theorem on double integrals), we obtain

$$\theta_X(\tau^{-1}(\mathcal{V}) \cap c^{-1}(H)) = P(\tau^{-1}(\mathcal{V}))\mu(H)$$

Now, if two random sets X_1 and X_2 are equal up to a random translation, for all $\mathcal{V}$, $P_1(\tau^{-1}(\mathcal{V})) = P_2(\tau^{-1}(\mathcal{V}))$, so that $\theta_{X_1}(\tau^{-1}(\mathcal{V}) \cap c^{-1}(H)) = \theta_{X_2}(\tau^{-1}(\mathcal{V}) \cap c^{-1}(H))$ and as a consequence $\theta_{X_1} = \theta_{X_2}$.

Combining these two last propositions yields the if part of the characterization theorem: $X_1 = X_2$ up to a random translation implies $D_{X_1} = D_{X_2}$.

3. Only if Part of the Theorem

Let $X_1 = (\mathcal{F},\sigma_f,P_1)$ and $X_2 = (\mathcal{F},\sigma_f,P_2)$ be two random a.s. compact sets. Let X_1' and X_2' be their associated centered sets. We denote $P_1' = P_1 \bigcirc \tau^{-1}$ and $P_2' = P_2 \bigcirc \tau^{-1}$. We have seen in the last section that $\theta_{X_1} = \theta_{X_1'}$ and $\theta_{X_2} = \theta_{X_2'}$.

The only if part of Theorem 3.1 is then proved if we prove:

Proposition 3.8. If θ_{X_1} and θ_{X_2} are proportional, then $X_1' = X_2'$.

Note that only the proportionality of θ_{X_1} and θ_{X_2} is needed.

Proof. The proof is based on the identity if X is centered

$$\theta_X(\mathcal{V} \cap c^{-1}(H)) = \int_H P(\mathcal{V}_{-x})\, dx \tag{5.10}$$

Suppose X centered.

$$\begin{aligned}\theta_X(\mathcal{V} \cap c^{-1}(H)) &= \int_{\mathbf{R}^2} P((\mathcal{V} \cap c^{-1}(H))_x)\, dx \\ &= \int_{\mathbf{R}^2} P((\mathcal{V} \cap c^{-1}(H))_{-x})\, dx \\ &= \int_{\mathbf{R}^2} P(\mathcal{V}_{-x} \cap c^{-1}(H_{-x}))\, dx \\ &= \int_{\mathbf{R}^2} P(\mathcal{V}_{-x} \cap c^{-1}(H_{-x}) \cap c^{-1}(\{0\}))\, dx\end{aligned} \tag{5.11}$$

because P is concentrated on $c^{-1}(\{0\})$. But

$$c^{-1}(H_{-x}) \cap c^{-1}(\{0\}) = \begin{cases}\emptyset & \text{if } x \notin H \\ c^{-1}(\{0\}) & \text{if } x \in H\end{cases}$$

Replacing $c^{-1}(H_{-x}) \cap c^{-1}(\{0\})$ in Eq. (5.11) yields

$$\theta_X(\mathcal{V} \cap c^{-1}(H)) = \int_H P(\mathcal{V}_{-x} \cap c^{-1}(\{0\}))\, dx = \int_H P(\mathcal{V}_{-x})\, dx$$

So, if $a_1\theta_{X_1} = a_2\theta_{X_2}$ (i.e., θ_{X_1} and θ_{X_2}, proportional), then for all Borelian sets H

$$a_1 \int_H P_1'(\mathcal{V}_{-x})\, dx = a_2 \int_H P_2'(\mathcal{V}_{-x})\, dx$$

It follows that for almost all x, $a_1P_1'(\mathcal{V}_{-x}) = a_2P_2'(\mathcal{V}_{-x})$. To show that this equality holds for all x, it suffices to notice

$$\begin{aligned}a_1P_1'(\mathcal{V}) &= a_1P_1'(\tau^{-1}(\mathcal{V})) = a_1P_1'(\tau^{-1}(\mathcal{V})_{-x}) \\ &\overset{a.s.}{=} a_2P_2'(\tau^{-1}(\mathcal{V})_{-x}) = a_2P_2'(\tau^{-1}(\mathcal{V})) = a_2P_2'(\mathcal{V})\end{aligned}$$

But an almost sure equality between constant functions is a usual equality. So $a_1P_1'(\mathcal{V}) = a_2P_2'(\mathcal{V})$. Taking $\mathcal{V} = \mathcal{F}$ yields $a_1 = a_2$. Then $P_1' = P_2'$ and the two random sets X_1 and X_2 are equal up to a random translation.

B. Inverse Theorem in the Deterministic Case

Theorem 3.1 is trivially translated into the deterministic case; it suffices to omit the mean values in the equations. However, we now prove that an adequate sequence of compact sets in the plane is sufficient to test if a given point x belongs to X when X is centered in the following sense: it has its right lower point at the origin. X can then be reconstructed using this procedure.

We denote by $\check{D}_X(B) = D_X(\check{B}) = \mu(X \oplus B)$, where μ is the surface area in the $\mathbf{R}^2$ plane. The right lower point $p = (p_1,p_2)$ of a set X is defined by

$$\begin{cases} p_2 = \min\{x_2, (x_1,x_2) \in X\} \\ p_1 = \max\{x_1, (x_1,p_2) \in X\} \end{cases} \tag{5.12}$$

In this section, we say that X is *centered* if its right lower point is at the origin. $S(u)$ is the square of edge length u having its right lower point at the origin.

Define sets $A(n,p,x)$ and $B(n,p,x)$ as depicted in Figure 11. Then

Proposition 3.9. Let X be a compact centered set.

$$x \in X \Leftrightarrow \forall p,\ \exists n_0,\ \forall n > n_0,$$
$$\check{D}_X(S(1/p) \oplus A(n,p,x)) - \check{D}_X(S(1/p) \oplus B(n,p,x)) = 0$$

We assume that the reconstructed X is centered. So, in the proposition, only $x = (x_1,x_2)$ with $x_2 > 0$ or $x_1 < 0$ and $x_2 = 0$ are required. The proof of the proposition needs the following lemma:

Lemma 3.10. Let f_u be defined by

$$f_u(\varepsilon) = \sup\{x_1,\ (x_1,\ x_2) \in X \oplus S(u),\ 0 \leq x_2 \leq \varepsilon\} \tag{5.13}$$

Then $f_u = f_0$ and $f_0(\varepsilon) \downarrow 0$ whe $\varepsilon \downarrow 0$.

Proof. We first prove that $f_u = f_0$. $f_u(\varepsilon) = \sup\{x_1 + u_1, (x_1,x_2) \in X, (u_1,u_2) \in S(u), 0 \leq x_2 + u_2 \leq \varepsilon\}$. But, according to the definition, $S(u) = \{(u_1,u_2), -u \leq u_1 \leq 0, 0 \leq u_2 \leq u\}$ so that $f_u(\varepsilon) = \sup\{x_1, (x_1,x_2) \in X, (u_1,u_2) \in S(u), 0 \leq x_2 \leq \varepsilon\} = f_0(\varepsilon)$. f_u decreases as ε decreases toward 0.

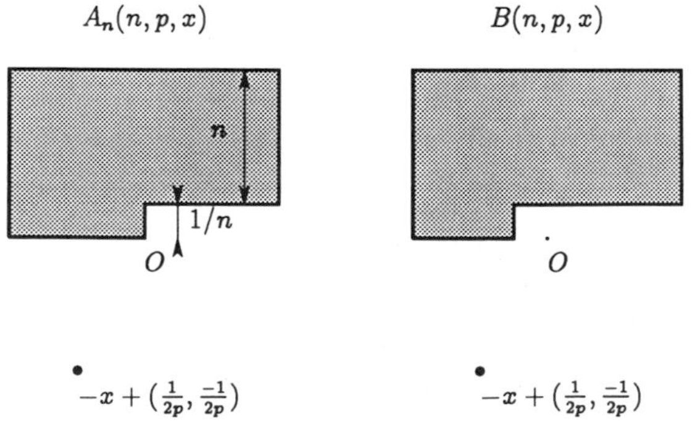

Figure 11. Sequence of sets allowing reconstruction of a set from the surface area of their dilations.

We now prove $f_0(\varepsilon) \downarrow 0$ when $\varepsilon \downarrow 0$. If not, let $\varepsilon_n \to 0$ and $f_0(\varepsilon_n) \geq \alpha > 0$. As X is compact, $f_0(\varepsilon_n) < M < \infty$, so that $f_0(\varepsilon_n)$ has a cluster point $h \geq \alpha$. Take ε_{n_k} such that $f_0(\varepsilon_{n_k}) \to h$. Then $(f_0(\varepsilon_{n_k}), \varepsilon_{n_k}) \to (h,0) \in X$ because X is closed. This is in contradiction with the definition of the right lower point at the origin.

We are now able to prove the proposition.

Proof. The proof is illustrated in Figure 12. Suppose $x \in X$. Let $p > 0$. We choose n_0 verifying three constraints: $n_0 \geq \mathrm{Diam}(X)$, $n_0 \geq 2p$, and $1/2p \geq f_0(1/n_0)$. n_0 verifying this last constraint is ensured by Lemma 3.10. Take $n \geq n_0$. If we discard point $-x + (1/2p, -1/2p)$ in both sets $A(n,p,x)$ and $B(n,p,x)$, the two sets

$$(X \oplus S(1/p)) \oplus (A(n,p,x)\backslash\{-x + (1/2p, -1/2p)\})$$

and

$$(X \oplus S(1/p)) \oplus (B(n,p,x)\backslash\{-x + (1/2p, -1/2p)\})$$

are identical except in the horizontal strip $0 \leq x_2 \leq 1/n$, where the right border of $(X \oplus S(1/p)) \oplus A(n,p,x)$ is shifted by vector $(1/n,0)$. More precisely, the difference is located in the rectangle $[-1/n, f_0(1/n)] \times [0,1/n]$, which is included in the rectangle $[-1/2p, 1/2p] \times [0,1/2p]$. The difference in surface area is $1/n^2$. If we take into account point $-x+(1/2p, -1/2p)$ and translate $X \oplus S(1/p)$ at this location, $x \in X$ implies that the square $[-1/2p, 1/2p] \times [-1/2p, 1/2p]$ is totally covered, so that $(X \oplus S(1/p)) \oplus A(n,p,x) = (X \oplus S(1/p)) \oplus B(n,p,x)$.

Now suppose $x \notin X$. We denote by $C(u)$ the square $[-u/2, u/2] \times [-u/2, u/2]$.

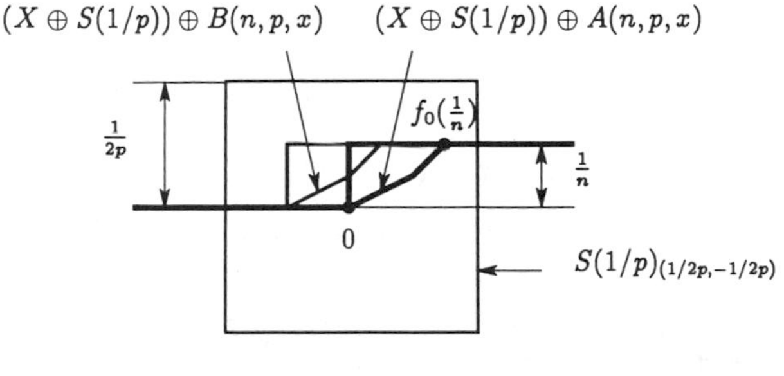

$x \in X \implies S(1/p)_{(1/2p,-1/2p)} \subset X_{-x+(1/2p,-1/2p)} \oplus S(1/p)$

$x \notin X \implies S(1/p)_{(1/2p,-1/2p)} \cap X_{-x+(1/2p,-1/2p)} \oplus S(1/p) = \emptyset$

Figure 12. Why $A(n,p,x)$ and $B(n,p,x)$ allow testing $x \in X$.

Because X is closed, there exists p such that $C(4/p)_x \cap X = \emptyset$. Then

$$X \oplus S(1/p) \cap C(2/p)_x = \emptyset$$

$$\Rightarrow (X \oplus S(1/p))_{-x+(1/2p,-1/2p)} \cap C(2/p)_{(1/2p,-1/2p)} = \emptyset$$

$$\Rightarrow (X \oplus S(1/p))_{-x+(1/2p,-1/2p)} \cap C(1/p) = \emptyset$$

For any n such that $n \geq 2p$ and $1/2p \geq f_0(1/n)$, the two sets $(X \oplus S(1/p)) \oplus A(n,p,x)$ and $(X \oplus S(1/p)) \oplus B(n,p,x)$ differ in rectangle $[-1/n, f_0(1/n)] \times [0,1/n]$, as explained in the previous paragraph, and the difference in surface area is $1/n^2$.

The utility of the small dilation by $S(1/p)$ is due to the fact that lines of zero thickness may exist in X and will not be sufficient to cover the region around the origin where the dilations by $A(n,p,x)$ and $B(n,p,x)$ may differ.

IV. CONCLUSION

We have studied two inverse problems related to the morphological framework:

The geometric covariogram (generic polygonal case)
The surface area of dilations by any compacts

Construction procedures for a compact polygon giving rise to these measurements have been exposed.

In the case of the geometric covariogram, more work has to be performed in the general case (sets without corner points on their boundary) and its extension to functions (autocorrelation function).

Many other inverse problems need more investigation, like the surface area of erosions or openings by circles of different radii. Experimental results show the discrimination power of such measurements [14,15], but no one knows their theoretical limitations.

V. NOTATION

$\mathcal{F}$	set of all topologically closed sets
$\mathcal{K}$	set of all topologically compact sets
$\mathcal{F}_K$	set of all closed sets hitting K
σ_f	σ-algebra generated by the $\mathcal{F}_K$, $K \in \mathcal{K}$
$(\mathcal{F},\sigma_f,P)$	random closed set [5]
$\oplus$	Minkowski addition $A \oplus B = \{a + b, a \in A \text{ and } b \in B\}$
B_x	Translation of B at point x: $B_x = B \oplus \{x\}$
$\check{X}$	symmetry of X about the origin: $\check{X} = -X = \{-x, x \in X\}$
X_B	Opening of X by B, $X_B = \cup\{B_x, B_x \subset X\}$
$[a,b]$	line segment from a to b
0	origin of the $\mathbf{R}^2$ plane

REFERENCES

1. Casasent, D., Optical pattern recognition and artificial intelligence: a review, in *Optical and Digital Pattern Recognition*, SPIE vol. 754, 1987.
2. Coster, M., and Chermant, J. L., *Précis d'analyse d'images*, CNRS, Paris, 1985.
3. Duda, R. O., and Hart, P. E., *Pattern Classification and Scene Analysis*, Wiley, New York, 1973.
4. Matheron, G., *Eléments pour une théorie des milieux poreux*, Masson, Paris, 1967.
5. Matheron, G., *Random Sets and Integral Geometry*, Wiley, New York, 1975.
6. Matheron, G., Le covariogramme géométrique des compacts convexes de $\mathbf{R}^2$. Technical Report 2/86, Centre de Géostatistique, Ecole des Mines de Paris, February 1986.
7. Mattioli, J., Squelette, érosion et fonction spectrale par érosion d'une forme binaire planaire, Rapport Interne L.C.R., ASRF-91-8, 1991.
8. Nagel, W., The uniqueness of a planar convex polygon when its set covariance is given, Technical report N/89/17, Friedrich-Schiller-Universität Jena, July 1989.
9. Prêteux, F., and Schmitt, M., Analyse et synthèse de fonctions booléennes: théorèmes de caractérisation et démonstrations, Technical Report 34/87/MM, CGMM, Ecole des Mines, 1987.
10. Prêteux, F., and Schmitt, M., Boolean texture analysis and synthesis, in *Image Analysis and Mathematical Morphology*, vol. 2, *Theoretical Advances* (J. Serra, ed.) Academic Press, London, 1988.
11. Schmitt, M., A step towards the statistical inference of the Boolean model and the Boolean function, in *Acta Stereol.* vol. 8/2, *4th European Congress for Stereology*, 1989, pp. 623–628.
12. Schmitt, M., Estimation of the density in a stationary Boolean model, *J. Appl. Probability*, *28*, 702–708 (1991).
13. Schmitt, M., On two inverse problems in mathematical morphology, in *SPIE*: *Image Algebra and Morphological Image Processing II*, vol. 1568, San Diego, July 23–24, 1991, pp. 283–297.
14. Schmitt, M., and Mattioli, J., Reconnaissance de formes planaires par morphologie mathématique et réseaux de neurones, *Rev. Tech. Thomson*, *22*, *1*, 573–610 (1990).
15. Schmitt, M., and Mattioli, J., Shape recognition combining mathematical morphology and neural networks, in *SPIE*: *Application of Artificial Neural Network*, Orlando, Florida, April 1–5 1991.
16. Simon, J. C., *La Reconnaissance des Formes par Algorithmes*, Masson, Paris, 1985.
17. Stoyan, D., Kendall, W. S., and Mecke, J., *Stochastic Geometry and Its Applications*, Wiley, New York, 1987.

Chapter 6

Graph Morphology in Image Analysis

Henk Heijmans

Centre for Mathematics and Computer Science
Amsterdam, The Netherlands

Luc Vincent

Xerox Imaging Systems
Peabody, Massachusetts

I. INTRODUCTION

Mathematical morphology can be considered as a set-based approach for the analysis of images [23,24,17]. One of its underlying ideas is to use so-called *structuring elements* to define neighborhoods of points. Recently it has been recognized that these ideas more generally apply to any space V that has the structure of a vector space, or at least a group [11]. In that case one can define a neighborhood of a point $x \in V$ as

$$N(x) = \{x + a \mid a \in A\}$$

where A is the structuring element. In this chapter we describe how many of the concepts of "classical morphology" (i.e., the case where $V = \mathbf{R}^d$ and where $+$ denotes vector addition) can be extended to spaces of images modeled by graphs.

A graph consists of a collection of points, called *vertices*, and a binary relation between them: two vertices either are related or they are not related. This relation is usually represented by a subset $E \subseteq V \times V$ called the *edges*; v and w are related if and only if $(v, w) \in E$. Graphs play an important role in many branches of mathematics and computer science. In the context of image analysis they are often used as a geometric representation of the scene under study. In this case the vertices correspond to the objects in the scene and the edges describe the (neighboring) relations between these objects. In this chapter, the images we consider consist of an underlying graph and a scalar function defined on the set of vertices. In Section II we recall some basic notation and terminology. In Section III we explain what we mean by a graph representation of an image, and we introduce the notion of a graph operator.

Using the neighboring relations between vertices we are able to propose a large class of morphological operations on a graph. Beyond the basic operations (erosions and dilations, openings and closings), this class embraces almost all of the classical morphological transformations (distance function, skeletons, geodesic transformations, watersheds, etc.); see Section IV. In Section V, we present a number of examples of graph representation of images. There we introduce the Delaunay graph, the Gabriel graph, and the relative neighborhood graph. Furthermore, we illustrate some of the classical morphological transformations for graph-based images.

The previously mentioned class of transformations can be extended by introducing the notion of *structuring graph*. Exactly like structuring elements in classical morphology, structuring graphs act as probes to extract structural information from graphs. They have a simple structure and are relatively small compared to the graph that is to be transformed. The structuring graph is used to construct a neighborhood function on the vertices by relating individual vertices to each other whenever they belong to a local instantiation of the structuring graph. This is explained in Section VI. Then, in Section VII we use these neighborhood functions to define dilations and erosions. Subsequently, Section VIII deals with openings, closings, and other filters. Finally, Section IX is devoted to brief notes on implementation of morphological graph operations and to some concluding comments.

Let us close this introduction with the following important remark:

> Although the theory is presented in the framework of gray-level graphs, all the drawings are "binary" for the sake of clarity and simplicity.

II. MORPHOLOGY FOR FUNCTIONS: CONCEPTS AND BASIC RESULTS

In this section we give a brief overview of morphology for gray-level functions. For general results on mathematical morphology we refer to Serra's books [23,24]. A systematic exposition on gray-level morphology can be found in [23, Chapter 12] and [10].

Although we shall often mean by a gray-level some continuous or discrete numerical value, it may also represent a vector in color space. The only restriction we have to make on the set of gray levels T is that it possesses a complete lattice structure. Recall that T is called a complete lattice if T is a partially ordered set in which every subset S has a least bound $\bigvee S$ called the *supremum* of S, and a greatest lower bound $\bigwedge S$, called the *infimum* of S; see Birkhoff [5].

Let V be an arbitrary set and define Fun(V) to be the space of all functions $f: V \to T$, where the gray-level set T has a complete lattice structure. If we take

$T = \{0, 1\}$, then Fun(V) is the space of all binary images on V, also represented by $\mathcal{P}(V)$, the power set of V. Other choices for T, sometimes found in the literature, are $T = \{0, 1, 2, \ldots, m\}$, $T = \overline{\mathbf{Z}} = \mathbf{Z} \cup \{-\infty, \infty\}$, $T = \overline{\mathbf{R}} = \mathbf{R} \cup \{-\infty, \infty\}$, $T = [0, 1]$, $T = [0, \infty]$. With the pointwise ordering

$$f \leq g \qquad \text{if } f(v) \leq g(v) \text{ for } v \in V$$

the space Fun(V) becomes a complete lattice. In fact, Fun(V) inherits the complete lattice structure from T. In this chapter we will always assume that $T = \{0, 1, \ldots, m\}$, but we point out that most results carry over to the case where T is an arbitrary complete lattice.

In morphology we are interested in operators mapping the image space into itself.

Definition 2.1. Let ψ be an operator on Fun(V). We say that ψ is
(a) increasing if $f \leq g$ implies that $\psi(f) \leq \psi(g)$,
(b) an erosion if $\psi(\bigwedge_{i\in I} f_i) = \bigwedge_{i\in I} \psi(f_i)$ for an arbitrary family $\{f_i \mid i \in I\}$,
(c) a dilation if $\psi(\bigvee_{i\in I} f_i) = \bigvee_{i\in I} \psi(f_i)$ for an arbitrary family $\{f_i \mid i \in I\}$,
(d) extensive if $\psi(f) \geq f$ for every f,
(e) antiextensive if $\psi(f) \leq f$ for every f,
(f) idempotent if $\psi^2 = \psi$,
(g) a (morphological) filter if ψ is increasing and idempotent,
(h) an opening if ψ is increasing, antiextensive, and idempotent,
(i) a closing if ψ is increasing, extensive, and idempotent.

An important result in morphology says that dilations and erosions always occur in pairs. To any dilation δ there corresponds a unique erosion ε (and vice versa) such that

$$\delta(f) \leq g \Leftrightarrow f \leq \varepsilon(g) \qquad \text{for } f, g \in \text{Fun}(V) \tag{6.1}$$

If ε, δ are operators on Fun(V) such that (6.1) holds, then ε is an erosion, δ is a dilation, and the pair (ε,δ) is called an adjunction. We say that ε and δ are each other's adjoints. If ε and δ are adjoint, then

$$\varepsilon\delta\varepsilon = \varepsilon \qquad \text{and} \qquad \delta\varepsilon\delta = \delta \tag{6.2}$$

There exists yet another duality relation between dilations and erosions. We denote by $t^* = m - t$. Furthermore, we define the "negative" f^* of the function f by

$$f^*(x) = (f(x))^* = m - f(x) \tag{6.3}$$

If ψ is an operator on Fun(V), then the dual operator ψ^* is defined as

$$\psi^*(f) = (\psi(f^*))^* \tag{6.4}$$

Note that this method carries over to any gray-level space T for which there exists an order-reversing bijection. On $T = [0, \infty]$ one may, for example, define

$t^* = 1/t$. The mapping $f \to f^*$ is called a *dual automorphism*. For more details we refer to [10, Section III]. It is easy to see that ψ is increasing if and only if ψ^* is. If δ is a dilation, then δ^* is an erosion and vice versa. For openings and closings there exist similar duality relations.

An important class of function operators is formed by the so-called flat operators [10,25]. Here we shall not give a formal treatment of such operators, but give only a brief sketch of the underlying idea. By a *flat operator* we mean an operator on Fun(V) that is derived from an operator on the power set $\mathcal{P}(V)$ by thresholding. Starting with a function f, one obtains a family of sets X_t by thresholding the function at gray level t, that is, $X_t = \{v \in V \mid f(v) \geq t\}$. One then applies the set operator ψ to this family and uses the transformed family to construct $\psi(f)$. In this chapter we shall deal exclusively with flat operators. In the literature many different names have been proposed for these operators, such as FSP filters (FSP = fuction set processing) [15] or stack filters [35].

Let us conclude this section with some statements concerning flat dilations and erosions. In a sense that we shall not make precise here, the only way to define flat dilations and erosions is by considering neighborhood functions. A neighborhood function on V is a mapping $N : V \to \mathcal{P}(V)$. To any neighborhood function, there corresponds a reciprocal neighborhood function N given by

$$\check{N}(v) = \{w \in V \mid v \in N(w)\} \tag{6.5}$$

Furthermore, with any neighborhood function one may associate an erosion and a dilation, adjoint to each other, given by

$$\begin{aligned} \delta(f)(v) &= \sup\{f(w) \mid w \in \check{N}(v)\} \\ \varepsilon(f)(v) &= \inf\{f(w) \mid w \in N(v)\} \end{aligned} \tag{6.6}$$

Let $\check{\varepsilon}$ and $\check{\delta}$ be the erosion and dilation corresponding to the reciprocal neighborhood $\check{N}$. One can show that

$$\varepsilon^* = \check{\delta}, \qquad \delta^* = \check{\varepsilon} \tag{6.7}$$

One can also show that every flat dilation and erosion is of the form (6.6).

III. BINARY AND GRAY-LEVEL GRAPHS

In the previous section we outlined the theory for morphological operators on the function lattice Fun(V) where V is an arbitrary set. In this section we are interested in the case that V is the vertex set of a graph. There exist many good textbooks on graphs; we refer in particular to the monograph of Berge [2].

In this section, by graph we always mean a nonoriented graph without loops and multiple edges. A graph G is a mathematical structure consisting of a set of vertices V and edges E. We denote this as $G = (V, E)$. Since edges are supposed to be simple, they may be represented as a pair of vertices (v, w), denoting that v and w are neighbors. Our assumption that G is undirected can be made explicit by putting $(v, w) = (w, v)$. Let $G = (V, E)$ and $G' = (V', E')$ be two graphs. We say that G is a *subgraph* of G' if $V \subseteq V'$ and $E \subseteq E'$. In literature, the word subgraph is often used in a more restricted sense [2]. By a *homomorphism* from G to G' we mean a one-to-one mapping $\theta : V \to V'$ with the property that $(v, w) \in E$ implies that $(\theta(v), \theta(w)) \in E'$. In that case we say that G and G' are *homomorphic* and write $G \tilde{\subset} G'$. If the homomorphism θ is onto (and hence a bijection), it is called an *isomorphism*. The graphs G and G' are called *isomorphic* if they are related by an isomorphism. We denote this as $G \simeq G'$. An isomorphism from the graph G to itself is called a *symmetry* of G. We denote by Sym(G) the family of all symmetries of G. Obviously, this family forms a group called the *symmetry group* of G. The identity mapping id, defined by $\mathrm{id}(v) = v$, is contained in Sym(G) and is called the *trivial symmetry* of G (see Figure 1).

Let $G = (V, E)$ be a graph, and let $f \in \mathrm{Fun}(V)$. Then we call f a gray-level graph. If the gray-level set is $\{0, 1\}$, f is called a binary graph. Sometimes, if we want to emphasize the role of the underlying graph G, we write $(f \mid G)$ instead of f. If $(f \mid G)$ is a gray-level graph and $\tau \in \mathrm{Sym}(G)$, then we define $(\tau f \mid G)$ by $\tau f(v) = f(\tau^{-1}v)$, for $v \in V(G)$. Here $V(G)$ denotes the vertex set associated with the graph G.

Definition 3.1. A graph operator is a mapping which assigns to any graph $G = (V, E)$ an operator $\psi(\cdot \mid G)$ on the function space Fun(V). A graph operator is called flat if every $\psi(\cdot \mid G)$ is a flat operator. The graph operator ψ will be called G-increasing if ψ increases in G, that is, $\psi(X \mid G) \subseteq \psi(X \mid G')$ for $G \subseteq G'$ and $X \subseteq V(G)$. G-decreasingness of ψ is defined analogously. The graph operator ψ is called increasing if for any graph G the operator $\psi(\cdot \mid G)$ is increasing.

We say that ψ is a *graph erosion* if $\psi(\cdot \mid G)$ is an erosion on Fun(V) for every graph $G = (V, E)$. Analogously, we define graph dilations, openings, closings, filters, etc. A *graph neighborhood function* is a mapping N that, for every graph G, defines a neighborhood function on the vertex set of G. A graph operator is called *symmetry-preserving* if $\psi(\tau f \mid G) = \tau\psi(f \mid G)$, for $f \in \mathrm{Fun}(V)$, $\tau \in$

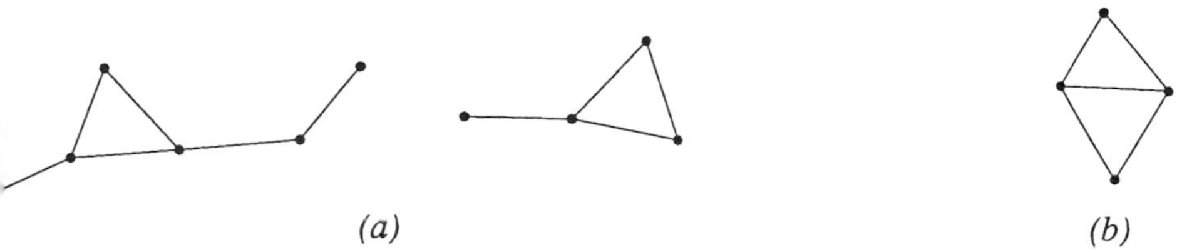

Figure 1. The left graph (a) has only trivial symmetry, whereas the symmetry group of the right graph (b) contains three elements (including the trivial symmetry).

Sym(G), and any graph $G = (V, E)$. Note that this last definition is the analogue of translation invariance in classical morphology. There is a one-to-one correspondence between graph neighborhood functions and flat graph adjunctions. If N is a graph neighborhood function that satisfies $N(\tau v \mid G) = \tau N(v \mid G)$ for $\tau \in$ Sym(G), then the resulting adjunction is symmetry preserving. To illustrate some of these abstract definitions we will present an example; this example is studied thoroughly in the following section.

Example 3.2. We define a graph neighborhood function N in the following way. If $G = (V, E)$ is some graph and $v \in V$, then we define $N(v \mid G)$ as the set containing v as well as all neighbors of v, that is, $N(v \mid G) = \{w \in V \mid (v, w) \in E\} \cup \{v\}$. It is obvious that N is symmetry preserving; that is, $N(\tau v \mid G) = \tau N(v \mid G)$ for every $\tau \in$ Sym(G). As a consequence, the erosion and dilation associated to this neighborhood function are symmetry preserving.

Throughout the remainder of this chapter we shall, whenever no confusion arises, suppress the argument G in the notation of both a gray-level graph and a graph operator.

IV. A SPECIAL CASE: NONSTRUCTURED GRAPH OPERATORS

Before the introduction of the notion of s-graph in [13], the only case of graph morphology that had ever been studied is what could be referred to as nonstructured graph morphology [30,29,32]. As we shall see (Section VI), these nonstructured graph operators turn out to be a special case of the structured ones.

In the present framework, as explained in detail in [30], the morphological operations are directly derived from the distance induced by the set of edges E on the set of vertices V. The graph distance d_G between two vertices v and w is given by the minimal number of edges to cross to go from v to w or, alternatively, by the length of the shortest paths connecting v to w in E. A collection of edges $\pi = (v_1, v_2, \ldots, v_k)$ is called a *path* between v and w if $v_1 = v$, $v_k = w$, and $(v_i, v_{i+1}) \in E$ for $i = 1, \ldots, k$. The *length* of the path is $l(\pi) = k - 1$. So we may write

$$d_G(v, w) = \inf\{l(\pi) \mid \pi \text{ is a path joining } v \text{ and } w \text{ in } E\} \tag{6.8}$$

Since the graph structure under study is not necessarily connected, it may well happen that *no* path connects v to w. In this case, we conventionally put the distance between these two vertices equal to ∞. So, strictly speaking, d_G is not a metric.

Given a vertex $v \in V$ and an integer $n \in \mathbf{Z}^+$, the ball $B_n(v)$ centered at v with radius n is given by

$$B_n(v) = \{v' \in V \mid d_G(v, v') \leq n\}$$

Following Serra [24, Chapters 1–2] and the example mentioned at the end of the previous section, we can now choose the $B_n(v)$'s as neighborhood functions (also called structural mappings, or sometimes structuring functions). These functions associate with each vertex the neighborhood it addresses in a dilation or erosion operation. Note that when $n = 1$, B_1 is exactly the neighborhood function N introduced in Example 3.2.

Dilations and erosions with respect to these neighborhood functions are then defined as follows:

Definition 4.1. Given a gray-level graph f on $G = (V, E)$, the dilation $\delta^{(n)}(f)$ and the erosion $\varepsilon^{(n)}(f)$ of size $n \geq 0$ of f are the gray-level graphs given by

$$\forall v \in V \qquad \begin{cases} \delta^{(n)}(f) = \max\{f(v') \mid v' \in B_n(v)\} \\ \varepsilon^{(n)}(f) = \min\{f(v') \mid v' \in B_n(v)\} \end{cases} \tag{6.9}$$

In the sequel, these operations shall be called dilation and erosion "of size n." Intuitively, just as in classical morphology, dilations and erosions are defined as local maxima and minima, respectively. One of the main differences is that, in the present case, the number of vertices in a given neighborhood (ball) $B_n(v)$ is highly dependent on the vertex v. Indeed, such properties as translation invariance are meaningless in a graph.

When $n = 1$ the resulting operations are called elementary dilation and erosion and are simply denoted δ and ε. As in classical morphology, we have the following properties:

$$\delta^{(n)} = \underbrace{\delta \circ \delta \circ \cdots \circ \delta}_{n \text{ times}}, \tag{6.10}$$

$$\varepsilon^{(n)} = \underbrace{\varepsilon \circ \varepsilon \circ \cdots \circ \varepsilon}_{n \text{ times}} \tag{6.11}$$

Thus, operations involving large neighborhoods can be decomposed into a succession of elementary operators. This is taken into account for the actual implementation of nonstructured graph dilations and erosions (see Section IX).

To illustrate the effect of these operations on graphs, we shall use a binary example: suppose that the gray-level graph f under study takes its values in $\{0, 1\}$, and let $n \geq 0$ be an integer. In this case, performing a dilation of size n of f comes down to giving value 1 to each vertex v with value 0 (i.e., such that $f(v) = 0$) having a vertex with value 1 in its neighborhood $B_n(v)$:

$$f(v) = 0 \quad \text{and} \quad \exists v' \in B_n(v), \qquad f(v') = 1 \Rightarrow \delta^{(n)}(f)(v) = 1 \tag{6.12}$$

$$f(v) = 1 \Rightarrow \delta^{(n)}(f)(v) = 1 \tag{6.13}$$

By duality, eroding f amounts to giving value 0 to each 1-vertex v having a 0-vertex in its neighborhood $B_n(v)$. Figure 2 illustrates the effect of an elementary

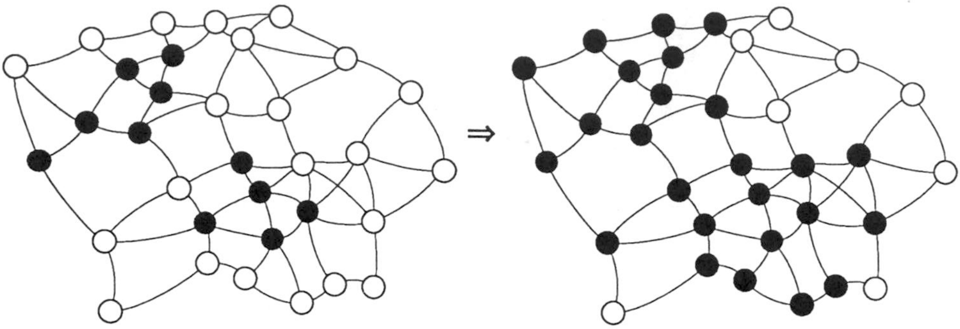

Figure 2. Nonstructured elementary dilation of a binary graph.

dilation of a binary graph. In this figure, the vertices with value one are represented in black.

In the present "nonstructured framework," beyond these basic operations and their associated openings and closings, the underlying distance d_G is particularly interesting because it allows us to define numerous more advanced transformations [28,30]. We cite among others,

- Distance functions
- Granulometries
- Skeletons, skeleton by influence zones (SKIZ)
- Catchment basins, watersheds
- Geodesic operators, reconstruction

These graph transformations, described and illustrated in [30], turn out to be particularly interesting in practice, as will be illustrated in the next section for distance functions, granulometries, and watersheds.

V. GRAPH REPRESENTATION OF IMAGES, EXAMPLES OF APPLICATION OF NONSTRUCTURED GRAPH TRANSFORMATIONS

A. Modeling of Neighborhood Relationships

At this point one may ask, why define all these morphological transformations? What concrete objects will they be applied to? What kind of problems will they help solve? It is now time to address these issues.

The initial motivation of graph morphology was the study of neighborhood relationships within populations of objects [30]. The idea is to model a set of objects V as the vertices of a graph and to process this graph via morphological transformations to extract useful information. For example, in histology, assuming that the objects of V represent cells in a tissue, it seems reasonable to model such a population as the vertices of a neighborhood graph [30,32]. This graph provides plausible relationships between cells and can be chosen to be relatively independent of the deformations of the tissue itself. Using tools described in the previous section, one can then try to characterize quantitatively such notions as

- Average number of neighbors of a cell.
- "Isolated" cells of a given type (with respect to the underlying graphs).
- Size distribution of the clusters of cells of a given type, or cells sharing common characteristics. Such analyses can be done in the graph itself, which means that the actual distances between cells do not matter; only the graph distance is accounted for, through the use of granulometries on the graph.
- Average distance in the graph between two cells of a given type, closest distance between a cell of type A and a cell of type B, etc. (use of distance functions on graphs).

In fact, graph morphology operations have already been successfully used in histology for the study of germinal centers [22]. The same kind of approach can be used in various problems involving the quantitative description of spatial relationships between objects.

B. Definition of Appropriate Neighborhood Graphs

For the category of aforementioned neighborhood problems, the first step is to define and construct a neighborhood graph from a two-dimensional (or even n-dimensional) population of objects. This initial population is often available under the form of a discrete binary image whose connected components represent the objects under study. It usually results from a previous segmentation stage.

This modeling purpose is generally best served by the neighborhood graphs of the Delaunay triangulation family (see, for example, [20]), namely the *Delaunay triangulation* (DT) itself, the *Gabriel graph* (GG) [8], and the *relative neighborhood graph* (RNG) [27]. Indeed, these graphs do not depend on any parameter such as a maximal distance between objects or a minimal number of neighbors, a property that is very useful in practice [32]. This property implies in particular that these graphs are independent of any scaling and can therefore be used equally well for various kinds of populations. In addition, DTs, GGs and RNGs are connected graphs, are planar, and are included in one another, thus enabling a modeling of neighborhood relationships of increasing strength.

These graphs are defined from the well-known Voronoï diagram (see, e.g., [20, § 5.5]). Let us assume for simplicity that the objects to be modeled are points $p_1, p_1, \ldots, p_n$ in the continuous place $\mathbf{R}^2$. Recall that the *Voronoï polygon* associated with p_i, denoted $V(p_i)$, is given by

$$V(p_i) = \{p \in \mathbf{R}^2 \mid \forall j \neq i,\ d(p,p_i) < d(p,p_j)\} \tag{6.14}$$

The set of the boundaries of these Voronoï polygons is called the Voronoï diagram (see Figure 3).

This definition easily extends to the discrete framework and to the case where the objects to be modeled are no longer isolated points. One often speaks then of influence zones and skeleton by influence zones (SKIZ).

The definitions of DT, GG, and RNG follow straightforwardly. Let $V = \{p_i \mid i = 1, \ldots, n\}$ be the initial set of points.

Definition 5.1. The Delaunay triangulation of V is the graph $\mathcal{G}_{\mathrm{dt}} = (V, E_{\mathrm{dt}})$ such that E_{dt} is the set of the point pairs (p_i,p_j) whose associated Voronoï polygons are adjacent, i.e., share an edge.

When V does not contain any cocircular 4 points, one can show that DT is effectively a triangulation that is connected and planar.

To define the Gabriel graph and the relative neighborhood graph, it is convenient to start from two regions associated with a pair (p,q) of points in the plane: $D(p,q)$ denotes the *closed* disk having $[p,q]$ as a diameter and $Cr(p,q)$ is the intersection of the two *open* disks of radius $\overline{pq}$ respectively centered in p and q, sometimes called the *crescent*. These notions are illustrated in Figure 4. Note that $D(p,q)\backslash\{p,q\} \subseteq Cr(p,q)$.

Definition 5.2. The Gabriel graph $G_{\mathrm{gg}} = (V, E_{\mathrm{gg}})$ of V and the relative neighborhood graph $G_{\mathrm{rng}} = (V, E_{\mathrm{rng}})$ of V are such that

E_{gg} is the set of the point pairs (p_i,p_j), with $p_k \notin D(p_i,p_j)$, if $k \neq i, j$.
E_{rng} is the set of the point pairs (p_i,p_j), with $p_k \notin Cr(p_i,p_j)$, if $k \neq i, j$.

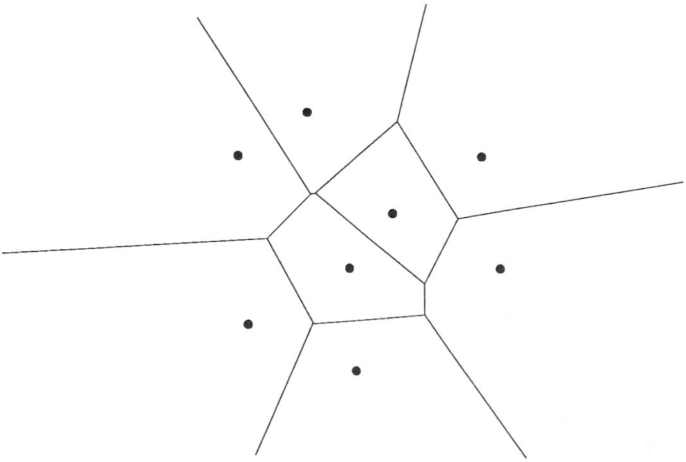

Figure 3. Example of Voronoï diagram.

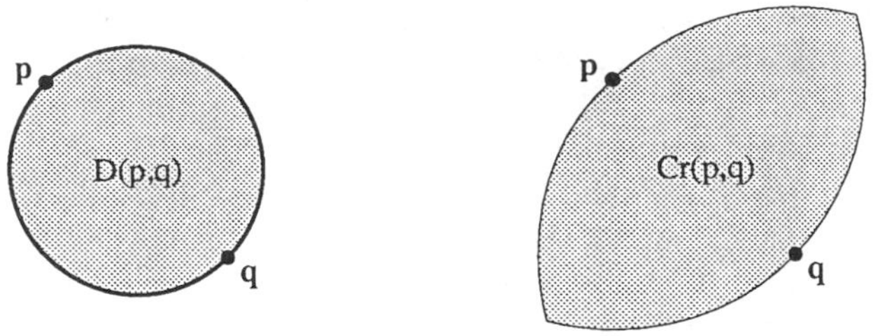

Figure 4. Regions on which the definitions of Gabriel graphs and relative neighborhood graphs are based.

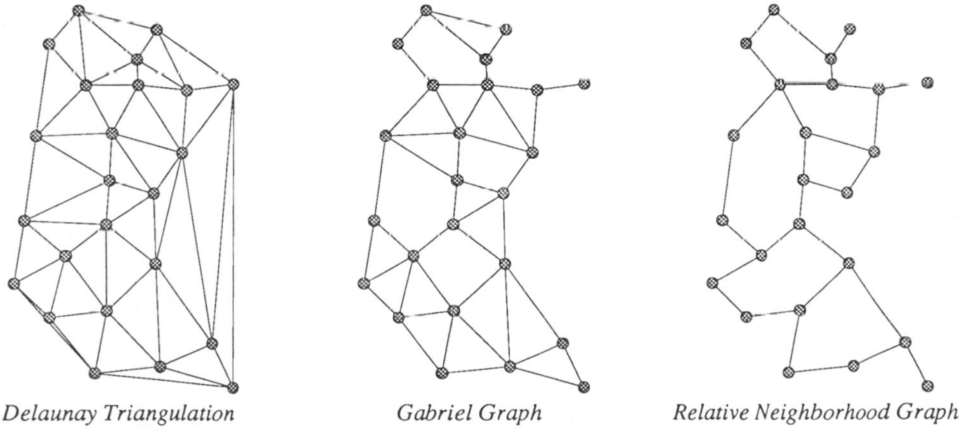

Figure 5. Three different neighborhood graphs stemming from the Voronoï diagram.

These three different graphs are illustrated by Figure 5. Obviously, the following inclusion relations hold:

$$E_{\text{rng}} \subseteq E_{\text{gg}} \subseteq E_{\text{dt}} \tag{6.15}$$

Definitions 5.1 and 5.2 can be extended to connected components of arbitrary size and shape in the plane [32, pp. 119–120].

C. Computation of DT, GG, and RNG

The typical algorithms for determining these neighborhood graphs in practice rely on computational geometry techniques [20]. When the connected components (objects) of the population are relatively circular or small in comparison to the distance between them, they can be assimilated to isolated points. One can then first apply some well-known Voronoï diagram algorithms, like the "divide-

and-conquer" approach described in [20, Chapter 5] or incremental techniques such as that proposed in [6]. These algorithms run in at most $O(n \log(n))$ time, n being the number of points. DT and GG can then be derived in a straightforward (and linear) manner. For RNG, the situation is much more complex, but it can anyhow be derived in $O(n \log(n))$ from the Voronoï diagram [26].

However, regardless of the computational efficiency of these methods, they are rather limited by the fact that they work only with isolated points as input data. Indeed, when the objects of the population are arbitrary connected components of a discrete binary image, they are to be modeled by polygons and the complexity of the associated algorithms becomes horrifying! Therefore, it seems much more appropriate to make use of digital techniques introduced in [30,28] and detailed in [32]. Let us now briefly describe and illustrate these algorithms.

For this purpose, we shall start from Figure 6a, whose connected components will be the vertices of our graphs. The successive steps of the algorithm are as follows:

1. Labeling of the connected components of the original image: each of them will be assigned a unique number. This labeling can be accomplished extremely efficiently by using, e.g., the first-in-first-out (FIFO)–based algorithms described in [34].
2. Computation of the labeled influence zones of these connected components. Roughly speaking, the labels associated with each component are propagated in the image until they completely fill up the remaining space, yielding

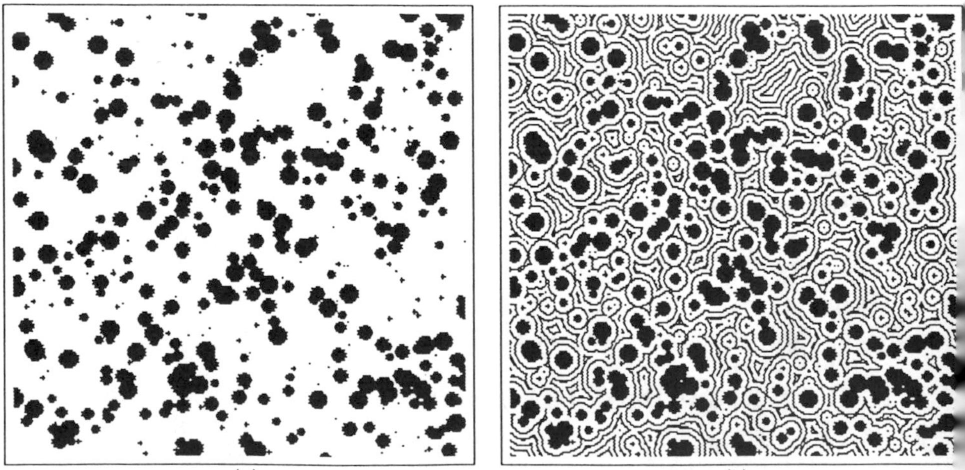

(a) *(b)*

Figure 6. (a) Original binary image with several connected components and (b) level lines of the distance function of its background.

a digital equivalent of the Voronoï diagram. This computation has to be performed as accurately as possible (see [32, Chapter 5]), and discrete distances such as the city block, chamfers, octagonal, or hexagonal distances [7] are generally not good enough. Here we shall make use of an exact Euclidean distance function algorithm described in [33] (see Figure 6b). The associated labeled influence zones are displayed in Figure 7a. The boundaries of these zones constitute the actual SKIZ (see Figure 7b).

3. Contour tracking of the influence zones. By tracking the contours of zone with label i, one gets successively all the labels of the neighboring zones. This allows an easy computation of the "discrete" Delaunay triangulation of our initial binary image. By using some additional constraints detailed in [32], one gets the Gabriel graph in the same way. The Delaunay triangulation and Gabriel graph corresponding to Figure 6a are displayed in Figure 8. Similar techniques also enable us to derive discrete RNGs starting from arbitrary connected components.

D. Examples of Application

Besides the histology applications mentioned earlier, graph morphology has been used for very different problems. To give the reader some flavor we shall briefly mention some of them below.

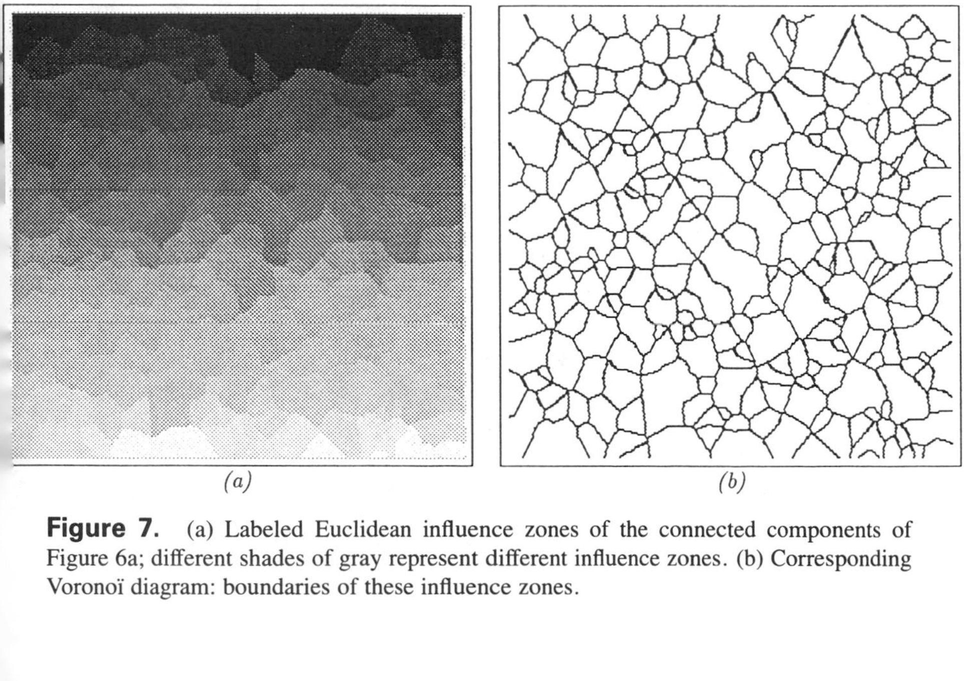

(a) (b)

Figure 7. (a) Labeled Euclidean influence zones of the connected components of Figure 6a; different shades of gray represent different influence zones. (b) Corresponding Voronoï diagram: boundaries of these influence zones.

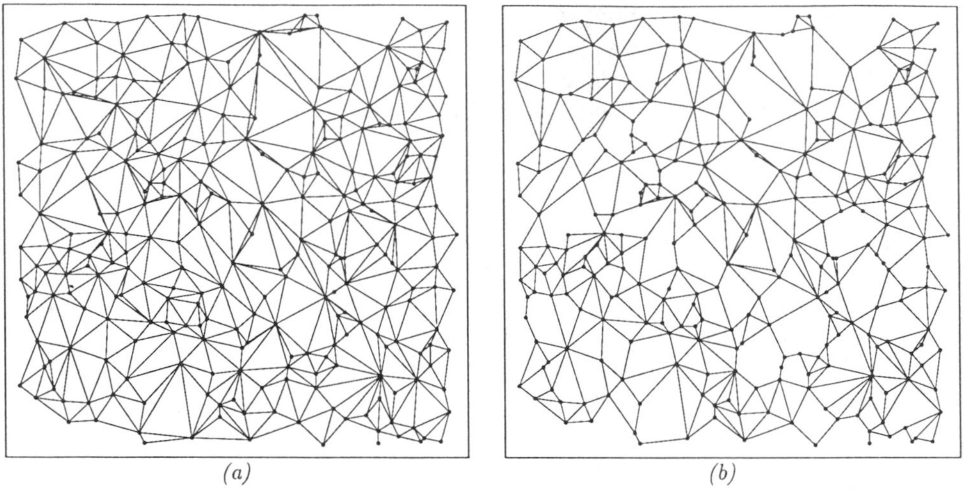

Figure 8. (a) Delaunay triangulation and (b) Gabriel graph corresponding to Figure 6a.

1. Fracture Simulations in Porous Media

Graph morphology provides nice tools for the study of heterogeneous media at a macroscopic level, based on information on their microstructure [14]. Typically, one starts from a digitized picture of a medium and models its microstructure as a graph. This graph can be either of DT or GG type (see previous section) or based on the underlying discrete grid, depending on the kind of information that is to be extracted.

Here we are concerned with the study of crack propagation in propous media (see [31]). We assume that we have a binary picture of the medium (pores, value 1; matrix, i.e., medium itself, value 0; see Figure 9a) and that a traction is exerted on it. Under this force, cracks will appear in the material, and we are interested in finding out what the crack paths look like, what their lengths are, etc. The assumption underlying our approach is that crack paths tend to go preferentially through the pores of the material. Moreover, assuming a crack has reached a particular pore p, its next propagation step will most probably be one of the pores in the immediate neighborhood of p. Therefore, we first model the set of pores as the vertices of a neighborhood graph and choose a specific vertex or set of vertices V_I as crack initiations. Then we actually simulate the propagation of a crack in the graph starting at the vertices of V_I.

The graph that seems most relevant to this kind of problem is the Gabriel graph described in the previous section. For the actual simulation, one uses distance functions on this graph (see [14,31]): the graph distance function asso-

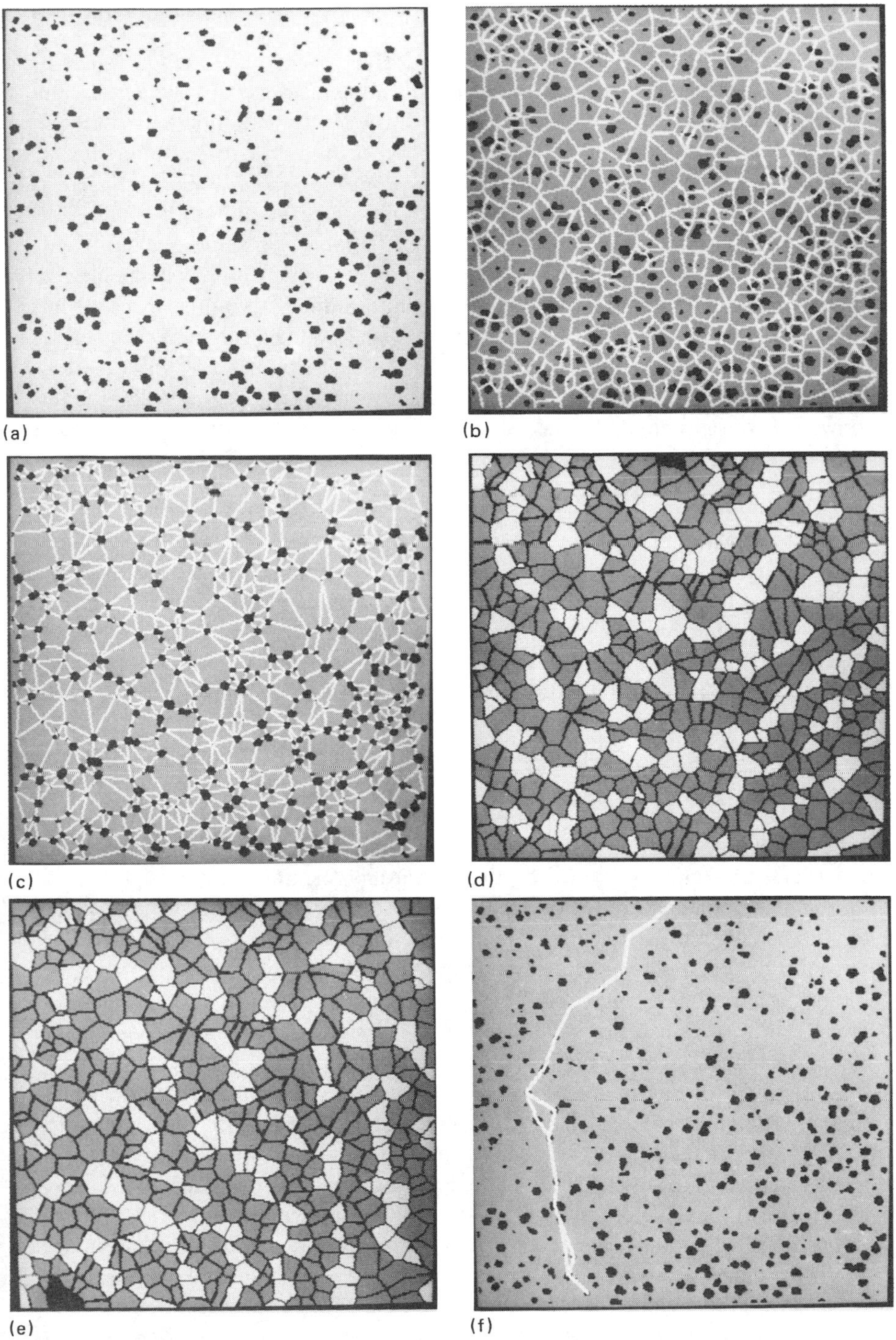

Figure 9. Using distance functions on a Gabriel graph to extract possible fracture lines in a porous material.

ciated with the set V_I is first determined; that is, to each vertex (pore), one assigns its distance to the closest crack initiation. This allows one to find the vertices V_E at one edge of the material that are first reached by a "wave" starting from V_I. Next, a backward distance function (back-propagation of a wave starting from V_E) allows one to determine the actual crack paths between V_I and V_E.

This technique has been tested on concrete examples as well as artificially designed random media and gives very promising results. It is illustrated in Figure 9 for the case of a medium made of graphite nodules (the pores) disconnected from a pig iron matrix. The Gabriel graph of the nodules (Figure 9c) is derived from the Voronoï diagram of Figure 9b. The forward and backward distance functions are then displayed modulo 2, each vertex being represented as its associated Voronoï zone, for the sake of clarity. Lastly, the extracted crack paths are shown in Figure 9f.

2. Hierarchical Representation and Segmentation of Images

Image segmentation is one of the most common problems in the field of image processing. A task that is often related to image segmentation is concerned with the "hierarchization" of an image, that is, the production of a series of images with decreasing level of detail: between two successive images of the series, details of least importance are suppressed while the important features are preserved. These two related issues can be approached in a common way by means of watersheds on images and graphs [29,32].

As explained in further detail in [29], the watershed transformation, whose use is more and more common in image analysis, associates with every minimum of a picture its *catchment basin*, that is, its influence zone. Computing the watersheds of the gradient of an image I allows one to decompose I into regions, each of which corresponds to a perceptually relevant feature. Unfortunately, due to noise, one often observes an oversegmentation; the regions into which the picture has to be decomposed are fragmented, sometimes very badly. For example, Figure 10b has been obtained by computing the catchment basins of the gradient of Figure 10a and assigning to each basin the mean gray level of the corresponding pixels in the original image. This image is often referred to as a mosaic image. The oversegmentation of Figure 10b can be clearly noted.

To get rid of this problem while producing a series of images with decreasing level of detail, the method originally proposed in [29] considers the adjacency graph of the catchment basins. A morphological gradient of this graph G is easily produced, for example, by computing the transform $\max(\delta(G) - G, G - \varepsilon(G))$. Then determining the watersheds of this graph allows one to merge catchment basins into catchment basins of second order. The resulting image, called a mosaic image of order 2 (see Figure 10c), has less detail than the previous one but the main features have been preserved. The procedure can be iterated to produce mosaic images of order 3, 4, etc. Unlike the classical Gaussian pyramids the present method has the advantage that it avoids blurring effects and preserves the most significant contours at best.

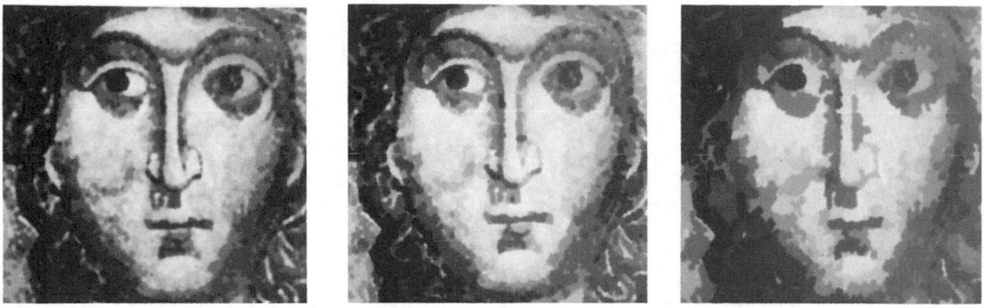

(a) (b) (c)

Figure 10. Successive image simplification by watersheds on graphs.

3. Study of Cornea Cell Populations

This is one of the medical applications in which graph morphology is currently under investigation. Images of the type shown in Figure 11* represent populations of cornea cells. The cells have roughly polygonal shapes, and the main problem is to determine the distribution of the number of edges of each cell. Beyond that, one would also like to answer such questions as: Do the small cells tend to have small cells as neighbors? What is the size of clusters of cells of a given type? The same questions can also be asked for the cells with few (or with many) edges.

To address these issues, we use the adjacency graph of the cells, which is obtained after watershed segmentation of the image in Figure 11 and contour tracking of each extracted region. As can be seen in Figure 12, this graph is a triangulation. Determination of the number of neighbors of a given vertex yields the number of edges of the associated cell in a straightforward manner. One can then assign to each cell its number of edges, thus producing a gray-level graph. Granulometries on this graph then provide useful information on the repartition of these cells in the tissue. The same analyses can be performed by assigning each cell its size or any other relevant parameter [30].

VI. STRUCTURING GRAPHS AND NEIGHBORHOOD FUNCTIONS

The basic idea underlying classical morphology is to extract information from an image by probing it at any position with some small geometric shape called a structuring element. Using operations related to the partial order of the underlying image space (e.g., supremum, infimum), one may construct a large class of image operators that are translation invariant. This approach easily carries over to gray-level graphs if one introduces the concept of a *structuring graph* or *s-graph*.

*Example provided by Dr. Barry Masters, USUHS, Bethesda, Maryland.

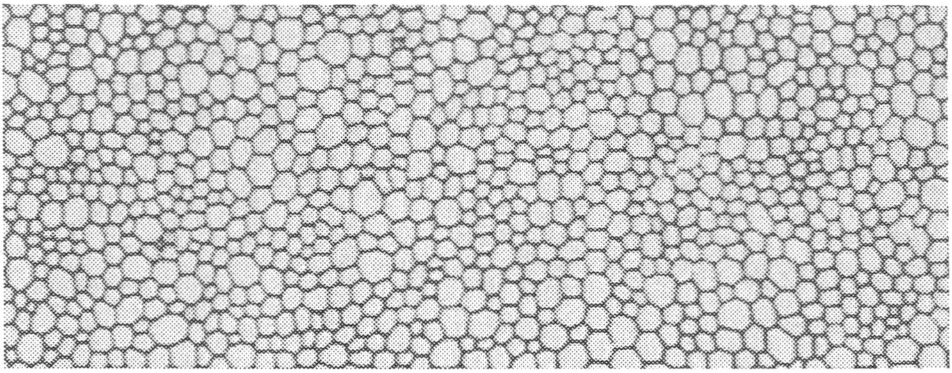

Figure 11. Population of cornea cells at high magnification.

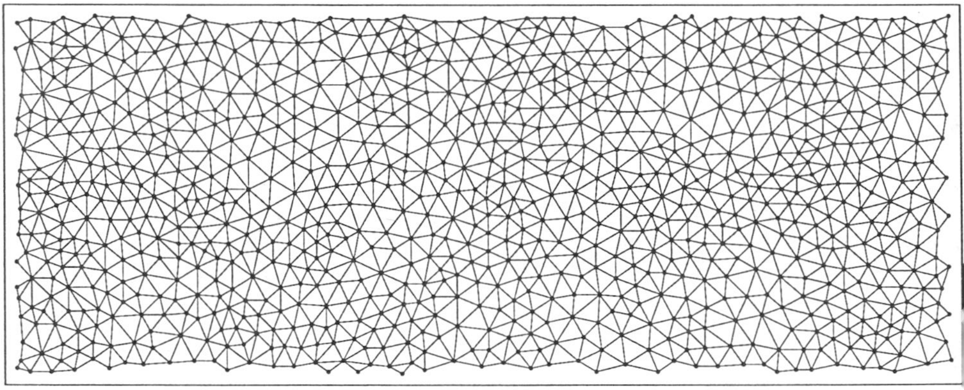

Figure 12. Adjacency graph corresponding to Figure 11.

Definition 6.1. An s-graph $\mathcal{A}$ consists of a graph $G_{\mathcal{A}} = (V_{\mathcal{A}}, E_{\mathcal{A}})$ and two nonempty subsets $B_{\mathcal{A}}, R_{\mathcal{A}} \subseteq V_{\mathcal{A}}$, respectively called the *buds* and the *roots*.

Matching an s-graph $\mathcal{A}$ to the graph G at vertex v amounts to finding a homomorphism θ mapping $G_{\mathcal{A}}$ into G such that $v \in \theta(R_{\mathcal{A}})$. Such a mapping θ is called an *embedding of* $\mathcal{A}$ *into G at v*. (*Note:* We point out that we use the word "matching" in a different meaning than is usual in the current literature on graphs.)

We can use an s-graph $\mathcal{A}$ to construct for any given graph $G = (V, E)$ a neighborhood function $N_{\mathcal{A}}$ on $\mathcal{P}(V)$ as follows:

$$N_{\mathcal{A}}(v \mid G) = \bigcup \{\theta(B_{\mathcal{A}}) \mid \theta \text{ is an embedding of } \mathcal{A} \text{ into } G \text{ at } v\} \tag{6.16}$$

Here the second argument G indicates the dependence on the underlying graph G. It is obvious that

$$N_{\mathcal{A}}(\tau v \mid G) = \tau N_{\mathcal{A}}(v \mid G) \qquad \text{for every } \tau \in \mathrm{Sym}(G)$$

In Figure 13 we have illustrated the concept of an s-graph and the corresponding neighborhood function. In this figure and the following ones, roots of s-graphs are designated by arrows and buds are drawn in bold. Comparing this construction of a neighborhood function with classical translation-invariant morphology, where the neighborhoods are translates of a small set called the structuring element, the roots of the s-graph correspond to the origin of the structuring element (note that an s-graph may have more than one root) and the buds to the points of the structuring element. An important difference from classical morphology, however, is that for graphs the neighborhood structure may differ at each vertex so that the s-graph must prescribe the structure near a vertex.

Figure 13 shows that the neighborhood determined by an s-graph $\mathcal{A}$ depends on several factors. Adding points to the bud set (s-graph $\mathcal{B}$) or to the root set (s-graph $\mathcal{C}$) or decreasing the underlying graph (s-graph $\mathcal{D}$) has the effect that the neighborhood increases. This motivates us to define a partial order $\leq$ on the collection of all s-graphs which formalizes this observation. For two s-graphs $\mathcal{A}$,

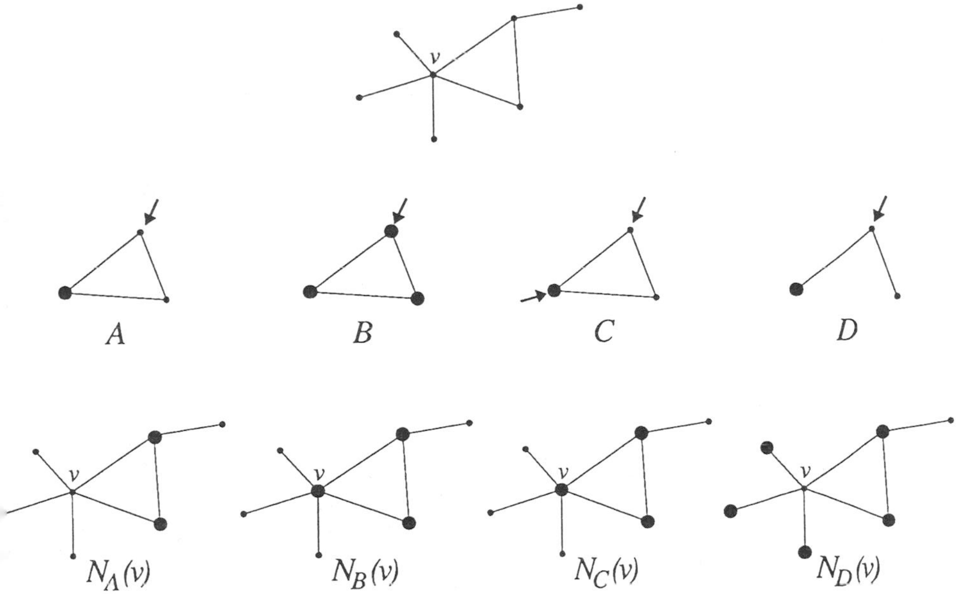

Figure 13. s-graphs A, B, C, D and their corresponding neighborhood functions at vertex v.

$\mathcal{B}$ we say that $\mathcal{A} \leq \mathcal{B}$ (stated as $\mathcal{A}$ is more selective than $\mathcal{B}$) if $N_{\mathcal{A}}(v \mid G) \subseteq N_{\mathcal{B}}(v \mid G)$ for any graph G and any vertex v on that graph. If $\mathcal{A} \leq \mathcal{B}$ and $\mathcal{B} \leq \mathcal{A}$, then $\mathcal{A}$ and $\mathcal{B}$ are equivalent and we write $\mathcal{A} \equiv \mathcal{B}$. In the example depicted in Figure 13 we have $\mathcal{A} \leq \mathcal{B}$, $\mathcal{B} \equiv \mathcal{C}$, $\mathcal{A} \leq \mathcal{D}$. In [13] we have shown the following result.

Proposition 6.2. Let $\mathcal{A}$, $\mathcal{B}$ be s-graphs. Then we have $\mathcal{A} \leq \mathcal{B}$ if and only if

(i) $G_{\mathcal{B}} \,\tilde{\subset}\, G_{\mathcal{A}}$

(ii) $N_{\mathcal{A}}(v \mid G_{\mathcal{A}}) \subseteq N_{\mathcal{B}}(v \mid G_{\mathcal{A}})$, for any $v \in R_{\mathcal{A}}$.

In particular, $\mathcal{A} \equiv \mathcal{B}$ if and only if $G_{\mathcal{A}} \simeq G_{\mathcal{B}}$ and $N_{\mathcal{A}}(v \mid G_{\mathcal{A}}) = N_{\mathcal{B}}(v \mid G_{\mathcal{A}})$ for any $v \in R_{\mathcal{A}}$.

In section II we have seen that to any neighborhood function N on the set V there corresponds a unique reciprocal neighborhood function $\check{N}$. So if $\mathcal{A}$ is an s-graph and $G = (V, E)$ a graph, then there exists a reciprocal neighborhood $\check{N}$ of the neighborhood function $N_{\mathcal{A}}(\cdot \mid G)$. One may wonder if there exists an s-graph $\mathcal{B}$ such that the reciprocal neighborhood function of $N_{\mathcal{A}}(\cdot \mid G)$ equals $N_{\mathcal{B}}(\cdot \mid G)$ for any graph G. In [13] we have shown that one can give an affirmative answer to this question by defining the so-called reciprocal s-graph $\check{\mathcal{A}}$, as follows:

$$G_{\check{\mathcal{A}}} = G_{\mathcal{A}}, \qquad B_{\check{\mathcal{A}}} = R_{\mathcal{A}}, \qquad R_{\check{\mathcal{A}}} = B_{\mathcal{A}}$$

see Figure 14a. Then we have the relation

$$\check{N}_{\mathcal{A}}(v \mid G) = N_{\check{\mathcal{A}}}(v \mid G)$$

If the s-graph $\mathcal{A}$ coincides with its reciprocal, or more precisely if $\mathcal{A} \equiv \check{\mathcal{A}}$, then we say that $\mathcal{A}$ is *symmetric*. Some examples can be found in Figure 14b.

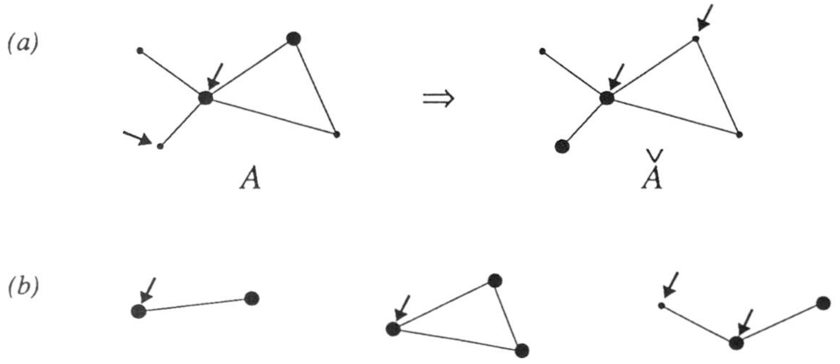

Figure 14. (a) An s-graph A and its reciprocal A; (b) symmetric s-graphs.

VII. DILATIONS AND EROSIONS

In Section II we indicated how to define gray-level dilations and erosions using neighborhood functions. In combination with the construction method for neighborhood functions from s-graphs described in the previous section, we have found a systematic way to build graph dilations and graph erosions from s-graphs.

Let $\mathcal{A}$ be an s-graph and let $N_{\mathcal{A}}$ be its corresponding neighborhood function. Consider an arbitrary graph $G = (V,E)$. Then $\delta_{\mathcal{A}}$ and $\varepsilon_{\mathcal{A}}$, given by

$$\begin{aligned}\delta_{\mathcal{A}}(f)(v) &= \sup\{f(w) \mid w \in N_{\check{\mathcal{A}}}(v \mid G)\} \\ \varepsilon_{\mathcal{A}}(f)(v) &= \inf\{f(w) \mid w \in N_{\mathcal{A}}(v \mid G)\}\end{aligned} \tag{6.17}$$

for $f \in \text{Fun}(V)$, define a graph dilation and a graph erosion, respectively, and the pair $(\varepsilon_{\mathcal{A}},\delta_{\mathcal{A}})$ forms an adjunction. Furthermore, both operators are symmetry preserving. They are illustrated in Figure 15.

Recall from Section II that the dual of an operator ψ on Fun(V) is defined as $\psi^*(f) = (\psi(f^*))^*$, where $f^* = m - f$ is the negative of f. Since the mapping $f \rightarrow f^*$ turns suprema into infima and vice versa, one may conclude that the dual

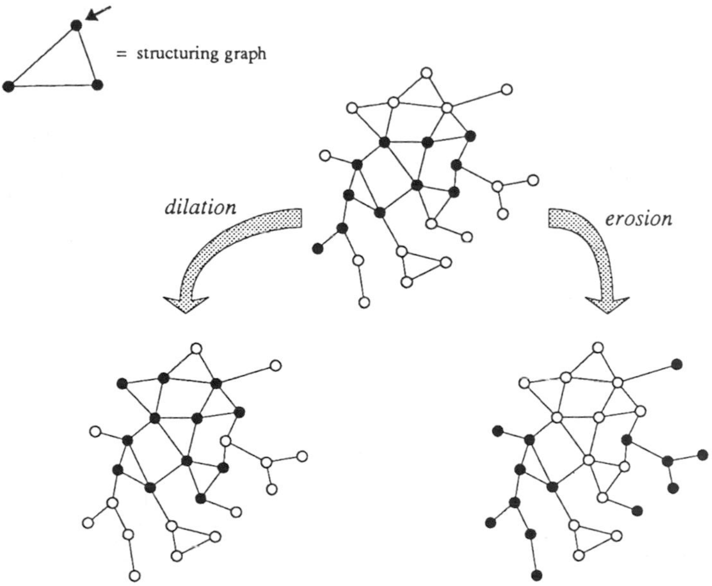

Figure 15. Dilation and erosion of a graph with respect to an s-graph.

of a dilation is an erosion and, conversely, that the dual of an erosion is dilation. In [13] we have established the following result.

Proposition 7.1. For any s-graph $\mathcal{A}$ we have

$$\delta_{\mathcal{A}}^* = \varepsilon_{\check{\mathcal{A}}} \qquad \text{and} \qquad \varepsilon_{\mathcal{A}}^* = \delta_{\check{\mathcal{A}}}$$

We point out that a similar property holds in classical morphology. Although the resemblances between graph morphological operators and the classical translation-invariant morphological operators are striking, there are also some important differences. In particular, it is well known that any translation-invariant dilation δ on the binary image space $\mathcal{P}(\mathbf{R}^d)$ is a Minkowski addition, that is, $\delta(X) = X \oplus \delta(\{0\})$. A similar result holds for gray-level dilations. Unfortunately, there exists no graph analogue of this fact. This means in particular that compositions or suprema of dilations using one structuring graph cannot be obtained using only one (larger) s-graph. This is due to the fact that the local graph structure near a vertex may be very diverse, and therefore the neighborhood determined by an s-graph depends not only on the number of buds and roots but also on the structure of the s-graph itself. This is quite different from classical morphology, where this local structure is independent of the position and therefore plays no role.

As a second distinction between graph and classical morphology, we note that Matheron's theorem, which, in the classical case, states that every increasing translation-invariant operator can be decomposed as an intersection of dilations or as a union of erosions, does not have an analogue in graph morphology. This follows immediately from the following considerations. The s-graph construction of a neighborhood function on the vertex set of a graph is not the most general method for obtaining neighborhood functions that are invariant under the symmetries of the graph. In fact, the s-graph approach requires a certain amount of local structure to be present near a vertex. One may construct more general neighborhood functions by requiring in addition that the local structure contents near a vertex does not exceed a certain amount. For example, we may define

$$N(v) = \{v\} \cup \{w \in V \mid (v, w) \in E\} \qquad \text{if } v \text{ has at most two neighbors}$$

and

$$N(v) = \{v\} \qquad \text{otherwise}$$

Such a construction gives rise to graph neighborhood functions (and hence graph dilations) that are not G-increasing in general; see Figure 16. In particular, the operators resulting from this construction are symmetry preserving but cannot be written as an infimum of graph dilations of the form $\delta_{\mathcal{A}}$.

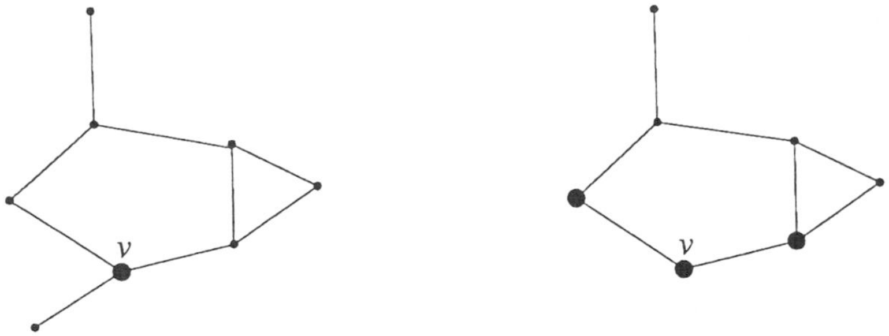

Figure 16. A neighborhood function that is not G-increasing; see text.

VIII. OPENINGS, CLOSINGS, AND OTHER FILTERS

An operator is called a *morphological filter* if it is increasing and idempotent. Idempotence is an important property for an operator because it means that repeated application of such an operator has no further effect on the outcome. In a sense, one could argue that any operator (morphological or otherwise) designed to clean noise from an image has to be applied repeatedly until the result remains constant. In practice, such an iterative procedure results in idempotent operators; see Heijmans and Serra [12]. A formal theory for morphological filters has been designed by Matheron [16]. In this section we shall apply some of his results to the framework of gray-level graphs. Although we shall mainly be concerned with openings and closings and construction methods for such operators, we will consider alternating sequential filters at the end.

Openings and closings lie at the heart of the theory of morphological filters. Here we shall only consider openings. The corresponding results for closings follow easily by duality in the following way: if ψ is an opening then ψ^* is a closing and vice versa. We refer to [13, Remark 6.5] for some difficulties concerning the definition of closings in the graph framework.

A well-known construction of openings is to compose an erosion and its adjoint dilation. For instance, $\delta_{\mathcal{A}}\varepsilon_{\mathcal{A}}$ is an opening for any s-graph $\mathcal{A}$. In Figure 17 we have depicted an example of an opening obtained in this way.

A second way to build openings, closely related to the previous one, also uses s-graphs. We use the following notation: if $G = (V, E)$ is a graph, $X \subseteq V$ and $t \in T$, then we define $\mathcal{F}_t(X)$ as the gray-level graph that equals t at the vertices in X and inf T elsewhere. Let $\mathcal{A}$ be an s-graph. We define the graph operator $\alpha_{\mathcal{A}}$ by

$$\alpha_{\mathcal{A}}(f) = \bigvee\{\mathcal{F}_t(\theta(B_{\mathcal{A}})) \mid G_{\mathcal{A}} \xrightarrow{\theta} G \text{ and } \mathcal{F}_t(\theta(B_{\mathcal{A}})) \leq f\}$$

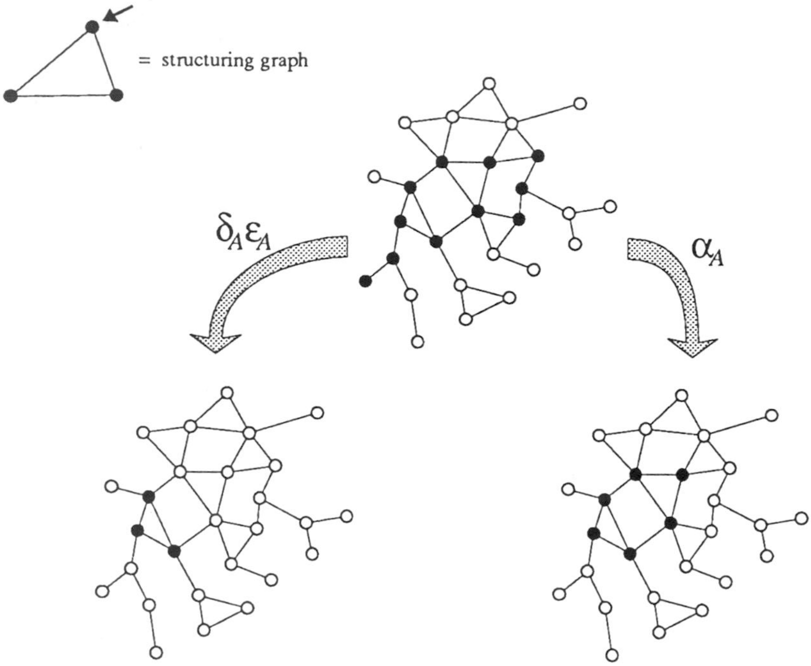

Figure 17. Two different openings of a binary graph with respect to a given *s*-graph.

This expression looks rather complex at first sight, but it becomes much simpler for binary graphs $(X \mid G)$ where $X \subseteq V$. In that case $\alpha_{\mathcal{A}}(X \mid G)$ is the union of all embedded bud sets $\theta(B_{\mathcal{A}})$ where θ is an embedding of $\mathcal{A}$ into G (at an arbitrary vertex) such that $\theta(B_{\mathcal{A}}) \subseteq X$.

It is obvious that $\alpha_{\mathcal{A}}$ is

1. an opening
2. a flat operator (i.e., it commutes with thresholding: see Section II)
3. symmetry preserving
4. $G-$ increasing

The opening $\alpha_{\mathcal{A}}$ is called a *structural graph opening*. This terminology stems from [21], where it has been shown that under rather mild assumptions (including translation invariance) structural openings form the basis for the collection of all openings. In [13] this result has been extended to the case of graphs.

Proposition 8.1. Let α be a graph opening that is flat, symmetry preserving and $G-$ increasing. Then α can be decomposed as a supremum of structural graph openings.

One can also show that (see [13]) that

$$\delta_{\mathcal{A}}\varepsilon_{\mathcal{A}} \leq \alpha_{\mathcal{A}}$$

An example where this inequality is strict is depicted in Figure 17b.

So far we have seen two ways to construct openings: the first is by composition of an erosion and its adjoint dilation, the second by definition of structural openings. Another powerful tool for building openings is provided by the so-called *inf-overfilters* (again, the corresponding results for closings follow by duality; here one must introduce the concept of a *sup-underfilter*). An increasing operator ψ is called an inf-overfilter if

$$\psi(\mathrm{id} \wedge \psi) = \psi$$

It is obvious that every extensive operator is an inf-overfilter. Furthermore, one can easily show that the class of inf-overfilters is closed under suprema and self-composition. Our interest in inf-overfilters stems from the fact that id $\wedge\ \psi$ is an opening if ψ is an inf-overfilter. Now, if (ε, δ) is an adjunction, and if δ' is a dilation such that $\delta' \geq \delta$, then $\delta'\varepsilon$ is an inf-overfilter. Namely,

$$\delta'\varepsilon \geq \delta'\varepsilon(\mathrm{id} \wedge \delta'\varepsilon) = \delta'(\varepsilon \wedge \varepsilon\delta'\varepsilon) \geq \delta'(\varepsilon \wedge \varepsilon\delta\varepsilon) = \delta'\varepsilon$$

The considerations above are valid on arbitrary complete lattices. Here we shall apply them to our graph framework. Let $\mathcal{A}$, $\mathcal{B}$ be two s-graphs such that $\mathcal{A} \leq \mathcal{B}$. Then $\delta_{\mathcal{A}} \leq \delta_{\mathcal{B}}$. Now the abstract theory gives us that $\delta_{\mathcal{B}}\varepsilon_{\mathcal{A}}$ is an inf-overfilter and hence that id $\wedge\ \delta_{\mathcal{B}}\varepsilon_{\mathcal{A}}$ is an opening; see Figure 18 for an example.

In classical morphology there exists yet another way to define openings. Take A to be a symmetric structuring element (i.e., $x \in A$ iff $-x \in A$) that does not contain the origin, and define

$$\alpha(X) = (X \oplus A) \cap X$$

One can easily show that α defines an opening. If A is a ring-shaped set (annulus) centered about the origin, then α removes isolated particles; see Figure 5.2 of [24]. For this reason α is called an *annular opening*. We can generalize this notion to graph morphology in the following way. Recall from Section VI that an s-graph is called symmetric if $\check{\mathcal{A}} = \mathcal{A}$

Proposition 8.2. Let $\mathcal{A}$ be a symmetric s-graph; then id $\wedge\ \delta_{\mathcal{A}}$ is an opening.

Proof. Though it is not difficult to give a direct proof for gray-level graphs, the demonstration for binary graphs is much more transparent. Since we are dealing exclusively with flat operators, a restriction to binary graphs means no loss of generality. Furthermore, we will suppress the argument G in the notation.

For a binary graph $X \subseteq V$ the dilation can be written as

$$\delta_{\mathcal{A}}(X) = \bigcup_{x \in X} N_{\mathcal{A}}(x)$$

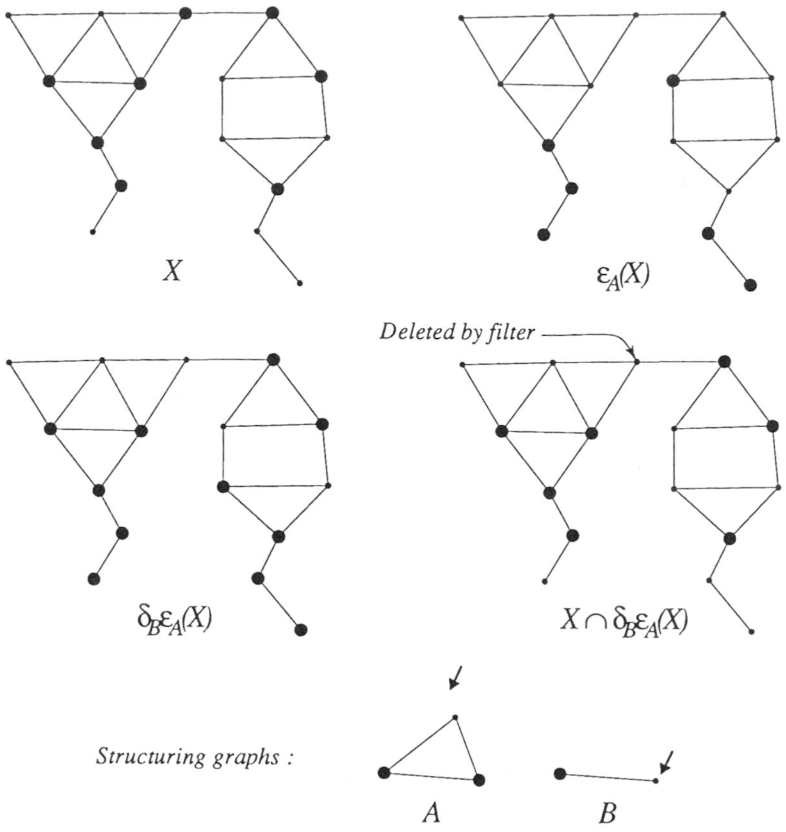

Figure 18. Example of an inf-overfilter.

It is obvious that id $\wedge\ \delta_{\mathcal{A}}$ is antiextensive. To get idempotence it suffices to show that

$$(\mathrm{id} \wedge \delta_{\mathcal{A}})(\mathrm{id} \wedge \delta_{\mathcal{A}}) \geq \mathrm{id} \wedge \delta_{\mathcal{A}},$$

or equivalently that

$$\delta_{\mathcal{A}}(\mathrm{id} \wedge \delta_{\mathcal{A}}) \geq \mathrm{id} \wedge \delta_{\mathcal{A}}$$

Let $X \subseteq V$ and $Y = X \cap \delta_{\mathcal{A}}(X)$. We show that

$$\delta_{\mathcal{A}}(Y) \supseteq Y$$

Let $y \in Y$. Then $y \in \delta_{\mathcal{A}}(X)$ and so $y \in N_{\mathcal{A}}(x)$ for some $x \in X$. Since $\mathcal{A}$ is symmetric we get $x \in N_{\mathcal{A}}(y) \subseteq \delta_{\mathcal{A}}(X)$, since $y \in X$. Therefore, $x \in X \cap \delta_{\mathcal{A}}(X) = Y$, and with $y \in N_{\mathcal{A}}(x)$ this yields $y \in \delta_{\mathcal{A}}(Y)$, which was to be proved.

An example of an annular opening is depicted in Figure 19.

A class of morphological filters which turned out to be quite successful for the cleaning of noisy images are the so-called alternating sequential filters. For a full account of the underlying theory we refer to [24,25]. Here we only sketch the underlying idea. Let $\alpha_1, \alpha_2, \alpha_3, \ldots$ be a sequence of openings, and $\phi_1, \phi_2, \phi_3, \ldots$ a sequence of closings such that

$$\alpha_i\alpha_j = \alpha_j\alpha_i = \alpha_j \qquad \text{and} \qquad \phi_i\phi_j = \phi_j\phi_i = \phi_j \qquad \text{if } j \geq i \tag{6.18}$$

The families α_i and ϕ_i may be chosen independently; often, however, they are taken to be each other's dual. The operators

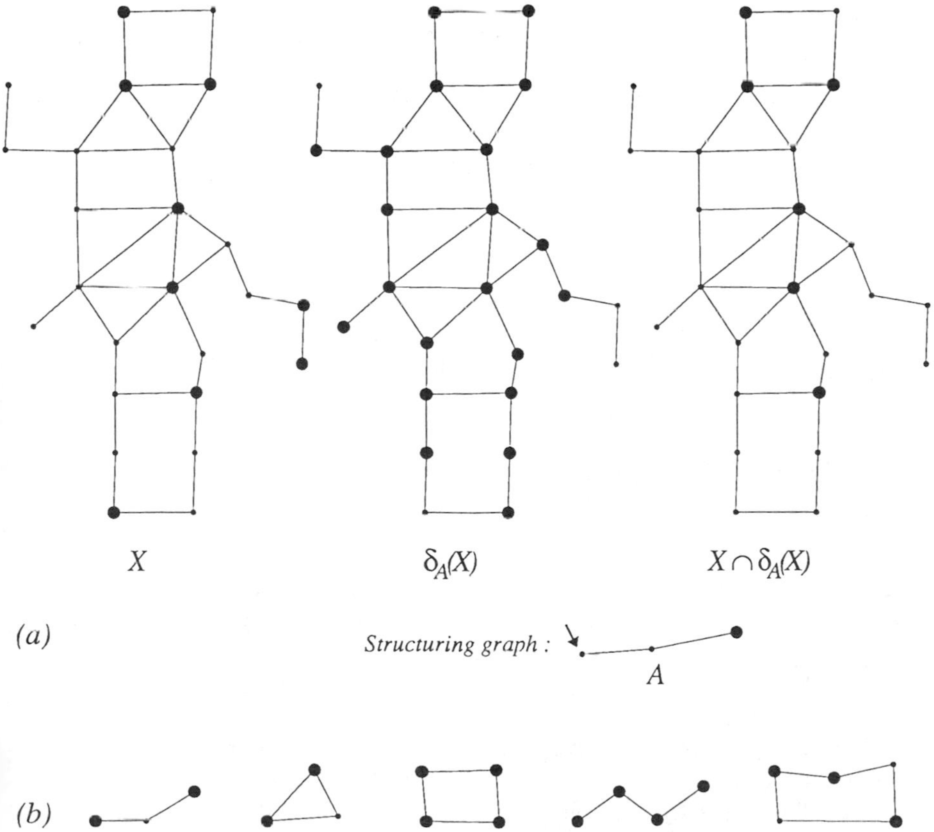

Figure 19. (a) Example of an annular opening. (b) Some typical invariants of the annular opening; note that every invariant is a "union of translates" of the invariant at the left.

$$M_i = \alpha_i\phi_i\alpha_{i-1}\phi_{i-1}\cdots\alpha_1\phi_1, \qquad N_i = \phi_i\alpha_i\phi_{i-1}\alpha_{i-1}\cdots\phi_1\alpha_1$$

are called alternating sequential filters or AS filters. In [25] two other AS filters have been introduced. A systematic way to obtain α_i and ϕ_i is to put

$$\alpha_i = \delta^i\varepsilon^i \qquad \text{and} \qquad \phi_i = \varepsilon^i\delta^i$$

where (ε, δ) is an adjunction and ε^i denotes the ith power of ε.

The effect of an AS filter is that it successively removes larger and larger noise particles. Furthermore, it treats fore- and background in a more or less similar way. In graph morphology we can construct AS filters by choosing structural openings $\alpha_{\mathcal{A}_i}$ and their dual closings $\phi_{\mathcal{A}_i}$ that satisfy the semigroup property (6.18). At this point it is important to recall the following results from [13]. Let $\mathcal{A}$, $\mathcal{B}$ be s-graphs; we say that $\mathcal{B}$ is *$\mathcal{A}$-open* if for every $v \in B_{\mathcal{B}}$ there is an embedding of $\mathcal{A}$ into $G_{\mathcal{B}}$ at v.

Proposition 8.3. Let $\mathcal{A}$ $\mathcal{B}$ be s-graphs. The equalities

$$\alpha_{\mathcal{A}}\alpha_{\mathcal{B}} = \alpha_{\mathcal{B}}\alpha_{\mathcal{A}} = \alpha_{\mathcal{B}}$$

hold if and only if $\mathcal{B}$ is $\mathcal{A}$-open.

Now take a sequence of s-graphs $\mathcal{A}_i$ such that $\mathcal{A}_{i+1}$ is $\mathcal{A}_i$-open. See Figure 20 for a number of such sequences. Then $\mathcal{A}_j$ is $\mathcal{A}_i$-open if $j \geq i$. Let $\alpha_i = \alpha_{\mathcal{A}_i}$ and $\phi_i = \phi_{\mathcal{A}_i}$ be the corresponding structural openings (respectively, closings). Then the operators M_i and N_i defined above are AS filters.

Proposition 8.3 also lays the foundation for the definition of granulometries that in turn yield size distributions. We can think of a granulometry as a collection of openings α_i $(i \geq 1)$ that satisfies the first condition in Eq. (6.18).

IX. CONCLUDING REMARKS

Before we conclude this chapter, let us give a few hints on the implementation of morphological transformations on graphs. We already dealt briefly with the obtention of neighborhood graphs from binary images (see Section V). Transforming such objects morphologically first involves encoding them in an appropriate way. This is achieved via a data structure derived from the adjacency matrix of the graph [2] and is detailed in [28,30]. Given a vertex, this structure allows direct access to its neighbors.

For the algorithms themselves, two cases have to be considered: structuring and nonstructuring graphs. The first case is by far the more difficult one; what one has to do basically is to find all the different ways of matching the structuring graphs within the graph to be transformed. This is achieved by scanning the search tree as efficiently as possible. The implementation of nonstructured graph operators is much easier. Indeed, as mentioned in Section IV, all the nonstruc-

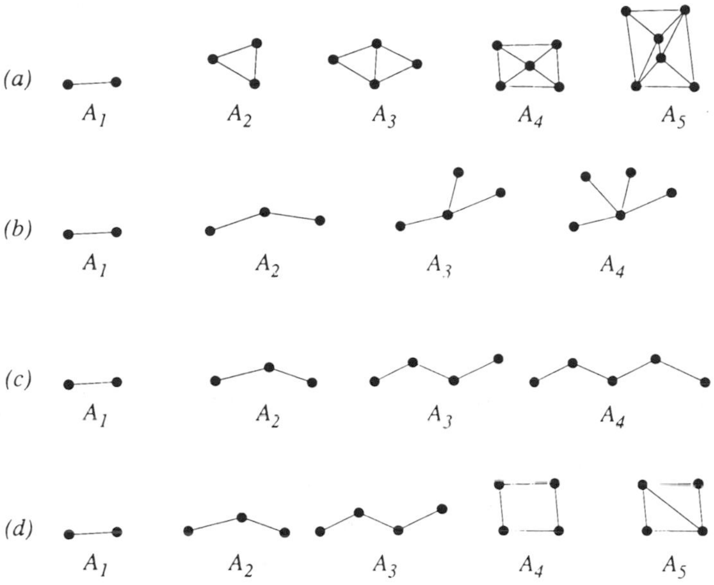

Figure 20. Families of s-graphs to be used as AS filters and granulometries on graphs. Note that A_{i+1} does not necessarily have more buds than A_i.

tured graph morphology operations are based on the *distance* induced by the set of edges E on the vertex set V. Therefore, *breadth-first scannings* will be at the basis of most algorithms. They are implemented via the use of a *queue* of vertices, that is, FIFO structure (see Figure 21). This is explained in further detail in [28,32,34].

For example, to determine the distance function of a graph—that is, to assign to a vertex its distance to a particular set W of vertices—one starts from the vertices of W and does a breadth-first scanning. In this procedure, the vertices at distance 1 are first met, then those at distance 2, etc., until stability is reached. This algorithm—as well as many others described in [32]—is particularly efficient because each vertex is considered a minimal number of times. It was used for the first example presented in Section V. Another example of a distance function, on a Delaunay triangulation this time, is presented in Figure 22.

Graph morphology provides a collection of morphological tools for the investigation of populations of objects for which neighborhood relations are of interest. Here objects may be physical or biological objects, such as the nuclei in a microscopic image of some cell tissue, but they may also refer to a symbolic description of a scene. As to the latter case, one may think of the situation in which the objects represent the intensity extrema of a scene (see [18]). One can

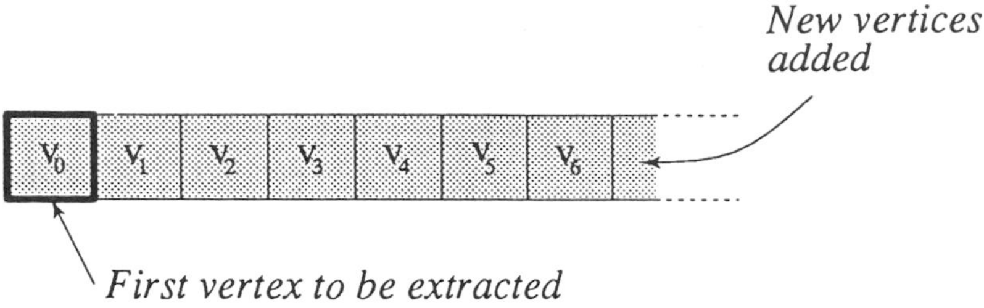

Figure 21. How a queue (FIFO structure) of vertices works.

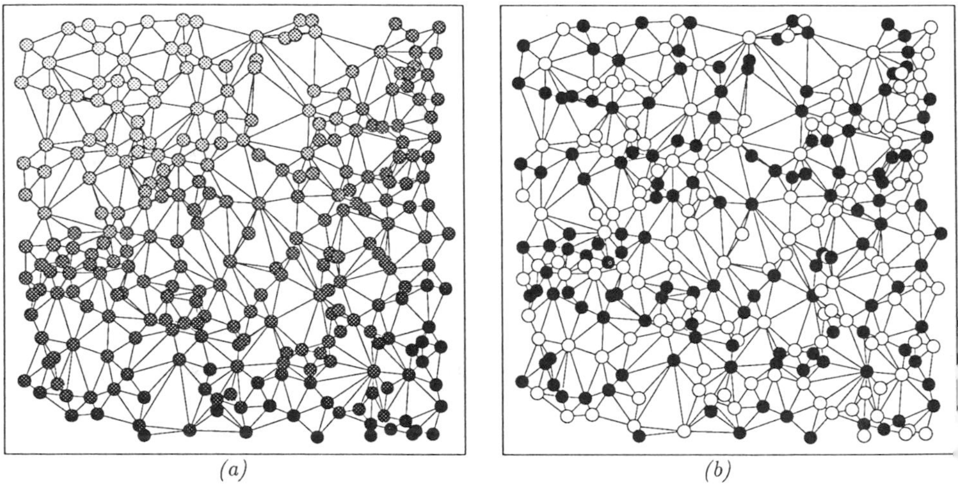

Figure 22. Example of a distance function on the Delaunay triangulation of Figure 8a. In (a), darker vertex values represent larger distances; in (b), the distance function is displayed modulo 2.

model spatial relationships between objects by different types of graphs. In this chapter we have discussed three such graphs, namely the Delaunay triangulation, the Gabriel graph, and the relative neighborhood graph.

Although the intrinsic structure of a gray-level graph is much poorer than that of the Euclidean space $\mathbf{R}^d$ or the discrete space $\mathbf{Z}^d$, it is still rich enough to define a large class of morphological operators for such images that inherit many of the nice properties of their classical analogues. Thereto we have introduced the concept of a structuring graph that generalizes the structuring element in classical morphology. If the structuring graph consists of only two vertices connected by an edge, then the resulting graph operators are of the so-called nonstructured type. They are easy to implement but still allow the construction of a large collection of morphological algorithms. Structured graph operations, which use more information about the local graph structure near a vertex, allow a much larger collection of morphological operations (see, e.g., the sequence of structuring graphs depicted in Figure 20, which can be used to define granulometries and AS filters), but they are definitely harder to implement. As often in science, the future will tell us to which extent these structured and nonstructured graph operations turn out to be useful in practical applications.

Acknowledgments

The work of L. Vincent was supported in part by Dassault Electronique and the National Science Foundation under grant MIPS-86-58150, with matching funds from DEC and Xerox.

REFERENCES

1. van Antwerpen, G., and van Munster, R. J., *TCL-IMAGE User's Manual*, Institute of Applied Physics, TNO, Delft, The Netherlands, 1989.
2. Berge, C., *Graphs*, 2nd rev. ed., North-Holland, Amsterdam, 1985.
3. Beucher, S., and Lantuéjoul, Ch., Use of watersheds in contour detection, in *Proceedings of the International Workshop on Image Processing, Real-Time Edge and Motion Detection/Estimation*, Rennes, France, 1979.
4. Beucher, S., Segmentation d'images et morphologie mathématique, Ph.D. dissertation, Ecole Nationale Supérieure des Mines de Paris, France, 1990.
5. Birkhoff, G., *Lattice Theory*, 3rd ed., Colloquium Publications, vol. 25, American Mathematical Society, Providence, Rhode Island, 1984.
6. Boissonat, J. D., Shape reconstructions from planar cross-sections, Technical Report INRIA, Le Chesnay, France, 1986.
7. Borgefors, G., Distance transformations in digital images, *Comput. Vision Graphics Image Process.*, *34*, 334–371 (1986).
8. Gabriel, K. R., and Sokal, R. R., A new statistical approach to geographic variations analysis, *Syst. Zool.*, *18*, 259–278 (1969).
9. Gondran, M., and Minoux, M., *Graphs and Algorithms*, Wiley, Chichester, 1984.
10. Heijmans, H. J. A. M., Theoretical aspects of gray-level morphology, *IEEE Trans. Pattern Anal. Machine Intell.*, *13*, 568–582 (1991).
11. Heijmans, H. J. A. M., and Ronse, C., The algebraic basis of mathematical morphology. Part I. Dilations and erosions, *Comput. Vision Graphics Image Process.*, *50*, 245–295 (1990).
12. Heijmans, H. J. A. M., and Serra, J., Convergence, continuity and iteration in mathematical morphology, *J. Visual Commun. Image Represent.*, in press.
13. Heijmans, H. J. A. M., Nacken, P., Toet, A., and Vincent, L., Graph morphology, *J. Visual Commun. Image Representation*, in press.
14. Jeulin, D., Vincent, L., and Serpe, G., Propagation algorithms on graphs for phys-

ical applications, Internal report CMM, Ecole Nationale Supérieure des Mines de Paris, France, 1990.
15. Maragos, P., and Schafer, R. W., Morphological filters. Part I. Their set-theoretic analysis and relations to linear shift-invariant filters, *IEEE Trans. Acoust. Speech Signal Process.*, *35*, 1153–1169 (1987).
16. Matheron, G., Filters and lattices, Internal report CGMM, Ecole Nationale Supérieure des Mines de Paris, France, 1989.
17. Matheron, G., *Random Sets and Integral Geometry*, Wiley, New York, 1975.
18. Nacken, P., Hierarchical image structure description based on intensity extrema, in *Geometrical Problems of Image Processing*, Georgenthal (DDR) (U. Eckhardt, A. Hübler, W. Nagel, and G. Werner, eds.), Akademie Verlag, Berlin, 1991, pp. 99–106.
19. O, Y.-L., and Toet, A., Mathematical morphology in hierarchical image representation, in *Proceedings, NATO ASI on the Formation, Handling and Evaluation of Medical Images*, A. E. Todd-Pokropek and M. A. Viergevere (eds), NATO ASI Series F, Springer, Heidelberg, 447–462, 1991.
20. Preparata, F. P., and Shamos, M. I., *Computational Geometry: An Introduction*, Springer-Verlag, New York, 1985.
21. Ronse, C., and Heijmans, H. J. A. M., The algebraic basis of mathematical morphology. Part II. Openings and closings, *Comput. Vision Graphics Image Process. Image Understand.*, *54*, 384–400, 1991.
22. Raphaël, M., Vincent, L., Raymond, E., Grimaud, M., and Meyer, F., Germinal center (GC) analysis with graphs and mathematical morphology, in *Proceedings, Congress of the International Society for Analytical Cytology*, Bergen, Norway, in press.
23. Serra, J., *Image Analysis and Mathematical Morphology*, Academic Press, London, 1982.
24. Serra, J., (ed.), *Image Analysis and Mathematical Morphology*, Part II, *Theoretical Advances*, Academic Press, London, 1988.
25. Serra, J., and Vincent, L., An overview of morphological filtering, *IEEE Trans. Circuits Syst. Signal Process.*, *11*, 47–100, 1992.
26. Supowit, K. J., The relative neighbourhood graph with an application to minimum spanning trees, *J. ACM*, *30*(3), 428–447 (1983).
27. Toussaint, G. T., The relative neighborhood graph of a finite planar set, *Pattern Recogn.*, *12*, 1324–1347 (1980).
28. Vincent, L., Mathematical morphology on graphs, in *Proceedings SPIE, Visual Communications and Image Processing 88*, Cambridge, Massachusetts, 1988, pp. 95–105.
29. Vincent, L., Mathematical morphology for graphs applied to image description and segmentation, in *Proceedings Electronic Imaging West 89*, Pasadena, 1989, vol. 1, pp. 313–318.
30. Vincent, L., Graphs and mathematical morphology, *Signal Process.*, *16*, 365–388 (1989).
31. Vincent, L., and Jeulin, D., Minimal paths and crack propagation simulations, in *Acta Stereologica 8, Proceedings 5th European Congress for Stereology*, 1989, pp. 487–494.

32. Vincent, L., Algorithmes morphologiques à base de files d'attente et de lacets. Extension aux graphes, Ph.D. dissertation, Ecole Nationale Supérieure des Mines de Paris, France, 1990.
33. Vincent, L., Exact Euclidean distance function by chain propagations, Technical Report 91-4, Harvard Robotics Laboratory, Cambridge, February 1991; *Proc. IEEE Comput. Vision Pattern Recogn.*, in press.
34. Vincent, L., Morphological algorithms, in *Mathematical Morphology in Image Processing* (E. Dougherty, ed.), Marcel Dekker, New York, 1991.
35. Wendt, P. D., Coyle, E. J., and Gallagher, N. C., Stack filters, *IEEE Trans. Acoust. Speech Signal Process.*, *34*, 898–911 (1986).

Chapter 7

Mathematical Morphology with Noncommutative Symmetry Groups

Jos B.T.M. Roerdink

Centre for Mathematics and Computer Science, Amsterdam, The Netherlands

Mathematical morphology as originally developed by Matheron and Serra is a theory of set mappings, modeling binary image transformations, that are invariant under the group of Euclidean translations. Because this framework turns out to be too restricted for many practical applications, various generalizations have been proposed. First, the translation group may be replaced by an arbitrary commutative group. Second, one may consider more general object spaces, such as the set of all convex subsets of the plane or the set of gray-level functions on the plane, requiring a formulation in terms of complete lattices. So far, symmetry properties have been incorporated by assuming that the allowed image transformations are invariant under a certain commutative group of automorphisms on the lattice. In this chapter we embark on another generalization of mathematical morphology by dropping the assumption that the invariance group is commutative. To this end we consider an arbitrary homogeneous space (the plane with the Euclidean translation group is one example, the sphere with the rotation group another), that is, a set $\mathcal{X}$ on which a transitive but not necessarily commutative transformation group Γ is defined. As our object space we then take the Boolean algebra $\mathcal{P}(\mathcal{X})$ of all subsets of this homogeneous space. First we consider the case in which the transformation group is simply transitive or, equivalently, the basic set $\mathcal{X}$ is itself a group, so that we may study the Boolean algebra $\mathcal{P}(\Gamma)$. The general transitive case is subsequently treated by embedding the object space $\mathcal{P}(\mathcal{X})$ into $\mathcal{P}(\Gamma)$, using the results for the simply transitive case and translating the results back to $\mathcal{P}(\mathcal{X})$. Generalizations of dilations, erosions, openings, and closings are defined and several representation theorems are proved. For clarity of exposition as well as to emphasize the connection with classical Euclidean morphology, we have restricted ourselves to the case of Boolean lattices, which is appropriate for binary image transformations.

I. INTRODUCTION

Mathematical morphology was originally developed at the Paris School of Mines as a set-theoretical approach to image analysis [17,29]. It has a strong algebraic component, studying image transformations with a simple geometric interpretation and their decomposition and synthesis in terms of set operations. Other aspects are the probabilistic one, modeling (images of) samples of materials by random sets, and the integral geometric one, which is concerned with image functionals. Although the main object of our present study is the algebraic approach, we emphasize that our primary motivation comes from the geometric side, in the sense that various image transformations used in mathematical morphology today (dilations, erosions, openings, closings) have a straightforward geometric analogue in a more general context. It is then a natural question to ask whether a corresponding algebraic description can be found. From a practical point of view the importance of such an algebraic decomposition theory no doubt derives from the fact that it enables fast and efficient implementations on digital computers and special image analysis hardware. Because we will not deal with such questions here, we refer the reader to [7] for an elementary introduction to Euclidean morphology with emphasis on implementation.

In the original approach of Matheron and Serra [17,29], a two-dimensional image of, let us say, a planar section of a porous material is modeled as a subset X of the plane. In order to reveal the structure of the material, the image is probed by translating small subsets B, called *structuring elements*, of various forms and sizes over the image plane and recording the locations h where certain relations (e.g., "B_h included in X" or "B_h hits X") between the image X and the translate B_h of the structuring element B over the vector h are satisfied (see Figure 1a). In this way one can construct a large class of image transformations that are compatible with translations of the image plane or, to put it differently, are invariant under the Euclidean translation group. The underlying idea here is that the form or shape of objects in the image does not depend on the relative location with respect to an arbitrary origin and that therefore the transformations performed on the image should respect this. Notice that the basic object of study, the "object space," is not the reference space (the plane in our example) itself but the collection of subsets of this reference space and the transformations defined on this collection of subsets.

In practice, one encounters various situations where this framework is too restrictive. One of the earliest examples is mentioned in Serra's book [29, p. 17],

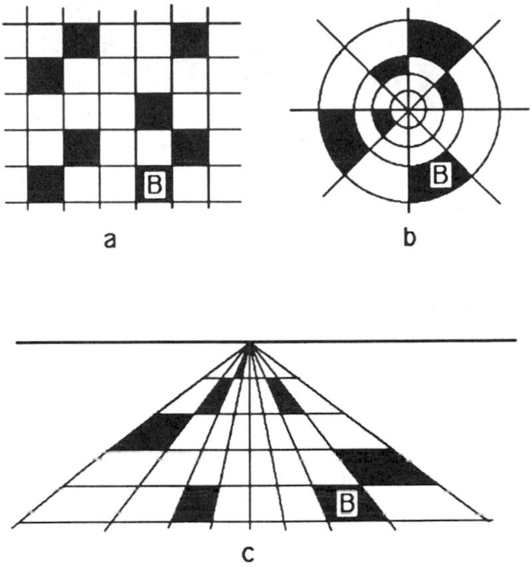

Figure 1. Copies (dark) of a structuring element B under (a) Euclidean translation, (b) rotations and scalar multiplication, and (c) perspective transformation.

where a photograph is shown of the trees in a forest, taken by putting the camera at ground level and aiming toward the sky. Such photographs are used to measure the amount of sunshine in the woods. The resulting image shows clear radial symmetry with intrinsic origin (the projection point of the zenith). It is clear that in this case we need image transformations that are adapted to the symmetries of this polar structure. It turns out that in fact one obtains a straightforward generalization of Euclidean morphology by replacing the Euclidean translations by an arbitrary *abelian* (*commutative*) group [11,25]. In the case of the example mentioned above, this would be the group generated by rotations and multiplications with respect to the origin. Here the size of the structuring element increases with increasing distance from the origin (Figure 1b). Another example occurs in the analysis of traffic scenes, where the goal is to recognize the shape of automobiles with a camera on a bridge overlooking a highway [3]. In this case the size of the structuring element has to be adapted according to the law of perspective (Figure 1c). It is not difficult to show that in this case there is again invariance under a commutative group. Notice that in the two examples just mentioned we have a variable structuring element as a function of position. This has been taken as the starting point by Serra and others to introduce arbitrary assignments of subsets to each point of the plane and define dilations and erosions accordingly, completely giving up invariance under a symmetry group (see also Section IV.A). However, in the examples just given the situation is different in the sense that there is a definite group connecting structuring elements at different locations, although their sizes differ. Actually, a metrical concept like "size" does not enter at all into the definition of the classical morphological operations. Only the group property of the Euclidean vector addition is involved, which explains why an extension to arbitrary groups is possible. In fact, we will argue that without a concept of invariance (under a group, or otherwise), one cannot even give a meaningful answer to the question when sets at different locations are "of the same shape" or not.

Instead of changing the symmetry group of the object space, one may generalize the object space itself. For example, instead of all subsets of the plane one may want to study a smaller collection, such as the open or closed sets or the convex sets. In that case the original approach is no longer valid because the union of an arbitrary collection of closed or convex sets is not necessarily closed or convex, the intersection of an arbitrary collection of open sets is not necessary open, etc. These difficulties can be overcome by taking as the object space a so-called *complete lattice*, that is, an ordered set $\mathcal{L}$ such that any subset of $\mathcal{L}$ has a supremum (smallest upper bound) and infimum (greatest lower bound), generalizing the set operations of union and intersection. This is the approach initiated by Serra and Matheron [30,31], as well as Heijmans [11]. A general study of this topic has been made by Heijmans and Ronse [12,27]. If one does not assume any invariance property one can only prove generalities. But again invariance under a group of automorphisms of the lattice may be introduced, as in [11,12,27], where so far the assumption made is always that the group is *commutative*. This enables a complete characterization of dilations, erosions, openings, closings, increasing transformations, etc. Another situation in which a lattice formulation is in order arises when one wants to go from binary images with their Boolean image algebra to gray-level images, that is, *functions*, defined on the basic reference space. Following Sternberg [32], one has introduced the so-called *umbras* to deal with this case [29,31,33]. After introducing an extra dimension for the function values, one performs the binary Euclidean operations in this enlarged space and translates the results back to the original space. However, for a mathematically satisfactory approach complete lattices are required; see Ronse [26].

In this chapter we want to generalize morphology by dropping the assumption that the invariance group is commutative. To this end we consider an arbitrary *homogeneous space*, a set $\mathcal{X}$ on which a transitive but not necessarily commutative group Γ of invertible transformations is defined. Here *transitive* means that for any pair of points in the set there is a transformation in the group that maps one point on the other. If this mapping is unique we say that the transformation group is *simply transitive* or *regular*. As the object space of interest from a morphological point of view, we take the Boolean algebra of all subsets of this homogeneous space.

We present two examples for basic motivation. First of all, one may extend Euclidean morphology in the plane by including rotations. This case has been extensively discussed in [24]. In many situations one does not want to distinguish between rotated versions of the same object. In that case it is appropriate to use the full Euclidean group of motions (the group generated by translations and rotations) as (noncommutative) invariance group. This is, for example, the basic assumption made in integral geometry to give a complete characterization (Hadwiger's theorem) of functionals of compact, convex sets in $\mathbf{R}^n$ [10]. As our second basic example we mention the sphere with its symmetry group of three-dimensional rotations, again a nonabelian group. Various motivations can be given here. First, the earth is spherical to a good approximation and this has to be taken into account when analyzing pictures taken by weather satellites. Second, pictures of virus particles show them to be nearly spherical with antibodies attached randomly to the surface, and a morphological description of the particle distribution on the surface is of interest. Third, from a theoretical point of view we observe that integral geometry and geometric probability on the sphere have been well investigated in the past [18,28]. Since there is a clear connection between these fields on the one hand and mathematical morphology on the other (see Serra [29, Chapters 4, 13]), it is of interest to develop morphology for the sphere as well. Here we can do no more than indicate how the sphere fits into our general framework, but clearly this case is important enough to warrant an in-depth study. A more detailed investigation of this case is presented in [23]. Another area of possible research is the question of how to take the projective geometry of the imaging process into account, because clearly the symmetry of a two-dimensional plane is not the same as the symmetry of the three-dimensional world of which it is a projection.

We first develop the theory for simply transitive transformation groups (all abelian transitive transformation groups fall in this category). It is easy to see that in this case there is a one-to-one correspondence between elements of $\mathscr{X}$ and those of Γ: let ω (the "origin") be an arbitrary point in $\mathscr{X}$ and associate to any $x \in \mathscr{X}$ the unique transformation in Γ that maps ω to x. Then a bijection between $\mathscr{X}$ and Γ is obtained. So in the simply transitive case we can assume without loss of generality that $\mathscr{X}$ coincides with the group Γ. This will be taken as the starting point in Section III, where we study the Boolean algebra $\mathscr{P}(\Gamma)$ of subsets of an arbitrary group Γ. Of course, this is precisely the situation in Euclidean morphology, where the group is that of the Euclidean translations, the only difference being that in the present case the group may be noncommutative.

Subsequently we will consider the general transitive case, enabling us to analyze the examples mentioned above (the Euclidean plane with the translation-rotation group, the sphere with the rotation group) as particular cases. It turns out that the general case can be handled by embedding the object space of interest (the set $\mathscr{P}(\mathscr{X})$ of subsets of $\mathscr{X}$) into another one (the set $\mathscr{P}(\Gamma)$ of subsets of Γ), which has a simply transitive transformation group. So the results for the latter case, although rather technical, have to be developed first in depth. The theory is illustrated by various examples. The possibility of an extension to non-Boolean lattices will be considered in future work.

Now some remarks about the organization of this chapter. In Section II we first review Euclidean morphology together with some lattice-theoretical concepts; then the concept of homogeneous spaces is introduced and we give some background material needed in the following. In Section III we generalize Euclidean morphology to the Boolean lattice of all subsets of an arbitrary group, ordered by set inclusion. In particular, we generalize the classical Minkowski set operations, as well as dilations, erosions, openings, and closings that are "translation invariant" in a generalized sense, that is, invariant under certain automorphisms induced by the transformation group. A complete characterization of these operations is given and we also prove a general representation theorem for translation-invariant mappings, generalizing earlier results of Matheron [17] and Banon and Barrera [1]. We point out the connection to the theory of residuated lattices and ordered semigroups [4,5]. Section IV then develops the general transitive case. Some interesting differences from the simply transitive case show up. We also introduce a concept of "shape" that explicitly depends on the symmetry group involved. Section V contains a discussion and we point out the possible relevance for some applications.

This chapter essentially contains the material of [21,22], apart from some minor additions and slight changes of notation. To enhance readability we have deferred all the proofs to an appendix. Definitions, theorems, etc. have been consecutively numbered in each section, e.g. Remark 2.1 is followed by Theorem 2.2, etc. The end of a remark is indicated by the symbol □.

II. PRELIMINARIES

In this section we first outline some elementary concepts and results from classical Euclidean morphology (Section II.A), followed by a few general lattice-theoretical concepts that are needed below (Section II.B). Then we introduce the concept of homogeneous spaces in Section II.C.

A. Euclidean Morphology

Let E be the Euclidean space $\mathbf{R}^n$ or the discrete grid $\mathbf{Z}^n$. By $\mathcal{P}(E)$ we denote the set of all subsets of E ordered by set-inclusion, henceforth called the object space. A binary image can be represented as a subset X of E. Now E is a commutative group under vector addition: we write $x + y$ for the sum of two vectors x and y, and $-x$ for the inverse of x. Then we can define the following elementary algebraic operations:

that is, dilating an image by A gives the same result as eroding the background by $\check{A}$. To any mapping $\psi : \mathcal{P}(E) \rightarrow \mathcal{P}(E)$ we associate the *dual* mapping $\psi' : \mathcal{P}(E) \rightarrow \mathcal{P}(E)$ defined by $\psi'(X) = \{\psi(X^c)\}^c$. To avoid confusion with other forms of duality to be discussed below, we will refer to ψ' as the *Boolean dual* of ψ.

Remark 2.1. Matheron and Serra define the Minkowski subtraction of X by A as follows: $X \ominus A = \bigcap_{a \in A} X_a$. Then one has to write $X \ominus \check{A}$ in Eq. (7.6). The advantage of this definition is that the duality relation (7.8) does not involve a reflection of the structuring element. But it complicates the expression of adjunctions (see below), which is a notion persisting in lattices without complementation. □

Two characteristic properties of dilation are:

$$\text{Distributivity w.r.t. union:} \quad \left(\bigcup_{i \in I} X_i\right) \oplus A = \bigcup_{i \in I} (X_i \oplus A) \tag{7.9}$$

$$\text{Translation invariance:} \quad (X \oplus A)_h = X_h \oplus A \tag{7.10}$$

Similar properties hold for the erosion with intersection instead of union. A consequence of the distributivity property is that dilation and erosion are *increasing* mappings, that is, mappings such that for all $X, Y \in \mathcal{P}(E)$, $X \subseteq Y$ implies that $\psi(X) \subseteq \psi(Y)$.

Other important increasing transformations are the opening and closing by a structuring element A (the closing is defined slightly differently in [17,29]):

$$\text{Opening:} \quad X \bigcirc A := (X \ominus A) \oplus A = \bigcup_{h \in E} \{A_h : A_h \subseteq X\} \tag{7.11}$$

$$\text{Closing:} \quad X \bullet A := (X \oplus A) \ominus A = \bigcap_{h \in E} \{(\check{A}^c)_h : (\check{A}^c)_h \supseteq X\} \tag{7.12}$$

The opening is the union of all the translates of the structuring element that are included in the set X. Opening and closing are related by Boolean duality: $(X^c \bigcirc A)^c = X \bullet \check{A}$. A more general definition of dilations, erosions, openings, and closings will be given in the next subsection in the framework of complete lattices.

We end this review of Euclidean morphology by presenting a theorem by Matheron [17], which gives a characterization in the Euclidean case of translation-invariant increasing mappings.

Theorem 2.2. A mapping $\psi : \mathcal{P}(E) \rightarrow \mathcal{P}(E)$ is increasing and translation-invariant if and only if ψ can be decomposed as a union of erosions or, alternatively, as an intersection of dilations:

$$\psi(X) = \bigcup_{A \in \mathcal{V}(\psi)} X \ominus A = \bigcap_{A \in (\psi')} X \oplus \check{A}$$

where $\mathcal{V}(\psi) = \{A \in \mathcal{P}(E) : o \in \psi(A)\}$ is the kernel of ψ, and ψ' is the Boolean dual of ψ.

B. Lattice-Theoretical Concepts

The object spaces of interest in mathematical morphology are not restricted to Boolean algebras. For example, if one is interested in convex subsets of the plane or gray-level images, one has to introduce the notion of complete lattices. This approach has been initiated by Serra [30], Serra et al. [31], and Heijmans and Ronse [12,27]. Although the present generalization of mathematical morphology is confined to Boolean lattices, it is nevertheless advantageous to summarize a few lattice-theoretical concepts that will be needed below. The reader may want to skip this subsection at first reading and refer back to it later. For a full discussion, see [12,27]. A general introduction to lattice theory is given by Birkhoff [4].

A complete lattice $(\mathcal{L}, \leq)$ is a partially ordered set $\mathcal{L}$ with order relation $\leq$, a supremum or join operation written $\bigvee$, and an infimum or meet operation written $\bigwedge$, such that every (finite or infinite) subset of $\mathcal{L}$ has a supremum (smallest upper bound) and an infimum (greatest lower bound). In particular, there exist two universal bounds, the least element written $O_{\mathcal{L}}$ and the greatest element $I_{\mathcal{L}}$. In the case of the power lattice $\mathcal{P}(E)$ of all subsets of a set E, the order relation is set-inclusion $\subseteq$, the supremum is the union $\cup$ of sets, the infimum is the intersection $\cap$ of sets, the least element is the empty set $\emptyset$, and the greatest element is the set E itself. An *atom* is an element X of a lattice $\mathcal{L}$ such that for any $Y \in \mathcal{L}$, $O\mathcal{L} \leq Y \leq X$ implies that $Y = O\mathcal{L}$ or $Y = X$. A complete lattice $\mathcal{L}$ is called *atomic* if every element of $\mathcal{L}$ is the supremum of the atoms less than or equal to it. It is called *Boolean* if (1) it satisfies the distributivity laws $X \bigvee (Y \bigwedge Z) = (X \bigvee Y) \bigwedge (X \bigvee Z)$ and $X \bigwedge(Y \bigvee Z) = (X \bigwedge Y) \bigvee (X \bigwedge Z)$ for all $X, Y, Z \in \mathcal{L}$, and (2) every element X has a unique complement X^c, defined by $X \bigvee X^c = I_{\mathcal{L}}$, $X \bigwedge X^c = O_{\mathcal{L}}$. The power lattice $\mathcal{P}(E)$ is an atomic complete Boolean lattice, and conversely any atomic complete Boolean lattice has this form.

Since we are interested in image transformations, a main object of study is the set $\mathbb{O} := \mathcal{L}^{\mathcal{L}}$ of all maps (operators) on $\mathcal{L}$, that is, mappings $\psi: \mathcal{L} \to \mathcal{L}$. Operators are generally written in Greek letters, with γ, ϕ, δ, ε being reserved for openings, closings, dilations, and erosions. The identity operator $X \mapsto X$ is written $\mathrm{id}\mathcal{L}$. The composition of two operators ψ_1 and ψ_2 is defined by $\psi_1\psi_2(X) = \psi_1(\psi_2(X))$, $X \in \mathcal{L}$. Instead of $\psi\psi$ we write ψ^2.

The power lattice $\mathbb{O}$ inherits the complete lattice structure of $\mathcal{L}$. The ordering, supremum, and infimum in $\mathbb{O}$ are denoted by $\leq$, $\bigvee$, $\bigwedge$ as well, and for any subset $\mathfrak{Q} \subseteq \mathbb{O}$ they are defined by

Similarly, the *structural closing* $\phi_B^{\mathbf{T}}$ by the structuring element B is defined by the formula

$$\phi_B^{\mathbf{T}}(X) = \bigwedge\{\tau(B): \tau \in \mathbf{T},\ \tau(B) \geq X\}, \qquad X \in \mathcal{L} \tag{7.17}$$

As the name suggests, structural openings and closings are defined in terms of a single structuring element. Notice that (7.11) is a structural opening by the structuring element A and (7.12) is a structural closing by the structuring element $\check{A}^c$.

An important result is the following characterization of **T**-openings [12]:

Proposition 2.10. Let γ be a **T**-opening on $\mathcal{L}$. Then γ is a supremum of structural **T**-openings, that is,

$$\gamma(X) = \bigvee\{\gamma_B^{\mathbf{T}}(X): B \in \mathcal{B}\}, \qquad X \in \mathcal{L} \tag{7.18}$$

where $\mathcal{B}$ is the domain of invariance of γ. The subset $\mathcal{B} \subseteq \mathcal{L}$ in this formula may be replaced by any subset $\mathcal{B}'$ that generates $\mathcal{B}$ under **T**-translations and infinite suprema.

In the Euclidean case, a structural opening by B is also a morphological opening: $\gamma_B^{\mathbf{T}}(X) = \delta_B\varepsilon_B(X) = (X \ominus B) \oplus B$. The corresponding representation (7.18) of Euclidean openings on $\mathcal{P}(E)$ as a union of morphological openings was originally proved by Matheron [17].

C. Homogeneous Spaces

In this subsection we introduce the concept of a homogeneous space and give a brief account of prerequisites for later use. For a general introduction we refer the reader to [2,19,20,34].

Let $\mathcal{X}$ be a non-empty set. A bijection $\mathcal{X} \to \mathcal{X}$ is called a *permutation* of $\mathcal{X}$. By $\mathrm{Sym}_{\mathcal{X}}$ we denote the group of all permutations of $\mathcal{X}$. If $\mathcal{X}$ is a finite set of n elements, we write $\mathcal{S}_n$ instead of $\mathrm{Sym}_{\mathcal{X}}$. A subgroup Γ of $\mathrm{Sym}_{\mathcal{X}}$ is called a *permutation group* or *transformation group* on $\mathcal{X}$. We also say that Γ is a *group action* on $\mathcal{X}$ or that Γ *acts on* $\mathcal{X}$. Each element $g \in \Gamma$ is a mapping $\mathcal{X} \to \mathcal{X} : x \mapsto g(x)$, satisfying

$$(i)\quad gh(x) = g(h(x)), \qquad (ii)\quad e(x) = x$$

where e is the unit element of Γ (i.e., the identity mapping $x \mapsto x$, $x \in \mathcal{X}$), and gh denotes the product of two group elements g and h. The *inverse* of an element $g \in \Gamma$ will be denoted by g^{-1}. Usually we will also write gx instead of $g(x)$.

The permutation group Γ is called *transitive on* $\mathcal{X}$ if for each $x, y \in \mathcal{X}$ there is a $g \in \Gamma$ such that $gx = y$, and *simply transitive* or *regular* when this element g is *unique*. We will sometimes write *multitransitive* to mean "transitive, but not simply transitive" (the more natural phrase "multiply transitive" is avoided because it has a special technical meaning in group theory [20,34]).

Definition 2.11. A *homogeneous space* or a Γ-*set* is a pair $(\Gamma, \mathscr{X})$ where Γ is a permutation group acting transitively on $\mathscr{X}$.

Remark 2.12. In the references cited above, homogeneous spaces are defined in terms of group actions or, equivalently, permutation representations of an abstract group Γ on a set $\mathscr{X}$. If the representation of this abstract group is faithful, then Γ is isomorphic to a group of permutations on $\mathscr{X}$. In that case the above definition applies, which is more intuitive and sufficient for our purposes. □

The following result is standard in group theory [2,20,34].

Lemma 2.13. Any transitive abelian permutation group Γ is simply transitive.

Therefore our extension of mathematical morphology has to deal with two classes beyond the commutative case: the noncommutative simply transitive case and the noncommutative multitransitive case. The situation is summarized in Table 1.

If Γ acts on $\mathscr{X}$, the *stabilizer* or *isotropy group* of $x \in \mathscr{X}$ is the subgroup $\Gamma_x := \{g \in \Gamma : gx = x\}$. Stabilizers of different points form *conjugated* subgroups: $\Gamma_{gx} = g\Gamma_x g^{-1}$. Let $\omega \in \mathscr{X}$ be an arbitrary but fixed point of $\mathscr{X}$, henceforth called the *origin*. The stabilizer Γ_ω will be noted by Σ from now on:

$$\Sigma := \Gamma_\omega = \{g \in \Gamma : g\omega = \omega\} \tag{7.19}$$

Definition 2.14. The canonical projection π_ω is the mapping $\pi_\omega : \Gamma \mapsto \mathscr{X}$ given by $\pi_\omega(g) = g\omega$.

Define an equivalence relation on Γ as follows $g \sim h \Leftrightarrow h^{-1}g \in \Sigma$. So two elements g and h of the group are equivalent if $g\omega = h\omega$, that is, if they map the origin to the same point of $\mathscr{X}$. If g_x is an arbitrary element of Γ that maps the origin to x, then one easily sees that the collection of all elements equivalent to g_x is the subset $g_x\Sigma := \{g_x s : s \in \Sigma\}$ of Γ. This set is called a *left coset* with respect to the subgroup Σ. The collection of equivalence classes is called the *left coset space* associated to Σ and is denoted by Γ/Σ (read "Γ modulo Σ").

From the above it follows that there is a bijection between $\mathscr{X}$ and the coset space Γ/Σ: each point $x \in \mathscr{X}$ is identified with the coset $g_x\Sigma$. So instead of the pair $(\Gamma,\mathscr{X})$ we might as well study $(\Gamma,\Gamma/\Sigma)$. If Γ acts regularly on $\mathscr{X}$, the stabilizer reduces to the unit element of the group: $\Sigma = \{e\}$. So in that case we are left with the pair (Γ,Γ), which will be studied in Section III.

Table 1. Classification of Transformation Groups and the Associated Morphologies

	Simply Transitive	Multitransitive
Commutative	Euclidean morphology (Section II.A)	—
Noncommutative	Morphology on groups (Section III)	Morphology on homogeneous spaces (Section IV)

$$g(G \cup H) = (gG) \cup (gH),$$
$$g(G \cap H) = (gG) \cap (gH), \qquad (gG)^c = gG^c$$

and similarly for right translations. So the sets $\Gamma^\lambda := \{\lambda_g : g \in \Gamma\}$ and $\Gamma^\rho := \{\rho_g : g \in \Gamma\}$ are both automorphism groups of $\mathcal{P}(\Gamma)$.

Remark 3.1. Notice that $\lambda_g\lambda_h = \lambda_{gh}$, $\rho_g\rho_h = \rho_{hg}$, so Γ_λ is isomorphic to Γ under the correspondence $g \leftrightarrow \lambda_g$, and Γ^ρ is isomorphic to Γ under the correspondence $g \leftrightarrow \rho_g^{-1}$. This is related to the concept of the *dual* Γ^* of a group Γ, which is obtained by defining a dual product "*" in Γ by $g * h = hg$. It is easy to see that the groups Γ^λ and Γ^ρ are dual. So we only need to give proofs for invariance with respect to left translations, say. The right-invariant counterparts then follow by group duality. For easy reference we nevertheless give most results in left- and right-invariant form. □

A simple yet fundamental observation is that left and right translations commute. Summarizing:

Lemma 3.2. Let Γ be a group. Then the groups Γ^λ and Γ^ρ of left and right translations are: (1) automorphism groups of the lattice $\mathcal{P}(\Gamma)$ and (2) isomorphic to Γ. Moreover, left and right translations commute: $\gamma_g\rho_h = \rho_h\gamma_g$ for all g, $h \in \Gamma$.

Finally, we define left and right translation-invariant mappings.

Definition 3.3. A mapping $\psi : \mathcal{P}(\Gamma) \to \mathcal{P}(\Gamma)$ is called *left translation-invariant* when, for all $g \in \Gamma$, $\lambda_g\psi = \psi\lambda_g$ (i.e., $\psi(gG) = g\psi(G)$, $\forall G \in \mathcal{P}(\Gamma)$). Similarly, a mapping $\psi : \mathcal{P}(\Gamma) \to \mathcal{P}(\Gamma)$ is called *right translation-invariant* when, for all $g \in \Gamma$, $\rho_g\psi = \psi\rho_g$ (i.e., $\psi(Gg) = (\psi(G))g$, $\forall G \in \mathcal{P}(\Gamma)$).

For brevity we will speak of *left-invariant* or *λ-mappings* and *right-invariant* or *ρ-mappings*.

B. Generalization of the Minkowski Operations

Since Γ is a group, we can use the group operation to define a multiplication on subsets of Γ, which leads to the generalization of the Minkowski addition.

Definition 3.4. Let G,H be subsets of the group Γ. The *product of G by H*, denoted by $G \overset{\Gamma}{\oplus} H$, is the subset of Γ defined by

$$G \overset{\Gamma}{\oplus} H = \{gh: g \in G, h \in H\} \tag{7.21a}$$
$$G \overset{\Gamma}{\oplus} \emptyset = \emptyset \overset{\Gamma}{\oplus} G = \emptyset \tag{7.21b}$$

Here we have explicitly indicated the dependence of the product $\overset{\Gamma}{\oplus}$ on the group Γ. It is immediate that, with e the unit element of Γ, $G \overset{\Gamma}{\oplus} \{e\} = \{e\} \overset{\Gamma}{\oplus} G = G$. Notice that, in general, the product operation is noncommutative, that is:

$$G \overset{\Gamma}{\oplus} H \neq H \overset{\Gamma}{\oplus} G \tag{7.22}$$

Remark 3.5. We notice in passing that $\mathcal{P}(\Gamma)$ is a *monoid* under the multiplication $\overset{\Gamma}{\oplus}$, that is, a semigroup with unit element $\{e\}$. Since $\mathcal{P}(\Gamma)$ is a complete lattice as well, and the multiplication $\overset{\Gamma}{\oplus}$ is distributive over unions (see the next proposition), we have an example here of a so-called *complete lattice-ordered monoid* or *cl-monoid*; see Birkhoff [4] or Blyth and Janowitz [5]. □

We can write (7.21a) in the alternative forms

$$G \overset{\Gamma}{\oplus} H = \bigcup_{g \in G} gH = \bigcup_{h \in H} Gh \tag{7.23}$$

The similarity with the Minkowski addition (7.1) is clear. Next we generalize the Minkowski subtraction.

Definition 3.6. Let G, H be subsets of the group Γ. The *left residual* of G by H, denoted by $G \overset{\lambda}{\ominus} H$, is the subset of Γ defined by

$$G \overset{\lambda}{\ominus} H = \{g \in \Gamma: gH \subseteq G\} \tag{7.24a}$$

The *right residual* of G by H, denoted by $G \overset{\rho}{\ominus} H$, is the subset of Γ defined by

$$G \overset{\rho}{\ominus} H = \{g \in \Gamma: Hg \subseteq G\} \tag{7.24b}$$

Remark 3.7. The above definition of residuals is standard in the theory of *residuated semigroups*. The left residual of G by H is characterized by the property that it is the largest subset K of Γ such that when multiplied on the right by H it is included in G:

(i) $(G \overset{\lambda}{\ominus} H) \overset{\Gamma}{\oplus} H \subseteq G$
(ii) $K \overset{\Gamma}{\oplus} H \subseteq G \Rightarrow K \subseteq G \overset{\lambda}{\ominus} H$

with a similar statement for right residuals; see Birkhoff [4] or Blyth and Janowitz [5]. Definition 3.6 also applies if Γ is just a semigroup instead of a group. Of course, the fact that we assume Γ to be a group enables us to derive more specific results. As far as notation is concerned, in residuation theory one usually writes GH, $G \cdot\!.\, H$, $G \,.\!\cdot H$ instead of $G \overset{\Gamma}{\oplus} H$, and $G \overset{\lambda}{\ominus} H$, and $G \overset{\rho}{\ominus} H$, respectively. With our choice of notation we maintain some resemblance to the symbols $\oplus$, $\ominus$ used in Euclidean morphology. □

Using the group nature of Γ, we easily derive the following equivalences:

$$\begin{aligned} gH \subseteq G &\Leftrightarrow gh \in G,\ \forall h \in H \Leftrightarrow g \in Gh^{-1},\ \forall h \in H \\ &\Leftrightarrow g \in \bigcap_{h \in H} Gh^{-1} \end{aligned}$$

Hence,

$$G \overset{\lambda}{\ominus} H = \bigcap_{h \in H} Gh^{-1}, \qquad G \overset{\rho}{\ominus} H = \bigcap_{h \in H} h^{-1}G \tag{7.25}$$

$$\tilde{\delta}_H^\lambda(G) = G \overset{\Gamma}{\oplus} H, \qquad \tilde{\delta}_H^\rho(G) = H \overset{\Gamma}{\oplus} G \tag{7.27}$$

That these mappings are dilations (i.e., commute with arbitrary unions; see Definition 2.3), is readily proved by extending Proposition 3.8(a) to distributivity with respect to infinite unions. The reason for the terminology is that left (right) dilations are left (right) translation invariant; see Proposition 3.8(g).

Next we show that left and right dilations can be decomposed in terms of the automorphisms of the lattice $\mathcal{P}(\Gamma)$. From (7.23) is it immediate that

$$\tilde{\delta}_H^\lambda(G) = \bigcup_{h\in H} Gh = \bigcup_{g\in G} gH \tag{7.28a}$$

$$\tilde{\delta}_H^\rho(G) = \bigcup_{h\in H} hG = \bigcup_{g\in G} Hg \tag{7.28b}$$

Defining the union and intersection of left and right translations pointwise (i.e., by the ordering inherited from $\mathcal{P}(\Gamma)$; see Section II.B), (7.28) can be written in operator form as

$$\tilde{\delta}_H^\lambda = \bigcup_{h\in H} \rho_h, \qquad \tilde{\delta}_H^\rho = \bigcup_{h\in H} \lambda_h \tag{7.29}$$

Since left and right translations commute, we see that $\tilde{\delta}_H^\lambda$ commutes with left translations and $\tilde{\delta}_H^\rho$ commutes with right translations. Below we will show that all left- and right-invariant dilations have this form. In a similar way we define left- and right-invariant erosions.

Definition 3.13. Let $H \in \mathcal{P}(\Gamma)$. The *left erosion* $\tilde{\varepsilon}_H^\lambda$ and *right erosion* $\tilde{\varepsilon}_H^\rho$ by the structuring element H are the mappings $\mathcal{P}(\Gamma) \to \mathcal{P}(\Gamma)$ defined by

$$\tilde{\varepsilon}_H^\lambda(G) = G \overset{\lambda}{\ominus} H, \qquad \tilde{\varepsilon}_H^\rho(G) = G \overset{\rho}{\ominus} H \tag{7.30}$$

We also write *λ-dilation*/*λ-erosion* instead of left dilation/left erosion, with a similar convention for the right-invariant counterparts.

Again we decompose left and right erosions in terms of left and right translations. Just as there are two equivalent forms for the left and right dilation (7.28), one can derive two forms for the erosions. To see this take the complement of (7.28a), which by (7.27) equals the complement of $G \overset{\Gamma}{\oplus} H$:

$$\bigcap_{h\in H} G^c h = \bigcap_{g\in G} gH^c = (G \overset{\Gamma}{\oplus} H)^c = G^c \overset{\lambda}{\ominus} \check{H}$$

where we have used Lemma 3.10(g). Since this formula holds for arbitrary $G, H \in \mathcal{P}(\Gamma)$ we find (the proof for the right erosion is analogous)

$$\tilde{\varepsilon}_H^\lambda(G) = \bigcap_{h\in H} Gh^{-1} = \bigcap_{g\in G^c} g\hat{H} \tag{7.31a}$$

$$\tilde{\varepsilon}_H^\rho(G) = \bigcap_{h\in H} h^{-1}G = \bigcap_{g\in G^c} \hat{H}g \tag{7.31b}$$

where, as before, $\hat{H} = \check{H}^c$. In operator form,

$$\tilde{\varepsilon}^{\lambda}_{H} = \bigcap_{h \in H} \rho_h^{-1}, \qquad \tilde{\varepsilon}^{\rho}_{H} = \bigcap_{h \in H} \lambda_h^{-1} \tag{7.32}$$

The following lemma shows that as soon as we have proved a result for left-invariant dilations, there is a corresponding result for right-invariant dilations, as well as for left- or right-invariant erosions. First we need a definition.

Definition 3.14. Let $\psi : \mathcal{P}(\Gamma) \to \mathcal{P}(\Gamma)$ be an arbitrary mapping. The *Boolean dual* ψ' of ψ is the mapping defined by $\psi'(G) = (\psi(G^c))^c$. The reflection $\check{\psi}$ of ψ is the mapping defined by $\check{\psi}(G) = (\psi(\check{G}))^{\vee}$. The *dual reflection* of ψ is the mapping $\hat{\psi}$ defined by $\hat{\psi}(G) = (\psi(\hat{G}))^{\wedge}$.

Lemma 3.15. Let $\psi : \mathcal{P}(\Gamma) \to \mathcal{P}(\Gamma)$ be an *arbitrary mapping*. Then,

(a) $(\psi')' = (\check{\psi})^{\vee} = (\hat{\psi})^{\wedge} = \psi$.

(b) $(\check{\psi})' - (\psi')^{\vee}$.

(c) ψ is an increasing λ-mapping $\Leftrightarrow \psi'$ is an increasing λ-mapping; ψ is a dilation $\Leftrightarrow \psi'$ is an erosion. In particular, $(\tilde{\delta}^{\lambda}_{H})' = \tilde{\varepsilon}^{\lambda}_{\hat{H}}$.

(d) ψ is right-invariant $\Leftrightarrow \check{\psi}$ is left-invariant.
In particular, $(\lambda_h)^{\vee} = \rho_h^{-1}$, $(\tilde{\delta}^{\lambda}_{H})^{\vee} = \tilde{\delta}^{\rho}_{\check{H}}$, $(\tilde{\varepsilon}^{\lambda}_{H})^{\vee} = \tilde{\varepsilon}^{\rho}_{\check{H}}$.

(e) $(\tilde{\delta}^{\rho}_{h})^{\vee} = \tilde{\varepsilon}^{\rho}_{\hat{H}}$.

Remark 3.16. Here is an example of how this lemma can be used. Suppose the following statement has been proved: ψ increasing $\Rightarrow \psi'$ increasing. To show the converse, apply this statement to ψ'. Then we find: ψ' increasing $\Rightarrow \psi''$ increasing, but since the complementation operator is an involution ($\psi'' = \psi$) the proof is complete. In a similar way we can use results for left-invariant dilations to derive counterparts for right-invariant dilations (using $(\check{\psi})^{\vee} = \psi$) or for right-invariant erosions (using $(\hat{\psi})^{\wedge} = \psi$). □

Next we make a few remarks about adjunctions. By Proposition 3.8(e) we have the equivalences

$$\tilde{\delta}^{\lambda}_{H}(G) \subseteq K \Leftrightarrow G \subseteq \tilde{\varepsilon}^{\lambda}_{H}(K) \tag{7.33a}$$

$$\tilde{\delta}^{\rho}_{H}(G) \subseteq K \Leftrightarrow G \subseteq \tilde{\varepsilon}^{\rho}_{H}(K) \tag{7.33b}$$

We call $(\tilde{\varepsilon}^{\lambda}_{H}, \tilde{\delta}^{\lambda}_{H})$ a *left-invariant adjunction* (λ-*adjunction*) and similarly we call $(\tilde{\varepsilon}^{\rho}_{H}, \tilde{\delta}^{\rho}_{H})$ a *right-invariant adjunction* (ρ-*adjunction*). In particular all the properties of adjunctions as summarized in Lemma 2.6 hold for these adjunctions. So $\tilde{\varepsilon}^{\rho}_{H}$ is the supper adjoint of $\tilde{\delta}^{\lambda}_{H}$, $\tilde{\delta}^{\lambda}_{H}$ is the lower adjoint of $\tilde{\varepsilon}^{\lambda}_{H}$, etc. Lemma 3.15(c–e) expresses the relation between the duality by complementation, reflection and adjoint pairs.

creasing) on $\mathcal{P}(E)$, where E denotes Euclidean space. Following the simplified proof in [13], we extend this result here to the case $\mathcal{P}(\Gamma)$ with Γ a noncommutative group, getting as a by-product a generalization of Matheron's theorem. We only formulate the left translation-invariant case. The right translation-invariant case is obtained by left-right symmetry.

Define, for $F, G, H \in \mathcal{P}(\Gamma)$, the *left wedge transform* of G by the pair (F, H) by

$$G \overset{\lambda}{\oslash} (F, H) := \{g \in \Gamma: gF \subseteq G \subseteq gH\}$$
$$= (G \overset{\lambda}{\ominus} F) \cap (G^c \overset{\lambda}{\ominus} H^c)$$

where the second line follows from the definition Eq. (7.24a) of the left residual. In the Euclidean case, this operation is a slight modification of the hit-or-miss transform [29]. Clearly, the mapping $G \mapsto G \overset{\lambda}{\oslash} (F, H)$ is left translation-invariant. Two cases are of special interest:

1. $G \overset{\lambda}{\oslash} (F, \Gamma) = G \overset{\lambda}{\ominus} F$
2. $G \overset{\lambda}{\oslash} (\emptyset, H) = G^c \overset{\lambda}{\ominus} H^c$

Define also the "interval" $[F, H]$ between two arbitrary sets F and H as

$$[F, H] = \{G \in \mathcal{P}(\Gamma) : F \subseteq G \subseteq H\}$$

Clearly, $[F, H]$ and $G \overset{\lambda}{\oslash} (F, H)$ are both empty if $F \not\subseteq H$.

Definition 3.22. Let ψ be a mapping on $\mathcal{P}(\Gamma)$, with kernel $\mathcal{V}(\psi)$ given by Definition 3.19. The *bikernel* of ψ is defined by

$$\mathcal{W}(\psi) = \{(F, H) \in \mathcal{P}(\Gamma) \times \mathcal{P}(\Gamma): [F, H] \subseteq \mathcal{V}(\psi)\}$$

If ψ is increasing and F is an element of $\mathcal{V}(\psi)$, then the whole interval $[F, H]$ is included in $\mathcal{V}(\psi)$ if $H \supseteq F$. Similarly, if ψ is decreasing and $H \in \mathcal{V}(\psi)$, then $[F, H]$ is included in $\mathcal{V}(\psi)$ if $F \subseteq H$. Hence

$$\psi \text{ increasing}, \qquad F \in \mathcal{V}(\psi) \Rightarrow (F, \Gamma) \in \mathcal{W}(\psi) \tag{7.37a}$$
$$\psi \text{ decreasing}, \qquad H \in \mathcal{V}(\psi) \Rightarrow (\emptyset, H) \in \mathcal{W}(\psi) \tag{7.37b}$$

Now we can state:

Theorem 3.23. Representation of translation-invariant mappings. The mapping $\psi : \mathcal{P}(\Gamma) \to \mathcal{P}(\Gamma)$ is left translation-invariant if and only if

$$\psi(G) = \bigcup_{(F,H)\in\mathcal{W}(\psi)} G \overset{\lambda}{\oslash} (F, H) \tag{7.38}$$

Corollary 3.24. Representation of monotone translation-invariant mappings. If $\psi : \mathcal{P}(\Gamma) \to \mathcal{P}(\Gamma)$ is an increasing λ-mapping it can be decomposed as a union of λ-erosions, or an intersection of λ-dilations:

$$\psi(G) = \bigcup_{F \in \mathcal{V}(\psi)} G \overset{\lambda}{\ominus} F = \bigcap_{F \in \mathcal{V}(\psi')} G \overset{\Gamma}{\oplus} \check{F} \tag{7.39a}$$

where ψ' is the Boolean dual of ψ. If $\psi : \mathcal{P}(\Gamma) \to \mathcal{P}(\Gamma)$ is a decreasing λ-mapping, it can be similarly decomposed:

$$\psi(G) = \bigcup_{H \in \mathcal{V}(\psi)} G^c \overset{\lambda}{\ominus} H^c = \bigcap_{H \in \mathcal{V}(\psi')} G^c \overset{\Gamma}{\oplus} \hat{H} \tag{7.39b}$$

Recall from Section II.B that the domain of invariance of a mapping $\psi : \mathcal{P}(\Gamma) \to \mathcal{P}(\Gamma)$ is the subset of $\mathcal{P}(\Gamma)$ defined by $\mathrm{Inv}(\psi) = \{G \in \mathcal{P}(\Gamma) : \psi(G) = G\}$.

Theorem 3.25. Representation of openings. A mapping $\psi : \mathcal{P}(\Gamma) \to \mathcal{P}(\Gamma)$ is a left-invariant opening if and only if ψ has the representation

$$\psi\,(G) = \bigcup_{H \in \mathcal{B}} \tilde{\gamma}_H^{\lambda}(G) \tag{7.40}$$

for some subset $\mathcal{B}$ of the lattice $\mathcal{P}(\Gamma)$, with $\gamma_H^{\lambda}(G) = (G \overset{\lambda}{\ominus} H) \overset{\Gamma}{\oplus}$ H. Moreover, $\mathrm{Inv}(\psi)$ is the class of sets generated by $\mathcal{B}$ under left translations and infinite unions, and any subset $\mathcal{B}$ which generates $\mathrm{Inv}(\psi)$ in this way defines the same opening ψ.

IV. MATHEMATICAL MORPHOLOGY ON SPACES WITH A TRANSITIVE GROUP ACTION

In this section we study a homogeneous space $(\Gamma, \mathcal{X})$, where Γ is a group acting transitively on $\mathcal{X}$. The object space of interest is again the Boolean lattice $\mathcal{P}(\mathcal{X})$ of all subsets of $\mathcal{X}$, ordered by set inclusion. In the following we first informally sketch the basic idea of our construction of morphological operations on this homogeneous space with full invariance under the acting group Γ (Section IV.A). Then we outline in Section IV.B a general strategy of handling the transitive case by making use of the results for the simply transitive case developed above. In Section IV.C we define a "lift" and "projection" between the lattices $\mathcal{P}(\mathcal{X})$ and $\mathcal{P}(\Gamma)$ and state several properties of the associated operators π and ϑ (to be defined). Two examples are given that allow an easy visualization of the results. Section IV.D contains the characterization of set mappings on $\mathcal{P}(\mathcal{X})$ (adjunctions, openings/closings, translation-invariant mappings). Section IV.E briefly describes the situation where the group Γ has a subgroup Δ that acts transitively on $\mathcal{X}$. In this context we also introduce a concept of "shape" that explicitly takes into account the group which is involved.

A. The Basic Idea

It may be helpful to the reader if we first give some motivation and sketch the main ingredients entering into our generalization of mathematical morphology.

structuring element $Y \subseteq \mathcal{X}$. To be more precise, let ω be an arbitrary point of $\mathcal{X}$, called the "origin" (see Section II.C). The choice of ω is immaterial (this is precisely what *homogeneous* means). Since Γ is simply transitive on $\mathcal{X}$, there exists for each $x \in \mathcal{X}$ a unique group element g_x that maps ω to x. Let Y be a fixed subset of $\mathcal{X}$. Then the mappings given by (7.47) and (7.48) with $\gamma(x) = g_x Y$ are a dilation and erosion, respectively, which in addition are Γ-invariant. For an illustration see Figure 2b, where we cover the plane with copies of a triangle and take the Euclidean translation group as the acting group Γ. More examples in this category were presented in Figure 1.

Finally, we return to the example of the sphere, which belongs to the multi-transitive case. Now there exists more than one rotation that "moves" Y from an initial position to an arbitrary point x. Following the construction above, we simply attach to x *all* the rotated sets gY, where g runs over the complete collection of rotations that move the origin to x, and repeat this process for all $x \in \mathcal{X}$. Now define

$$\delta(X) := \bigcup_{x \in X} \bigcup_{\{g \in \Gamma \,:\, g\omega = x\}} gY \tag{7.49}$$

It is plausible, and will be proved below, that in this way one indeed obtains a dilation δ that is rotation invariant. Moreover, we will show that all rotation-invariant dilations are of this form. The adjoint erosion of (7.49) is formed by associating to a subset X the collection of points $y \in \mathcal{X}$ such that $gY \subseteq X$ for *all* rotations $g \in \Gamma$ that move the origin to y. More details on the spherical case can be found in [23].

The construction sketched above for the sphere can in fact be generalized to any homogeneous space $(\Gamma, \mathcal{X})$, for example, to the plane with the translation-rotation group $E^+(3)$ as the acting group. For the latter case a sketch of the Γ-dilation by a structuring element Y consisting of a line segment is given in Figure 2c. From this figure it is clear that the line segment may be replaced by a disk, which is rotation invariant. More generally we will see below that dilations/erosions on any homogeneous space can be reduced to a form involving structuring elements that are invariant under all elements of Γ that leave the origin ω invariant.

In contrast to the Euclidean case, the dilations and erosions on the lattice $\mathcal{P}(\mathcal{X})$ constructed above are not the building blocks for other morphological transformations such as the opening (7.42). For this purpose we have to introduce dilations and erosions between the distinct lattices $\mathcal{P}(\mathcal{X})$ and $\mathcal{P}(\Gamma)$. Before we can explain this, we need to develop the ideas sketched above in full detail, to which we proceed now.

B. General Strategy

From now on we will write $\mathcal{L}$ instead of $\mathcal{P}(\mathcal{X})$ and $\tilde{\mathcal{L}}$ instead of $\mathcal{P}(\Gamma)$. It will be convenient to introduce a notation that clearly distinguishes between subsets of

$\mathcal{X}$ and subsets of Γ, as well as between mappings on $\mathcal{L}$ and mappings on $\tilde{\mathcal{L}}$; see Table 2.

Elements of the set $\mathcal{X}$ will be denoted by lower case letters x, y, z, x′, y′, z′; subsets of $\mathcal{X}$ by the corresponding capitals X, Y, Z, X′, Y′, Z′. For the group Γ we use the notation of the previous section, i.e. g, h, k, g′, h′, k′ for the group elements and G, H, K, G′, H′, K′ for the subsets of Γ. Mappings $\tilde{\mathcal{L}} \to \tilde{\mathcal{L}}$ will be denoted by a tilde, e.g., $\tilde{\psi}$, to distinguish the from mappings $\psi : \mathcal{L} \to \mathcal{L}$

As explained in Section II.C, the fact that Γ is a group acting on $\mathcal{X}$ means that each $g \in \Gamma$ defines a mapping $\mathcal{X} \to \mathcal{X} : x \mapsto gx$. This mapping can be extended to subsets of $\mathcal{X}$ as in Section III.A: define, for each $g \in \Gamma$, a mapping γ_g by

$$\gamma_g\colon \mathcal{L} \to \mathcal{L}, \qquad \gamma_g(X) := \{gx\colon x \in X\} \tag{7.50}$$

where instead of $\gamma_g(X)$ we usually write gX. We call the mapping (7.50) translation by g. On $\tilde{\mathcal{L}} = \mathcal{P}(\Gamma)$ we have the left and right translations,

$$\lambda_g\colon \tilde{\mathcal{L}} \to \tilde{\mathcal{L}}, \qquad \lambda_g(H) = gH := \{gh\colon h \in H\} \tag{7.51a}$$

$$\rho_g\colon \tilde{\mathcal{L}} \to \tilde{\mathcal{L}}, \qquad \rho_g(H) = Hg := \{hg\colon h \in H\} \tag{7.51b}$$

It is easy to check that γ_g commutes with unions, intersections, and complements on $\mathcal{L}$ for each $g \in \Gamma$, just as is the case for λ_g and ρ_g on $\tilde{\mathcal{L}}$. So the group Γ induces:

1. An automorphism group of $\mathcal{L} = \mathcal{P}(\mathcal{X})$, acting transitively on $\mathcal{X}$
2. Two mutually commuting automorphism groups of $\tilde{\mathcal{L}} = \mathcal{P}(\Gamma)$, acting *simply* transitively on Γ

In order to avoid a proliferation of symbols we will replace the notation (7.43) for the action of a subset G of Γ on a subset X of $\mathcal{X}$ by

$$GX := \bigcup_{g \in G} gX \tag{7.52}$$

We define a mapping on $\mathcal{L}$ to be *translation-invariant* or a Γ*-mapping* when it commutes with Γ-translations; see (7.41). On $\tilde{\mathcal{L}}$ we have left and right translation invariance; see Section III.A. We also need to study set mappings between $\mathcal{L}$ and $\tilde{\mathcal{L}}$. To define translation invariance for such mappings, we use *left* translations on

Table 2. Notation for the Elements and Subsets of $\mathcal{X}$ and Γ

Reference space	Elements	Subsets	Lattice
$\mathcal{X}$	x, y, z, x', y', z'	X, Y, Z, X', Y', Z'	$\mathcal{L} = \mathcal{P}(\mathcal{X})$
Γ	g, h, k, g', h', k'	G, H, K, G', H', K'	$\tilde{\mathcal{L}} = \mathcal{P}(\Gamma)$

lattice as the set $\mathscr{X}$ and where the group Γ is the hexagonal group $\mathscr{H}$ (see Example 2.19). The allowed rotations are over an integer multiple of $\pi/3$ and a coset is represented by the six unit vectors corresponding to the allowed rotations. On the hexagonal lattice, subsets of $\mathscr{X}$ are indicated by collections of heavy dots and subsets of $\mathscr{H}$ by heavy dots with one or more unit vectors attached to them.

Example 4.4. The rotation group on the sphere [23]. Consider the unit 2-sphere $\mathscr{X} = S^2$, acted upon by the three-dimensional rotation group $\Gamma = \mathrm{SO}(3)$; see Example 2.20. As in the previous example, define a pointer p to be a pair $(x, \mathbf{v})$, where $x \in S^2$ and $\mathbf{v}$ a unit tangent vector at the base-point x. Choose the north pole $\mathcal{N}$ as the origin of the sphere, and define a base-pointer b to be an arbitrarily chosen (fixed) pointer with base-point $\mathcal{N}$. Then again any pointer p represents a unique rotation, i.e., the one that maps b to p. The stabilizer Σ and the left coset $g_x\Sigma$ are represented by the collection of unit tangent vectors attached to $\mathcal{N}$ and x, respectively; see Figure 5, where we have drawn six representative pointers belonging to $g_x\Sigma$.

Having introduced this geometric picture, it is now easy to visualize the action of the mappings ϑ and π defined above. Any subset G of Γ is represented by a set of pointers and π maps G to the set of base-points of the pointers in G (Figure 6a). Conversely, ϑ maps a subset X of $\mathscr{X}$ to the set of pointers in Γ that have their base-points in X (Figure 6b).

In a moment we will list several properties of the lift ϑ and the projection π that will enable us to settle the case of a transitive group on $\mathscr{X}$ by making use of the results for the simply transitive case. But first we need to introduce a special dilation and corresponding erosion.

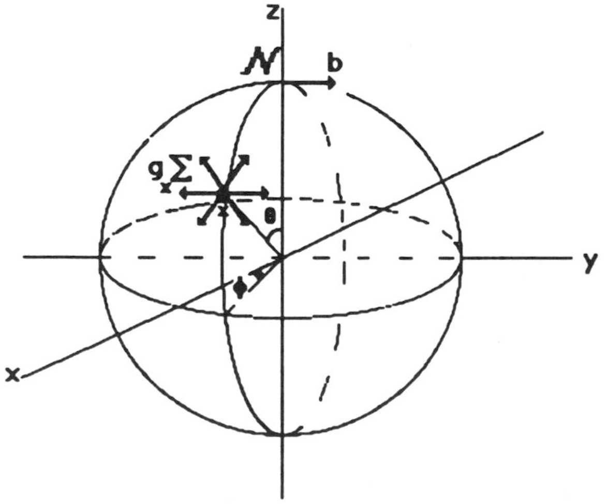

Figure 5. Representation of elements from the rotation group SO(3). b, base-pointer; $g_x\Sigma$, left coset representing all rotations that map $\mathcal{N}$ to x.

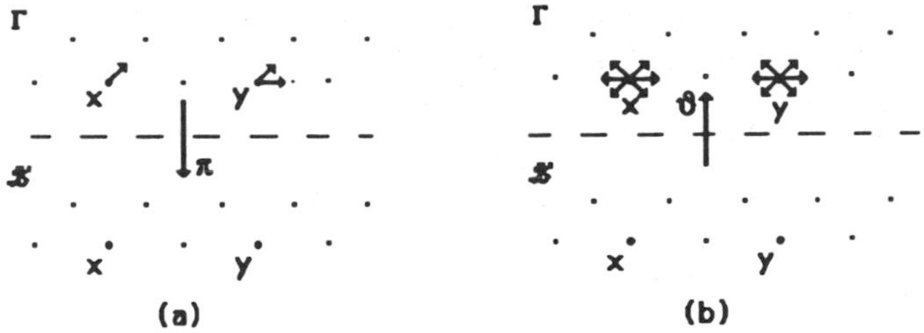

Figure 6. (a) Action of π on a subset of Γ; (b) action of θ on a subset of $\mathcal{X}$.

Definition 4.5. The dilation $\tilde{\delta}_\Sigma : \tilde{\mathcal{L}} \to \tilde{\mathcal{L}}$ and erosion $\tilde{\varepsilon}_\Sigma : \tilde{\mathcal{L}} \to \tilde{\mathcal{L}}$ by the stabilizer Σ are the mappings defined by

$$\tilde{\delta}_\Sigma(G) = G \overset{\Gamma}{\oplus} \Sigma, \qquad \tilde{\varepsilon}_\Sigma(G) = G \overset{\lambda}{\ominus} \Sigma \tag{7.58}$$

Here we have used the set-product and left residual on $\tilde{\mathcal{L}} = \mathcal{P}(\Gamma)$; see Section III.B. Clearly, $(\tilde{\varepsilon}_\Sigma, \tilde{\delta}_\Sigma)$ is a λ-adjunction on $\tilde{\mathcal{L}}$. Furthermore, we have

Lemma 4.6. The *adjunction* $(\tilde{\varepsilon}_\Sigma, \tilde{\delta}_\Sigma)$ satisfies

$$\tilde{\varepsilon}_\Sigma = \tilde{\varepsilon}_\Sigma^2 = \tilde{\delta}_\Sigma \tilde{\varepsilon}_\Sigma \tag{7.59a}$$

$$\tilde{\delta}_\Sigma = \tilde{\delta}_\Sigma^2 = \tilde{\varepsilon}_\Sigma \tilde{\delta}_\Sigma \tag{7.59b}$$

This lemma says that $\tilde{\varepsilon}_\Sigma$ is not only an erosion but also a (morphological) opening, and $\tilde{\delta}_\Sigma$ is not only a dilation but also a (morphological) closing.

The effect of the closing $\tilde{\delta}_\Sigma$ on a subset G of Σ is to make G "Σ-closed," that is, invariant under right multiplication by Σ. To put it differently, $\tilde{\delta}_\Sigma$ augments G by all group elements that are equivalent to some $g \in G$ (recall from Section II.C that two group elements $g, h \in \Gamma$ are called equivalent if $g\omega = h\omega$). Pictorially, any pointer $p = (x, \mathbf{v})$ is extended to the set of pointers $p\Sigma := \{(x, \mathbf{v}') : \mathbf{v}' \in S^1\}$. Similarly, the opening $\tilde{\varepsilon}_\Sigma$ extracts from a subset G of Γ all the "complete cosets" present in G, that is, the subset $G^* \subseteq G$ which is such that if $g \in G^*$, all the elements equivalent to g are also in G^*.

We also need to introduce a modified projection as follows.

Definition 4.7. Let π be the projection (7.54) and $\tilde{\varepsilon}_\Sigma$ the erosion (7.59a). Then $\pi_\Sigma : \tilde{\mathcal{L}} \to \mathcal{L}$ is the *modified projection* defined by

$$\pi_\Sigma = \pi \tilde{\varepsilon}_\Sigma \tag{7.60}$$

It follows from this definition that $\pi_\Sigma(G) = \{g\omega : g\Sigma \subseteq G\}$, so the projection π_Σ maps $G \in \tilde{\mathcal{L}}$ to the subset of $\mathcal{L}$ consisting of only those base-points of pointers in G to which a complete set of unit vectors is attached.

The next proposition contains a collection of properties of the operators ϑ and π.

$$\varepsilon(X) = \varepsilon_Y^\Gamma(X) = \bigcap_{g \in \vartheta(X^c)} g\,\hat{Y}^* \tag{7.63b}$$
$$= \{y \in \mathscr{X}: gY \subseteq X \text{ for all } g \in \vartheta_\omega(y)\}$$

where $\check{Y}^* = \pi(\check{\vartheta}(Y))$ and $\hat{Y}^* = (\check{Y}^*)^c$. Here $\vartheta_\omega(y)$ is the coset representing y, that is, the collection of all group elements that map ω to y.

Here we have written $\check{\vartheta}(X)$ instead of $(\vartheta(X))^\vee$, and $Y \Uparrow X$ (read: Y hits X) is a shorthand notation for $Y \cap X \neq \emptyset$.

The proposition above shows that any dilation on $\mathscr{L}$ can be reduced to a dilation δ_Y^Γ involving a Σ-*invariant* structuring element Y; the same is true for erosions. Also, in (7.63a) we may replace "some" by "all": since $\check{Y}$ is Σ-invariant (easy to show), $g\check{Y}^*$ will hit X for *all* $g \in \vartheta_\omega(y)$ as soon as it hits X for *some* element g_y of this coset. For example, in the case of the Euclidean motion group acting on the plane, it would be natural to take for g_y the Euclidean translation τ_y that maps ω to y. Also, (7.63b) may equivalently be written as $\varepsilon_Y^\Gamma(X) = \{y \in \mathscr{X}: g_y\overline{Y} \subseteq X\}$.

A second consequence of the corollary is that the morphological opening $\varepsilon_Y^\Gamma\delta_Y^\Gamma$ and closing $\delta_Y^\Gamma\varepsilon_Y^\Gamma$ associated to an adjunction $(\varepsilon_Y^\Gamma, \delta_Y^\Gamma)$ with Y an arbitrary subset of $\mathscr{X}$ are also equivalent to the morphological opening or closing by the structuring element $\overline{Y}$. This raises the question of how to decompose Γ-openings that are not reducible to openings by a Σ-invariant structuring element. Consider the *structural* opening and closing by a subset Y of $\mathscr{X}$ defined by

$$\gamma_Y^\Gamma(X) = \bigcup_{g\in\Gamma}\{gY: gY \subseteq X\} \tag{7.64a}$$

$$\phi_Y^\Gamma(X) = \bigcap_{g\in\Gamma}\{gY: gY \supseteq X\} \tag{7.64b}$$

In other words, $\gamma_Y^\Gamma(X)$ is the union of all translates gY of Y that are included in X. For example, let X be a union of line segments of varying sizes in the plane and Y a line segment of size L with center at the origin. Take Γ equal to the translation-rotation group $\mathcal{M} := E^+(3)$. Then $\gamma_Y^{\mathcal{M}}(X)$ consists of the union of all segments in X of size L or larger, but $\delta_Y^{\mathcal{M}}\varepsilon_Y^{\mathcal{M}}(X) = \gamma_{\Sigma Y}^{\mathcal{M}}(X) = \emptyset$, since $\overline{Y}$ is a disk of radius $L/2$ and does not fit anywhere in X; see Figure 7.

So in general we can not build the opening γ_Y^Γ from a Γ-erosion ε_Y^Γ on $\mathscr{P}(\mathscr{X})$ followed by a Γ-dilation δ_Y^Γ on $\mathscr{P}(\mathscr{X})$. However, Theorem 2.7 of [5] guarantees that, given an opening or closing ψ on a lattice $\mathscr{L}$, there exists another lattice $\mathscr{L}'$ such that ψ can be decomposed into erosions and dilations between $\mathscr{L}$ and $\mathscr{L}'$. In the present case the situation is clarified by the next proposition.

Proposition 4.14. Decomposition of structural openings and closings. The structural opening $\gamma_Y^\Gamma : \mathscr{L} \to \mathscr{L}$ defined by (7.64a) is the projection of the Γ-opening $\tilde{\gamma}_{\vartheta(Y)}$ on $\tilde{\mathscr{L}}$, that is,

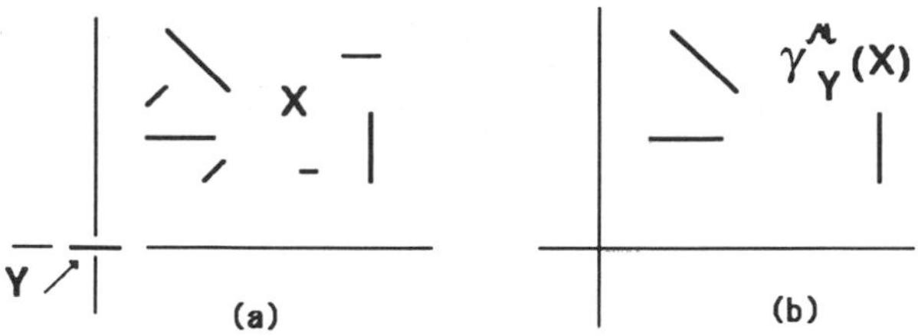

Figure 7. (a) X, a subset of the plane consisting of line segments; Y, a segment of size L at the origin. (b) opening $\gamma_Y^{\mathcal{M}}(X)$ by the Euclidean motion group $\mathcal{M}$. The opening $\delta_Y^{\mathcal{M}}\varepsilon_Y^{\mathcal{M}}(X) = \gamma_Y^{\mathcal{M}}(X)$ is empty.

$$\gamma_Y^\Gamma(X) = (\pi\tilde{\delta}_{\vartheta(Y)}\tilde{\varepsilon}_{\vartheta(Y)}\vartheta)(X) = \pi[\{\vartheta(X) \overset{\lambda}{\ominus} \vartheta(Y)\} \overset{\Gamma}{\oplus} \vartheta(Y)] \tag{7.65}$$

with $(\tilde{\varepsilon}_{\vartheta(Y)}, \tilde{\delta}_{\vartheta(Y)})$ the left-invariant adjunction $\tilde{\mathcal{L}}$ with structuring element $\vartheta(Y)$. Equivalently, γ_Y^Γ is the product of a Γ-erosion $\varepsilon^\uparrow : \mathcal{L} \to \tilde{\mathcal{L}}$ followed by a Γ-dilation $\delta^\downarrow : \tilde{\mathcal{L}} \to \mathcal{L}$, where $(\varepsilon^\uparrow, \delta^\downarrow) := (\tilde{\varepsilon}_{\vartheta(Y)}\vartheta, \pi\tilde{\delta}_{\vartheta(Y)})$ is a Γ-adjunction between $\mathcal{L}$ and $\tilde{\mathcal{L}}$.

Similarly, the structural closing $\phi_Y^\Gamma : \mathcal{L} \to \mathcal{L}$ defined by (7.64b) has the decomposition

$$\phi_Y^\Gamma(X) = (\pi_\Sigma\tilde{\varepsilon}_{\check{\vartheta}(Y^c)}\tilde{\delta}_{\check{\vartheta}(Y^c)}\vartheta)(X) = \pi[\{\vartheta(X) \overset{\Gamma}{\oplus} \check{\vartheta}(Y^c)\} \overset{\lambda}{\ominus} \check{\vartheta}(Y^c)] \tag{7.66}$$

Equivalently, ϕ_Y^Γ has the decomposition $\phi_Y^\Gamma = \varepsilon^\downarrow\delta^\uparrow$ where $(\varepsilon^\downarrow, \delta^\uparrow) := (\pi_\Sigma\tilde{\varepsilon}_{\check{\vartheta}(Y^c)}, \tilde{\delta}_{\check{\vartheta}(Y^c)}\vartheta)$ is a Γ-adjunction between $\tilde{\mathcal{L}}$ and $\mathcal{L}$

Remark 4.15. We could have introduced the following generalization of the Minkowski operations on $\mathcal{P}(\mathcal{X})$:

$$X \overset{\Gamma}{\oplus} Y := \pi[\vartheta(X) \overset{\Gamma}{\oplus} \vartheta(Y)], \qquad X \overset{\lambda}{\ominus} Y := \pi_\Sigma[\vartheta(X) \overset{\lambda}{\ominus} \vartheta(Y)]$$

where $\overset{\Gamma}{\oplus}$ and $\overset{\lambda}{\ominus}$ denote the set-product and left-residual on $\mathcal{P}(\Gamma)$, respectively. But since the basic building blocks for structural Γ-openings and Γ-closings are not the dilation $X \overset{\Gamma}{\oplus} Y$ and erosion $X \overset{\lambda}{\ominus} Y$ on $\mathcal{L}$, but dilations/erosions between $\mathcal{L}$ and $\tilde{\mathcal{L}}$, we have refrained from doing this. □

By Proposition 2.10 (see also [27]), every Γ-opening $\gamma : \mathcal{L} \to \mathcal{L}$ is a union of structural openings γ_Y^Γ, where Y ranges over a subset $\mathcal{Y} \subseteq \mathcal{L}$. Combining this with Proposition 4.14, we therefore can decompose any Γ-opening into Γ-openings of the form $\pi\tilde{\delta}_{\vartheta(Y)}\tilde{\varepsilon}_{\vartheta(Y)}\vartheta$. The formulation of the precise result, paralleling Theorem 3.25, is left to the reader.

Finally, we discuss the characterization of increasing Γ-mappings. We first need:

where $\mathcal{W}(\psi) = \{(W,Y) \in \mathcal{P}(\mathcal{X}) \times \mathcal{P}(\mathcal{X}): [W,Y] \subseteq \mathcal{V}(\psi)\}$ with $\mathcal{V}(\psi)$ the kernel of ψ.

E. The Role of Symmetry Groups in Shape Description

In this final subsection we make a few succinct remarks that are relevant in the present context and on which further research is needed.

A first remark concerns the problem how to define "shape," which is known to present great difficulties and is a recurring theme in the image processing literature. Often, shape is defined as referring to those properties of geometric figurs which are invariant under the Euclidean similarity group [14]. Intuitively, one first has to bring figures to a standard location, orientation, and scale before being able to "compare" them. Now it is not necessary to restrict oneself to the similarity group, although in the absence of any form of group invariance there is no way at all to compare figures. In the present context the following definition seems appropriate.

Definition 4.20. Let $\mathcal{X}$ be a set, Γ a group acting on $\mathcal{X}$. Two subsets X, Y of $\mathcal{X}$ are said to have *the same shape with respect to* Γ, or *the same* Γ*-shape*, if they are Γ-equivalent, meaning that there is a $g \in \Gamma$ such that $Y = gX$. If no such $g \in \Gamma$ exist, X and Y are said to have *different* Γ*-shape*.

In essence this definition goes back to F. Klein's Erlanger Programm (1872), which considers geometry to be the study of transformation groups and the properties invariant under these groups [15]. So in Euclidean morphology, all translates of a set X by the Euclidean translation group $\mathcal{T}$ have the same $\mathcal{T}$-shape. Adding rotations to get the Euclidean motion group $\mathcal{M}$, rotated versions of X or its translates have the same $\mathcal{M}$-shape as X. Extreme cases are (1) $\Gamma = \{\text{id}\}$, so that all sets have different shape, and (2) $\Gamma = \text{Sym}_{\mathcal{X}}$, in which case all sets with the same cardinality have the same shape.

A related observation is the following: if the group Γ contains two subgroups Δ_1 and Δ_2 such that any $g \in \Gamma$ has a unique decomposition $g = d_1 d_2$ ($d_1 \in \Delta_1$, $d_2 \in \Delta_2$), where Δ_1 acts itself transitively on $\mathcal{X}$, we can accordingly decompose Γ-mappings. For example, in the case of the Euclidean motion group $\mathcal{M}$, each $g \in \mathcal{M}$ has the form $g = tr$, where $t \in \mathcal{T}$ and $r \in \mathcal{R}$, with $\mathcal{T}$ and $\mathcal{R}$ the two-dimensional translation and rotation group, respectively. It is easy to see that in this case every $\mathcal{M}$-dilation δ has the form $\delta(X) = \delta_B^{\mathcal{M}}(X) := X \oplus B$, where B is an $\mathcal{R}$-invariant structuring element and $\oplus$ denotes the Euclidean Minkowski addition (7.1). Also, we have the following decomposition of the structural $\mathcal{M}$-opening $\gamma_Y^{\mathcal{M}}$ by the structuring element Y:

$$\begin{aligned}\gamma_Y^{\mathcal{M}}(X) &:= \bigcup_{g \in \mathcal{M}} \{gY: gY \subseteq X\} = \bigcup_{r \in \mathcal{R}} \bigcup_{t \in \mathcal{T}} \{trY: trY \subseteq X\} \\ &= \bigcup_{r \in \mathcal{R}} \gamma_{rY}^{\mathcal{T}}(X),\end{aligned}$$

where $\gamma_{rY}^{\mathcal{T}}(X) = [X \ominus (rY)] \oplus (rY)$ is the Euclidean opening by the (not necessarily $\mathcal{R}$-invariant) structuring element rY. This shows that one can perform the opening $\gamma_Y^{\mathcal{M}}$ by carrying out the Euclidean openings $\gamma_{rY}^{\mathcal{T}}$ for all rotated versions of the structuring element Y and taking the union of the results. On the sphere such a decomposition cannot be found, since there is no proper subgroup of SO(3) that acts transitively on the sphere [2, Chapter 1.8].

A decomposition that is possible for any homogeneous space $(\Gamma, \mathcal{X})$ is the following. Consider the partitioning into cosets of the group Γ by the stabilizer Σ, and choose, for all $x \in \mathcal{X}$, a representative g_x for the coset associated to x. Let Δ denote the collection (in general not a group) of all representatives:

$$\Delta = \{g_x : x \in \mathcal{X}\}$$

Then the dilation and erosion by the structuring element Y can be written

$$\delta_Y^\Gamma(X) = \bigcup_{g \in \vartheta(X)} gY = \bigcup_{x \in X} \bigcup_{s \in \Sigma} g_x sY = \bigcup_{x \in X} g x \bar{Y} =: \delta_{\bar{Y}}^\Delta(X) \tag{7.68a}$$

$$\begin{aligned} \varepsilon_Y^\Gamma(X) &= \{y \in \mathcal{X} : gY \subseteq X \ \forall g \in \vartheta_\omega(y)\} \\ &= \{y \in \mathcal{X} : g_y \bar{Y} \subseteq X\} =: \varepsilon_{\bar{Y}}^\Delta(X) \end{aligned} \tag{7.68b}$$

where, for any $Z \subseteq \mathcal{X}$,

$$\delta_Z^\Delta(X) := \bigcup_{x \in \mathcal{X}} g_x Z, \qquad \varepsilon_Z^\Delta(X) := \{y \in \mathcal{X} : g_y Z \subseteq X\} \tag{7.69}$$

So Γ-dilations/Γ-erosions by the structuring element Y are identical to dilations/erosions "with respect to Δ" by the structuring element $\bar{Y}$.

In the case of structural opening by Y we have

$$\begin{aligned} \gamma_Y^\Gamma(X) &= \bigcup_{g \in \Gamma} \{gY : gY \subseteq X\} = \bigcup_{s \in \Sigma} \bigcup_{x \in \mathcal{X}} \{g_x sY : g_x sY \subseteq X\} \\ &= \bigcup_{s \in \Sigma} \gamma_{sY}^\Delta(X) \end{aligned} \tag{7.70}$$

where γ_Z^Δ is the opening defined by

$$\gamma_Z^\Delta(X) := \bigcup_{x \in \mathcal{X}} \{g_x Z : g_x Z \subseteq X\} \tag{7.71}$$

In general (i.e., if Z is not Σ-invariant) neither the dilation δ_Z^Δ/erosion ε_Z^Δ nor the opening γ_Z^Δ possesses invariance properties, unless Δ is a group (i.e., a subgroup of Γ). In that case γ_Z^Δ is invariant under Δ-translations (e.g., $\delta_Z^\Delta(gX) = g\delta_Z^\Delta(X)$ for all $g \in \Delta$), but not under translations by elements $s \in \Sigma$. An example has been given above for the Euclidean motion group, where $\Delta = \mathcal{T}$ and $\Sigma = \mathcal{R}$.

V. DISCUSSION AND EXAMPLES

In this chapter we have generalized Euclidean morphology to arbitrary homogeneous spaces $(\mathscr{X}, \Gamma)$, where the group Γ acting on $\mathscr{X}$ is not necessarily commutative. The case where Γ acts simply transitively on $\mathscr{X}$, considered in Section III, leads to the study of transformations of subsets of an arbitrary group that are invariant under either left or right group translations. The general case where Γ acts transitively on $\mathscr{X}$ has been treated in Section IV by (1) mapping the subsets of $\mathscr{X}$ to subsets of Γ, (2) using the results for the simply transitive case, and (3) projecting back to the original space. The main result is that the scope of mathematical morphology is widened to situations where a noncommutative group is involved. Examples are the translation-rotation group acting on the plane, the rotation group acting on the sphere, or a subgroup of the symmetric group S_n acting on a finite set of n points. Although the emphasis of our work has clearly been on the mathematical framework, we want to finish by mentioning a few areas of possible practical relevance. As indicated in the introduction, we expect that other applications will be found as well.

A first example is that of the search for structures in a graph; see [31, p. 90]. As in Example 2.23, let E' be a set of vertices of a graph and E the complete graph generated by E'. Then $\mathscr{P}(E)$ is the set of subgraphs of E. The group Γ acting on E is generated by the set of all bijections of E' to itself. Let B be an arbitrary subgraph of E, which plays the role of the structuring element. Then the opening γ_B^Γ applied to a graph $X \subseteq E$ is the union of all the subgraphs of X that are isomorphic to B; see Figure 4.9 of [31].

A second example, which occurs in the problem of motion planning for robots, has been considered in great detail in [24]. The problem is to find a path for a robot moving in a plane E with obstacles. Since a robot has a finite size, one can find allowed positions of the (arbitrarily chosen) center of the robot by an erosion $\varepsilon_B^{\mathscr{T}}$ of the obstacle-free space E_{free}, where the structuring element B is the robot itself and $\mathscr{T}$ is the Euclidean translation group. Here we assume that only translations of the robot are allowed. Equivalently, one may perform the dilation $\delta_{\check{B}}^{\mathscr{T}}$ by the reflected set $\check{B}$ on the set E_{ob} of obstacles to find the *forbidden* positions of the center of the robot. In this connection we refer to related work by Ghosh [6], on spatial planning and other problems using the classical Minkowski operations (i.e., only Euclidean translations allowed).

If the robot has rotational degrees of freedom, one has to perform dilations with all rotated versions of the robot (Verwer [35]). In the framework of this chapter, the situation can be described as follows: given a set B (the robot), find the collection of all locations in the plane and all orientations of the robot at those locations such that the displaced robot fits into the obstacle-free space E_{free}. In the terminology of Section IV.C, this is precisely the erosion $\varepsilon_B^{\mathscr{M}\uparrow}(E_{\text{free}}) := \{(t, r) \in$

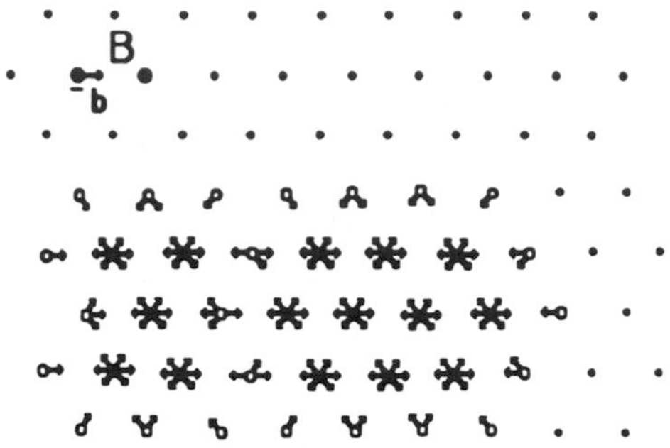

Figure 8. The forbidden positions and orientations for a robot with translational and rotational degrees of freedom are found by the dilation $\delta^{\mathcal{H}}_{B\uparrow}(E_{\text{ob}})$ of the obstacles by the robot B, where $\mathcal{H}$ is the hexagonal group. Heavy dots, the obstacle space; arrows attached to heavy and open dots, the dilated set $\delta^{\mathcal{H}}_{B\uparrow}(E_{\text{ob}})$ of forbidden states. The underlining in B indicates the origin and b is the base pointer (taken from [24]).

'positions' by taking the complement of $\varepsilon^{\mathcal{M}}_{\mathcal{M}} \uparrow (E_{\text{free}})$, which equals the dilation (see [24]) $\delta^{\mathcal{M}}_{B} \uparrow (E_{\text{ob}}) = \bigcup_{b \in E_{ob}} \tau_b \check{\vartheta}(B)$, where τ_b is the unique Euclidean translation that maps the origin to b. An example for the hexagonal grid is given in Figure 8.

ACKNOWLEDGMENT

We thank Dr. C. Ronse (Philips, Brussels) and Dr. H. J. A. M. Heijmans (CWI, Amsterdam) for a critical reading of the original manuscripts and many useful suggestions.

APPENDIX: PROOFS

Proof of Proposition 3.8. In cases where pairs of statements occur which differ only by left-right symmetry, we prove only one of them. In all proofs we use without comment that translations commute with unions and intersections.

(a) $$G \overset{\Gamma}{\oplus} (H \cup K) = \bigcup_{g \in G} g(H \cup K) = \bigcup_{g \in G} (gH \cup gK)$$
$$= (\bigcup_{g \in G} gH) \bigcup (\bigcup_{g \in G} gK) = (G \overset{\Gamma}{\oplus} H) \cup (G \overset{\Gamma}{\oplus} K),$$
which proves the left distributivity of the set product.

(b) Using that multiplication in a group is associative, we find
$$(G \overset{\Gamma}{\oplus} H) \overset{\Gamma}{\oplus} K = \bigcup_{g \in G, h \in H, k \in K} (gh)k = \bigcup_{g \in G, h \in H, k \in K} g(hk)$$
$$= G \overset{\Gamma}{\oplus} (H \overset{\Gamma}{\oplus} K)$$

(c) $(G \cap H) \overset{\lambda}{\ominus} K = \bigcap_{k\in K}(G \cap H)k^{-1} = \bigcap_{k\in K}(Gk^{-1}) \cap (Hk^{-1})$

$= (\bigcap_{k\in K} Gk^{-1}) \bigcap (\bigcap_{k\in K} Hk^{-1})$

$= (G \overset{\lambda}{\ominus} K) \cap (H \overset{\lambda}{\ominus} K).$

(d) $G \overset{\lambda}{\ominus} (H \cup K) = \bigcap_{m\in H\cup K} Gm^{-1} = (\bigcap_{m\in H} Gm^{-1}) \bigcap (\bigcap_{m\in K} Gm^{-1})$

$= (G \overset{\lambda}{\ominus} H) \cap (G \overset{\lambda}{\ominus} K).$

(e) $G \overset{\Gamma}{\oplus} H \subseteq K \Leftrightarrow \forall h \in H\colon Gh \subseteq K \Leftrightarrow \forall h \in H : G \subseteq Kh^{-1}$

$\Leftrightarrow G \subseteq \bigcap_{h\in H} Kh^{-1} = K \overset{\lambda}{\ominus} H$. Similarly,

$G \overset{\Gamma}{\oplus} H \subseteq K \Leftrightarrow \forall g \in G\colon gH \subseteq K \Leftrightarrow \forall g \in G : H \subseteq g^{-1}K$

$\Leftrightarrow H \subseteq \bigcap_{g\in G} g^{-1}K = K \overset{\rho}{\ominus} G.$

(f) $(G \overset{\lambda}{\ominus} H) \overset{\lambda}{\ominus} K = \bigcap_{k\in K} (G \overset{\lambda}{\ominus} H)k^{-1} = \bigcap_{k\in K} (\bigcap_{h\in H} Gh^{-1})k^{-1}$

$= \bigcap_{h\in H, k\in K} G(kh)^{-1} = \bigcap_{m\in K \overset{\Gamma}{\oplus} H} Gm^{-1} = G \overset{\lambda}{\ominus} (K \overset{\Gamma}{\oplus} H).$

In a similar way one proves that $(G \overset{\rho}{\ominus} H) \overset{\rho}{\ominus} K = G \overset{\rho}{\ominus} (H \overset{\Gamma}{\oplus} K)$. Finally,

$(G \overset{\rho}{\ominus} H) \overset{\lambda}{\ominus} K = \bigcap_{h\in H, k\in K} h^{-1}Gk^{-1} = (G \overset{\lambda}{\ominus} K) \overset{\rho}{\ominus} H.$

(g) Follows from (b) and the identities (7.26).

(h) $H \overset{\lambda}{\ominus} (gK) = \bigcap_{m\in gK} Hm^{-1} = \bigcap_{k\in K} H(gk)^{-1} = \bigcap_{k\in K} (Hk^{-1})g^{-1}$

$= (\bigcap_{k\in K} Hk^{-1})g^{-1} = (H \overset{\lambda}{\ominus} K)g^{-1}$. The other results are proved similarly.

Proof of Lemma 3.10. We only prove (g)–(i). The other items are obvious.

(g) $(G \overset{\Gamma}{\oplus} H)^c = (\bigcup_{h\in H} Gh)^c = \bigcap_{h\in H} G^c h = G^c \overset{\lambda}{\ominus} \check{H}$. Also,

$(G \overset{\Gamma}{\oplus} H)^c = (\bigcup_{g\in G} gH)^c = \bigcap_{g\in G} gH^c = H^c \overset{\rho}{\ominus} \check{G}.$

(h) $(G \overset{\lambda}{\ominus} H)^\vee = (\bigcap_{h\in H} Gh^{-1})^\vee = \bigcap_{h\in H} h\check{G} = \check{G} \overset{\rho}{\ominus} \check{H}.$

(i) Follows from (e) and (g).

Proof of Lemma 3.15. Items (a) and (b) follow from Lemma 3.10(a,b).

(c) Let ψ be increasing. Then if $G \subseteq H$, $G^c \supseteq H^c$, so $\psi(G^c) \supseteq \psi(H^c)$. Therefore if $G \subseteq H$, then $\psi'(G) = (\psi(G^c))^c \subseteq (\psi(H^c))^c = \psi'(H)$, hence ψ' is increasing. The converse is proved similarly. Also, let ψ be left-invariant, $g \in \Gamma$, $H \subseteq \Gamma$. Then $\psi'(gH) = \{\psi((gH)^c)\}^c = \{\psi(gH^c)\}^c = \{g\psi(H^c)\}^c = g\{\psi(H^c)\}^c = g\psi'(H)$, hence ψ' is left-invariant. Next, let ψ be a dilation. Then $\psi'(\cap X_i) = \{\psi((\cap X_i)^c)\}^c = \{\psi(\cup X_i^c)\}^c = \{\cup \psi(X_i^c)\}^c = \cap\{\psi(X_i^c)\}^c = \cap\, \psi'(X_i)$, hence ψ' is an erosion; the reverse implication is proved similarly. Finally, $(\tilde{\delta}^{\lambda}_{H})'(G) = (\tilde{\delta}^{\lambda}_{H}(G^c))^c = (G^c \overset{\Gamma}{\oplus} H)^c = G \overset{\lambda}{\ominus} \check{H} = \tilde{\varepsilon}^{\lambda}_{\check{H}}(G)$, where we have used Lemma 3.10(g).

(d) Let ψ be right-invariant, $g \in \Gamma$, $H \subseteq \Gamma$. Then $\check{\psi}(gH) = \{\psi((gH)^{\vee})\}^{\vee} = \{\psi(\check{H}g^{-1})\}^{\vee} = \{\psi(\check{H})g^{-1}\}^{\vee} = g\{\psi(\check{H})\}^{\vee} = g\check{\psi}(H)$, where we used Lemma 3.10(d). So we have shown that if ψ is right-invariant, $\check{\psi}$ is left-invariant. The reverse statement is proved similarly.
Also,

$$(\lambda_g)^{\vee}(H) = \{\lambda_g(\check{H})\}^{\vee} = \{g\check{H}\}^{\vee} = Hg^{-1} = \rho_g^{-1}(H)$$

and

$$(\tilde{\delta}^{\lambda}_{H})^{\vee} = (\bigcup_{h\in H} \rho_h)^{\vee} = \bigcup_{h\in H} (\rho_h)^{\vee} = \bigcup_{h\in H} \lambda_h^{-1} = \tilde{\delta}^{\rho}_{\check{H}}$$

The result for the erosion follows in the same way.
(e) Follows from (c) and (d).

Proof of Proposition 3.17. We only prove the first and third formulas. From the defining Eq. (7.24a) of the left residual we have $(G \overset{\lambda}{\ominus} H) \overset{\Gamma}{\oplus} H = (\bigcup_{g\in\Gamma} \{g : gH \subseteq G\}) \overset{\Gamma}{\oplus} H = \bigcup_{g\in\Gamma}\{gH : gH \subseteq G\}$, which proves the result for the left-invariant opening. Using Boolean duality, we have $\bigcap_{g\in\Gamma}\{g\hat{H} : g\hat{H} \supseteq G\} = \bigcap_{g\in\Gamma}\{g\hat{H} : g(\hat{H})^c \subseteq G^c\} = (\bigcup_{g\in\Gamma}\{g\check{H} : g\check{H} \subseteq G^c\})^c = ((G^c \overset{\lambda}{\ominus} \check{H}) \overset{\Gamma}{\oplus} \check{H})^c = (G \overset{\Gamma}{\oplus} H) \overset{\lambda}{\ominus} H$, proving the third line.

Proof of Proposition 3.18. We have seen above that (7.36) is a λ-adjunction. Therefore it remains to prove the "only if" part. So assume that (ε, δ) is a λ-adjunction. Let $H = \delta(\{e\})$, where e is the unit element of Γ. Then, for each $g \in \Gamma$,

$$\delta(\{g\}) = \delta(\lambda_g\{e\}) = \lambda_g\delta(\{e\}) = \lambda_g(H)$$

Hence, for each $G \in \mathcal{P}(\Gamma)$,

$$\delta(G) = \delta(\bigcup_{g\in G} \{g\}) = \bigcup_{g\in G} \delta(\{g\}) = \bigcup_{g\in G} \lambda_g(H) = G \overset{\Gamma}{\oplus} H = \tilde{\delta}^{\lambda}_{H}(G)$$

proving that each left-invariant dilation has the form as in (7.36). To complete the proof, observe that if ε is a λ-erosion, then its lower adjoint δ is a λ-dilation, so $\delta = \tilde{\delta}^{\lambda}_{H}$ for some $H \in \mathcal{P}(\Gamma)$, whose unique upper adjoint is $\tilde{\varepsilon}^{\lambda}_{H}$. Hence $\varepsilon = \tilde{\varepsilon}^{\lambda}_{H}$.

Proof of Theorem 3.23. It is clear that ψ as given by (7.38) is a left-invariant mapping, since it is a union of such mappings. Conversely, let ψ be a left-invariant mapping. We show that ψ has the form (7.38). Given $G \in \mathcal{P}(\Gamma)$, let $Z = \bigcup_{(F,H) \in \mathcal{W}(\psi)} G \overset{\lambda}{\oslash} (F, H)$. We show that $\psi(G) = Z$.

(a) $\psi(G) \supseteq Z$: Let $g \in G \overset{\lambda}{\oslash} (F, H)$ for some $(F, H) \in \mathcal{W}(\psi)$. Then $gF \subseteq G \subseteq gH$, hence $F \subseteq g^{-1}G \subseteq H$ and so $g^{-1}G \in [F, H] \subseteq \mathcal{V}(\psi)$ by assumption on (F, H). It follows that $e \in \psi(g^{-1}G) = g^{-1}\psi(G)$, where e is the identity of Γ and we used left invariance of ψ. Therefore $g \in \psi(G)$, hence $\psi(G) \supseteq Z$.

(b) $\psi(G) \subseteq Z$: Let $g \in \psi(G)$. Then, using left invariance, $e \in g^{-1}\psi(G) = \psi(g^{-1}G)$, hence $g^{-1}G \in \mathcal{V}(\psi)$ and therefore $(g^{-1}G, g^{-1}G) \in \mathcal{W}(\psi)$. Combining this with the obvious fact that $G \overset{\lambda}{\oslash} (g^{-1}G, g^{-1}G) \supseteq \{g\}$, we conclude that $g \in Z$ and so $\psi(G) \subseteq Z$.

Proof of Corollary 3.24. By application of the above theorem to an increasing λ-mapping, and using (7.37a) combined with the obvious fact that $G \overset{\lambda}{\oslash} (F, H)$ is increasing in H, we have

$$\psi(G) = \bigcup_{F \in \mathcal{V}(\psi)} G \overset{\lambda}{\oslash} (F, \Gamma) = \bigcup_{F \in \mathcal{V}(\psi)} G \overset{\lambda}{\ominus} F$$

To prove the representation as an intersection of dilations, observe that the dual mapping ψ' of ψ is itself left-invariant and increasing; see Lemma 3.15. So, applying the decomposition just proved to ψ', we get

$$\psi'(G) = \bigcup_{F \in \mathcal{V}(\psi')} G \overset{\lambda}{\ominus} F$$

Now we take again the Boolean dual of ψ', using Lemma 3.10(g) and the fact that $\psi'' = \psi$ to find

$$\psi(G) = (\bigcup_{F \in \mathcal{V}(\psi')} G^c \overset{\lambda}{\ominus} F)^c = \bigcap_{F \in \mathcal{V}(\psi')} G \overset{\Gamma}{\oplus} \check{F}$$

This completes the proof for increasing λ-mappings. The proof for decreasing λ-mappings is analogous.

Proof of Theorem 3.25. We only have to prove the "only if" part, since a union of λ-openings is a λ-opening (see Section II.B). So assume that ψ is a λ-opening. Applying Proposition 2.10 of Section II.B with $\mathbf{T} = \Gamma^\lambda$, one finds that ψ has the form (7.40) with $\tilde{\gamma}^\lambda_H$ the structural λ-opening by the structuring element H. Since from Proposition 3.17, $\tilde{\gamma}^\lambda_H = \tilde{\delta}^\lambda_H \tilde{\varepsilon}^\lambda_H$, the proof is complete.

Proof of Lemma 4.6. Since Σ is a group, $\Sigma \overset{\Gamma}{\oplus} \Sigma = \Sigma = \Sigma \overset{\lambda}{\ominus} \Sigma$, so

$$G \overset{\lambda}{\ominus} \Sigma = G \overset{\lambda}{\ominus} (\Sigma \overset{\Gamma}{\oplus} \Sigma) = (G \overset{\lambda}{\ominus} \Sigma) \overset{\lambda}{\ominus} \Sigma$$

where we used Proposition 3.8(f). Also,

$$\forall s \in \Sigma, \qquad \Sigma s = \Sigma \Rightarrow G \overset{\lambda}{\ominus} \Sigma = \bigcup_{s\in\Sigma} G \overset{\lambda}{\ominus} (\Sigma s) = \bigcup_{s\in\Sigma} (G \overset{\lambda}{\ominus} \Sigma)s$$
$$= (G \overset{\lambda}{\ominus} \Sigma) \overset{\Gamma}{\oplus} \Sigma$$

This proves (7.59a). To prove (7.59b), apply the Boolean duality relation Lemma 3.10(g) to (7.59a).

Proof of Proposition 4.8. We prove (a) to (i), but not necessarily in the stated order.

(a): Since π and ϑ are extensions of the point mappings π_ω and ϑ_ω to sets, they are increasing. So we will prove that π and ϑ are Γ-invariant, which then automatically implies that π_Σ is an increasing Γ-mapping, since $\tilde{\varepsilon}_\Sigma$ is increasing and translation-invariant. Let $g_0 \in \Gamma$; then, for any $G \in \tilde{\mathcal{L}}$, $\pi(g_0G) = \{g\omega : g \in g_0G\} = \{g_0g'\omega : g' \in G\} = g_0\{g'\omega : g' \in G\} = g_0\pi(G)$. Similarly, one proves the Γ-invariance of ϑ.

(b) and the first part of (c): These follow from the fact that π is the extension of a point mapping $\pi_\omega : \Gamma \to \mathcal{X}$ to subsets of Γ. The second part of (c) is proved below.

(e), first part: Using (a) and the fact that $\vartheta(\{\omega\}) = \Sigma$, one has for any $G \in \tilde{\mathcal{L}}$, $X \in \mathcal{L}$,

$$\pi\vartheta(X) = \pi\,\{g \in \Gamma : \pi_\omega(g) \in X\} = \{\pi_\omega(g) : g \in \Gamma,\ \pi_\omega(g) \in X\} = X$$
$$\vartheta\pi(G) = \vartheta\pi\left[\bigcup_{g\in G} \{g\}\right] = \bigcup_{g\in G} \vartheta(\{g\omega\}) = \bigcup_{g\in G} g\Sigma = G \overset{\Gamma}{\oplus} \Sigma = \tilde{\delta}_\Sigma(G)$$

(d): From (e), first part, we have $\vartheta = \vartheta\pi\vartheta = \tilde{\delta}_\Sigma\vartheta$, and, using Lemma 4.6, $\tilde{\varepsilon}_\Sigma\vartheta = \tilde{\varepsilon}_\Sigma\tilde{\delta}_\Sigma\vartheta = \tilde{\delta}_\Sigma\vartheta$.

(e), second part: From (d), $\pi_\Sigma\vartheta = \pi\tilde{\varepsilon}_\Sigma\vartheta = \pi\vartheta = \mathrm{id}_{\mathcal{L}}$, $\vartheta\pi_\Sigma = \vartheta\pi\tilde{\varepsilon}_\Sigma = \tilde{\delta}_\Sigma\tilde{\varepsilon}_\Sigma = \tilde{\varepsilon}_\Sigma$.

(c), second part: Since $\vartheta\pi_\Sigma = \tilde{\varepsilon}_\Sigma$, with $\tilde{\varepsilon}_\Sigma$ an erosion, we have, using (b), $\vartheta\pi_\Sigma(\bigcap_{i\in I} G_i) = \bigcap_{i\in I} \vartheta\pi_\Sigma(G_i) = \vartheta\left[\bigcap_{i\in I} \pi_\Sigma(G_i)\right]$ for any index set I. Operating on both sides of this quality by π and using $\pi\vartheta = \mathrm{id}_{\mathcal{L}}$, we get $\pi_\Sigma(\bigcap_{i\in I} G_i) = \bigcap_{i\in I} \pi_\Sigma(G_i)$.

(f,$\Rightarrow$): Follows from (a).

(f,$\Leftarrow$): Let $\vartheta(X) \subseteq \vartheta(Y)$. Then, since π is increasing and $\pi\vartheta = \mathrm{id}_{\mathcal{L}}$, $\pi\vartheta(X) \subseteq \pi\vartheta\,(Y)$, so $X \subseteq Y$.

(g): First, $(\vartheta(X))^c = \{g \in G\colon \pi_\omega(g) \in X\}^c = \{g \in G : \pi_\omega(g) \notin X\} = \{g \in G : \pi_\omega(g) \in X^c\} = \vartheta(X^c)$. Using this identity and the fact that $(G \overset{\Gamma}{\oplus} \Sigma)^c = G \overset{\lambda}{\ominus} \check{\Sigma}$

$= G \overset{\lambda}{\ominus} \Sigma$ (since Σ is a group, $\check{\Sigma} = \Sigma$), we find $(\pi(G))^c = \pi\vartheta[\pi(G)]^c = \pi[\{\vartheta\pi(G)\}^c] = \pi[\tilde{\delta}_\Sigma(G)]^c = \pi\tilde{\varepsilon}_\Sigma(G^c) = \pi_\Sigma(G^c)$.

(h,$\Rightarrow$): $\pi(G) \subseteq X \Rightarrow G \subseteq \vartheta\pi(G) \subseteq \vartheta(X)$, since $\vartheta\pi = \tilde{\delta}_\Sigma$ is a closing and ϑ is increasing.

(h,$\Leftarrow$): $G \subseteq \vartheta(X) \Rightarrow \pi(G) \subseteq \pi\vartheta(X) = X$, since π is increasing and $\pi\vartheta = \mathrm{id}_{\mathscr{L}}$.

(i,$\Rightarrow$): $\vartheta(X) \subseteq G \Rightarrow X = \pi_\Sigma\vartheta(X) \subseteq \pi_\Sigma(G)$, since π_Σ is increasing and $\pi_\Sigma\vartheta = \mathrm{id}_{\mathscr{L}}$.

(i,$\Leftarrow$): $X \subseteq \pi_\Sigma(G) \Rightarrow \vartheta(X) \subseteq \vartheta\pi_\Sigma(G) \subseteq G$, for $\vartheta\pi_\Sigma = \tilde{\varepsilon}_\Sigma$ is an opening and ϑ is increasing.

Proof of Lemma 4.9. (a) and (a′) are obvious. We prove (b) and the part of (c) and (c′) concerning openings. The other entries can be proved similarly.

(b): Let ε be an erosion on $\mathscr{L}$ with adjoint δ. Then $\vartheta\varepsilon\pi_\Sigma$ is an erosion on $\tilde{\mathscr{L}}$, and, since the adjoint of a product is the product of the adjoints in reverse order (see Lemma 2.6(f)), the corresponding dilation is $\vartheta\delta\pi$.

(c): Let γ be an opening. Then $\tilde{\gamma} := \vartheta\gamma\pi_\Sigma$ is increasing, antiextensive (since $\vartheta\gamma\pi_\Sigma \subseteq \vartheta\pi_\Sigma = \tilde{\varepsilon}_\Sigma \subseteq \mathrm{id}_{\tilde{\mathscr{L}}}$) and idempotent, for $\tilde{\gamma}^2 = \vartheta\gamma\pi_\Sigma\vartheta\gamma\pi_\Sigma = \vartheta\gamma^2\pi_\Sigma = \vartheta\gamma\pi_\Sigma = \tilde{\gamma}$. Hence $\tilde{\gamma}$ is an opening.

(c′): Let $\tilde{\gamma}$ be an opening. Then $\gamma := \pi\tilde{\gamma}\vartheta$ is increasing and antiextensive, since $\gamma = \pi\tilde{\gamma}\vartheta \subseteq \pi\vartheta = \mathrm{id}_{\mathscr{L}}$. Therefore $\gamma^2 \subseteq \gamma$, but also $\gamma^2 = \pi\tilde{\gamma}\vartheta\pi\tilde{\gamma}\vartheta = \pi\tilde{\gamma}\tilde{\delta}_\Sigma\tilde{\gamma}\vartheta \supseteq \pi\tilde{\gamma}^2\vartheta = \pi\tilde{\gamma}\vartheta = \gamma$, hence $\gamma^2 = \gamma$. So γ is an opening.

Proof of Proposition 4.12. We give the proof in operator form. Since $(\tilde{\varepsilon}_H, \tilde{\delta}_H)$ is a Γ-adjunction on $\tilde{\mathscr{L}}$, the pair (ε, δ) as defined in (7.61) forms a Γ-adjunction on $\mathscr{L}$ by Lemma 4.9(a′,b′). Conversely, let (ε, δ) form a Γ-adjunction on $\mathscr{L}$. Then by Lemma 4.9(a,b), the pair $(\vartheta\varepsilon\pi_\Sigma, \vartheta\delta\pi)$ is a Γ-adjunction on $\tilde{\mathscr{L}}$. Hence we know from Section III.D that $\tilde{\delta} := \vartheta\delta\pi = \tilde{\delta}_H$, $\tilde{\varepsilon} := \vartheta\varepsilon\pi_\Sigma = \tilde{\varepsilon}_H$ for some $H \in \tilde{\mathscr{L}}$, where $(\tilde{\varepsilon}_H, \tilde{\delta}_H)$ is the left-invariant adjunction (7.36) on $\tilde{\mathscr{L}}$ with structuring element H. Using that $\pi\vartheta = \pi_\Sigma\vartheta = \mathrm{id}_{\mathscr{L}}$, we find that $\delta = \pi\tilde{\delta}_H\vartheta$, $\varepsilon = \pi_\Sigma\tilde{\varepsilon}_H\vartheta$. This proves (7.61). Now $\pi = \pi\tilde{\delta}_\Sigma$, hence $\delta = \pi\tilde{\delta}_\Sigma\tilde{\delta}_H\vartheta = \pi\tilde{\delta}_{H\oplus\Sigma}\vartheta = \pi\tilde{\delta}_{\vartheta\pi(H)}\vartheta$, and $\vartheta = \tilde{\varepsilon}_E\vartheta$, hence $\varepsilon = \pi_\Sigma\tilde{\varepsilon}_H\tilde{\varepsilon}_\Sigma\vartheta = \pi_\Sigma\tilde{\varepsilon}_{H\oplus\Sigma}\vartheta = \pi_\Sigma\tilde{\varepsilon}_{\vartheta\pi(H)}\vartheta$. Writing Y instead of $\pi(H)$, we thus have found $\delta = \delta_Y^\Gamma := \pi\tilde{\delta}_{\vartheta(Y)}\vartheta$, $\varepsilon = \varepsilon_Y^\Gamma := \pi_\Sigma\tilde{\varepsilon}_{\vartheta(Y)}\vartheta$.

Since $\pi = \pi\tilde{\delta}_\Sigma$, $\vartheta = \tilde{\delta}_\Sigma\vartheta$, one has that $\delta = \pi\tilde{\delta}_\Sigma\tilde{\delta}_H\tilde{\delta}_\Sigma\vartheta = \pi\tilde{\delta}_{\Sigma\oplus H\oplus\Sigma}\vartheta = \pi\tilde{\delta}_{\vartheta(\bar{Y})}\vartheta$, and from $\pi_\Sigma = \pi_\Sigma\tilde{\varepsilon}_\Sigma$, $\vartheta = \tilde{\varepsilon}_\Sigma\vartheta$, one has $\varepsilon = \pi\tilde{\varepsilon}_\Sigma\tilde{\varepsilon}_H\tilde{\varepsilon}_\Sigma\vartheta = \pi\tilde{\varepsilon}_{\Sigma\oplus H\oplus\Sigma}\vartheta = \pi\tilde{\varepsilon}_{\vartheta(\bar{Y})}\vartheta$, where as before $Y = \pi(H)$. It is clear that nothing changes when we replace Y by $\bar{Y}$. This completes the proof.

Proof of Corollary 4.13. From Proposition 4.12 we have that any adjunction has the form $(\varepsilon, \delta) = (\varepsilon_Y^\Gamma, \delta_Y^\Gamma)$ where

$$\delta_Y^\Gamma(X) = \pi[\vartheta(X) \overset{\Gamma}{\oplus} \vartheta(Y)] = \bigcup_{g \in \vartheta(X)} gY$$

$$\varepsilon_Y^\Gamma(X) = \pi_\Sigma[\vartheta(X) \overset{\lambda}{\ominus} \vartheta(Y)] = \bigcap_{g \in \vartheta(X^c)} g\hat{Y}^*$$

Here $\hat{Y}^* = \pi\Sigma[(\check{\vartheta}(Y))^c]$ and we have used the form of adjunctions on $\tilde{\mathscr{L}}$ in (7.28) and (7.32). Using Proposition 4.8g we also have $\hat{Y}^* = [\pi(\check{\vartheta}(Y))]^c = (\check{Y}^*)^c$ with $\check{Y}^* = \pi(\check{\vartheta}(Y))$. This proves the first part.

Second, using Proposition 4.12 again, $\delta_Y^\Gamma(X) = \pi[\vartheta(X) \overset{\Gamma}{\oplus} \vartheta(\bar{Y})] = \pi\{g \in \Gamma\colon g\check{\vartheta}(\Sigma Y) \Uparrow \vartheta(X)\} = \pi\{g \in \Gamma\colon g(\check{\vartheta}(Y) \overset{\Gamma}{\oplus} \Sigma) \Uparrow \vartheta(X)\} = \pi\{g \in \Gamma\colon g\vartheta(\pi\check{\vartheta}(Y)) \Uparrow \vartheta(X)\} = \pi\{g \in \Gamma\colon g\check{Y}^* \Uparrow X\} = \{y \in \mathscr{X}\colon g\check{Y}^* \Uparrow X \text{ for some } g \in \vartheta_\omega(y)\}$. Here we have made use of the geometric interpretation of dilations on $\mathscr{P}(\Gamma)$ (see Remark 3.11) and the obvious equivalence $\vartheta(X) \Uparrow \vartheta(Y) \Leftrightarrow X \Uparrow Y$. Finally, for the erosion we have $\varepsilon_Y^\Gamma(X) = \pi_\Sigma[\vartheta(X) \overset{\lambda}{\ominus} \vartheta(Y)] = \pi_\Sigma\{g \in \Gamma\colon gY \subseteq X\} = \{y \in \mathscr{X}\colon gY \subseteq X$ for all $g \in \vartheta_\omega(y)\}$.

Proof of Proposition 4.14. By explicit computation, we find

$$\begin{aligned}
\pi\tilde{\delta}_{\vartheta(Y)}\tilde{\varepsilon}_{\vartheta(Y)}\vartheta(X) &= \pi\,[(\vartheta(X) \overset{\lambda}{\ominus} \vartheta(Y)) \overset{\Gamma}{\oplus} \vartheta(Y)] = (\vartheta(X) \overset{\lambda}{\ominus} \vartheta(Y))Y \\
&= \left[\bigcup_{g \in \Gamma} \{g\colon g\vartheta(Y) \subseteq \vartheta(X)\}\right]Y = \left[\bigcup_{g \in \Gamma} \{g\colon gY \subseteq X\}\right]Y \\
&= \bigcup_{g \in \Gamma} \{gY\colon gY \subseteq X\}
\end{aligned}$$

Here we have used the notation (7.52) for the action of a subset G of Γ on a subset X or $\mathscr{X}$. This proves (7.65). Second, for the closing ϕ_Y^Γ we have

$$\begin{aligned}
\phi_Y^\Gamma(X) &= [\gamma_{Y^c}(X^c)]^c = [\pi\tilde{\delta}_{\vartheta(Y^c)}\tilde{\varepsilon}_{\vartheta Y^c}\vartheta(X^c)]^c \\
&= \pi_\Sigma\tilde{\varepsilon}_{\check{\vartheta}(Y^c)}\,\tilde{\delta}_{\check{\vartheta}(Y^c)}\,\vartheta(X) = \pi\tilde{\varepsilon}_{\check{\vartheta}(Y^c)}\tilde{\delta}_{\check{\vartheta}(Y^c)}\vartheta(X)
\end{aligned}$$

where we used Proposition 4.8(g) and Boolean duality for adjunctions on $\tilde{\mathscr{L}}$, as well as the fact that $\pi_\Sigma\tilde{\varepsilon}_{\check{\vartheta}(Y^c)} = \pi\tilde{\varepsilon}_{\check{\vartheta}(Y^c)}$, which follows from the Σ-invariance of $\check{\vartheta}(Y^c)$.

Proof of Lemma 4.17. $\mathscr{V}(\tilde{\psi}) = \{G \in \tilde{\mathscr{L}}\colon e \in \tilde{\psi}(G)\} = \{G \in \tilde{\mathscr{L}}\colon \pi(\{e\}) \in \psi(\pi(G))\} = \{G \in \tilde{\mathscr{L}}\colon \omega \in \psi(\pi(G))\} = \{G \in \tilde{\mathscr{L}}\colon \pi(G) \in \mathscr{V}(\psi)\}$, where we have used that $e \in \tilde{\psi}(G) = \vartheta\psi(\pi(G)) \Leftrightarrow \pi(\{e\}) \in \psi(\pi(G))$.

Proof of Theorem 4.18. Clearly, ψ as given by (7.67) is an increasing Γ-mapping. Conversely, assume that ψ is an increasing Γ-mapping. Then $\tilde{\psi} = \vartheta\psi\pi$ is an increasing λ-mapping (Lemma 4.9). Hence

$$\tilde{\psi}(G) = \bigcup_{H \in \mathscr{V}(\tilde{\psi})} \tilde{\varepsilon}_H(G)$$

where $\tilde{\varepsilon}_H(G) = G \overset{\lambda}{\ominus} H$ is the left-invariant erosion on $\mathcal{L}$. Now, from Proposition 4.8(e), $\psi = \pi\tilde{\psi}\vartheta$, so

$$\psi(X) = \pi\tilde{\psi}\vartheta(X) = \pi\left[\bigcup_{H\in\mathcal{V}(\tilde{\psi})} \tilde{\varepsilon}_H\vartheta(X)\right] = \bigcup_{H\in\mathcal{V}(\tilde{\psi})} \pi\tilde{\varepsilon}_H\vartheta(X)$$

Since $\tilde{\varepsilon}_H\vartheta = \tilde{\varepsilon}_H\tilde{\varepsilon}_\Sigma\vartheta = \tilde{\varepsilon}_{H\overset{\Gamma}{\oplus}}\mathrm{Sq}_{\,=\;;\vartheta\pi(H)}\vartheta$, we have from Lemma 4.17

$$\psi(X) = \bigcup_{Y\in\mathcal{V}(\psi)} \bigcup_{H:\pi(H)=Y} \pi\tilde{\varepsilon}_{\vartheta\pi(H)}\vartheta(X) = \bigcup_{Y\in\mathcal{V}(\psi)} \pi\tilde{\varepsilon}_{\vartheta(Y)}\vartheta(X)$$

which completes the proof of (7.67a). Equation (7.67b) follows by applying (7.67a) to the Boolean dual ψ'.

REFERENCES

1. Banon, G. J. F., and Barrera, J., Minimal representation for translation-invariant set mappings by mathematical morphology. *SiAM J. Appl. Math.*, *51*, 1782–1798, 1991.
2. Berger, M., *Geometry I*, Springer-Verlag, Berlin, 1989.
3. Beucher, S., Blosseville, J. M., and Lenoir, F., Traffic spatial measurements using video image processing, in *SPIE Cambridge 87 Symposium on Advances in Intelligent Robotics Systems*, November 1987.
4. Birkhoff, G., *Lattice Theory*. Colloquium Publications, Vol. 25, 3rd ed., American Mathematical Society, Providence, Rhode Island,1984.
5. Blyth, T. S., and Janowitz, M. F., *Residuation Theory*, Pergamon Press, Oxford, 1972.
6. Ghosh, P. K., A solution of polygon containment, spatial planning, and other related problems using Minkowski operations. *Comput. Vision Graphics Image Process.*, *49*, 1–35, 1990.
7. Giardina, C. R., and Dougherty, E. R., *Morphological Methods in Image and Signal Processing*, Prentice-Hall, Englewood Cliffs, New Jersey, 1988.
8. Gierz, G, Hofmann, K. H., Keimel, K., Lawson, J. D., Mislove, M., Scott, D. S., *A Compendium of Continuous Lattices*, Springer-Verlag, Berlin, 1980.
9. Gouzènes, L., Strategies for solving collision-free trajectories problems for mobile and manipulator robots, *Int. J. Robotics Res.*, *3*, 51–65 (1984).
10. Hadwiger, H., *Vorlesungen über Inhalt, Oberfläche, und Isoperimetric*, Springer-Verlag, Berlin, 1957.
11. Heijmans, H. J. A. M., Mathematical morphology: an algebraic approach. *CWI Newslett.*, *14*, 7–27, 1987.
12. Heijmans, H. J. A. M., and Ronse, C., The algebraic basis of mathematical morphology. Part I. Dilations and erosions. *Comput. Vision Graphics Image Process.*, *50*, 245–295, 1989.
13. Heijmans, H. J. A. M., and Serra, J., Convergence, continuity and iteration in mathematical morphology. To appear in *J. Vis. Comm. Image Repr.*
14. Kendall, D., Shape manifolds, procrustean metrics, and complex projective spaces. *Bull. London Math. Soc.*, *16*, 81–121, 1984.

15. Klein, F., Vergleichende Betrachtungen über neuere geometrische Forschungen, *Ges. Math. Abh.*, *1*, 460–497, 1872.
16. Maragos, P., Affine morphology and affine signal models, in *Proceedings, SPIE Conference on Image Algebra and Morphological Image Processing*, San Diego, July 1990.
17. Matheron, G., *Random Sets and Integral Geometry*, Wiley, New York, 1975.
18. Miles, R. E., Random points, sets and tesselations on the surface of a sphere, *Sankhya*, *A33*, 145–174 (1971).
19. Miller, W. *Symmetry Groups and Their Applications*, Academic Press, New York, 1972.
20. Robinson, D. J. S., *A Course in the Theory of Groups*, Springer-Verlag, New York, 1982.
21. Roerdink, J. B. T. M., Mathematical morphology on homogeneous spaces. Part I: The simply transitive case, Report AM-R8924, Centre for Mathematics and Computer Science, Amsterdam, 1989.
22. Roerdink, J. B. T. M., Mathematical morphology on homogeneous spaces. Part II: The transitive case, Report AM-R9006, Centre for Mathematics and Computer Science, Amsterdam, 1990.
23. Roerdink, J. B. T. M., Mathematical morphology on the sphere, in Proceedings, SPIE Conference on Visual Communications and Image Processing '90, Lausanne, pp. 263–271, 1990.
24. Roerdink, J. B. T. M., On the construction of translation and rotation invariant morphological operators, in *Morphology: Theory and Applications* (R. Haralick, ed.), Oxford Univ. Press, in press.
25. Roerdink, J. B. T. M., and Heijmans, H. J. A. M., Mathematical morphology for structures without translation symmetry, *Signal Process.*, *15*, 271–277 (1988).
26. Ronse, C., Why mathematical morphology needs complete lattices, submitted.
27. Ronse, C., and Heijmans, H. J. A. M., The algebraic basis of mathematical morphology. Part II: Openings and closings, *Comput. Vision Graphics Image Process.: Image Understanding, 54*, 74–97 (1991).
28. Santalo, L. A., *Integral Geometry and Geometric Probability*, Addison-Wesley, Reading, Massachusetts, 1976.
29. Serra, J., *Image Analysis and Mathematical Morphology*, Academic Press, London, 1982.
30. Serra, J., Éléments de Théorie pour l'Optique Morphologique, Ph.D. thesis, Université P. and M. Curie, Paris,1986.
31. Serra, J., ed., *Image Analysis and Mathematical Morphology*, vol. 2: *Theoretical Advances*, Academic Press, London, 1988.
32. Sternberg, S. R., Cellular computers and biomedical image processing, in *Biomedical Images and Computers* (J. Sklansky and J. C. Bisconte, eds.), Lecture Notes in Medical Informatics, vol. 17, Springer-Verlag, Berlin, 1982, pp. 294–319.
33. Sternberg, S. R., Grayscale morphology, *Comput. Vision Graphics Image Process.*, *35*, 333–355 (1986).
34. Suzuki, M., *Group Theory*, Springer-Verlag, Berlin, 1982.
35. Verwer, B. J. H., Distance transforms for path planning, Report, Technical University Delft, 1987.
36. Weyl, H., *Symmetry*, Princeton University Press, Princeton, New Jersey, 1952.

Chapter 8

Morphological Algorithms

Luc Vincent

Xerox Imaging Systems
Peabody, Massachusetts

I. INTRODUCTION

This chapter is concerned with the efficient implementation of "low-level" morphological transformations [25,38]. The qualifier "low-level" used here means that we deal with the implementation of transformations which serve as elementary bricks when solving practical image analysis problems. This does not mean that these transformations are simple, or cannot be decomposed into simpler ones; on the contrary, some of the operations considered in this chapter (e.g., skeletons, watersheds, propagation functions) are complex, both to define and to compute! However, from a user's perspective, these transformations share the characteristics of being easily and intuitively understandable; for example, watersheds extract from a gray-level image the crest lines that are located between the minima, top-hat transformations extract thin and light (or dark) regions, and skeletons reduce binary shapes to their medial axes.

Solving a moderately complex image analysis application by morphological methods often involves the concatenation of several tens or hundreds of low-level transformations [44]. This is the reason why each of these elementary bricks should be implemented as efficiently as possible. This task can be approached via various algorithmic techniques, the majority of which are described in this chapter. Each category of techniques is characterized by its advantages and drawbacks and illustrated using transformations for which it is particularly suited. For more details, see [34,47,37].

The present section is first concerned with the notation that will be used throughout the chapter. A particular transformation, the *distance function*, is also recalled, since it is used as a leitmotiv to illustrate how the described families of algorithms work. The characteristics one expects morphological algorithms to be equipped with are then briefly discussed.

Section II is devoted to the most classical morphological algorithms, namely the *parallel* ones. As explained below, these algorithms turn out to be rather inefficient on conventional computers. A first step toward the implementation of fast morphological algorithms was made by introducing the *sequential* methods. They are presented and illustrated in Section III.

Although sequential techniques serve very well for the computation of transformations such as distance function, granulometry function, or geodesic reconstruction (see Section III), they remain inefficient in many cases, since they involve numerous scannings of the entire image. To get rid of this problem, new scanning techniques have been introduced; the algorithms relying on them are such that, throughout the computation of a given morphological transform, only those pixels likely to be modified are taken into account. Such algorithms are based on contours and can be divided up into two families: the chains and loops propagation algorithms [34], which constitute the topic of Section IV, and the queue algorithms [47], discussed in Section V. Both families have recently been introduced in the morphology world and constitute one of the best possible choices for implementing complex transformations on conventional computers. Not only do these methods lead to faster algorithms, they are also extremely flexible and usually produce more accurate results. We shall illustrate their use by the computation of such transformations as propagation functions, Euclidean distance functions, skeletons, and watersheds. Lastly, the conclusion summarizes the qualities and drawbacks of these categories of algorithms and provides some guidelines as to what methods should be used for a given purpose.

A. Discrete Images and Grids, Notation

In the following, we consider binary and gray-scale images I as mappings from a rectangular domain $D_I \subset \mathbf{Z}^2$ into $\mathbf{Z}$. A binary image may take only values 0 and 1 and is often reduced to the set of its *feature pixels*, that is, pixels with value 1. Many of the algorithms described below extend to n-dimensional spaces, but for simplicity they will always be presented for two-dimensional images.

The underlying grid $G \subset \mathbf{Z}^2 \times \mathbf{Z}^2$ defines the neighborhood relations between pixels. G is usually a square grid (of 4- or 8-connectivity) or a hexagonal one (see Figure 1). $N_G(p)$ denotes the set of the neighbors of a pixel $p \in \mathbf{Z}^2$ according to grid G:

$$N_G(p) = \{q \in \mathbf{Z}^2 \mid (p,q) \in G\}$$

The discrete distance associated with G is denoted d_G: $d_G(p,q)$ is the minimal length of the paths of G connecting p to q.

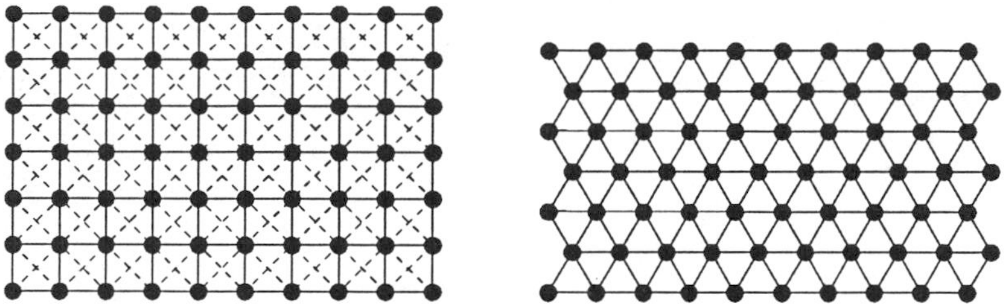

(a) *(b)*

Figure 1. Square (a) and hexagonal (b) digital grids. The former one can be considered in either 4- or 8-connectivity.

In the present chapter, we mostly use the hexagonal grid (6-connectivity). Indeed, the discrete distance it induces, called *hexagonal distance* and denoted d_6, is more isotropic than the city-block distance d_4 or the chessboard distance d_8, respectively induced by the square grid in 4- and 8-connectivity [8]. More importantly, the hexagonal grid is a *triangulation* and thus satisfies the *digital Jordan property* [34, page 61] according to which every nondegenerate simple loop separates the digital plane $\mathbf{Z}^2$ into two different connected components. As illustrated by Figure 2, this is not true for square grids and causes endless practical difficulties. For example, when dealing with square grids, consistency makes it often necessary to use 8-connectivity for the objects and 4-connectivity for the background (or vice versa) [47]. For these reasons, morphologists often prefer the hexagonal grid. Its elementary vectors are denoted $\mathbf{u}_0, \mathbf{u}_1, \ldots, \mathbf{u}_5$ and are illustrated by Figure 3. Note, however, that all the algorithms described below extend to any kind of discrete grid.

The algorithms themselves are described in a pseudocode that bears similarities to C and Pascal. It makes use of a certain number of keywords and symbols, which are summarized in Table 1. Some shortcuts like:

Repeat until stability { . . .

or

For every pixel p' *in* $N_G(p)$ { . . .

will also be used. Instructions specific to the type of image scanning used by the algorithm being discussed will be introduced as needed.

B. The Distance Function

Throughout the chapter, a particular transformation called the *distance function* [32,8] is used to illustrate the four families of algorithms described. This trans-

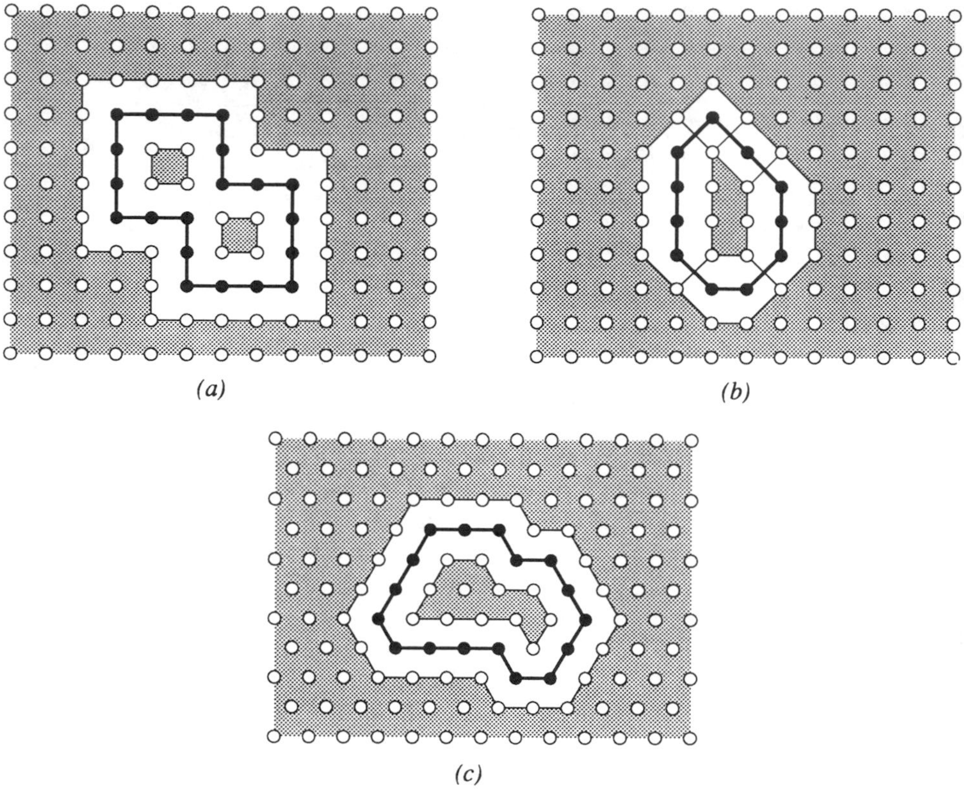

Figure 2. Square grids of 4- or 8-connectivity do not satisfy the digital Jordan property, in contrast to the hexagonal grid (c). Indeed, the simple nondegenerate loop drawn in (a) (4-connectivity) separates the discrete plane into three connected components, whereas that of (b) (8-connectivity) does not separate anything!

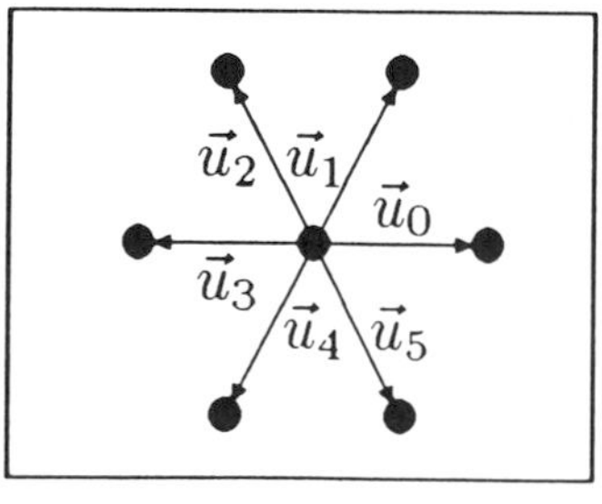

Figure 3. The six elementary vectors of the hexagonal grid.

Table 1. Symbols and Keywords of the Pseudocode Used for Algorithm Descriptions

$=, \neq, <, >, \leq, \geq$	Comparisons of values
$\leftarrow$	Assignment
{, }	Beginning and end of group of instructions
;	End of instruction
/*,*/	Beginning and end of comments
If, then, else	Logical tests
For . . . to; while; repeat until	Classical loops
True, false	Logical values

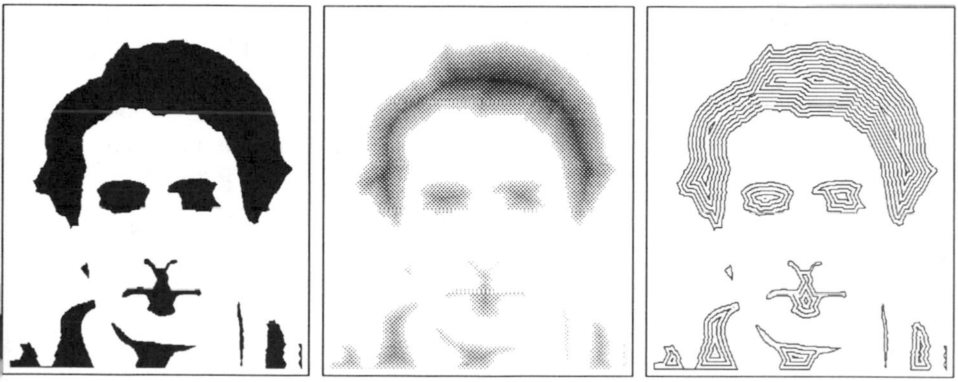

(a) (b) (c)

Figure 4. Example of hexagonal distance function. (a) Original image; (b) distance function; (c) level lines of this function.

formation is indeed very typical and gives rise to several different implementations. The distance function dist_X of a set $X \subset \mathbf{Z}^2$ associates with each pixel of X its distance to the background:

$$\text{dist}_X \begin{cases} X \to \mathbf{Z} \\ p \mapsto \min\{d_G(p,q) \mid q \notin X\} \end{cases} \tag{8.1}$$

The distance function d_I of a binary image I is equivalent to that of its set of *feature pixels*, that is, pixels with value 1. In addition, we put conventionally $\forall p \in D_I$, $I(p) = 0 \Rightarrow \text{dist}_I(p) = 0$. An example of distance function is shown in Figure 4.

C. Estimating the Quality of a Morphological Algorithm

The performance of a morphological algorithm may be defined using three main criteria: speed, accuracy, and flexibility.

1. Speed

This is a crucial issue in the field of image analysis. Indeed, on the one hand, an application program is often designed to be used routinely, either on a large amount of data (e.g., in medicine) or daily (e.g., in quality control). It is then unacceptable for the execution time to be larger than a specified upper bound. On the other hand, even during the solution of a given image analysis problem, many different possibilities have to be considered; for each of them, many transformations have to be used, often repeatedly, with parameter adjustments, filter modifications, etc. It is therefore extremely important for the image analyst to have fast algorithms at his or her disposal; it considerably speeds up this development step and even makes it possible to explore ideas that could not be considered otherwise. For example until recently, the use of the watershed transformation [11,2] was impossible in practice because of its prohibitive computation time. However, the appearance of the most recent specialized architectures (e.g., the Quantimet 570 of Leitz) and algorithms [47,50] (see Section V) has moved it to one of the highest ranking of morphological segmentation tools.

2. Accuracy

An algorithm should of course give results that are as accurate as possible. In fact, most of the time, the result is expected to be totally exact. However, the definition of some transformations—like skeletons (see Section V)—is sometimes not well adapted to the discrete framework. The algorithms for computing such transformations should then be designed to produce results "as close as possible" to the continuous one. Morphological algorithms are also expected to avoid some of the aberrations associated with the use of discrete grids, like the "cone effect" (see Figure 5b). Lastly, one often tries to compute morphological transformations in an isotropic fashion. This involves resorting to discrete distances d closer to the Euclidean one than d_G (see Section IV).

3. Flexibility

By flexibility, we mean any of the following:

The algorithm is adaptable to other grids.
It works in both the Euclidean and geodesic [22] cases.
It allows one to produce several transforms close to one another.
It is adaptable to several metrics.

Flexible algorithms are very interesting in that they spare the energy of the programmer and reduce the implementation costs.

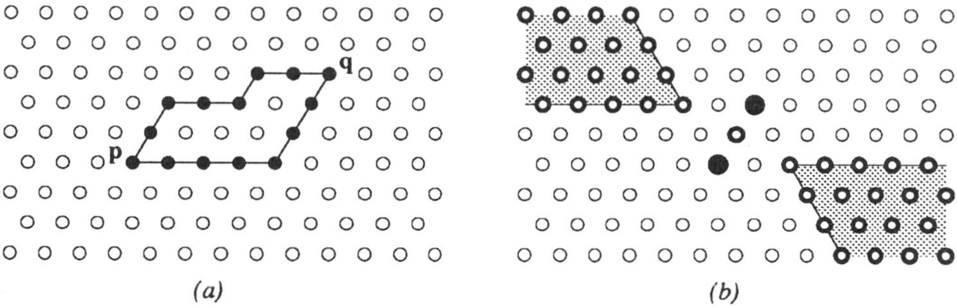

Figure 5. Some difficulties in discrete spaces. (a) There exist several paths of minimal length between two pixels. (b) The set of the pixels that are equidistant to two given connected components may be a thick area; here, according to the hexagonal distance, all the bold pixels (gray area) are equidistant to the two black ones.

However, most algorithms cannot be extremely fast, accurate, and flexible at the same time. Improving one of these characteristics is generally done to the detriment of the two remaining ones. For example, increasing the accuracy of an algorithm mostly requires additional tests and computations that affect its speed. On the other hand, morphological operations are not implemented in the same way on two different computers; on specialized architectures, complex transformations like skeletons and watersheds will be implemented using built-in thinning and thickening capabilities. However, such algorithms would be terribly slow on classical computers, where their execution times could be close to a couple of hours!

D. Image Structures and How to Access Them

The most common data structure to represent and process binary and gray-scale images is the two-dimensional array of pixels, all of which are either 1, 8, or 16 bits. This type of structure is very simple and enables fast access to the neighbors of a given pixel, no matter what discrete grid is used. Several attempts have been made in morphology to manipulate images stored using different data structures, e.g., quadtrees and octrees [33,4], interval coding [29], and structures stemming from computational geometry, like polygons [30]. However, as explained in [47, p. 22], none of these structures is really adapted to the implementation of morphological transformations.

This is the reason why, in the following, we consider only images stored as arrays of pixels. We always assume that the number of bits per pixel is sufficient and we do not account for edge effects. Indeed, as explained in [47, Chapter 2], they are usually easy to cope with by giving to the pixels of the frame a particular value, usually 0, $-\infty$, or $+\infty$. Here, the images under study are considered to be defined in the entire space $\mathbf{Z}^2$ and to take value 0 outside of their definition domain, unless otherwise mentioned. In addition, special data structures like loops or queues will be used to manipulate these arrays of pixels efficiently.

II. PARALLEL ALGORITHMS

This category of algorithms is the most common and classical one in the field of morphology. A parallel algorithm typically works as follows: given an input image I, the pixels of I are scanned and the new value of the current pixel p is determined from that of the pixels in a given neighborhood $N(p)$ of p. In doing this, the following constraint is satisfied:

The new pixel values are written in an output image J different from I

J is then copied into I, and additional image scannings are performed until a given criterion is fulfilled, or until stability is reached.

Since I is different from J, its pixels can actually be scanned in an arbitrary order. In particular, one can imagine parallelizing the processing on some image parts, or even on all pixels, as is done by some specialized architectures. Hereafter, a "parallel scanning" is introduced by a sentence like:

For every pixel p of D_I, *do* { . . .

The parallel algorithm to determine the distance function of a binary image I in grid G is given below in pseudocode.

Algorithm: Parallel Distance Function

- – input: I, binary image,
- – output: J, grayscale image defined on D_I; $J \neq I$.

•Repeat until stability {

For every pixel $p \in D_I$ { /* actual parallel scanning */

If $I(p) = 1$ then $J(p) \leftarrow \min\{I(q), q \in N_G(p)\} + 1$;

Copy image J in I; }

}

This algorithm is illustrated on Figure 6. One is easily convinced that the number of scannings it requires is proportional to the largest computed distance. More generally, parallel algorithms usually require a large number of complete image scannings, sometimes several hundreds! Therefore, although these algorithms are particularly suited to some architectures, they are definitely not adapted to conventional computers.

The basic parallel algorithms (dilations, erosions, distance function, etc.) easily extend to any grid and to n-dimensional images [16], but here again, prohib-

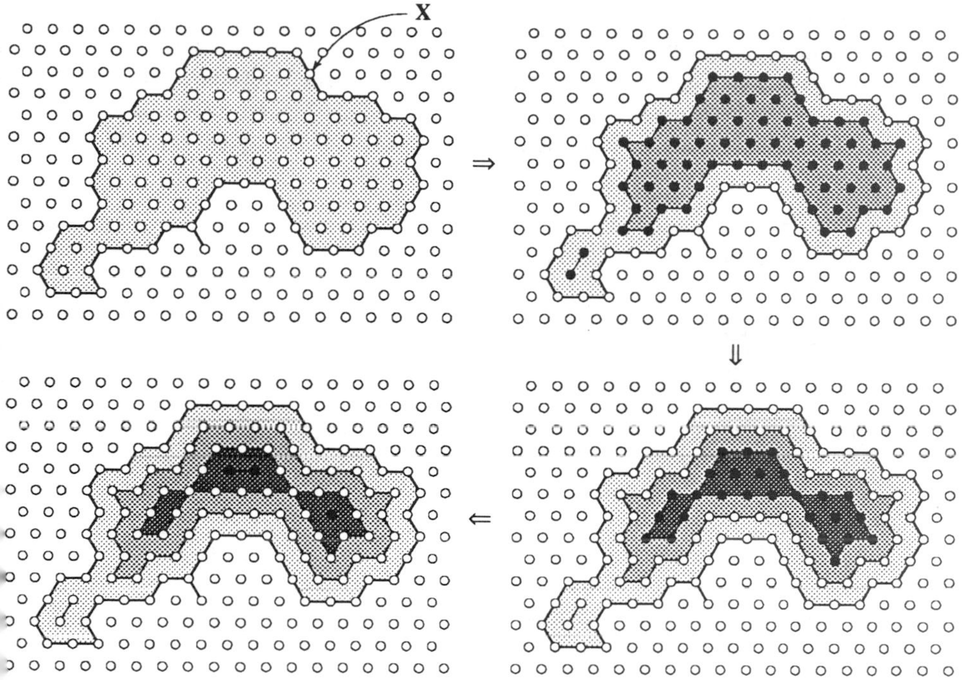

Figure 6. Successive steps involved in the parallel computation of a distance function.

itive execution times limit their practical interest. Moreover, the parallel computation of some more complex transformations like skeletons [9] and skeletons by influence zones (SKIZ) [21] is usually achieved via iterations of parallel thinnings or thickenings. These operations involve *structuring elements* [38], that is, pixel templates used as probes (see Golay's alphabet [17]). This is the reason why their adaptation to other grids requires cumbersome neighborhood analyses. This remark is even more true when it comes time to extend these algorithms to n-dimensional data! Furthermore, although some of them can bring Euclidean distances into play [54], in many cases the very local way parallel algorithms work leads to approximative transforms (e.g., for skeletons).

III. SEQUENTIAL ALGORITHMS

In an attempt to reduce the number of scannings required for the computation of an image transform, sequential or recursive algorithms have been proposed [31]. They rely on the following two principles:

1. The image pixels are scanned in a predefined order, generally raster (left to right and top to bottom) or antiraster.
2. The new value of the current pixel, determined from the values of the pixels in its neighborhood, is written directly in the same image, so that it is taken into account when determining the new values of the as yet unconsidered pixels.

Note that here, unlike the situation for parallel algorithms, the scanning order is essential. A number of transformations that can be obtained sequentially are described in [23]. In the following, a sequential scanning will be introduced by:

Scan D_I in raster order {

Let p be the current pixel; . . .

To compute a distance function sequentially, a raster scanning followed by an antiraster one are sufficient [32]: out of the original binary image I, the raster scanning creates an intermediate gray-level image I′, whose highest values are located in the lower right part of the connected components of I. Each feature pixel p of I is assigned the length of (one of) the shortest path P between p and the background, with the following constraint: every path element $\overline{xy}$ of P has either a strictly positive vertical component (i.e., $\overline{xy}$ is pointing upward) or a zero vertical component and a positive left component. This is why a second scanning, of antiraster type, is necessary to get from I' an actual distance function. This algorithm is given below for the hexagonal distance:

<u>Algorithm:</u> Sequential Distance Function

- input: I, binary image:

  /* The distance function is computed directly in I */

- Scan D_I in raster order {

  Let p be the current pixel;

  If $I(p) \neq 0$ then

  $I(p) \leftarrow \min\{I(p + \mathbf{u}_1) + 1, I(p + \mathbf{u}_2) + 1, I(p + \mathbf{u}_3) + 1\}$;

  }

- Scan D_I in anti-raster order {

  Let p be the current pixel;

  If $I(p) \neq 0$ then

  $I(p) \leftarrow \min\{I(p), I(p + \mathbf{u}_4) + 1, I(p + \mathbf{u}_5) + 1,$

  $I(p + \mathbf{u}_0) + 1\}$;

  }

In all cases, the above algorithm only requires *two* image scannings. In comparison to the parallel one, it thus constitutes a clear improvement! This is even more true since this sequential distance function algorithm has been integrated into the chip PIMM1 described in [20]. In fact, sequential algorithms often constitute one of the best possible choices. Some of the operations for whose implementation they are best suited are mentioned below. In addition, sequential algorithms such as the distance function one are easily extended to *n*-dimensional spaces [7] and can be adapted to better discrete distances [8,10].

A. Granulometry Function, Gray-Scale Dilations, and Erosions

Let I be a discrete binary image and let $(B_n)_{n \geq 0}$ be a family of structuring elements such that the B_n's are the homothetics of a given convex set B. The following equations hold:

$$\begin{aligned} & B_0 = \{\mathbf{o}\} \\ & B_1 = B \\ \forall_n \geq 1, \quad & B_{n+1} = B_n \oplus \mathrm{B} \end{aligned} \tag{8.2}$$

$\oplus$ denoting the Minkowski addition. Also, let γ_C denote the morphological opening with respect to structuring element C [39]. The *granulometry function* of I with respect to (B_n), denoted here by $g(I)$, associates with each pixel $p \in D_I$ the smallest integer k such that $\gamma_{B_k}(p) = 0$. In other words:

$$g(I) \begin{cases} D_I \rightarrow \mathbf{Z} \\ p \mapsto \min\{k \in \mathbf{Z}^+ \mid \gamma_{B_k}(p) = 0\} \end{cases} \tag{8.3}$$

Just as the distance function of a binary image I is the "pile" of its successive erosions, the granulometry function is nothing but the pile of its successive openings with respect to the B_n's. This means that by thresholding $g(I)$ at value k, one simply gets the binary opening of I with respect to element B_{k-1}. The histogram of $g(I)$ provides the granulometric analysis of I. Thus, although this transformation is not very exciting from a theoretical point of view, its great interest comes from the fact that it is possible to obtain it very quickly by using sequential methods. For example, the granulometry function shown in Figure 7 was obtained in 3 seconds on a Macintosh II. A very similar algorithm can be used to determine gray-scale dilations and erosions by the B_n's.

B. Morphological Shadowing

This is another case where sequential methods outperform all other techniques. The shadowing of a gray-scale image is realized through dilation by a ray of the three-dimensional space $\mathbf{Z}^2 \times \mathbf{Z}$. This is often interesting for visualization purposes and can be efficiently implemented thanks to a recursive algorithm de-

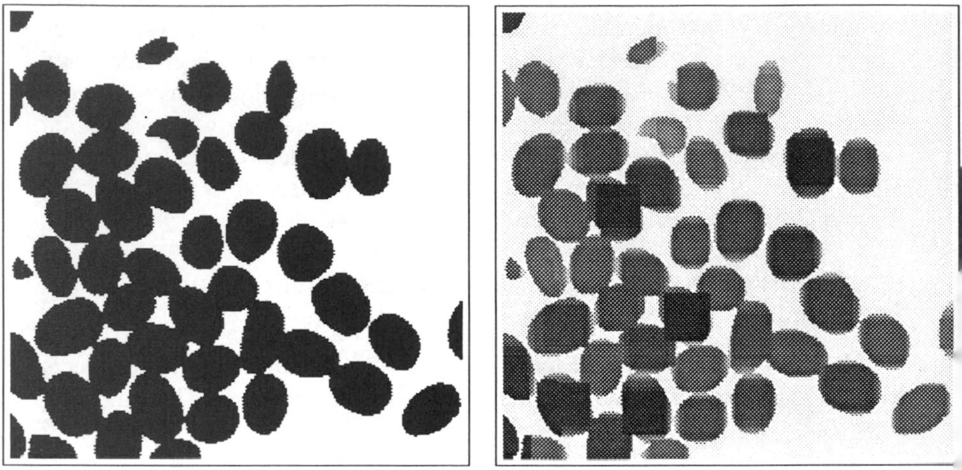

Figure 7. A binary image and its corresponding granulometry function with respect to a family of squares.

scribed in [45]. An example of morphological shadowing of a distance function is shown in Figure 8. One can notice in this example that the crest lines of the distance function have been highlighted, thereby leading to a family of methods for computing skeletons.

C. Geodesic Reconstruction

In geodesic morphology [22], a transformation called *reconstruction* turns out to be of immense interest. Given two (binary or gray-scale) images f and g such that $g \leq f$ (i.e., for every pixel p, $g(p) \leq f(p)$), the reconstruction $R_f(g)$ of f from g is obtained by dilating g geodesically under f until stability is reached. f is called the *mask* image, whereas g is the *marker*. More precisely, denote by B the elementary ball of the grid being used. For example, B is hexagon H in 6-connectivity, 5-pixel square S_1 in 4-connectivity, or 9-pixel square S_2 in 8-connectivity (see Figure 9). Let δ_B stand for the dilation with respect to B and $\wedge$ be the pointwise minimum. The reconstruction of f from g is obtained by iterating the following operation until stability is reached:

$$g \leftarrow \delta_B(g) \wedge f \tag{8.4}$$

In the binary case, reconstructing f from g allows us to extract those connected components of binary image f which contain at least a pixel of g [44]. This extends to the gray-scale case in terms of peaks: as illustrated by Figure 10, only the peaks of f that are marked by g are preserved through reconstruction. From

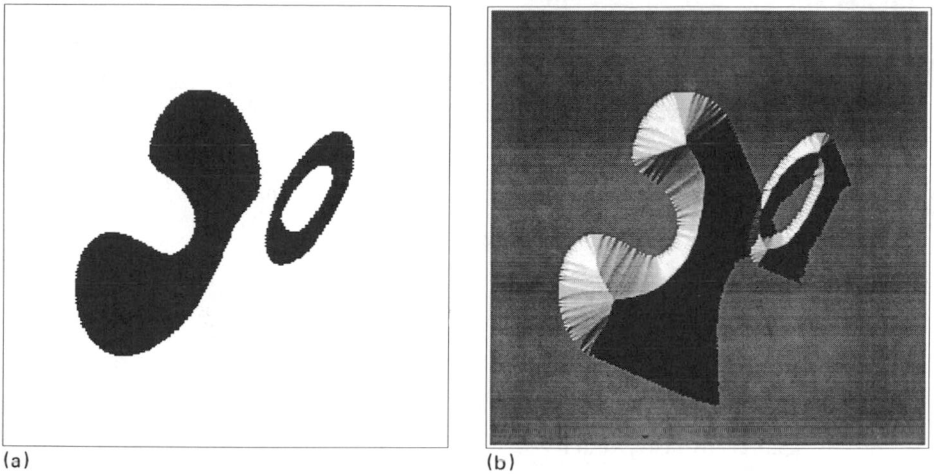

Figure 8. Shadowing of the distance function of a binary shape. (a) Original image; (b) distance function with artificial shadowing.

Figure 9. Elementary ball in 4-, 6-, and 8-connectivity, respectively.

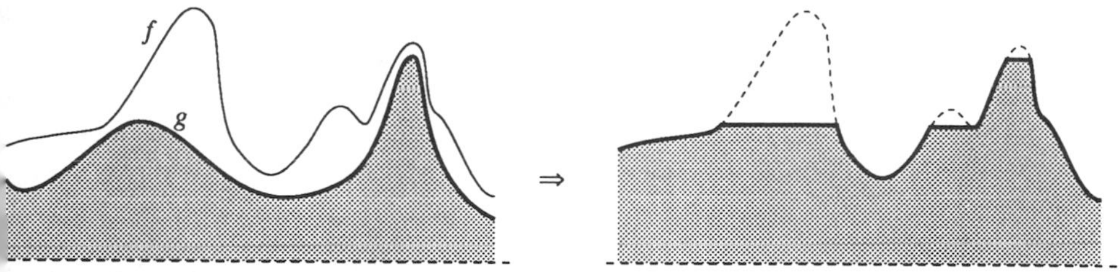

Figure 10. Gray-scale construction of f from g.

Eq. (8.4), it is straightforward to derive a parallel algorithm for binary and gray-scale reconstruction, but it is particularly inefficient on standard equipment. Here, it is much preferable to resort to a sequential algorithm [23,47]: like the sequential distance function algorithm previously described, the present one works by propagating information downward in a raster scanning and then upward in an antiraster scanning. Here, however, as explained later, these raster and antiraster scannings have to be iterated until stability is reached (see Figure 12). This algorithm is described below for the hexagonal case. It works for both the gray-scale and the binary case and usually only requires no more than 10 complete image scannings:

Algorithm: Sequential Reconstruction

- { –mask: I, binary or grayscale image,
  –marker: J, image defined on domain D_I.

/★ The reconstruction is determined in marker-image J ★/

/★ *Note*: we assume that $\forall p \in D_I$, $J(p) \leq I(p)$ ★/

- Repeat until stability is reached {

    Scan D_I in raster order {

      Let p be the current pixel;

      $J(p) \leftarrow (\max\{J(p), J(p + \mathbf{u}_1), J(p + \mathbf{u}_2), J(p + \mathbf{u}_3)\}) \wedge I(p)$;

    }

    Scan D_I in anti-raster order {

      Let p be the current pixel;

      $J(p) \leftarrow (\max\{J(p), J(p + \mathbf{u}_4), J(p + \mathbf{u}_5), J(p + \mathbf{u}_0)\}) \wedge I(p)$;

    }

}

As mentioned earlier, reconstruction is a particularly powerful morphological tool. Its several binary applications (filtering, hole filling, etc.) are rather well known, but it is even more useful in the gray-scale case [44,47]. For example, to extract the maxima of an image I, it suffices to reconstruct I from $I - 1$. By algebraic difference between I and the reconstructed function, one gets the desired maxima. Alternatively, the example of Figure 11 illustrates the use of gray-scale reconstruction for picture segmentation; Figure 11a is an image of blood vessels in the eye in which microaneurisms have to be detected. They are small compact light spots that are disconnected from the network of the (light) blood vessels. To extract them, the first step is to perform a series of openings of Figure 11a with respect to segments of different orientations. These segments are chosen to be longer than any possible aneurism, so that the aneurisms are removed by any such opening. On the other hand, since the blood vessels are elongated and light, there will be at least one orientation at which they are not completely removed by opening. After taking the supremum of these different openings, one gets Figure 11b, which is still an algebraic opening of Figure 11a [39]. It is used

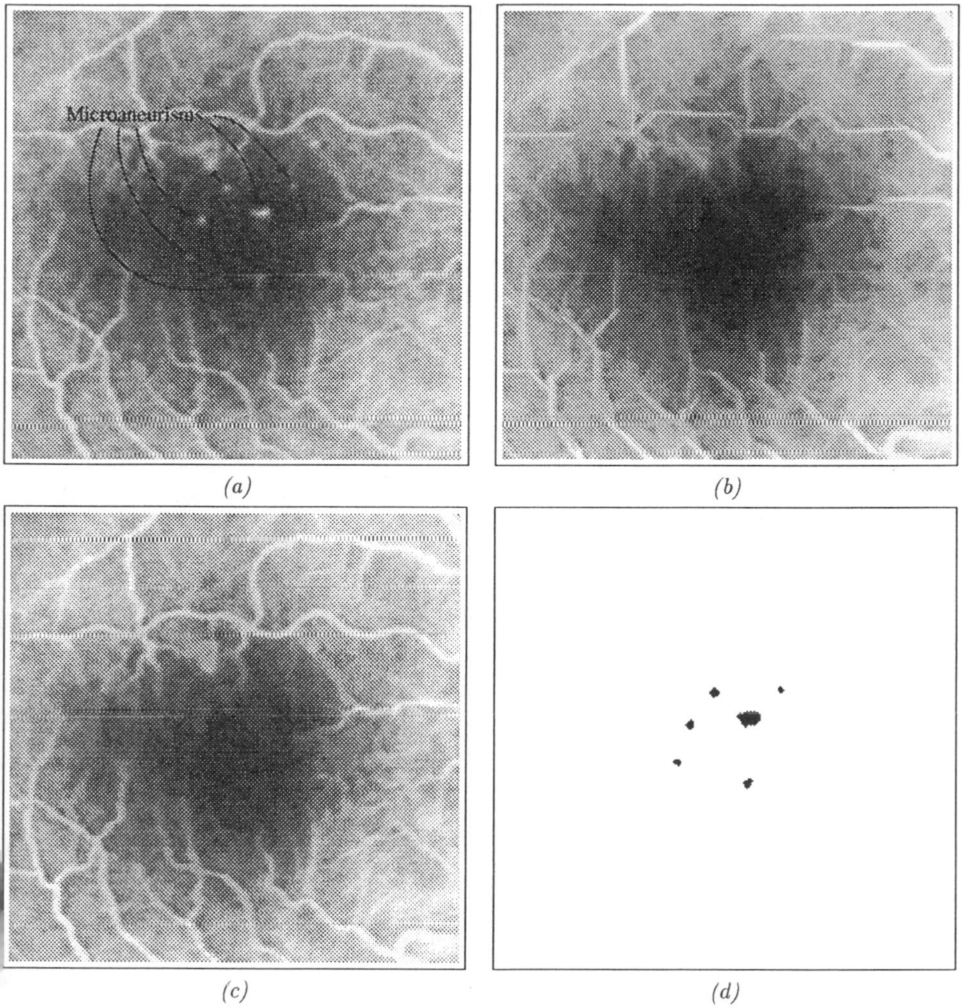

Figure 11. Use of gray-scale segmentation for image segmentation: (a) original image blood vessels; (b) supremum of openings by segments; (c) reconstructed image; (d) microaneurisms obtained by subtracting (c) from (a) and thresholding the result.

as marker to reconstruct the blood vessels entirely. Figure 11c is the result of the gray-scale reconstruction of Figure 11a from Figure 11b. Since the aneurisms are disconnected from the blood vessels, they have not been reconstructed! Thus, by algebraic difference between Figure 11a and Figure 11c, followed by thresholding, the microaneurisms shown in Figure 11d are easily extracted.

The above reconstruction algorithm also underscores some typical drawbacks of sequential algorithms. For example, in the binary case, when the mask is a "rolled-up" particle, the number of image scannings required for its reconstruction may be very important, as illustrated by Figure 12. However, only the values of a few pixels are actually modified after each scanning!

For this reason, a further step in the design of efficient morphological algorithms consists in *considering only the pixels whose value may be modified*. A first scanning is used to detect the pixels that are the process initiators and are typically located on the boundaries of the objects or regions of interest. Then, starting from these pixels, information is propagated only in the relevant image parts. The categories of algorithms described in the next two sections rely on these principles. They both require *random access* to the image pixels as well as to the neighbors of a given pixel.

IV. LOOP AND CHAIN ALGORITHMS

These methods were proposed in 1988 by M. Schmitt [34] and are based on the following simple remark:

> In a metric space (E,d), the boundary of the dilation $\delta(X)$ of a set X by an isotropic structuring element is a curve which is parallel to the boundary of X.

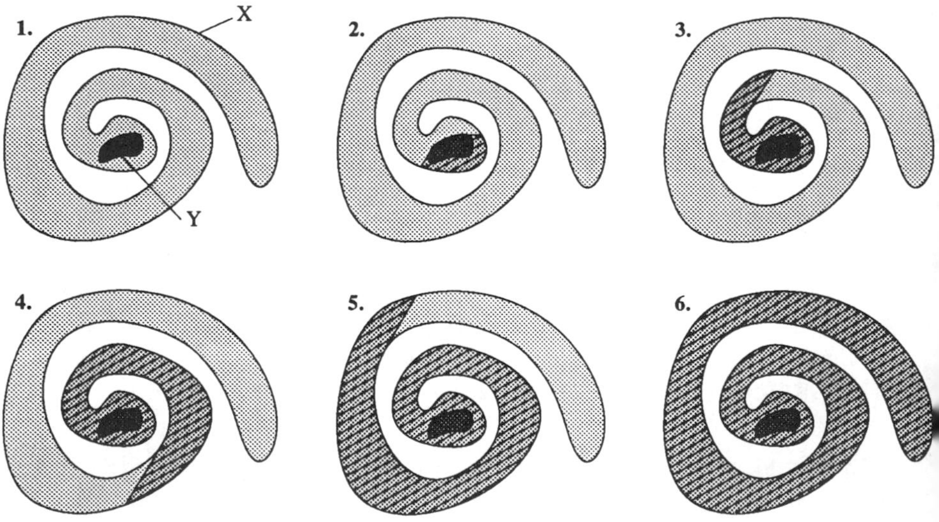

Figure 12. The sequential computation of a binary reconstruction in a rolled up mask may involve several complete image scannings; here, the hatched zones represent the pixels that have been modified after each step.

This is illustrated by Figure 13. Hence, if one is able to determine quickly the curves parallel to a given one, the calculation of isotropic dilations can be efficiently realized. This process can then be used to compute a large number of other morphological transformations that can be defined from isotropic dilations in an incremental fashion. Among others, distance functions, which are nothing but "piles" of erosions, are attainable this way.

The first step of these algorithms therefore consists in a tracking of the contours of the image I under study and in their encoding as Freeman loops [13]. A loop L is a data structure made of:

1. An origin pixel Or_L
2. A length $l(L)$
3. A list of $l(L)$ integers of the segment [0, 5], coding the elementary vectors $\mathbf{u}_0, \mathbf{u}_1, \ldots, \mathbf{u}_5$ of the hexagonal grid

The two extremities of a loop coincide. An example of a loop and of its encoding is shown in Figure 14.

Given a loop L coding the boundary of a set $X \subset \mathbf{Z}^2$, the dilated loop δL—coding the boundary of $\delta(X)$—is determined by means of *rewriting rules*. These rules allow one to derive from two successive contour elements of L a certain number (between 0 and 5) of contour elements of δL. The dilated loop is thus

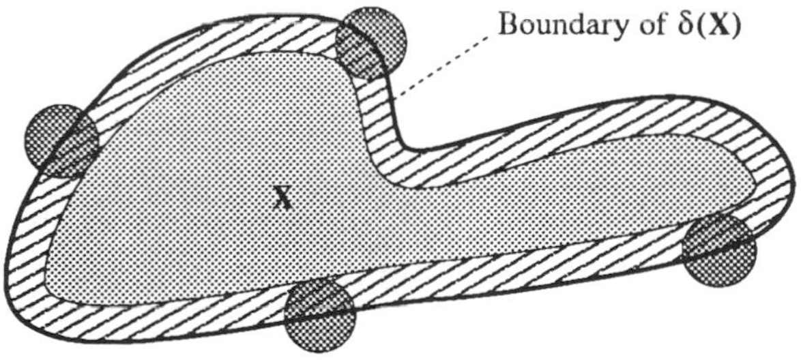

Figure 13. The boundary of the dilated set $\delta(X)$ is parallel to the boundary of X.

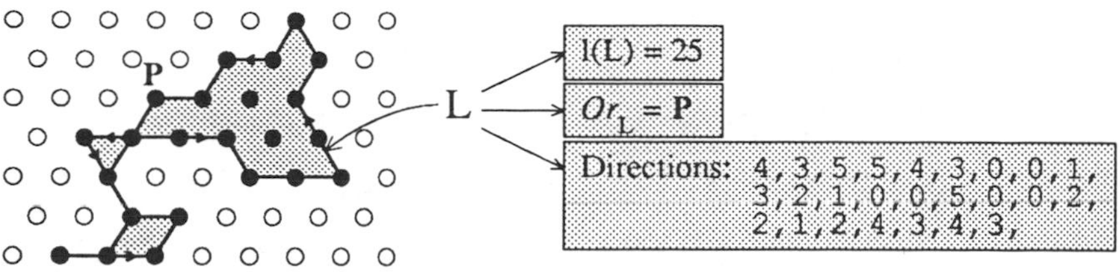

Figure 14. Example of a loop and of its encoding.

obtained from L in linear time with respect to $l(L)$. In the hexagonal case, there are exactly six rewriting rules (up to the six rotations), which are illustrated by Figure 15. Rule number 4 may seem useless but is, in fact, essential as soon as two successive dilations have been performed [34].

Now, once the dilated loops are determined, they must be written in the image and the corresponding pixels have to be given the appropriate value. For example, in the case of a binary dilation, the pixels corresponding to the dilated loops have to be assigned value 1. In fact, while the loops are written in the original image, one can detect if some of them intersect. For example, it may well happen that two loops coming from two different connected components intersect after a dilation step, as illustrated by Figure 16. In such cases, the overlapping parts become useless and can be cut. The remaining loop parts are called

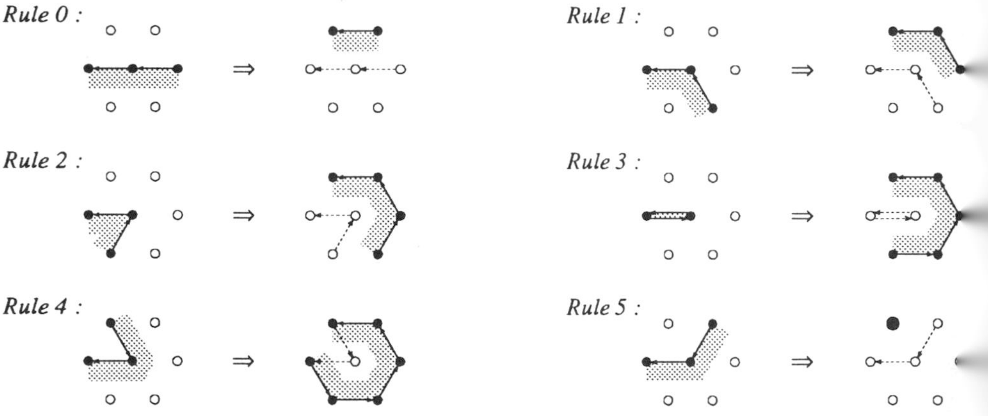

Figure 15. Rewriting rules allowing one to determine δL from L.

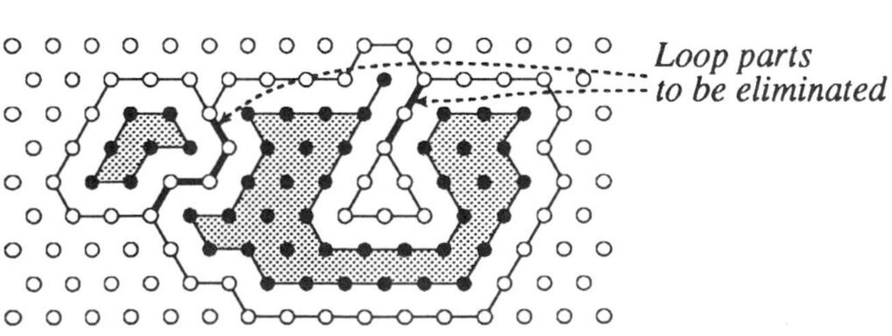

Figure 16. After dilation, certain loop parts must be eliminated; this is the adjustment step, in which chains are created. Here, the original (two) loops are located around the gray areas, and the parts of the dilated loops to be cut are drawn in a bold stroke.

chains; they are nothing but loops whose extremities do not coincide, and they are manipulated exactly as loops by using the rewriting rules of Figure 15. They can well be cut again at further steps. An example of a chain dilation is shown in Figure 17. The succession of operations described in this paragraph is referred to as *adjustment*. More precisely, during this adjustment step, one keeps only the chain or loop parts that are located in a given mask (set of pixels having a certain value) and gives them the appropriate value.

As an illustration, let us consider again the case of the hexagonal distance function. To determine it, we iterate dilations and adjustments of the chains until stability is reached. After each step, the value given to pixels in the "adjusted" chains is incremented by 1.

<u>Algorithm:</u> Distance Function by Chains and Loops

●input: I, binary image; /★ distance directly computed in I ★/
●Track the contours of I^C (complement image) and encode them as loops;
●dist ← 2; /★ variable containing the current distance ★/
●Repeat until there remain chains or loops {
  Dilate the chains and the loops;
  Adjust them in the mask $\{p \in D_I \mid I(p) = 1\}$, giving the corresponding pixels value dist;
  dist ← dist + 1;
}

Like almost all algorithms relying on this chain propagation principle, this one requires only two image scannings: one for the contour tracking step plus one scanning *of the feature pixels only* in the propagation step (in fact, here, to avoid an additional scanning, the algorithm is designed to yield $\text{dist}_I + 1$).

Chains and loops algorithms are thus extremely fast. Moreover, after the initial contour tracking is achieved, loops and chains may well be propagated *inside*

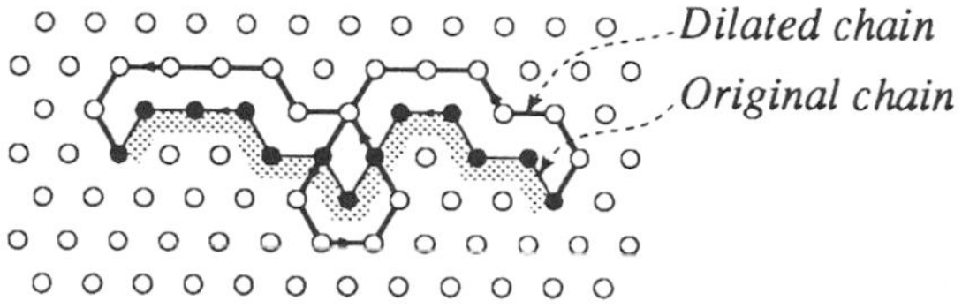

Figure 17. Dilation of a chain. Note that a loop has been created here, which will be eliminated during the adjustment step.

a given mask; for this reason, the present methods are particularly suited to the computation of binary geodesic transformations [34]. In this framework, they are probably the fastest techniques available.

A. Propagation Function

The previous remark is particularly true for the *propagation function* of a simply connected set, where chain propagation methods provide the only known efficient algorithm [35,24]. Recall that the propagation function p_x associates with each pixel of a connected set X its geodesic distance to the farthest pixel of X:

$$p_x \begin{cases} X \rightarrow \mathbf{Z}^+ \\ p \mapsto \sup\{d_X(p,q) \mid q \in X\} \end{cases} \tag{8.5}$$

The algorithm for computing p_X is detailed in [34] for the hexagonal case. It basically works through the determination of a supremum of geodesic distance functions, each of these being obtained via chain propagations. This transformation is illustrated by Figure 18 in the case of a 4-connected square grid. It has a very large number of practical applications, ranging from the extraction of extremities and geodesic centers [34] to the determination of antiskeletons [36].

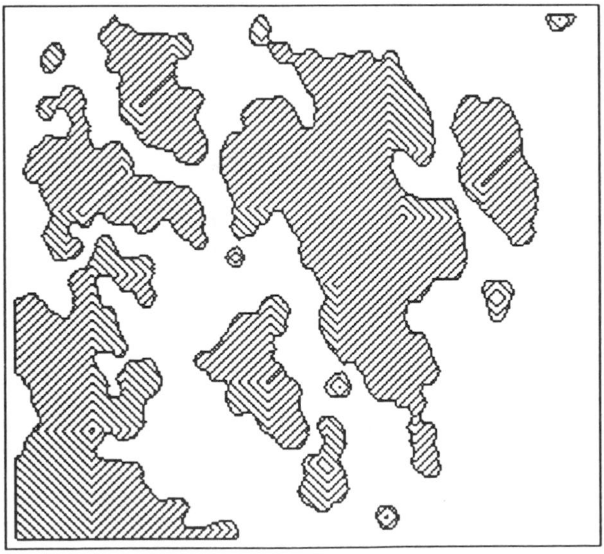

Figure 18. Level lines of the 4-connected propagation function of a binary image.

B. Euclidean Distance Function and Applications

Chains and loops algorithms are flexible in that they can also be adapted to better distances; for example, the dodecagonal one can be obtained by modifying the rewriting rules of Figure 15 [34, pages 86–89]. But it is even possible to adapt these algorithms to actual Euclidean distance [51]; the idea is to modify the chain and loop structure as well as the rewriting rules of Figure 15 in such a way that Euclidean distances are conveyed in the image by theses structures. Previous algorithms were of sequential type and yielded only more isotropic distances [8] or approximations of Euclidean distance functions [10]. An example of an exact Euclidean distance function is shown in Figure 19.

The same technique extends to the determination of Euclidean skeletons [27] and skeletons by influence zones [21]. In addition, Delaunay triangulations [30,6], Gabriel graphs [15], and relative neighborhood graphs [40] can be derived from these methods and obtained in arbitrary binary pictures [47,51]. Examples can be found in Chapter 6 of this book [18].

C. Morphological Transformations with Arbitrary Structuring Elements

Other extensions of the present loop-based methods include efficient algorithms for computing binary dilations, erosions, openings, and closings with structuring elements of arbitrary size and shape [48]. Here, chains and loops are no longer

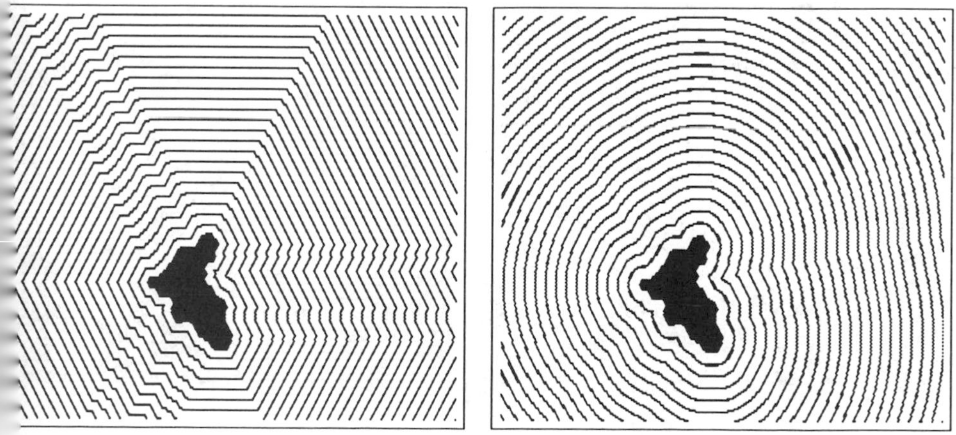

Figure 19. Comparison between hexagonal and Euclidean distance functions.

propagated in the image. Instead, the involved structuring element is encoded appropriately and propagated along the loops representing the set to be dilated or eroded. Combining an erosion and a dilation step allows us to determine openings and closings equally well, as illustrated by Figure 20.

To summarize, chains and loops are particularly efficient for the computation of binary morphological transformations and give rise to the interesting extensions described above. Unfortunately, they are not easy to adapt from one grid to another. Furthermore, they do not extend to multidimensional spaces. In these respects, the algorithms based on queues of pixels, which are discussed next, are much more general.

V. ALGORITHMS BASED ON QUEUES OF PIXELS

In this section, we again satisfy the principle according to which only the "interesting" image pixels are considered at each step. The image under study is regarded as a graph whose vertices are the pixels and whose edges are provided by the discrete grid G. Then, instead of loops and chains, we make use of a *queue of pixels* to perform breadth-first scannings of this graph. This idea has already proved to be particularly interesting in image analysis and morphology [52,47].

A queue is a first-in-first-out (FIFO) data structure, which means that the pix-

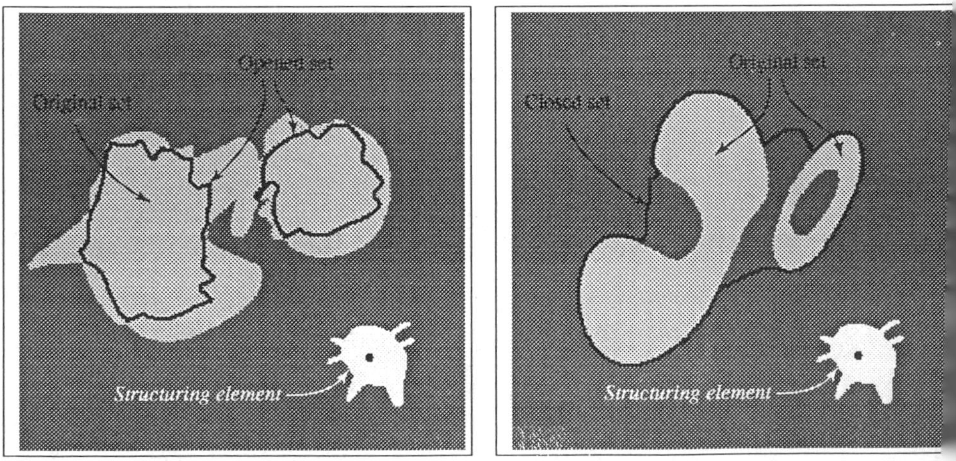

Figure 20. Binary opening and closing by an arbitrary (and weird!) structuring element.

els which are first put into it are those which can first be extracted. In other words, each new pixel included in the queue is put on one side, whereas a pixel being removed is taken from the other side (see Figure 21). In practice, a queue is simply a large enough array of pointers to pixels, on which three operations may be performed:

fifo-add(p): puts the (pointer to) pixel p into the queue
fifo-first( ): returns the (pointer to) pixel which is at the beginning of the queue and removes it
fifo-empty ( ): returns *true* if the queue is empty and *false* otherwise

The implementation of our distance function using this queue and the above operations is accomplished as follows:

<u>Algorithm:</u> Distance Function Using a Queue of Pixels

- input: I, binary image;
  /★ The distance function is computed in I directly ★/
- For every pixel $p \in D_I$, do {
  /★ detection of the pixels to be initially put on the queue ★/
  If $I(p) = 1$ and $\exists p' \in N_G(p), I(p') = 0$ {
    fifo_add(p);
    $I(p) \leftarrow 2$;
  }
  }
- While fifo_empty( ) = false {
  $p \leftarrow$ fifo_first( );
  For every $p' \in N_G(p)$ {

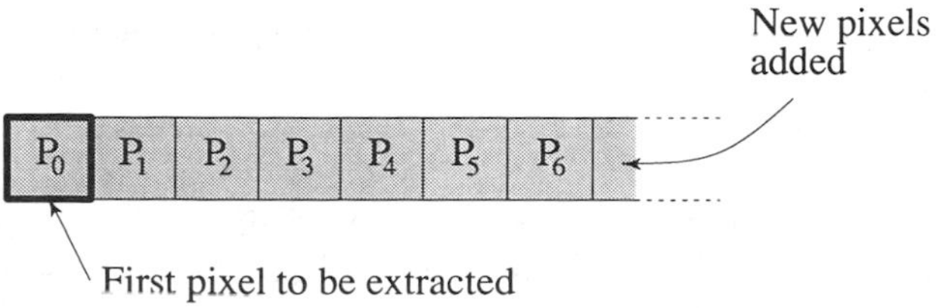

Figure 21. How a queue of pixels works.

```
            If I(p') = 1 {
                I(p') ← I(p) + 1;
                fifo_add(p');
            }
        }
    }
```

Here again, this algorithm actually yields $\text{dist}_I + 1$, a trick that avoids an additional image scanning.

Like the methods described in the previous section, queue-based algorithms are extremely efficient, in both the nongeodesic and the geodesic cases. The simplicity of the above distance function procedure is also interesting, and this characteristic is shared by most FIFO algorithms. They are thus more suited than the chain propagation ones to the development of procedures for computing complex transformations like gray-scale reconstruction (for an efficient alternative to the sequential algorithm described in Section III, see [47, chapter 6]), skeletons, and watersheds. The latter two are briefly described and illustrated below.

Moreover, contrary to the chain propagation algorithms, the present ones are extremely easy to adapt from one grid to another, since it suffices to modify the routine that generates the neighbors of a given pixel. Similarly, their extension to n-dimensional images and even to graphs is straightforward. They have been used to implement a large number of morphological transformations on graphs [42], which have already been used in physical applications [46,19] and are expected to be of great interest for complex picture segmentation tasks [43]. Several ideas and algorithms about mathematical morphology on graphs can be found in [41] or in chapter 6 of this book [18].

A. Skeletons

The skeleton transformation is widely used in morphological image processing. It was introduced by Blum in 1961 as the *medial axis transformation* [5]. The definition he proposed is based on the concept of grassfire: assuming a grassfire starting from the boundary of a set $X \subset \mathbf{Z}^2$ is propagating within it at uniform speed, the skeleton $S(X)$ of X is the set of the pixels where different firefronts meet. This is illustrated by Figure 22. A more formal definition of the skeleton was then proposed by Calabi and Harnett [9], based on the notion of maximal ball: the skeleton of X is defined as the locus of its maximal balls for the used metrics. One can show that this skeleton can be obtained as the set of local maxima of the distance function of X [47,49].

Unfortunately, a well-known result is that the skeleton by maximal balls, sometimes called the *true skeleton*, is not necessarily connected even though the original set is! However, to be useful in practice, skeletonization needs to be a

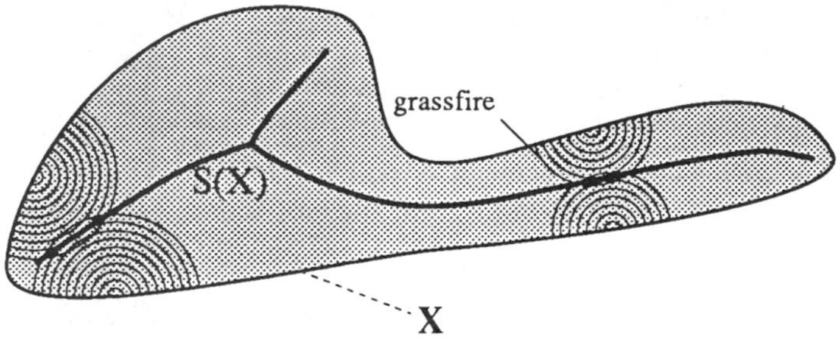

Figure 22. The skeleton can be viewed as the locus of the pixels where two firefronts stemming from the object's boundaries meet.

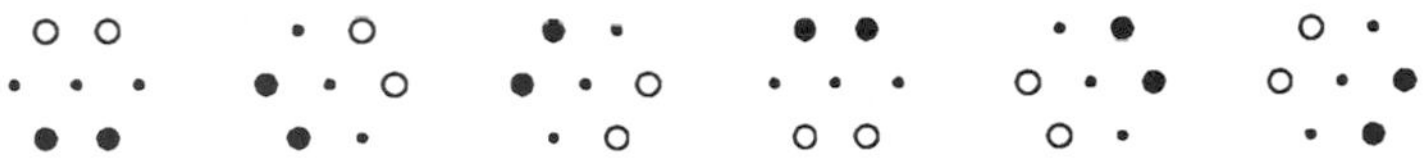

Figure 23. The two-phase structuring element L and its six rotations.

homotopic transformation [38]; roughly speaking, it needs to preserve the number of connected components and the number of holes in the original set, as well as the inclusion relationships between these components and holes. The practical problem with the implementation of skeletons consists therefore in extracting an object of unit width that would be as close as possible to the skeleton by maximal balls, while preserving the homotopy of the original set. The first methods proposed in the literature are part of what we referred to in this chapter as parallel algorithms (see Section II). They consist in removing successive peels of the set by means of homotopic thinnings until stability is reached [38]. For example, with hexagonal connectivity one generally performs homotopic thinnings with respect to structuring elements L shown in Figure 23 [17]. The main drawbacks of these techniques are their inefficiency and the fact that the resulting objects, though homotopic and of unit width, have not much in common with skeletons by maximal balls!

Sequential algorithms based on the crest lines of the distance function have also been proposed [26,28]. They are more efficient than the previous ones (they work in a fixed number of image scannings), produce accurate results, and can even be extended to Euclidean distances [27]. However, as explained in [49], these algorithms require cumbersome neighborhood analyses and their flexibility is rather poor. Other methods proposed in the literature include computational geometry–based algorithms [30] as well as contour-based techniques [1,53]. They are among the most efficient methods but are very complex, have little flexibility, and allow the determination of only one given type of skeleton [47].

Based on the above remarks, the skeleton algorithm detailed in [49] makes use of homotopic peelings, crest points, and contours: more precisely, starting from the boundaries of X, successive peelings are realized until stability (i.e., one-pixel thickness) is reached. These peelings—or grassfire propagation process—are efficiently implemented via a queue of pixels. At every step, the current pixel p may be removed (i.e., given value 0) if and only if one of the following conditions is fulfilled:

1. p does not belong to the skeleton by maximal balls. In other words, p is not a crest point (local maximum) of the distance function.
2. Removing p does not modify the homotopy locally.

The first condition ensures the accuracy of the result in that it will be a superset of the skeleton by maximal balls. The second one means that the resulting object is homotopic; the local homotopy checkings are realized via specially designed look-up tables [49]. In this skeletonization process, the pixels belonging to the skeleton by maximal balls play the role of *anchor points*: the firefronts stemming from the boundaries of X tend to anchor themselves on these particular points. An example of skeletonization based on these principles is shown in Figure 24.

Like every FIFO algorithm, the present one is particularly efficient because only the feature pixels are considered during the fire propagation step. For example, the skeletonization process illustrated by Figure 24 takes less than 1 second on a Sun Sparc Station 1. The resulting skeletons are also more accurate than with most other methods; they are skeletons by maximal balls to which connecting arcs of unit thickness have been added for homotopy preservation.

The algorithm works in the Euclidean case as well as in the geodesic one. Most interestingly, it allows us to calculate a whole range of different skeletonlike transformations whose computation is hardly possible otherwise; this is simply achieved by using different sets of anchor points. For example, by taking as anchor points the *regional* maxima of the distance function instead of its local maxima, one gets an object referred to in the literature as the *minimal skeleton*. Similarly, an empty set of anchor points results in a "homotopic marking" of the set. Using as anchor points only the local maxima of elevation greater than n yields a smoother skeleton called a *skeleton of order n* [49]. Any of these skeletons can then be postprocessed via prunings, themselves realized via FIFO algorithms. Figure 25 shows a sample of these possibilities. To summarize, the present queue-based skeleton algorithm is particularly efficient, accurate, and flexible.

B. Watersheds

In the last decade, increasing attention has been put on the watershed transformation as a tool for image segmentation [11,2,3,50]. It is defined for gray-scale images via the notion of a *catchment basin*: let us regard the image under study

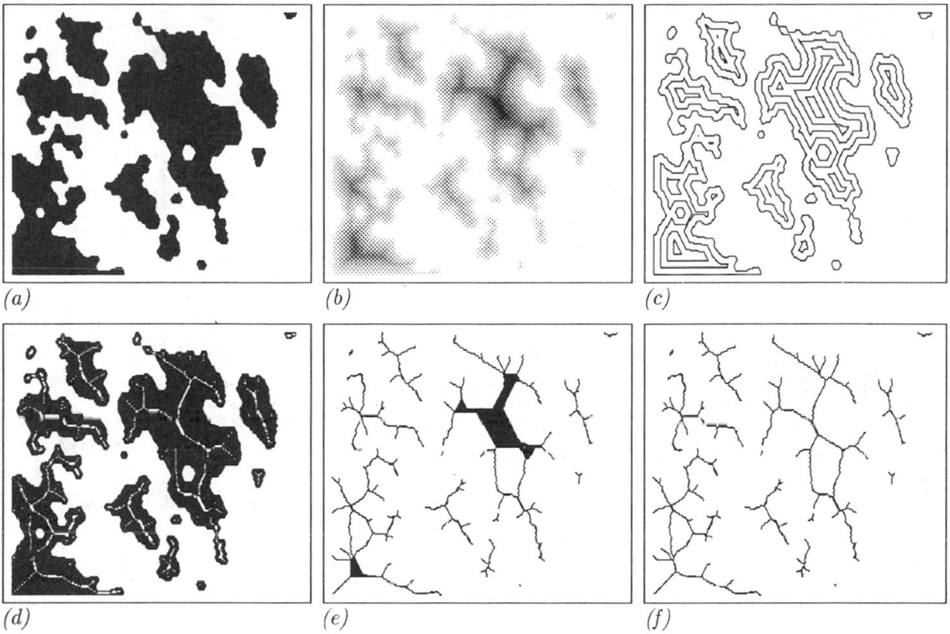

Figure 24. Construction of the standard skeleton. (a) Original image (256 × 256 pixels); (b) distance function; (c) level lines of this function; (d) skeleton by maximal balls; (e) propagation of the fire front; (f) final skeleton.

as a topographic relief (where the gray level of a pixel stands for its altitude) on which it is raining. A drop of water falling at a point p flows down along a steepest-slope path until it is trapped in a minimum m of the relief. The set $C(m)$ of the pixels such that a drop falling on them eventually reaches m is called the catchment basin associated with the minimum m. The set of the boundaries of the different catchment basins of an image constitutes its *watersheds*. These notions are demonstrated in Figure 26.

Here again, numerous techniques have been proposed to determine watersheds in digital pictures. The major ones are reviewed in [47,50]. One of the most interesting algorithms, originally proposed by Beucher, consists of "inverting" the watershed definition. Consider that the minima of the image—regarded here as a three-dimensional surface—have been pierced and that this image is slowly immersed into a lake. The water progressively floods the different catchment basins, and at some point water originating from two different minima will merge, thereby connecting the corresponding catchment basins. We prevent this by erecting dams at every place where connection would otherwise occur. Once the surface is totally immersed, the set of dams thus built corresponds to the watersheds of the initial image.

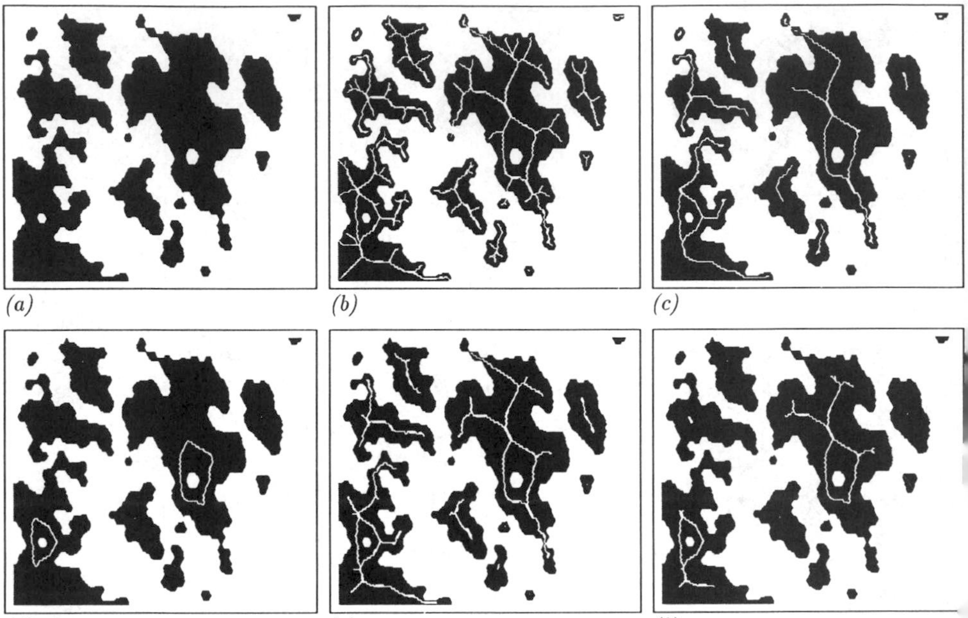

(a) (b) (c)

(d) (e) (f)

Figure 25. Various kinds of skeletons that may be efficiently determined using algorithms based on queues of pixels [49]. (a) Original image; (b) standard skeleton; (c) minimal skeleton; d) homotopic marking; (e) pruned skeleton (15 iterations); (f) skeleton of order 15. This example was produced using the hexagonal grid.

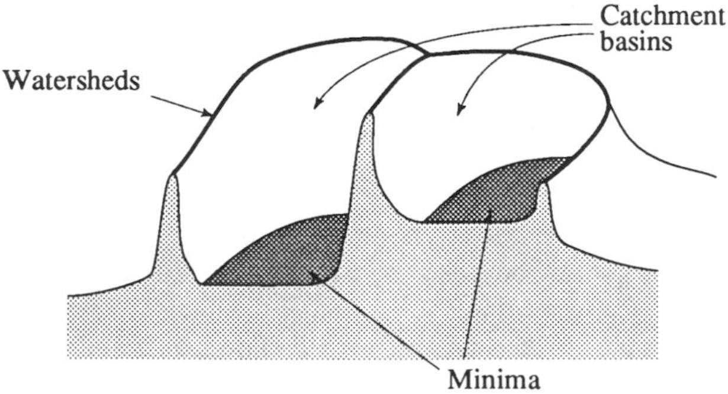

Figure 26. Minima, catchment basins, and watersheds.

This immersion and dam erection process can now be simulated by an algorithm. The most efficient algorithm described in the literature makes use of FIFO breadth-first scanning techniques for the actual flooding of the catchment basins [50]. A labeling of the catchment basins is also used, which automatically prevents the connection of two different basins. It has been shown that the results provided by this technique are more accurate that those of any other method. For example, the cone effect of Figure 5b is always avoided. Furthermore, just like almost all other FIFO algorithms, the present one extends to any grid and any dimension in a straightforward manner.

This algorithm dramatically reduces the computation times required for extracting watersheds. On conventional computers, previous approaches typically needed up to a couple of hours! The present one takes about 5 to 10 seconds on a Macintosh II, for a 256 × 256 image, thus opening the door to powerful segmentation methods in inexpensive software-based systems.

Let us conclude this section by an example of application of the present algorithm. We consider here Figure 27a; it is part of a series of successive images of the same scene, and the problem is to recover the motion of the camera. One of the approaches taken to solve this motion estimation problem consists of decomposing the images into regions and matching these regions over successive time frames [13]. Performing this decomposition by means of watershed techniques turns out to provide meaningful regions, which are then easily matched from one image to the next one [14]. Following the methodology described in [44], the watershed tool is applied to the morphological gradient of the original image (Figure 27c). In fact, to avoid oversegmentation, the watersheds of the gradient are constrained by a *marker image*. The markers are connected components of pixels belonging to each of the regions to be extracted and are obtained in this case from the intensity "domes" and "basins" of the original image (see Figure 27b). The result of this constrained watershed transformation is the highest watershed lines of the gradient, which are located between two markers (Figure 27d). The entire segmentation process takes approximately 5 seconds on a Sun Sparc Station 2.

VI. CONCLUSION AND SUMMARY

Table 2 summarizes the qualities and drawbacks of the families of algorithms that have been briefly reviewed in this chapter. Now, in practice, what algorithm should be chosen to implement a given transformation in a given environment? Of course, there is no absolute answer to this question. However, the following guidelines can be proposed:

Parallel algorithms should generally be avoided, unless running on specialized architectures.

(a) (b) (c) (d)

Figure 27. Example of a watershed segmentation: (a) original image; (b) marker image; (c) gradient image; (d) final segmentation obtained via watersheds of the gradient controlled by the markers.

Sequential algorithms constitute one of the best choices to implement gray-scale reconstructions, gray-scale dilations and erosions with some structuring elements, and granulometry functions.

Chain and loop methods should be chosen whenever binary transformations are concerned. They are indeed, the fastest in this case, particularly for geodesic

Table 2. Respective Qualities of the Various Families of Algorithms

Family of algorithms	Speed	Accuracy (if appropriate)	Adaptation to other grids	Development ease	Hardware imple-entation
Parallel	X	XX	XX	XXX	XXXX
Sequential	XX	XX	XX	XX	XXX
Loops and chains	XXXX	XXX	X	XXX	X
Queues	XXX	XXX	XXXX	XXXX	X

operations (reconstruction, geodesic distance function, hole filling, labeling, etc.) and provide the only known efficient propagation function algorithm.

FIFO algorithms will be preferred in all other cases and in particular in the following ones: geodesic transformations in square grids, n-dimensional or graph morphology, complex transformation like skeletons, SKIZ, and watersheds.

Clearly, the last two families are going to become increasingly important in the future. Regardless of accuracy and flexibility considerations, chain or queue algorithms are often faster on conventional computers than parallel algorithms on specialized hardwares! Now, a few years after the introduction of the parallel morphological algorithms, the first specialized hardwares were built on these principles. Between the publication of the sequential distance function algorithm (1968) and its first hardware implementation (1989), more than 20 years have elapsed. Let us hope that we will not have to wait 20 more years to see the first hardware realizing queue-based morphological operations. Indeed, this would probably allow us to compute complex morphological transformations in just several hundredths of a second.

ACKNOWLEDGMENTS

This work was supported in part by Dassault Electronique and the National Science Foundation under grant MIPS-86-58150, with matching funds from DEC and Xerox. Most of the examples have been produced using the software DIP Station on a Macintosh. The author would like to thank Dr. Peter Winthrop Hallinan for his help in reviewing this chapter.

REFERENCES

1. Arcelli, C., Cordella, L. P., and Levialdi, S., From local maxima to connected skeletons, *IEEE Trans. Pattern Anal. Machine Intelli.*, *2*, 134–143 (1981).

2. Beucher, S., and Lantuéjoul, Ch., Use of watersheds in contour detection, in *Proceedings of the International Workshop on Image Processing, Real-Time Edge and Motion Detection/Estimation*, Rennes, France, 1979.
3. Beucher, S., Segmentation d'images et morphologie mathématique, Ph.D. dissertation, Ecole Nationale Supérieure des Mines de Paris, 1990.
4. Bloch, I., Morphologie mathématique et représentation d'images par arbres, Internal Report CMM, School of Mines, Paris, 1987.
5. Blum, H., An associative machine for dealing with the visual field and some of its biological implications, in *Biological Prototypes and Synthetic Systems*, vol. 1, pp. 244–260, *Proceedings, 2nd Annual Bionics Symposium*, Cornell University, 1961 (E. E. Bernard and M. R. Kare, eds.), Plenum, New York, 1962.
6. Boissonat, J. D., Shape reconstructions from planar cross-sections, Technical Report INRIA, Le Chesnay, France, 1986.
7. Borgefors, G., Distance transformations in arbitrary dimensions, *Comput. Vision Graphics Image Process.*, *27*, 321–345 (1984).
8. Borgefors, G., Distance transformations in digital images, *Comput. Vision Graphics Image Process.*, *34*, 334–371 (1986).
9. Calabi, L., and Harnett, W. E., Shape recognition, prairie fires, convex deficiencies and skeletons, Parke Mathematical Laboratories Inc., Scientific Report No. 1, Carlisle, Massachusetts, 1966.
10. Danielsson, P. E., Euclidean distance mapping, *Comput. Graphics Image Process.*, *14*, 227–248 (1980).
11. Digabel, H., and Lantuéjoul, Ch., Iterative algorithms, in *Proceedings of the 2nd European Symposium on Quantitative Analysis of Microstructures in Material Science, Biology and Medicine*, Caen, France, Oct. 1977 (J.-L. Chermant, ed.), Riederer Verlag, Stuttgart, 1978, pp. 85–99.
12. Freeman, H., On the encoding of arbitrary geometric configurations, *IEEE Trans. Comput.*, *10*, 260–268 (1961).
13. Fuh, C.-S., and Maragos, P., Region based optical flow estimation, in *Proceedings IEEE Computer Vision and Pattern Recognition Conference '89*, San Diego, 1989, pp. 130–135.
14. Fuh, C.-S., Maragos, P., and Vincent, L., Region based approaches to visual motion correspondence, Technical Report HRL, Harvard University, Cambridge, Massachusetts, 1991.
15. Gabriel, K. R., and Sokal, R. R., A new statistical approach to geographic variations analysis, *Syst. Zool.*, *18*, 259–278 (1969).
16. Gesbert, S., Howard, V., Jeulin, D., and Meyer, F., The use of basic morphological operations for 3-D biological image analysis, *Trans. R. Microsc. Soc.*, *1* (1991).
17. Golay, M., Hexagonal pattern transforms, *Proc. IEEE Trans.* Comput., *C18*, (1969).
18. Heijmans, H. J. A. M., and Vincent, L., Graph morphology in image analysis, in *Mathematical Morphology in Image Processing* (E. Dougherty, ed.), Marcel Dekker, New York, 1991.
19. Jeulin, D., Vincent, L., and Serpe, G., Propagation algorithms on graphs for physical applications, Internal Report CMM, Ecole Nationale Supérieure des Mines de Paris, 1990.

20. Klein, J.-C., and Peyrard, R., PIMM1, an image processing ASIC based on mathematical morphology, in *IEEE ASIC Seminar and Exhibit*, Rochester, New York, 1989, pp. 25–28.
21. Lantuéjoul, Ch., Skeletonization in quantitative metallography, in *Issues of Digital Image Processing* (R. M. Haralick and J.-C. Simon, eds.), Sijthoff and Noordhoff, Groningen, The Netherlands, 1980.
22. Lantuéjoul, Ch., and Maisonneuve, F., Geodesic methods in image analysis, *Pattern Recogn.*, *17*, 117–187 (1984).
23. Laÿ, B., Recursive algorithms in mathematical morphology, in *Acta Stereol.*, vol. 6/III, *Proceedings of the 7th International Congress for Stereology*, Caen, France, 1987, pp. 691–696.
24. Maisonneuve, F., and Schmitt, M., An efficient algorithm to compute the hexagonal and dodecagonal propagation function, in *Acta Stereol.*, vol. 8/2, *Proceedings of the 5th European Congress for Stereology*, Freiburg-im-Breisgau, FRG, 1989, pp. 515–520.
25. Matheron, G., *Random Sets and Integral Geometry*, Wiley, New York, 1975.
26. Meyer, F., Skeletons and perceptual graphs, *Signal Process.*, *16*(4), 335–363 (1989).
27. Meyer, F., Euclidean digital skeletons, in *Proceedings of SPIE Visual Communications and Image Processing '90*, 1360, Lausanne, Switzerland, 1990.
28. Montanvert, A., Contribution au traitement de formes discrètes. Squelettes et codage par graphe de la ligne médiane, Ph.D. disscrtation, Université Scientifique, Technologique et Médicale de Grenoble, France, 1987.
29. Piper, L. J., and Tang, J.-Y., Erosion and dilation of binary images by arbitrary structuring elements using interval coding, *Pattern Recogn. Lett.*, 201–209 (April 1989).
30. Preparata, F. P., and Shamos, M. I., *Computational Geometry: An Introduction*, Springer-Verlag, New York, 1985.
31. Rosenfeld, A., and Pfaltz, J. L., Sequential operations in digital picture processing, *J. Assoc. Comput. Mach.*, *13*(4), 471–494 (1966).
32. Rosenfeld, A., and Pfaltz, J. L., Distance functions on digital pictures, *Pattern Recogn.*, *1*, 33–61 (1968).
33. Samet, H., The quadtree and other related hierarchical data structures, *ACM Comput. Surv.*, *16*(2), 87–260 (1984).
34. Schmitt, M., Des algorithmes morphologiques à l'intelligence artificielle, Ph.D. dissertation, Ecole Nationale Supérieure des Mines de Paris, 1989.
35. Schmitt, M., Geodesic arcs in non Euclidean metrics: application to the propagation function, *Rev. 'Intell. Artif.*, *13*(2), 43–76 (1989).
36. Schmitt, M., Anti-skeleton: some theoretical properties and applications, in *Proceedings, SPIE Visual Communication and Image Processing '90*, 1360, Lausanne, Switzerland, 1990, pp. 272–283.
37. Schmitt, M., and Vincent, L., *Morphology: Practical and Algorithmic Handbook for Image Analysis*, Cambridge University Press, London, in press.
38. Serra, J., *Image Analysis and Mathematical Morphology*, Academic Press, London, 1982.
39. Serra, J., *Image Analysis and Mathematical Morphology*, Part II: *Theoretical Advances*, Academic Press, London, 1988.

40. Toussaint, G. T., The relative neighborhood graph of a finite planar set, *Pattern Recogn.*, *12*, 1324–1347 (1980).
41. Vincent, L., Mathematical morphology on graphs, in *Proceedings*, *SPIE* Visual Communications and Image Processing 88, Cambridge, Massachusetts, 1988, pp. 95–105.
42. Vincent, L., Graphs and mathematical morphology, *Signal Process.*, *16*, 365–388 (1989).
43. Vincent, L., Mathematical morphology for graphs applied to image description and segmentation, in *Proceedings*, *Electronic Imaging West 89*, vol. 1, Pasadena, 1989, pp. 313–318.
44. Vincent, L., and Beucher, S., The morphological approach to segmentation: an introduction, Internal Report CMM, Ecole Nationale Supérieure des Mines de Paris, 1989.
45. Vincent, L., Morphological shading and shadowing algorithm, in *Proceedings*, *PIXIM 89*, *Computer Graphics in Paris*, Hermès, Paris, 1989, pp. 109–124.
46. Vincent, L., and Jeulin, D., Minimal paths and crack propagation simulations, in *Acta Stereol.*, vol. 8, *Proceedings*, *5th European Congress for Stereology*, 1989, pp. 487–494.
47. Vincent, L., Algorithmes morphologiques à base de files d'attente et de lacets. Extension aux graphes, Ph.D. dissertation, Ecole Nationale Supérieure des Mines de Paris, France, 1990.
48. Vincent, L., Morphological transformations of binary images with arbitrary structuring elements, *Signal Process.*, *22*(1), 3–23 (1991).
49. Vincent, L., Efficient computation of various types of skeletons, in *Proceedings*, *SPIE Medical Imaging V*, San Jose, California, 1991.
50. Vincent, L., and Soille, P., Watersheds in digital spaces: an efficient algorithm based on immersion simulations, *IEEE Trans. Pattern Anal. Machine Intell.*, *13*(6), 583–598 (1991).
51. Vincent, L., Exact Euclidean distance function by chain propagations, in *Proceedings*, *IEEE Computer Vision and Pattern Recognition '91*, Maui, Hawaii, 1991, pp. 520–525.
52. Van Vliet, L. J., and Verwer, B. J. H., A contour processing method for fast binary neighborhood operations, *Pattern Recogn. Lett.*, *7*, 27–36 (1988).
53. Xia, Y., Skeletonization via the realization of the fire front's propagation and extinction in digital binary shapes, *IEEE Trans. Pattern Anal. Machine Intell.*, *11*(10), 1076–1086 (1989).
54. Yamada, H., Complete Euclidean distance transformation by parallel operations, in *Proceedings*, *7th International Conference on Pattern Recognition*, Montreal, 1984, pp. 69–71.

Chapter 9

Discrete Half-Plane Morphology for Restricted Domains

Tapas Kanungo and Robert M. Haralick

University of Washington
Seattle, Washington

I. INTRODUCTION

Morphological operations, when performed on objects represented as sets of discrete points, are of $O(n^2)$ complexity, where n is the size of each set [6]. But when objects of interest are convex, or can be decomposed into convex objects, a more appropriate representation of the object is in terms of its boundary [1,2].

In this paper we extend our earlier work [7] and give two boundary representations for a class of two-dimensional binary shapes and define all the morphological operations—dilation, erosion, opening, closing, n-fold dilation, and n-fold erosion—in terms of these boundary representations. Further, we prove that each of these algorithms is $O(1)$ and hence independent of the size of the object. In addition, we prove that the results of these algorithms are equivalent to those obtained using the regular set-theoretic definitions. We also suggest how the algorithms can be extended for more complicated objects.

Morphological algorithms using boundary representations have been attempted by Ghosh [3,4] for polytopes in continuous domain. Xu [9] gave algorithms for decomposition of a class of binary convex shapes using boundary representation. There he used the notion of dilation of boundaries but did not prove it equivalent to the set-theoretic definitions. Our work sets the basic foundation by providing all the basic morphological operations in terms of the boundary representations.

This paper is organized as follows. In Section II we set the stage by giving the basic definitions and notation to be used. We introduce the B-code representation in Section III. In Section IV we formally characterize restricted domains in terms of B-codes and half-planes and show how to interconvert the representations. Morphology on restricted domains using B-codes and discrete half-planes is addressed in Section V. Here we start from the set, perform set morphology, and show that the same result is obtained using morphology using the half-plane representations. In Section VI we discuss the algorithms, evaluate their computational complexity, and compare them with the complexity of existing set-theoretic algorithms. Here we also walk through some examples of dilation and erosion of restricted domains. The open problems and work in progress are discussed in Section VII, and in Section VIII we give our overall conclusion.

II. PRELIMINARIES

In this section we define all the necessary terms and give the notation used in this chapter.

Images are represented as mathematical functions over some finite domain. One of its common forms is $f : (x, y) \rightarrow z$ where $x, y, z \in \mathbf{R}$, and $\mathbf{R}$ is the set of real numbers. The ordered pair (x, y) represents the spatial coordinates of a point with respect to some reference frame and z is the luminance value at that point. One way to represent such a function in a computer is by discretizing the space variables and the values the function takes.

The space domain can be discretized by tessellating it. A tessellation can be obtained by dividing the space into a set of nonintersecting domains such that the union of all the domains is the $\mathbf{R}^2$ space. Each of these domains is then represented by a point inside it. The most common technique for tessellating the $\mathbf{R}^2$ space is to make all the domains squares of the same size and represent them by their centers. To make the treatment simple, we will use squares of size 1×1, centered on the points $(i, j) \in \mathbf{Z}^2$, where $\mathbf{Z}$ is the set of integers. An illustration is given Figure 1. Note that the origin (0,0) represents the unit square centered on it; that is, the real axis passes through its center. Thus, the tessellation domains can be uniquely represented by the ordered pair $(i, j) \in \mathbf{Z}^2$. The ordered pairs will be interchangeably called *lattice points*, *points*, or *pixels*.

The function values are defined on the lattice points and can also be discretized. We will not go into the details of discretization of the function values, although the process is similar to that of space discretization. In this chapter we are interested in the case where the function takes binary values of 0 or 1 at the lattice points. Such images are called *discrete binary images* or *binary images*. Since the set of all the lattice points having the value 1 completely characterizes an image, the term binary image will be used to imply the set of all lattice points where the function value is 1. The terms *structuring element* and *shape* will also refer to sets of lattice points with function value 1.

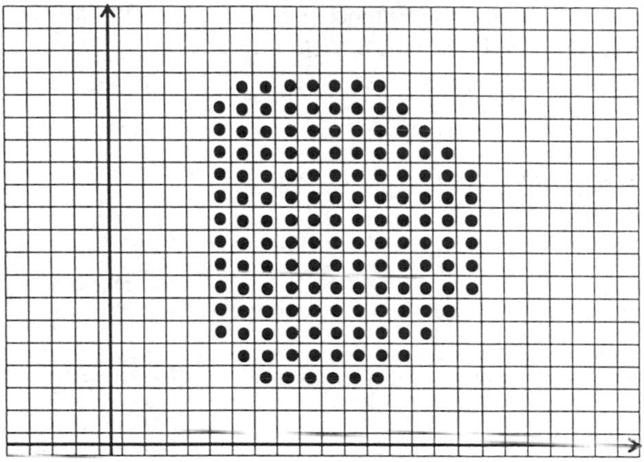

Figure 1. A binary image. Note that the lattice points are represented by the center of each pixel.

Definition 2.1. The **x**[ ] and **y**[ ] operators take a lattice point as an argument and return its x and y ordinates, respectively. Formally, let $p = (i, j) \in \mathbf{Z}^2$. Then $\mathbf{x} : \mathbf{Z}^2 \to \mathbf{Z}$ and $\mathbf{y} : \mathbf{Z}^2 \to \mathbf{Z}$ such that $\mathbf{x}[p] = i$ and $\mathbf{y}[p] = j$.

Definition 2.2. A point (x, y) is called a *foreground point* if its value is nonzero and a *background point* otherwise.

Definition 2.3. Two foreground points (i, j) and (k, l) are said to be *4-neighbors* if and only if (k, l) is an element of $\{(i + 1, j), (i - 1, j), (i, j), (i, j + 1), (i, j - 1)\}$.

Definition 2.4. Two foreground points (i, j) and (k, l) are said to be *8-neighbors* if and only if (k, l) is an element of $\{(i + 1, j), (i - 1, j), (i, j + 1), (i, j - 1), (i - 1, j - 1), (i + 1, j + 1), (i + 1, j - 1), (i - 1, j + 1)\}$.

Definition 2.5. Two foreground points x_i and x_j are said to be *4-connected* if and only if there is a sequence of foreground points $\{x_1 = x_1, x_2 \ldots, x_{n-1}, x_n = x_j\}$ such that x_k and x_{k+1} are 4-neighbors for all k in $\{1, \ldots, n\}$.

Definition 2.6. Two foreground points x_i and x_j are said to be *8-connected* if and only if there is a sequence of foreground points $\{x_1 = x_1, x_2 \ldots, x_{n-1}, x_n = x_j\}$ such that x_k and x_{k+1} are 8-neighbors for all k in $\{1, \ldots, n\}$.

Definition 2.7. A set of foreground points F is a *4-connected component* if for all x_i, $x_j \in F$, x_i and x_j arc 4-connected.

Definition 2.8. A set of foreground points F is an *8-connected component* if for all x_i, $x_j \in F$, x_i and x_j are 8-connected.

Definition 2.9. A foreground point of an 8-connected component is a *boundary* or *edge* point if one or more of its 4-neighbors is a background point.

Definition 2.10. A *discrete half-plane* is a set of lattice points $H \subset \mathbf{Z}^2$ defined as

$$H = \{(i, j) \mid a_0 i + b_0 j \leq c_0 \text{ such that } i, j, a_0, \mathrm{b}_0, \text{ and } c_0 \in \mathbf{Z}\}$$

Definition 2.11. A 4- or 8-connected component F is *convex discretely* if and only if all the lattice points lying inside or on the convex hull of F belong to F. This definition directly implies that a discretely convex connected component has no holes.

Next, we restate the definitions of the basic morphological operations based on the tutorial by Haralick, et al. [6].

Dilation is the morphological transformation that combines two sets using vector addition of set elements. If A and B are sets in $\mathbf{Z}^2$, the dilation of A by B is the set of all possible vector sums of pairs of elements, one coming from A and one coming from B.

Definition 2.12. The *dilation* of A by B is denoted by $A \oplus B$ and is defined by

$$A \oplus B = \{c \in \mathbf{Z}^2 \mid c = a + b \text{ for some } a \in A \text{ and } b \in B\}$$

Erosion is the morphological dual of dilation. If A and B are sets in $\mathbf{Z} \times \mathbf{Z}$, then the erosion of A by B is the set of all elements x for which $x + b \in A$ for every $b \in B$.

Definition 2.13. The *erosion* of A by B is denoted by $A \ominus B$ and is defined as

$$A \ominus B = \{x \in \mathbf{Z}^2 \mid x + b \in A \text{ for every } b \in B\}$$

Definition 2.14. The *opening* of a set B by a structuring element K is denoted by $B \bigcirc K$ and is defined as

$$B \bigcirc K = (B \ominus K) \oplus K$$

Definition 2.15. The *closing* of a set B by a set K is denoted by $B \bullet K$ and

$$B \bullet K = (B \oplus K) \ominus K$$

Definition 2.16. The *n-fold dilation* of a set B by a set A is denoted by $B \oplus (\oplus_n A)$ and is defined as

$$B \oplus (\underset{n}{\oplus} A) = B \oplus \overbrace{A \oplus A \oplus \cdots}^{n \text{ times}} \oplus A$$

Definition 2.17. The *n-fold erosion* of a set B by a set A is

$$B \ominus (\underset{n}{\oplus} A) = ((\cdots ((\overbrace{B \ominus A) \ominus A) \cdots}^{n \text{ times}}) \ominus A$$

III. BOUNDARY CODES

Line drawings have been commonly used to represent the boundaries of two-dimensional objects. In the case of discrete, binary images these line drawings of the object boundary can be represented in either of the following ways: (1) as a sequence of points, (2) by a chain code representation, or (3) as a sequence of line segments. A description of these methods can be found in [1,5].

The chain code representation as proposed by Freeman [2] does not incorporate the lengths of the edges in its notation. It nevertheless has a provision for a special token in the implementation that allows for the length of the edge to be stored. In this section, we discuss a notation for chain codes that requires explicit representation of the boundary edge lengths and directions. This boundary encoding scheme, referred to as B-code, uses a list data structure.

B-code is a representation scheme for connected components in terms of their boundary lattice points. Only a starting boundary point is represented explicitly, while the rest of the boundary points are represented in terms of successive displacements in one of possible eight directions. If the successive displacements happen to be in the same direction, they are encoded as the direction followed by the number of moves in that direction. The formal notation to represent a connected component A is given by

$$A = \langle (i_A, j_A) \mid (\mathbf{d}_l : n_l)(\mathbf{d}_{l+1} : n_{l+1}) \cdots (\mathbf{d}_m : n_m) \rangle \tag{9.1}$$

Here (i_A, j_A) is the starting boundary lattice point, and the ordered pairs following the vertical bar describe each successive displacement. The number of ordered pairs is equal to the number of changes in the direction of displacement. In the ordered pair $(\mathbf{d}_k : n_k)$, $\mathbf{d}_k \in \{\mathbf{d}_0, \mathbf{d}_1, \ldots, \mathbf{d}_7\}$ represents the direction of the displacement and the nonnegative integer n_i following the colon represents the number of successive moves in that direction. The directions $\mathbf{d}_0, \ldots, \mathbf{d}_7$ are the same as the chain code directions $0, \ldots, 7$, which correspond to angles of $\{0°,45°,90°,135°,180°,225°,270°,315°\}$, with respect to the positive x-axis: $\mathbf{d}_0 = (1,0)$, $\mathbf{d}_1 = (1,1)$, $\mathbf{d}_2 = (0,1)$, $\mathbf{d}_3 = (-1,1)$, $\mathbf{d}_4 = (-1,0)$, $\mathbf{d}_5 = (-1,-1)$, $\mathbf{d}_6 = (0,-1)$, and $\mathbf{d}_7 = (1,-1)$.

Figure 2 illustrates the relation between B-codes and chain codes. Figure 2a is a binary image. Figure 2b shows the boundary pixels explicitly represented by the corresponding chain code:

(5, 1) : 000001111222244442222244444456666666666677

Figure 2c shows the pixels explicitly represented by the corresponding B-code: $\langle (5,1) \mid (\mathbf{d}_0 : 5)(\mathbf{d}_1 : 4)(\mathbf{d}_2 : 4)(\mathbf{d}_4 : 4)(\mathbf{d}_2 : 5)(\mathbf{d}_4 : 6)(\mathbf{d}_5 : 5)(\mathbf{d}_4 : 6)(\mathbf{d}_5 : 5)$ $(\mathbf{d}_6 : 10)(\mathbf{d}_7 : 2)\rangle$. It can be seen that the B-code representation can be thought of as a runlength encoding of the chain code. Also, any binary image, simply or multiply connected, that can be encoded using the chain codes can be encoded using the B-codes too.

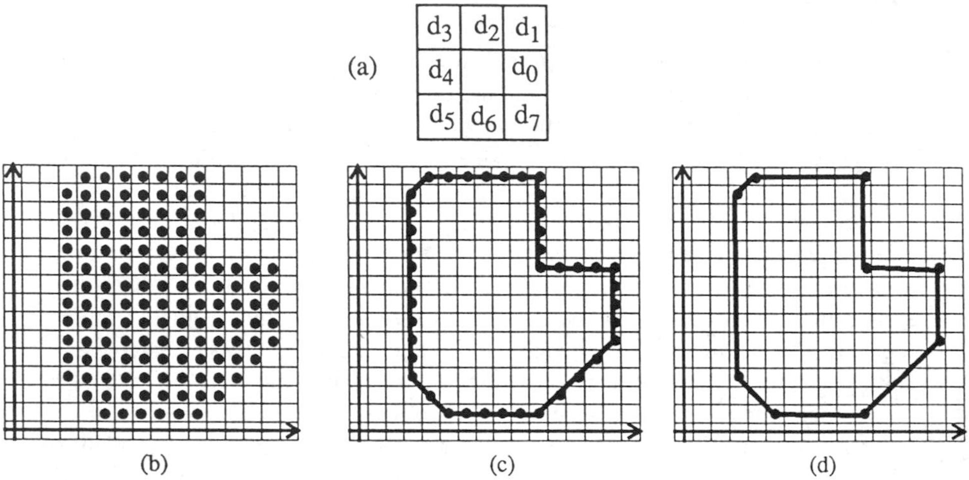

Figure 2. Example of B-coding of images. (a) Basic directions; (b) a binary image; (c) pixels explicitly represented by its chain code (5,1): 00000111122224444222224444445666666666677; (d) pixels explicitly represented by its B-code $\langle(5,1) \mid (\mathbf{d}_0 : 5)(\mathbf{d}_1 : 4)(\mathbf{d}_2 : 4)(\mathbf{d}_4 : 4)(\mathbf{d}_2 : 5)(\mathbf{d}_4 : 6)(\mathbf{d}_5 : 5)(\mathbf{d}_6 : 10)(\mathbf{d}_7 : 2)\rangle$.

IV. RESTRICTED DOMAINS

The class of objects we will decompose and work on will be discretely convex, four-connected sets all of whose boundaries are oriented at angles which are multiples of 45° and whose lengths are multiples of the pixel side lengths for 0° and 90° orientations and multiples of $\sqrt{2}$ times the pixel side length for 45° and 135° orientations. We will refer to the set of all objects belonging to this class as restricted domains.

Definition 4.1. A *restricted domain* is a discretely convex, four-connected shape whose convex hull has sides at angles that are multiples of 45° with respect to the positive x axis.

Some examples of restricted domains are given in Figure 3. In the following sections we will define the restricted domains in terms of their B-codes and present an equivalent representation in terms of half-planes.

A. B-Code Representation

1. Convention

Given a binary image of a restricted domain, A, we will represent it in the B-code form for further processing. The binary image of a restricted domain can be

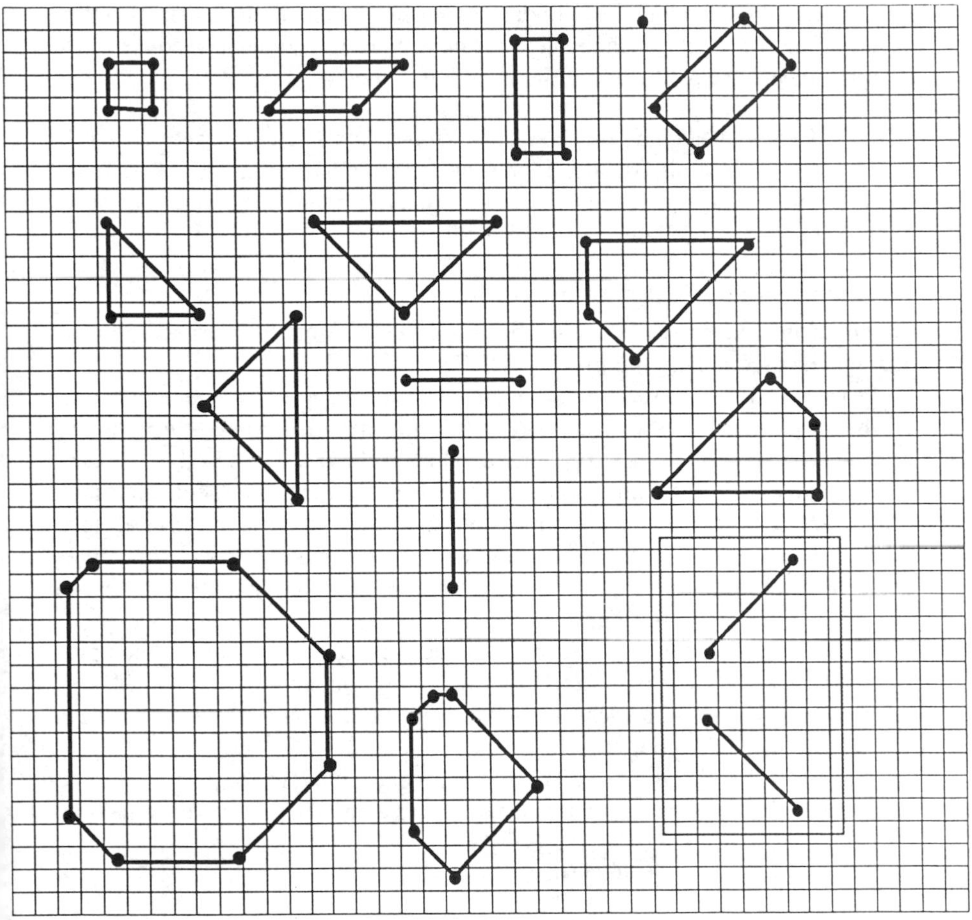

Figure 3. A few examples of restricted domains. Note that the diagonal lines in the box are not strictly restricted domains since they are not 4-connected.

represented in many ways using a B-code representation; i.e., the B-code representation is not unique. This is due to the fact that the only restriction on the starting point of a B-code representation is that it should be a vertex. Thus, there are as many B-code representations of a restricted domain as the number of vertices it has. To avoid ambiguity we will use the following convention:

> The starting point will always be the lowest and leftmost vertex of the restricted domain. The rest of the vertices are encoded by traversing around the

restricted domain along its boundary points in the counterclockwise direction, encoding the length of the edges that constitute A. The interior points of the set will be those on the left of the direction of motion.

The B-code obtained using this convention represents an equivalence class of B-codes—the class of all B-codes representing the considered restricted domain. Each B-code in the equivalence class is a rotated version of the other but representing the same set of lattice points nevertheless.

2. Properties of B-Coded Restricted Domains

In this section we present some useful properties of B-coded restricted domains that will be used in later proofs.

Property 4.1. Any restricted domain can be represented by a general B-code of the form $A = \langle (i,j) \mid (\mathbf{d}_0 : n_0)\ (\mathbf{d}_1 : n_1)\ (\mathbf{d}_2 : n_2)\ (\mathbf{d}_3 : n_3)\ (\mathbf{d}_4 : n_4)\ (\mathbf{d}_5 : n_5)\ (\mathbf{d}_6 : n_6)\ (\mathbf{d}_7 : n_7)\rangle$ by giving appropriate values to the n_i's. Thus in this representation there are always eight vertices, eight displacements, and the displacement angles are monotonically increasing from $\mathbf{d}_0$ to $\mathbf{d}_7$. If there is no displacement corresponding to one of the directions, the corresponding pair can be dropped from the B-code and the particular n_i is given a value zero. Note that in this case two vertices become coincident.

Given a closed contour, the net displacement on traversing its complete boundary is zero. Since the B-code of a restricted domain $A = \langle (i,j) \mid (\mathbf{d}_0 : n_0)\ (\mathbf{d}_1 : n_1) \cdots (\mathbf{d}_7 : n_7)\rangle$ represents a closed contour, it inherits the following two properties of a closed contour.

Property 4.2. The sum of displacements contributing to the positive x direction is equal to the sum of displacements contributing to the negative x direction:

$$n_0 + n_1 + n_7 = n_3 + n_4 + n_5 \tag{9.2}$$

Property 4.3. The sum of displacements contributing to the positive y direction is equal to the sum of displacements contributing to the negative y direction:

$$n_1 + n_2 + n_3 = n_5 + n_6 + n_7 \tag{9.3}$$

Property 4.4. Any B-code of the form $A = \langle (i,j) \mid (\mathbf{d}_0 : n_0)\ (\mathbf{d}_1 : n_1)\ (\mathbf{d}_2 : n_2)\ (\mathbf{d}_3 : n_3)\ (\mathbf{d}_4 : n_4)\ (\mathbf{d}_5 : n_5)\ (\mathbf{d}_6 : n_6)\ (\mathbf{d}_7 : n_7)$ whose n_i's satisfy the properties in Eqs. (9.2) and (9.3) is either a restricted domain or a line at 45° or 135°. The lines are special cases and are of the form $A = \langle (i,j) \mid (\mathbf{d}_1 : n_1)\ (\mathbf{d}_5 : n_5)\rangle$ and $A = \langle (i, j \mid \mathbf{d}_3 : n_3)\ (\mathbf{d}_7 : n_7)\rangle$. Details on this are given in the Appendix.

Given a B-code of a restricted domain $A = \langle (i,j) \mid (\mathbf{d}_0 : n_0)\ (\mathbf{d}_1 : n_1) \cdots (\mathbf{d}_7 : n_7)\rangle$, all the eight vertices of the polygon are uniquely defined and can be found in the following two ways.

Property 4.5. Let the vertex v_0 be the starting lattice point (i, j). The rest of the vertices are given recursively. Given the kth vertex v_k, the x and y coordinates of the $(k+1)$th vertex v_{k+1} are given by the recursive equations

$$\mathbf{x}[v_{k+1}] = \mathbf{x}[v_k] + n_k\mathbf{x}[d_k] \tag{9.4}$$

$$\mathbf{y}[v_{k+1}] = \mathbf{y}[v_k] + n_k\mathbf{y}[d_k] \tag{9.5}$$

for $0 \leq k \leq 6$. Here $\mathbf{x}[v_0] = i$ and $\mathbf{y}[v_0] = j$—the x and y coordinates of the starting point of the B-code.

The coordinates of the vertices of A can also be computed relative to the starting point of the restricted domain.

Property 4.6. The coordinates of the kth vertex v_k can be computed in terms of the starting location (i, j), and the lengths n_l, $0 \leq l \leq k$. Let $\mathbf{V}_x$, $\mathbf{V}_y$, $\mathbf{V}$, and $\mathbf{N}$ be the matrices

$$\mathbf{V}_x = \begin{bmatrix} \mathbf{x}[v_0] \\ \mathbf{x}[v_1] \\ \vdots \\ \mathbf{x}[v_7] \end{bmatrix}, \quad \mathbf{V}_y = \begin{bmatrix} \mathbf{y}[v_0] \\ \mathbf{y}[v_1] \\ \vdots \\ \mathbf{y}[v_7] \end{bmatrix}, \quad \mathbf{V} = \begin{bmatrix} \mathbf{V}_x \\ \mathbf{V}_y \end{bmatrix}, \quad \mathbf{N} = \begin{bmatrix} n_0 \\ n_1 \\ \vdots \\ n_7 \end{bmatrix} \tag{9.6}$$

Then,

$$\mathbf{V} = \begin{bmatrix} \mathbf{V}_x \\ \mathbf{V}_y \end{bmatrix} = \mathbf{P} \begin{bmatrix} i \\ j \\ \mathbf{N} \\ i \\ j \\ \mathbf{N} \end{bmatrix} \tag{9.7}$$

where

$$\mathbf{P} = \begin{bmatrix} \mathbf{P}_1 0 \\ 0\ \mathbf{P}_2 \end{bmatrix} \tag{9.8}$$

$$\mathbf{P}_1 = \begin{bmatrix} 1 & 0 & 0 & 0 & 0 & 0 & 0 & 0 & 0 & 0 \\ 1 & 0 & 1 & 0 & 0 & 0 & 0 & 0 & 0 & 0 \\ 1 & 0 & 1 & 0 & 1 & 0 & 0 & 0 & 0 & 0 \\ 1 & 0 & 1 & 0 & 1 & 0 & 0 & 0 & 0 & 0 \\ 1 & 0 & 1 & 0 & 1 & -1 & 0 & 0 & 0 & 0 \\ 1 & 0 & 1 & 0 & 1 & -1 & -1 & 0 & 0 & 0 \\ 1 & 0 & 1 & 0 & 1 & -1 & -1 & -1 & 0 & 0 \\ 1 & 0 & 1 & 0 & 1 & -1 & -1 & -1 & 0 & 0 \end{bmatrix} \tag{9.9}$$

$$
\mathbf{P}_2 = \begin{bmatrix}
0 & 1 & 0 & 0 & 0 & 0 & 0 & 0 & 0 & 0 \\
0 & 1 & 0 & 0 & 0 & 0 & 0 & 0 & 0 & 0 \\
0 & 1 & 0 & 1 & 0 & 0 & 0 & 0 & 0 & 0 \\
0 & 1 & 0 & 1 & 1 & 0 & 0 & 0 & 0 & 0 \\
0 & 1 & 0 & 1 & 1 & 1 & 0 & 0 & 0 & 0 \\
0 & 1 & 0 & 1 & 1 & 1 & 0 & 0 & 0 & 0 \\
0 & 1 & 0 & 1 & 1 & 1 & 0 & -1 & 0 & 0 \\
0 & 1 & 0 & 1 & 1 & 1 & 0 & -1 & -1 & 0
\end{bmatrix} \tag{9.10}
$$

B. Normalized Half-Plane Representation

Restricted domains can be represented in terms of the intersections of discrete half-planes. Let $A = \langle (i,j) \mid (\mathbf{d}_0 : n_0)(\mathbf{d}_1 : n_1)\cdots(\mathbf{d}_7 : n_7)\rangle$ be a restricted domain. Then the lattice points belonging to A can be defined in terms of intersections of eight discrete half-planes $\mathcal{H}_i$, $0 \leq i \leq 7$. Each of these half-planes $\mathcal{H}_i$ is a function of the basic directions of the displacement d_i and the vertices v_i of the restricted domain. Each discrete half-plane $\mathcal{H}_i$ is such that its boundary passes through the vertex v_i and its edge is along the direction d_i. The half-plane $\mathcal{H}_i$ represents all the points on the left and on the boundary while traversing in the direction d_i along the boundary. Therefore, a restricted domain $A = \langle (i,j) \mid (\mathbf{d}_0 : n_0)(\mathbf{d}_1 : n_1)\cdots(\mathbf{d}_7 : n_7)\rangle$ can be represented as

$$A = \mathcal{H}_0 \cap \mathcal{H}_1 \cap \cdots \cap \mathcal{H}_7 \tag{9.11}$$

where $\mathcal{H}_i$ is a discrete half-plane given by

$$\mathcal{H}_i = \left\{ p = (x,y) \in \mathbf{Z}^2 \text{ such that } \begin{vmatrix} x - \mathbf{x}[v_i] & y - \mathbf{y}[v_i] \\ \mathbf{x}[v_i] + \mathbf{x}[d_i] & \mathbf{y}[v_i] + \mathbf{y}[d_i] \end{vmatrix} \leq 0 \right\} \tag{9.12}$$

Figure 4 illustrates the half-plane concept. We can expand the above expression for the particular cases of $\mathcal{H}_i$, $0 \leq i \leq 7$. Substituting the expression for the vertex v_i of the restricted domain given in Eqs. (9.7) into the inequality (9.12), the inequalities for the half-planes $\mathcal{H}_0$ to $\mathcal{H}_7$ thus obtained are:

$$
\begin{aligned}
\mathcal{H}_0&: \quad (0)x + (-1)y \leq c_0 \\
\mathcal{H}_1&: \quad (1)x + (-1)y \leq c_1 \\
\mathcal{H}_2&: \quad (1)x + (0)y \leq c_2 \\
\mathcal{H}_3&: \quad (1)x + (1)y \leq c_3 \\
\mathcal{H}_4&: \quad (0)x + (1)y \leq c_4
\end{aligned} \tag{9.13}
$$

$$\mathcal{H}_5: \quad (-1)x + \quad (1)y \le c_5$$
$$\mathcal{H}_6: \quad (-1)x + \quad (0)y \le c_6$$
$$\mathcal{H}_7: \quad (-1)x + (-1)y \le c_7$$

where $x, y, c_i \in \mathbf{Z}$ and the c_i are given by the equations

$$\mathbf{C} = \mathbf{L}\begin{bmatrix} i \\ j \\ \mathbf{N} \end{bmatrix} \tag{9.14}$$

where

$$\mathbf{C} = \begin{bmatrix} c_0 \\ c_1 \\ c_2 \\ c_3 \\ c_4 \\ c_5 \\ c_6 \\ c_7 \end{bmatrix} \tag{9.15}$$

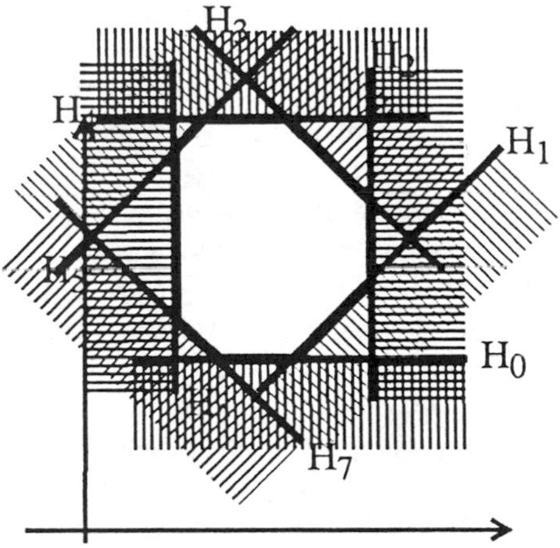

Figure 4. Restricted domains as intersections of half-planes. $\mathcal{H}_1 \ldots \mathcal{H}_7$. The unshaded half represents the half-plane. Here the intersection set is the unshaded central region.

$$\mathbf{L} = \begin{bmatrix} 0 & -1 & 0 & 0 & 0 & 0 & 0 & 0 & 0 & 0 \\ 1 & -1 & 1 & 0 & 0 & 0 & 0 & 0 & 0 & 0 \\ 1 & 0 & 1 & 1 & 0 & 0 & 0 & 0 & 0 & 0 \\ 1 & 1 & 1 & 2 & 1 & 0 & 0 & 0 & 0 & 0 \\ 0 & 1 & 0 & 1 & 1 & 1 & 0 & 0 & 0 & 0 \\ -1 & 1 & -1 & 0 & 1 & 2 & 1 & 0 & 0 & 0 \\ -1 & 0 & -1 & -1 & 0 & 1 & 1 & 1 & 0 & 0 \\ -1 & -1 & -1 & -2 & -1 & 0 & 1 & 2 & 1 & 0 \end{bmatrix} \tag{9.16}$$

To make the information more compact, we will use matrices to represent the system of linear inequalities in (9.13) as

$$\mathbf{M}p' \leq \mathbf{C} \tag{9.17}$$

where

$$\mathbf{M} = \begin{bmatrix} 0 & -1 \\ 1 & -1 \\ 1 & 0 \\ 1 & 1 \\ 0 & 1 \\ -1 & 1 \\ -1 & 0 \\ -1 & -1 \end{bmatrix} \tag{9.18}$$

and $p = (x, y)$ is a lattice point. Note that the inequalities (9.17) are considered row-wise.

The physical interpretation of the system of inequalities (9.17) is as follows. Consider eight half-planes passing through the origin, each one corresponding to a direction d_i, $0 \leq i \leq 7$. The half-planes are translated from the origin up, down, left, and right such that they pass through the corresponding vertices v_i. The intersections of these half-planes gives us the lattice points belonging to the restricted domain.

Notice that since the d_i's are fixed, the slope of the discrete half-planes are also fixed and hence the half plane $\mathcal{H}_i$ is uniquely represented by the corresponding c_i's. But a set of c_i's representing a restricted domain need not be unique. For example, in Figure 5, we see that the half-planes corresponding to two different sets of c_i's represent the same intersection set. This is because the half-plane $\mathcal{H}_1$ is *redundant* and can be translated to infinitely many locations without having any effect on the intersection set. All the possible sets of c_i's representing a given

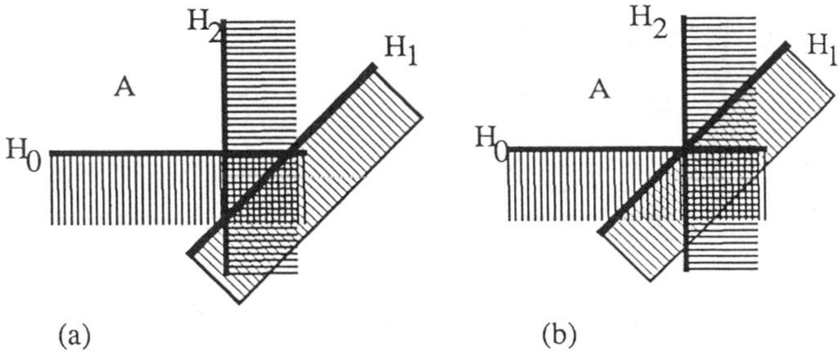

Figure 5. Unnormalized and normalized half-planes. The unshaded half represents the half-plane. The half-plane $\mathcal{H}_1$ in (a) is redundant and can be moved until it passes through the vertex of set A; this situation is shown in (b).

restricted domain form an equivalence class. This raises the question about a convention that we can follow such that an equivalence class of restricted domains can be represented through a unique c_i set. We notice that the c_i's that are obtained from the B-code representation using Eq. (9.14) always represent discrete half-planes passing through the vertices of the restricted domains. Those that are redundant—that is, those that correspond to a displacement of length zero along the d_i direction—also pass through a vertex even though they have potentially infinite possibilities. Thus, we will follow the convention that if a set of c_i's represents a restricted domain, it should be normalized such that all the half-planes pass through the vertices of the intersection set. Such a set of eight c_i's, represented using a vector **C** will be called the *normalized half-plane representation* of the restricted domain. The half-planes that are not redundant and form the sides of the polygon will be called *primary*.

Before we proceed further, we need to address the following issues:

1. Under what conditions the set of c_i's represents a nonempty set
2. Under what conditions the restricted domain represented by the set of c_i's is a normalized representation and, if it is not, how to normalize it

The c_i's represent a set of discrete half-planes. Hence, the set of points belonging to the intersection of these half-planes is not empty if and only if the set of points belonging to the intersection of any two of these half-planes is not empty. Figure 6a illustrates an example where the half-plane $\mathcal{H}_0$ is unnormalized. Since it should be moved such that it touches the intersection set, it is obvious that it should be moved to r, the intersection point of $\mathcal{H}_1$ and $\mathcal{H}_7$. The other possibilities could have been p, the intersection point of $\mathcal{H}_7$ and $\mathcal{H}_2$, or q, the intersection point of $\mathcal{H}_1$ and $\mathcal{H}_6$. Notice that p and q do not belong to the intersection set and they are below r. Figures 6b and c show examples where $\mathcal{H}_0$ has to be moved to p and q, respectively. Notice that in this case p is above q and r. And in Figure

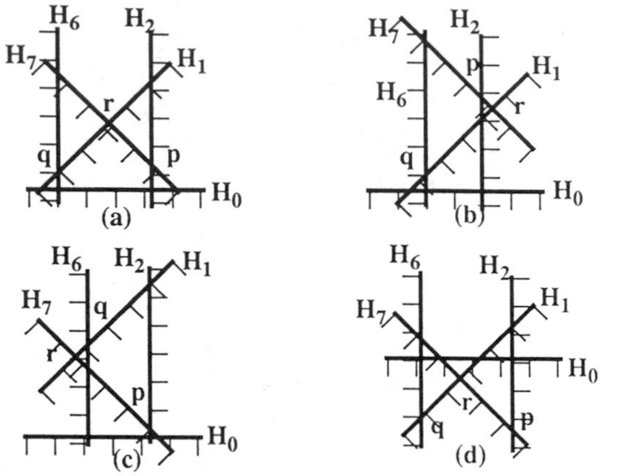

Figure 6. The normalization of $\mathcal{H}_0$—four cases. In (a), (b), and (c) the half-plane $\mathcal{H}_0$ is redundant and has to be normalized, that is, moved up so that it passes through r, p, and q, respectively. In (d) $\mathcal{H}_0$ is primary and cannot be moved.

6d, $\mathcal{H}_0$ is a primary half-plane and cannot be moved. In this case $\mathcal{H}_0$ is above p, q, and r. Thus the algorithm for normalization of $\mathcal{H}_0$ then becomes: find the intersection points p, q, and r, and update c_0 such that $\mathcal{H}_0$ passes through the one belonging to the set. In the case that $\mathcal{H}_0$ is primary, nothing should be done to c_0. Conveniently, the c_0 found in this way also forms a bound for the half-plane $\mathcal{H}_3$, that is, the half plane $\mathcal{H}_3$ cannot be below this level. In case it is, the intersection of the half-planes results in an empty set. Using the same argument for all other half-planes, it can be shown that a set of eight c_i's represents a nonempty set if and only if

$$\mathbf{C} \geq \mathbf{C}_{\text{bound}} = \max[\mathbf{G}_1\mathbf{C},\ \mathbf{G}_2\mathbf{C},\ \mathbf{G}_3\mathbf{C},\ -\lfloor \mathbf{G}_4\mathbf{C} \rfloor] \tag{9.19}$$

and a set of c_i's is normalized if

$$\mathbf{C} \geq \max[\mathbf{G}_1\mathbf{G}_2\mathbf{C},\ \mathbf{G}_1\mathbf{G}_3\mathbf{C},\ -\lfloor \mathbf{G}_1\mathbf{G}_4\mathbf{C} \rfloor] \tag{9.20}$$

where the 8×8 matrices $\mathbf{G}_1, \ldots, \mathbf{G}_4$ used in the algorithm are given below.

$$\mathbf{G}_1 = \begin{bmatrix} 0 & 0 & 0 & 0 & -1 & 0 & 0 & 0 \\ 0 & 0 & 0 & 0 & 0 & -1 & 0 & 0 \\ 0 & 0 & 0 & 0 & 0 & 0 & -1 & 0 \\ 0 & 0 & 0 & 0 & 0 & 0 & 0 & -1 \\ -1 & 0 & 0 & 0 & 0 & 0 & 0 & 0 \\ 0 & -1 & 0 & 0 & 0 & 0 & 0 & 0 \\ 0 & 0 & -1 & 0 & 0 & 0 & 0 & 0 \\ 0 & 0 & 0 & -1 & 0 & 0 & 0 & 0 \end{bmatrix} \tag{9.21}$$

$$\mathbf{G}_2 = \begin{pmatrix} 0 & 0 & -1 & 0 & 0 & -1 & 0 & 0 \\ 0 & 0 & 0 & -1 & 0 & 0 & -2 & 0 \\ 0 & 0 & 0 & 0 & -1 & 0 & 0 & -1 \\ -2 & 0 & 0 & 0 & 0 & -1 & 0 & 0 \\ 0 & -1 & 0 & 0 & 0 & 0 & -1 & 0 \\ 0 & 0 & -2 & 0 & 0 & 0 & 0 & -1 \\ -1 & 0 & 0 & -1 & 0 & 0 & 0 & 0 \\ 0 & -1 & 0 & 0 & -2 & 0 & 0 & 0 \end{pmatrix} \tag{9.22}$$

$$\mathbf{G}_3 = \begin{pmatrix} 0 & 0 & 0 & -1 & 0 & 0 & -1 & 0 \\ 0 & 0 & 0 & 0 & -2 & 0 & 0 & -1 \\ -1 & 0 & 0 & 0 & 0 & -1 & 0 & 0 \\ 0 & -1 & 0 & 0 & 0 & 0 & -2 & 0 \\ 0 & 0 & -1 & 0 & 0 & 0 & 0 & -1 \\ -2 & 0 & 0 & -1 & 0 & 0 & 0 & 0 \\ 0 & -1 & 0 & 0 & -1 & 0 & 0 & 0 \\ 0 & 0 & -2 & 0 & 0 & -1 & 0 & 0 \end{pmatrix} \tag{9.23}$$

$$\mathbf{G}y = \begin{pmatrix} 0 & 0 & 0 & 1/2 & 0 & 1/2 & 0 & 0 \\ 0 & 0 & 0 & 0 & 1 & 0 & 1 & 0 \\ 0 & 0 & 0 & 0 & 0 & 1/2 & 0 & 1/2 \\ 1 & 0 & 0 & 0 & 0- & 0 & 1 & 0 \\ 0 & 1/2 & 0 & 0 & 0 & 0 & 0 & 1/2 \\ 1 & 0 & 1 & 0 & 0 & 0 & 0 & 0 \\ 0 & 1/2 & 0 & 1/2 & 0 & 0 & 0 & 0 \\ 0 & 0 & 1 & 0 & 1 & 0 & 0 & 0 \end{pmatrix} \tag{9.24}$$

The lower ceilings come about because the 45° and 135° lines need not intersect at a lattice point. Notice that $\mathbf{G}_1^2 = \mathbf{I}$. Here the matrix multiplications find the intersection points. The max operation selects the one nearest to the set. Thus, both the issues mentioned above have been addressed.

Notice that when $c_1 = -c_5$ or $c_3 = -c_7$, we have diagonal lines at 45° or 135°, respectively. These are not strictly restricted domains since they are not 4-connected (but they are 8-connected). Thus since restricted domains are 4-connected, the following constraints should hold:

$$c_1 > -c_5 \qquad \text{and} \qquad c_3 > -c_7 \tag{9.25}$$

The algorithm *Normalize* given in Table 1 takes as input the **C** array of a restricted domain and returns the normalized **C** array if one exists, else it returns a NULL value. Since the algorithm has five multiplications of 8 × 8 matrices with 8 × 1 vectors, one lower ceiling of an 8 × 1 vector, one 8 × 1 vector comparison one row-wise max operation of four 8 × 1 vectors, and no loops, the algorithm is constant in time.

C. Conversion from Normalized Half-Plane to B-Code

Given the c_i's of a normalized restricted domain, we should be able to (1) find the vertices of a restricted domain in terms of the c_i's, (2) find the n_i's in terms of the c_i's, and (3) find the B-code representation of the restricted domain.

The vertices of the restricted domain can be computed by finding the intersections of the consecutive half-planes. They can be expressed in terms of the vector **C** as follows:

$$\mathbf{V} = \mathbf{D}\begin{bmatrix}\mathbf{C}\\ \mathbf{C}\end{bmatrix} \tag{9.26}$$

where

$$\mathbf{D} = \begin{bmatrix}\mathbf{D}_1 & 0\\ 0 & \mathbf{D}_2\end{bmatrix} \tag{9.27}$$

$$\mathbf{D}_1 = \begin{bmatrix} 1 & 0 & 0 & 0 & 0 & 0 & 0 & -1\\ -1 & 1 & 0 & 0 & 0 & 0 & 0 & 0\\ 0 & 0 & 1 & 0 & 0 & 0 & 0 & 0\\ 0 & 0 & 1 & 0 & 0 & 0 & 0 & 0\\ 0 & 0 & 0 & 1 & -1 & 0 & 0 & 0\\ 0 & 0 & 0 & 0 & 1 & -1 & 0 & 0\\ 0 & 0 & 0 & 0 & 0 & 0 & -1 & 0\\ 0 & 0 & 0 & 0 & 0 & 0 & -1 & 0 \end{bmatrix} \tag{9.28}$$

$$\mathbf{D}_2 = \begin{bmatrix} -1 & 0 & 0 & 0 & 0 & 0 & 0 & 0\\ -1 & 0 & 0 & 0 & 0 & 0 & 0 & 0\\ 0 & -1 & 1 & 0 & 0 & 0 & 0 & 0\\ 0 & 0 & -1 & 1 & 0 & 0 & 0 & 0\\ 0 & 0 & 0 & 0 & 1 & 0 & 0 & 0\\ 0 & 0 & 0 & 0 & 1 & 0 & 0 & 0\\ 0 & 0 & 0 & 0 & 0 & 1 & -1 & 0\\ 0 & 0 & 0 & 0 & 0 & 0 & 1 & -1 \end{bmatrix} \tag{9.29}$$

Table 1. The Algorithm for Normalizing Half-Planes

```
function Normalize(C) : ArrayObject

Input:
ArrayObject C;

begin
    C_bound := max[G_1C, G_2C, G_3C, − ⌊G_4C⌋];
    if (C < C_bound)
    then
        return NULL;
    else
        C := max[C, G_1G_2C, G_1G_3C, − ⌊G_1G_4C⌋];
        return C;
end Normalize;
```

Then n_i's can be computed by finding the distance between the two consecutive vertices v_{i+1} and v_i. Thus,

$$\mathbf{N} = \mathbf{QC} \tag{9.30}$$

where

$$\mathbf{Q} = \begin{bmatrix} -2 & 1 & 0 & 0 & 0 & 0 & 0 & 1 \\ 1 & -1 & 1 & 0 & 0 & 0 & 0 & 0 \\ 0 & 1 & -2 & 1 & 0 & 0 & 0 & 0 \\ 0 & 0 & 1 & -1 & 1 & 0 & 0 & 0 \\ 0 & 0 & 0 & 1 & -2 & 1 & 0 & 0 \\ 0 & 0 & 0 & 0 & 1 & 1 & 1 & 0 \\ 0 & 0 & 0 & 0 & 0 & 1 & -2 & 1 \\ 1 & 0 & 0 & 0 & 0 & 0 & 1 & -1 \end{bmatrix} \tag{9.31}$$

The B-code representation of the restricted domain is determined by v_0 and $\mathbf{N}$.

V. BOUNDARY CODE MORPHOLOGY FOR RESTRICTED DOMAINS

In this section we will give constant time algorithms for dilation, erosion, opening, closing, n-fold dilation, and n-fold erosion of restricted domains using their half-plane and B-code representations. We will show that the results obtained using these algorithms are equivalent to those obtained using regular morphology. If the input restricted domains are in their B-code representations or if the output restricted domains are needed in their B-code representation, the results of the previous section can be used for the interconversion between representations.

A. Dilation of Restricted Domains

Let A and B be two restricted domains given by the B-codes

$$A = \langle (i_A, j_A) \mid \mathbf{d}_0 : n_0^A)(\mathbf{d}_1 : n_1^A)\cdots(\mathbf{d}_7 : n_7^A)\rangle \tag{9.32}$$

$$B = \langle (i_B, j_B) \mid (\mathbf{d}_0 : n_0^B)(\mathbf{d}_1 : n_1^B)\cdots(\mathbf{d}_7 : n_7^B)\rangle \tag{9.33}$$

and their normalized half-plane representations be

$$A = \{a \in \mathbf{Z}^2 \mid \mathbf{M}a' \leq \mathbf{C}^A\} \tag{9.34}$$

$$B = \{b \in \mathbf{Z}^2 \mid \mathbf{M}b' \leq \mathbf{C}^B\} \tag{9.35}$$

where $\mathbf{C}^A$ and $\mathbf{C}^B$ are given by

$$\mathbf{C}^A = \mathbf{L} \begin{bmatrix} i^A \\ j^A \\ \mathbf{N}^A \end{bmatrix} \tag{9.36}$$

$$\mathbf{C}^B = \mathbf{L} \begin{bmatrix} i^B \\ j^B \\ \mathbf{N}^B \end{bmatrix} \tag{9.37}$$

$\mathbf{N}^A$ and $\mathbf{N}^B$ are 8×1 column vectors with the respective edge lengths as their elements, and $\mathbf{M}$ and $\mathbf{L}$ matrices are defined in Eqs. (9.18), and (9.16), respectively.

Lemma 5.1. The set C given by

$$C = \{c \in \mathbf{Z}^2 \mid \mathbf{M}c' \leq \mathbf{C}^C\} \tag{9.38}$$

where $\mathbf{C}^C = \mathbf{C}^A + \mathbf{C}^B$, is a restricted domain, and the vector $\mathbf{C}^C$ is a normalized half-plane representation of C.

Proof. From the discussion in Section IV.B and Eq. (9.19), the sufficient condition for $\mathbf{C}^C$ to be a restricted domain is that $\mathbf{C}^C \geq \mathbf{C}^C_{\text{bound}}$. Since A and B are restricted domains and $\mathbf{C}^A$ and $\mathbf{C}^B$ are normalized half-plane representations,

$$\mathbf{C}^A \geq \mathbf{C}^A_{\text{bound}} = \max[\mathbf{G}_1\mathbf{C}^A, \mathbf{G}_2\mathbf{C}^A, \mathbf{G}_3\mathbf{C}^A, -\lfloor \mathbf{G}_4\mathbf{C}^A \rfloor] \tag{9.39}$$

$$\mathbf{C}^B \geq \mathbf{C}^B_{\text{bound}} = \max[\mathbf{G}_1\mathbf{C}^B, \mathbf{G}_2\mathbf{C}^B, \mathbf{G}_3\mathbf{C}^B, -\lfloor \mathbf{G}_4\mathbf{C}^B \rfloor] \tag{9.40}$$

Thus, adding the above equations we get

$$\begin{aligned} \mathbf{C}^C &= \mathbf{C}^A + \mathbf{C}^B \\ &\geq \max[\mathbf{G}_1\mathbf{C}^A, \mathbf{G}_2\mathbf{C}^A, \mathbf{G}_3\mathbf{C}^A, -\lfloor \mathbf{G}_4\mathbf{C}^A \rfloor] \\ &\quad + \max[\mathbf{G}_1\mathbf{C}^B, \mathbf{G}_2\mathbf{C}^B, \mathbf{G}_3\mathbf{C}^B, -\lfloor \mathbf{G}_4\mathbf{C}^B \rfloor] \end{aligned} \tag{9.41}$$

But we know that max$[a,b]$ + max$[c,d] \geq$ max$[(a + c), (b + d)]$. Hence,

$$\mathbf{C}^C \geq \max[\mathbf{G}_1(\mathbf{C}^A + \mathbf{C}^B), \mathbf{G}_2(\mathbf{C}^A + \mathbf{C}^B), \mathbf{G}_3(\mathbf{C}^A + \mathbf{C}^B), -\lfloor \mathbf{G}_4(\mathbf{C}^A + \mathbf{C}^B) \rfloor] \quad (9.42)$$

Therefore, $\mathbf{C}^C$ represents a restricted domain. Similarly, we now show that $\mathbf{C}^C$ is a normalized half-plane representation. Since $\mathbf{C}^A$ and $\mathbf{C}^B$ are normalized representations, we have

$$\mathbf{C}^A \geq \max[\mathbf{G}_1\mathbf{G}_2\mathbf{C}^A, \mathbf{G}_1\mathbf{G}_3\mathbf{C}^A, -\lfloor \mathbf{G}_1\mathbf{G}_4\mathbf{C}^A \rfloor] \text{ and} \quad (9.43)$$

$$\mathbf{C}^B \geq \max[\mathbf{G}_1\mathbf{G}_2\mathbf{C}^B, \mathbf{G}_1\mathbf{G}_3\mathbf{C}^B, -\lfloor \mathbf{G}_1\mathbf{G}_4\mathbf{C}^B \rfloor] \quad (9.44)$$

As before, from the above equations we get

$$\mathbf{C}^A + \mathbf{C}^B \geq \max[\mathbf{G}_1\mathbf{G}_2(\mathbf{C}^A + \mathbf{C}^B), \mathbf{G}_1\mathbf{G}_3(\mathbf{C}^A + \mathbf{C}^B), -\lfloor \mathbf{G}_1\mathbf{G}_4(\mathbf{C}^A + \mathbf{C}^B) \rfloor] \quad (9.45)$$

Thus, $\mathbf{C}^C$ is normalized. Furthermore, since A and B are 4-connected restricted domains, we have from Eq. (9.25): $c_1^A > -c_5^A$, $c_3^A > -c_7^A$, $c_1^B > -c_5^B$, and $c_3^B > -c_7^B$. Manipulating, we get $c_1^A + c_1^B > c_5^A + c_5^B$ and $c_3^A + c_3^B > c_7^A + c_7^B$. Thus C is four-connected and $\mathbf{C}^C$ is a normalized half-plane representation of a restricted domain. Note that even if either A or B is a diagonal line, e.g., $c_1^A = -c_5^A$ or $c_1^B = -c_5^B$ for the case of 45° diagonal lines, the resultant shape is still a restricted domain (because the 4-connectivity constraints are still satisfied).

Lemma 5.2. The eight vertices $\mathbf{V}^C$ of C are the vector sums of the respective vertices $\mathbf{V}^A$ and $\mathbf{V}^B$ of A and B.

Proof. The vertices of C are given by

$$\mathbf{V}^C = \mathbf{D}\begin{bmatrix}\mathbf{C}^C \\ \mathbf{C}^C\end{bmatrix} \quad (9.46)$$

where $\mathbf{D}$ is the matrix given in Eq. (9.27). Since $\mathbf{C}^C = \mathbf{C}^A + \mathbf{C}^B$, we have

$$\mathbf{V}^C = \mathbf{D}\begin{bmatrix}\mathbf{C}^A + \mathbf{C}^B \\ \mathbf{C}^A + \mathbf{C}^B\end{bmatrix} = \mathbf{D}\begin{bmatrix}\mathbf{C}^A \\ \mathbf{C}^A\end{bmatrix} + \mathbf{D}\begin{bmatrix}\mathbf{C}^B \\ \mathbf{C}^B\end{bmatrix} \quad (9.47)$$

Hence,

$$\mathbf{V}^C = \mathbf{V}^A + \mathbf{V}^B \quad (9.48)$$

$$v_i^C = v_i^A + v_i^B, \qquad 0 \leq i \leq 7 \quad (9.49)$$

Thus the lemma is proved.

The dilation of two restricted domains can be performed by adding their respective $\mathbf{C}$ vectors.

Proposition 5.1. $A \oplus B$ is given by the restricted domain C whose normalized half plane representation is given by $\mathbf{C}^C = \mathbf{C}^A + \mathbf{C}^B$.

Proof. We will proceed by proving that (i) $A \oplus B \subseteq C$ and then (ii) $C \subseteq A \oplus B$.

(i) $A \oplus B \subseteq C$. Let $c \in A \oplus B$. Then, by definition of dilation, there exist $a \in A$ and $b \in B$ such that $c = a + b$. Since $a \in A$ and $b \in B$, from Eqs. (9.34) and (9.35) we have

$$\mathbf{M}a' \leq \mathbf{C}^A$$
$$\mathbf{M}b' \leq \mathbf{C}^B$$

Adding the above equations we get

$$\mathbf{M}a' + \mathbf{M}b' \leq \mathbf{C}^A + \mathbf{C}^B$$
$$\mathbf{M}(a + b)' \leq \mathbf{C}^A + \mathbf{C}^B$$

Hence $c \in C$. Therefore, $A \oplus B \subseteq C$.

(ii) $C \subseteq A \oplus B$. Let $c \in C$. We have to prove that there exist $a \in A$ and $b \in B$ such that $a + b = c$. Since $c \in C$, it satisfies the relation

$$\mathbf{M}c' \leq \mathbf{C}^A + \mathbf{C}^B \tag{9.50}$$

From Lemma 5.1 we know that C is a restricted domain and thus by definition it is discretely convex. Hence, c belongs to the convex hull of C, and it can be expressed as the convex combination of the vertices of C. Thus,

$$c = \sum_{0 \leq i \leq 7} \alpha_i v_i^C \tag{9.51}$$

where $\alpha_i \in \mathbf{R}$, $0 \leq \alpha_i \leq 1$, and $\Sigma\, \alpha_i = 1$. From Lemma 5.2 we get

$$c = \sum_{0 < i < 7} \alpha_i (v_i^A + v_i^B) \tag{9.52}$$

$$c = c_A + c_B \tag{9.53}$$

where $c_A = \Sigma_{0 \leq i \leq 7}\, \alpha_i v_i^A$ and $c_B = \Sigma_{0 \leq j \leq 7}\, \alpha_i v_i^B$. The first term, c_A, on the right-hand side of the above equation belongs to the convex hull of the restricted domain A, the second term, c_B, to the convex hull of the restricted domain B. Notice that Eq. (9.52) does not guarantee that c_A and c_B are lattice points; i.e., they need not belong to $\mathbf{Z}^2$. It just represents the fact the vector sum of two points c_A and c_B in $\mathbf{R}^2$ is the lattice point c in $\mathbf{Z}^2$. We will now show that we can always find a lattice point belonging to A, in the neighborhood of c_A, and another to B, in the neighborhood of c_B, such that their vector sum is the lattice point c. This is illustrated in Figure 7.

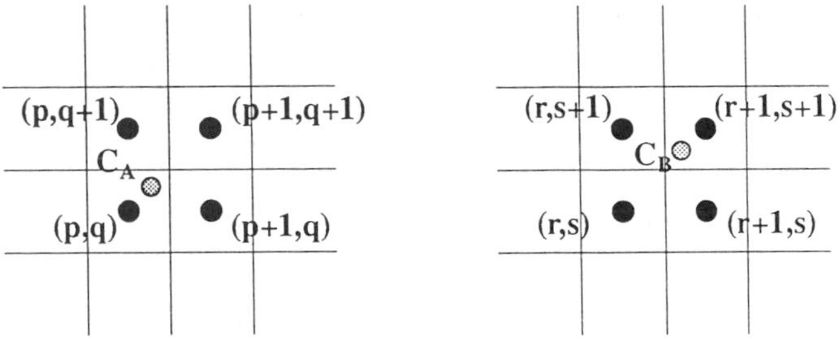

Figure 7. Relation between p, q, r, s, δ, and γ. We have to find one neighboring lattice point of c_A and another of c_B such that their vector sum is $c = (p + r + 1, q + s + 1)$.

Let $c = (l, m)$, $c_A = (p + \delta, q + \gamma)$, and $c_B = (r + 1 - \delta, s + 1 - \gamma)$ such that $l, m, p, q, r, s \in \mathbf{Z}$ and $0 \leq \delta, \gamma \leq 1$. Going back to Eq. (9.53) and replacing c, c^A, and c^B by their values, we get

$$\begin{aligned} c = (l,m) &= (p + \delta, q + \gamma) + (r + 1 - \delta, s + 1 - \gamma) \\ &= (p + r + 1, q + s + 1) \end{aligned} \tag{9.54}$$

It can be seen from the above equations that the point c_A lies between the four lattice points (p, q), $(p + 1, q)$, $(p, q + 1)$, and $(p + 1, q + 1)$. Similarly, the point c_B lies between the four lattice points (r, s), $(r + 1, s)$, $(r, s + 1)$, and $(r + 1, s + 1)$. We will prove that the vector sum of two of these eight points, one belonging to A and another belonging to B, is the lattice point $(l, m) = (p + r + 1, q + s + 1)$.

We can find out which of the four surrounding lattice points necessarily belong to the restricted domain A given c_A (and hence δ and γ). Depending on the values of δ and γ, the area between the lattice points surrounding c_A and c_B can be divided into several regions. The inclusion of a particular neighbor in the set A depends on where the point c_A falls. The attack has to be on a case-by-case basis.

Case (i)—c_A lies in the region defined by $\gamma > \delta$. We can see that in this case the neighboring lattice point $(p, q + 1)$ necessarily belongs to A. This is because if $(p, q + 1)$ did not belong to A, no convex combination of any subset of the other three neighboring lattice points could produce a c_A in the region defined by $\gamma > \delta$. From symmetry we can see that for $\gamma > \delta$, $(r + 1, s)$ necessarily belongs to the set B. Thus the desired lattice points are $(p, q + 1)$ and $(r + 1, s)$ since their vector sum is the lattice point $(p + 1, q + 1) = (l, m) = c$.

Similarly, for the cases (ii) $\gamma < \delta$, (iii)$\gamma > 1 - \delta$, and (iv) $\gamma < 1 - \delta$, we can find lattice points belonging to A and B such that their vector sum is the lattice point $(l, m) = c$.

The only region inside the square not yet considered is when $\delta = \gamma = 0.5$, which is the center of the square. When $\delta = \gamma = 0.5$, we fall into neither of the above categories and hence we have to treat this case separately. We notice that the lattice point $(p+0.5, q+0.5)$ can result from the convex combination if (i) all four neighboring lattice points belong to the set, (ii) only three of the neighboring lattice points belong to the set, and (iii) any two diagonally opposite lattice points belong to the set. We can eliminate the third case since it implies that A is a diagonal line and hence is not four connected and thereby contradicting our assumptions. Thus for the case when $\delta = \gamma = 0.5$, either three or four of the lattice points neighboring c_A necessarily belong to A. The same is true for the set B. Note that the lattice points belonging to A do not in any way constrain the ones belonging to B. It is easy to verify that given any three lattice points surrounding C_A and three lattice points surrounding c_B, we can always find two lattice points, one neighboring c_A and one neighboring c_B, such that their sum is $c = (l, m) = (p + r + 1, q + s + 1)$. In fact there are many such pairs.

Thus we have proved that $C \subset A \oplus B$. Hence $C = A \oplus B$.

We will now prove an important lemma that says that a dilation of A by B is just the addition of the respective side lengths and the starting points.

Lemma 5.3. If $C = A \oplus B$, then $(i_C, j_C) = (i_A, j_A) + (i_B, j_B)$ and $\mathbf{N}^C = \mathbf{N}^A + \mathbf{N}^B$.

Proof. Since $\mathbf{C}^C = \mathbf{C}^A + \mathbf{C}^B$, we have $\mathbf{N}^C = \mathbf{Q}\mathbf{C}^C = \mathbf{Q}(\mathbf{C}^A + \mathbf{C}^B) = \mathbf{N}^A + \mathbf{N}^B$. The rest of the lemma follows from the fact that $\mathbf{V}^C = \mathbf{V}^A + \mathbf{V}^B$.

B. Erosion of Restricted Domains

Let A and B be two restricted domains with normalized half-plane representations,

$$A = \{a \in \mathbf{Z}^2 \mid \mathbf{M}a' \leq \mathbf{C}^A\} \tag{9.55}$$

$$B = \{b \in \mathbf{Z}^2 \mid \mathbf{M}b' \leq \mathbf{C}^B\} \tag{9.56}$$

where $\mathbf{C}^A$ and $\mathbf{C}^B$ are given by

$$\mathbf{C}^A = \mathbf{L}\begin{bmatrix} i^A \\ j^A \\ \mathbf{N}^A \end{bmatrix} \tag{9.57}$$

$$\mathbf{C}^B = \mathbf{L}\begin{bmatrix} i^B \\ j^B \\ \mathbf{N}^B \end{bmatrix} \tag{9.58}$$

$\mathbf{N}^A$ and $\mathbf{N}^B$ are 8×1 column vectors with the respective edge lengths as their elements, $\mathbf{M}$ and $\mathbf{L}$ are the matrices defined in Eqs. (9.18) and (9.16), respectively.

The erosion of A by B can be performed by subtracting the $\mathbf{C}$ matrix of B from that of A. The resulting $\mathbf{C}$ matrix need not be a normalized half-plane representation—it has to be normalized using the algorithm given in the previous section. Furthermore, the erosion of a restricted domain with another need not produce a restricted domain. Consider, for example, the erosion of a rectangle by a rhombus where the sides of the rectangle and the rhombus are oriented at 45° and 135° and the side lengths of the rhombus equal the smaller of the two sides of the rectangle. It can easily be seen that the result of the erosion is a line oriented along the longer side of the rectangle, i.e., a line at 45° or 135°. Since lines at 45° and 135° are not 4-connected (but are 8-connected), they are not restricted domains. These special cases have to be considered separately.

Proposition 5.2. $A \ominus B$ is given by C, whose half-plane representation is given by

$$C = \{c \in \mathbf{Z}^2 \mid \mathbf{M}c' \leq \mathbf{C}^C\}$$

where $\mathbf{C}^C = \mathbf{C}^A - \mathbf{C}^B$. C can be either a restricted domain or a diagonal line.

Proof. We will proceed by proving that (i) $A \ominus B \subseteq C$ and then (ii) $C \subseteq A \ominus B$.

(i) $A \ominus B \subseteq C$: Let $c \in A \ominus B$. By definition of erosion, $c + b \in A$ for all $b \in B$. Therefore,

$$\mathbf{M}(c + b)' \leq \mathbf{C}^A \qquad \text{for all } b \in B \tag{9.59}$$

We will prove it by contradiction. Suppose that $c \notin C$. Thus,

$$\mathbf{M}c' \not\leq \mathbf{C}^C \tag{9.60}$$

That is, there exists at least one inequality in the system of inequalities (9.60) that does not hold. Without loss of generality, let i, $0 \leq i \leq 7$, be the number of the inequality that is not satisfied:

$$\mathbf{e}_i\mathbf{M}c' > \mathbf{e}_i\mathbf{C}^C \tag{9.61}$$

where $\mathbf{e}_i$ is a 1×8 row vector with 1 in the ith column and zeros elsewhere.

Since the vector $\mathbf{C}^B$ is a normalized half-plane representation of the restricted domain B, the ith vertex of B, v_i^B, lies on $\mathcal{H}_i^B$. Thus,

$$\mathbf{e}_i\mathbf{M}(v_i^B)' = \mathbf{e}_i\mathbf{C}^B \tag{9.62}$$

The system of inequalities (9.59) holds for every $b \in B$, and in particular it holds for the vertex v_i^B:

$$\mathbf{M}(c + v_i^B)' = \mathbf{M}c' + \mathbf{M}(v_i^B)' \leq \mathbf{C}^A \tag{9.63}$$

Substituting Eq. (9.62) in the ith row of the system of inequalities (9.63), we get

$$\mathbf{e}_i\mathbf{M}c' + \mathbf{e}_i\mathbf{C}^B \leq \mathbf{e}_i\mathbf{C}^A \tag{9.64}$$

Since $\mathbf{C}^C = \mathbf{C}^A - \mathbf{C}^B$, we have

$$\mathbf{e}_i \mathbf{M} c' \leq \mathbf{e}_i \mathbf{C}^C \qquad (9.65)$$

contradicting Eq. (9.61). Thus, $c \in C$, and $A \ominus B \subseteq C$.

(ii) $C \subseteq A \ominus B$: Let $c \in C$. We need to prove that for all $b \in B$, $c + b \in A$. Since $c \in C$, c satisfies the set of inequalities:

$$\mathbf{M} c' \leq \mathbf{C}^A - \mathbf{C}^B$$

Adding $\mathbf{C}^B$ to both sides, we have

$$\mathbf{M} c' + \mathbf{C}^B \leq \mathbf{C}^A$$

But, $\mathbf{M} b' \leq \mathbf{C}^B$ for all $b \in B$. Thus,

$$\mathbf{M} c' + \mathbf{M} b' \leq \mathbf{M} c' + \mathbf{C}^B \leq \mathbf{C}^A$$

Taking $\mathbf{M}$ as a common factor,

$$\mathbf{M}(c + b)' \leq \mathbf{C}^A \qquad \text{for all } b \in B$$

Therefore, $c + b \in A$ for all $b \in B$, and $C \subseteq A \ominus B$. C is a diagonal line when either $c_1^C = -c_5^C$ or $c_3^C = -c_7^C$.

C. Opening

Morphological opening of a binary set A by another binary set B is denoted by $A \bigcirc B$ and is defined as

$$A \bigcirc B = (A \ominus B) \oplus B \qquad (9.66)$$

Since dilations and erosions of restricted domains have been defined, the above definition of opening is also valid for restricted domains. The definition is also valid for the following more general cases when either A or B or both are not restricted domains: (i) $A \ominus B$ is a restricted domain and B is a line at 45°; (ii) $A \ominus B$ is a restricted domain and B is a line at 135°; (iii) $A \ominus B$ and B are lines at 135°; (iv) $A \ominus B$ and B are lines at 45°. Note that lines at 45° and 135° are not restricted domains since they are not 4-connected. The algorithm cannot be used if $A \ominus B$ and B are lines at 45° and 135°, respectively. This constraint is due to the fact that the dilation of a 45° line with a 135° line results in a rhombus-like shape with one-pixel holes; i.e., the shape is not filled. Thus the set theory dilation results in a shape that not filled but the half-plane and B-code dilation algorithms produce a shape that is filled.

D. Closing

Morphological closing of a binary set A by another binary set B is denoted by $A \bullet B$ and is defined as

$$A \bullet B = (A \oplus B) \ominus B \tag{9.67}$$

Since dilations and erosions of restricted domains have been defined, the above definition of opening is also valid for restricted domains. The definition is also valid for the following more general cases when either A or B or both are not restricted domains: (i) A is a line at 45° and B is a restricted domain; (ii) B is a line at 45° and A is a restricted domain; (iii) A and B are lines at 45°; (iv) A and B are lines at 135°. Note that lines at 45° and 135° are not restricted domains since they are not 4-connected. The algorithm cannot be used if A and B are lines at 45° and 135°, respectively. This constraint is due to the fact that the set theory dilation of a 45° line with a 135° line results in a rhombus-like shape with one-pixel holes; i.e., the shape is not filled. Thus the set theory dilation results in a shape that not filled but the half-plane and B-code dilation algorithms produce a shape that is filled.

VI. ALGORITHMS AND THEIR COMPLEXITY

In this section we give the algorithms for computing the dilation and erosion of restricted domains represented by half-planes. The algorithms for opening and closing can be easily obtained by applying the dilation and erosion algorithms in the appropriate order. The algorithms for n-fold dilation and n-fold erosion need one multiplication step, which we explain at the end of this section.

The following data structures are used in the algorithms:

ArrayObject is a data structure containing an array and its dimensions. In the algorithms the vectors associated with B-codes, half-planes, etc. are stored using this data structure type.

RDObject is a data structure used to represent restricted domains. It contains the three matrices **N**, **V**, and **C** associated with the restricted domain.

The procedure *DilateRDObject* takes as input two RDObject and outputs RDObject which is the dilation of the two input RDObject. The algorithm is given in Table 2.

The procedure *ErodeRDObject* takes as input two RDObject and outputs RDObject which is the erosion of the two input RDObject. The algorithm is given in Table 3. This procedure calls the normalization function which is given in Table 1. The function *Normalize* takes as input an ArrayObject containing the **C** array of a restricted domain and returns the normalized **C** array if one exists; else it returns a NULL value.

The n-fold dilation of a restricted domain B by a restricted domain A is $B \oplus (\oplus_n A)$, the dilation of B by the n-fold dilation of A. In the B-code domain it amounts to multiplying the side lengths of A by n and adding it to the starting point of B. If A and B have the side lengths given by the vectors $\mathbf{N}^A$ and $\mathbf{N}^B$, and

Table 2. The Algorithm for Dilation of Restricted Domains

```
procedure DilateRDObject (A, B, C)

Input:
RDObject A, B;
Output:
RDObject C;

begin
    N^C := N^A + N^B;
    (i_C, j_C) := (i_A, j_A) + (i_B, j_B);

end DilateRDObjects;
```

Table 3. The Algorithm for Erosion of Restricted Domains

procedure ErodeRDObject (A, B, C)

Input:
RDObject A, B;
Output:
RDObject C;

begin

$\mathbf{C}^C := \mathbf{C}^A - \mathbf{C}^B$;

$\mathbf{C}^C :=$ Normalize$(\mathbf{C}^C)$;

$\mathbf{N}^C := \mathbf{Q}\mathbf{C}^C$;

$$\mathbf{V}^C := \mathbf{D}\begin{bmatrix}\mathbf{C}^C\\\mathbf{C}^C\end{bmatrix};$$

end ErodeRDObjects;

starting points (i_A, j_A) and (i_B, j_B), then $(\oplus_n A)$ has side lengths given by the vector $n\mathbf{N}^A$ and starting point $n(i_A, j_A)$. It follows that $B \oplus (\oplus_n A)$ has side lengths given by the vector $\mathbf{N} = \mathbf{N}^B + n\mathbf{N}^A$ and the starting point $(i_B, j_B) + n(i_A, j_A)$. Dilation can also be performed by going into the discrete half-plane representation and adding the $\mathbf{C}$ vectors associated with A and B. Thus the $\mathbf{C}$ vector associated with $B \oplus (\oplus_n A)$ is $\mathbf{C} = \mathbf{C}^B + n\mathbf{C}^A$.

The n-fold erosion of a restricted domain B by a restricted domain A is $B \ominus (\oplus_n A)$, the erosion of B by the n-fold dilation of A. Let $\mathbf{C}^A$ and $\mathbf{C}^B$ be the vectors associated with A and B. Then $n\mathbf{C}^A$ is the vector associated with $(\oplus_n A)$. Thus in the half-plane domain the n-fold erosion of B by A amounts to $\mathbf{C}^B - n\mathbf{C}^A$.

The algorithm for the dilation of restricted domains given in Table 2 consists of 10 additions only. Hence it is a constant-time algorithm. Note that the time complexity is independent of the size of the structuring element. In conventional morphology this is not the case—the time complexity is $O(n^2)$, where n is the number of elements in each set.

The algorithm for an n-fold dilation of restricted domains consists of eight multiplications. Hence it is also a constant-time algorithm.

The erosion algorithm given in Table 3 consists of eight subtractions followed by the process of normalization. The normalization algorithm in Table 1 was shown to be constant in time. Thus, the erosion algorithm is a constant-time algorithm.

The n-fold erosion is represented in terms of n-fold dilation, and since the n-fold dilation algorithm is constant in time, the n-fold erosion algorithm is constant in time.

The algorithm for opening consists of two stages—an erosion stage followed by a dilation stage. Since erosion and dilation algorithms are constant in time, the algorithm for opening is also constant in time. Similarly, the algorithm for closing consists of two stages—a dilation stage followed by a erosion stage. Since erosion and dilation both are constant in time, the algorithm for closing is also constant in time.

A. Walkthrough

In this section we apply the algorithms on some typical restricted domains. Figure 8 shows an example of a dilation. Here $C = A \oplus B$ where

$$A = \langle (0,2) \mid (\mathbf{d}_1 : 3)(\mathbf{d}_3 : 2)(\mathbf{d}_4 : 1)(\mathbf{d}_6 : 5) \rangle$$
$$B = \langle (1,1) \mid (\mathbf{d}_1 : 1)(\mathbf{d}_3 : 1)(\mathbf{d}_5 : 1)(\mathbf{d}_7 : 1) \rangle$$

Thus,

$$(i_A, j_A) = (0,2) \quad \text{and} \quad \mathbf{N}^A = [0\ 3\ 0\ 2\ 1\ 0\ 5\ 0]'$$
$$(i_B, j_B) = (1,1) \quad \text{and} \quad \mathbf{N}^B = [0\ 1\ 0\ 1\ 0\ 1\ 0\ 1]'$$
$$\mathbf{N}^C = \mathbf{N}^A + \mathbf{N}^B = [0\ 4\ 0\ 3\ 1\ 1\ 5\ 1]'$$
$$(i_C, j_C) = (i_A, j_A) + (i_B, j_B) = (1,3)$$

Therefore,

$$C = \langle (1,3) \mid (\mathbf{d}_1 : 4)(\mathbf{d}_3 : 3)(\mathbf{d}_4 : 1)(\mathbf{d}_5 : 1)(\mathbf{d}_6 : 5)(\mathbf{d}_7 : 1) \rangle$$

Thus we started from the B-code representations of A and B. Then we added the respective side lengths and starting locations to get the B-code of C.

Figure 9 shows an erosion when the shape is not open under the structuring element. Here $C = A \ominus B$ where

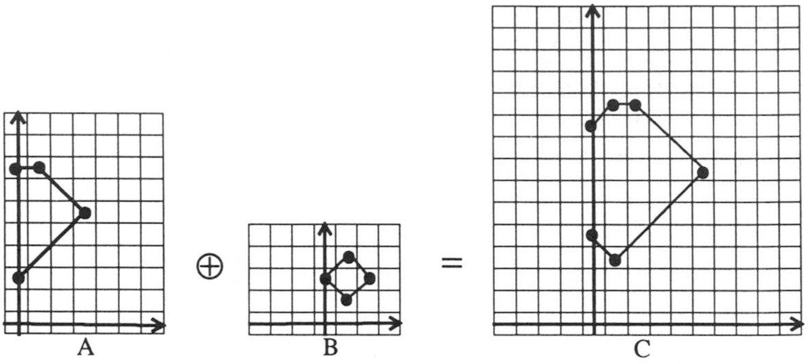

Figure 8. An example of dilation. C is obtained by dilating A by B. Here $A = \langle(0,2) \mid (\mathbf{d}_1 : 3)(\mathbf{d}_3 : 2)(\mathbf{d}_4 : 1)(\mathbf{d}_6 : 5)\rangle$, $B = \langle(1,1) \mid (\mathbf{d}_1 : 1)(\mathbf{d}_3 : 1)(\mathbf{d}_5 : 1)(\mathbf{d}_7 : 1)\rangle$, and $C = \langle(1,3) \mid (\mathbf{d}_1 : 4)(\mathbf{d}_3 : 3)(\mathbf{d}_4 : 1)(\mathbf{d}_5 : 1)(\mathbf{d}_6 : 5)(\mathbf{d}_7 : 1)\rangle$.

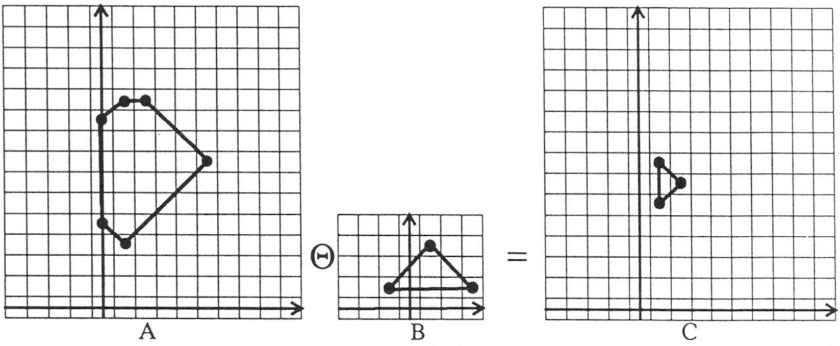

Figure 9. An example of erosion. C is obtained by eroding A by B. Here $A = \langle(1,3) \mid (\mathbf{d}_1 : 4)(\mathbf{d}_3 : 3)(\mathbf{d}_4 : 1)(\mathbf{d}_5 : 1)(\mathbf{d}_6 : 5)(\mathbf{d}_7 : 1)\rangle$, $B = \langle(-1,1) \mid (\mathbf{d}_0 : 4)(\mathbf{d}_3 : 2)(\mathbf{d}_5 : 2)\rangle$, and $C = \langle(1,5) \mid (\mathbf{d}_1 : 1)(\mathbf{d}_3 : 1)(\mathbf{d}_6 : 2)\rangle$.

$$A = \langle(1,3) \mid (\mathbf{d}_1 : 4)(\mathbf{d}_3 : 3)(\mathbf{d}_4 : 1)(\mathbf{d}_5 : 1)(\mathbf{d}_6 : {}_5)(\mathbf{d}_7 : 1)\rangle$$
$$B = \langle(-1,1) \mid (\mathbf{d}_0 : 4)(\mathbf{d}_3 : 2)(\mathbf{d}_5 : 2)\rangle$$

Thus,

$$(i_A, j_A) = (1,3) \quad \text{and} \quad \mathbf{N}^A = [0\ 4\ 0\ 3\ 1\ 1\ 5\ 1]'$$
$$(i_B, j_B) = (-1,1) \quad \text{and} \quad \mathbf{N}^B = [4\ 0\ 0\ 2\ 0\ 2\ 0\ 0]'$$
$$\mathbf{C}^A = \mathbf{L}[i_A\ j_A\ N^A]' = [-3\ \ -2\ 5\ 12\ 10\ 9\ 0\ \ -4]'$$
$$\mathbf{C}^B = \mathbf{L}[i_B\ j_B\ N^B]' = [-1\ 2\ 3\ 4\ 3\ 2\ 1\ 0]'$$

$$\mathbf{C}^C = \mathbf{C}^A - \mathbf{C}^B = [-2\ -4\ 2\ 8\ 7\ 7\ -1\ -4]'$$
$$\mathbf{C}^C = \text{Normalize}(\mathbf{C}^C) = [-5\ -4\ 2\ 8\ 7\ 6\ -1\ -6]'$$
$$\mathbf{N}^C = \mathbf{Q}\mathbf{C}^C = [0\ 1\ 0\ 1\ 0\ 0\ 2\ 0]'$$
$$\mathbf{V}^C = \mathbf{D}[\mathbf{C}^C\ \mathbf{C}^C]' = [\ 1\ 1\ 2\ 2\ 1\ 1\ 1\ 1\ 5\ 5\ 6\ 6\ 7\ 7\ 7\ 5]'$$
$$(i_C, j_C) = (\mathbf{V}^C[0], \mathbf{V}^C[8]) = (1,5)$$

Thus,

$$C = \langle(1,5) \mid (\mathbf{d}_1 : 1)(\mathbf{d}_3 : 1)(\mathbf{d}_6 : 2)\rangle$$

Thus we proceeded differently from the way we did for dilation. First we constructed the normalized half-plane representations of A and B and performed the erosion in this representation. Next we normalized the result. We then converted the result back into the B-code form.

VII. WORK IN PROGRESS

Many extensions of the work presented here are being tried out. Here we list a few of them.

Basic set theory operations with B-codes are an immediate task. Questions that come up are: is the B-coded shape A a subset of B? A superset? Or, does A intersect B or is the intersection empty? Is there an algorithm for finding these sets?

The algorithms presented in this chapter can be generalized for the case of any discrete, convex figure. In that case the polygon edges can be at any angle. These angles can be defined in terms of the basic angles that can be formed by a vector starting from the origin and ending on any pixel (m,n) such that m and n are coprime. The problem of holes in the dilation of shapes needs to be addressed.

Further, the algorithms have been extended to the case of continuous convex polyhedra. But B-code data structure cannot be used for representing the polyhedra. In fact, the polyhedra are represented as the intersection of n-dimensional continuous polyhedra.

The problem of decomposing nonconvex shapes, two- or three-dimensional, is difficult. One way to attack this problem is to represent the nonconvex shape as a union of restricted domains and then decompose each of the restricted domains of the union. Another approach is to represent a shape A as a union of disjoint sets A^1 and K^1 where K^1 is the largest restricted domain that is a subset of A and $A = A^1 - K^1$. This process can be repeated and the shape A can be represented as $A = K^1 \cup K^2 \cup \cdots \cup K^n$, where each restricted domain K^i can then be decomposed further.

Morphological dilation on nonconvex shapes will have to be carried out by first representing the shape as a union of restricted domains. Morphological erosion of nonconvex shapes can be done by representing the shapes as intersections of restricted domains and complements of restricted domains. How to decompose a nonconvex shape as a union of restricted domains and intersection of restricted domains is a problem. An algorithm for doing this has to be developed. Furthermore, in higher dimensions the representation scheme of nonconvex shapes has to be in terms of half-planes.

VIII. CONCLUSION

We defined restricted domains—a restricted class of two-dimensional shapes. Two boundary schemes for representing a restricted domains, the B-code and the discrete half-planes representation, were introduced. Morphological dilation, erosion, n-fold dilation, n-fold erosion, openings, and closings of restricted domains with structuring elements that are also restricted domains were expressed in terms of B-codes and half-planes. Algorithms for performing these operations were provided and were proved to be constant-time algorithms.

Suggestions have been made as to how the algorithms can be generalized to any arbitrary two- and three-dimensional convex shapes. Further work needs to be done in the direction of set operations on restricted domains. Finally, it is important to solve the difficult problem of decomposing nonconvex shapes into restricted domains so that the algorithms presented in this chapter can be used on general images.

REFERENCES

1. Ballard, D. H., and Brown, C. M., *Computer Vision*, Prentice-Hall, Englewood Cliffs, New Jersey, 1982.
2. Freeman, H., Computer processing of line-drawing images, *Comput. Surv.*, *6*(1), 57–97 (1974).
3. Ghosh, P. K., A mathematical model for shape description using Minkowski operations, *Comput. Vision Graphics Image Process.*, *44*, 239–269 (1988).
4. Ghosh, P. K., A solution of polygon containment, spatial planning, and other related problems using Minkowski operations, *Comput. Vision Graphics Image Process.*, *49*, 1–35 (1990).
5. Haralick, R. M., and Shapiro, L. G., *Machine Vision*, Addison-Wesley, Reading, Massachusetts, 1992.
6. Haralick, R. M., Sternberg, S. R., and Zhuang, X., Image analysis using mathematical morphology, *IEEE Trans. Pattern Anal. Machine Intell.*, *PAMI-9*(4), 532–550 (1987).
7. Kanungo, T., Haralick, R. M., and Zhuang, X., B-code dilation and decomposition of restricted convex shapes, in *Proceedings of SPIE*, vol. 1350: *Image Algebra and Morphological Image Processing*, San Diego, 1990.
8. Preparata, F. P., and Shamos, M. I., *Computational Geometry*, Springer-Verlag, New York, 1985.
9. Xu, J., The optimal implementation of morphological operations on neighbourhood connected array processors, in *Proc. of IEEE Conf. on Computer Vision and Pattern Recognition*, pp. 166–171, 1989.

APPENDIX: B-CODE DETAILS

For a B-code to be the representation of a restricted domain, the sequence of displacements in the directions d_i should be ordered monotonically increasing on the directions between d_0 and d_7. The maximum number of nonzero displacements along the boundary can be eight and the minimum zero, which corresponds to a singular point—the starting point. Thus the B-codes of restricted domains are of the forms $A = \langle (i, j) \mid (\mathbf{d}_0 : n_0)(\mathbf{d}_1 : n_1) \cdots (\mathbf{d}_7 : n_7) \rangle$. Note that two exceptions, corresponding to diagonal lines at 45° and 135° are of similar form: $A = \langle (i, j)\ \mathbf{d}_1 : n_1)(\mathbf{d}_5 : n_5) \rangle$ and $A = \langle (i, j) \mid (\mathbf{d}_3 : n_3)(\mathbf{d}_7 : n_7) \rangle$. These are not restricted domains since they are not 4-connected. To prove this property, we need the following two lemmas:

Lemma A.1. Let $A = \langle (i, j) \mid (\mathbf{d}_0 : n_0)(\mathbf{d}_1 : n_1) \cdots (\mathbf{d}_7 : n_7)$ be a B-code representing a shape with more than one point. Then there exists i, $0 \le i \le 3$, such that $n_i \neq 0$.

Proof. By contradiction: Let $n_0 = n_1 = n_2 = n_3 = 0$. Since $n_i \ge 0$ for $0 \le i \le 7$, from Eq. (9.3):

$$n_5 = n_6 = n_7 = 0$$

and from Eq. (9.2):

$$n_4 = 0$$

Hence, $n_i = 0$ for $0 \le i \le 7$, and the B-code A reduces to just one point—the starting point (i, j). This contradicts the hypothesis. Therefore, at least one of n_i with $0 \le i \le 3$ must be nonzero.

Lemma A.2. Let $A = \langle (i, j) \mid (\mathbf{d}_0 : n_0)(\mathbf{d}_1 : n_1) \cdots (\mathbf{d}_7 : n_7) \rangle$ be a B-code representing a shape with more than one point. Then, if $n_i \neq 0$, $0 \le i \le 7$, there exists j, $\mathrm{mod}(i + 1, 8) \le j \le \mathrm{mod}(i + 4, 8)$, such that $n_j \neq 0$.

Proof. The lemma can be proved using Eqs. (9.2), (9.3) and the fact that some $n_i \ge 0$ for $0 \le i \le 7$.

Case (1): $i = 0$. We have to prove that if $n_0 \neq 0$, then there exists j, $1 \le j \le 4$ such that $n_j \neq 0$. From Eq. (9.2)

$$n_3 + n_4 + n_5 > 0$$

and therefore at least one of n_3, n_4, or n_5 is nonzero.

1. $n_3 \neq 0$. Then $j = 3$.
2. $n_4 \neq 0$. Then $j = 4$.
3. $n_3 = n_4 = 0$, and $n_5 \neq 0$. From Eq. (9.3) we have

   $$n_1 + n_2 > 0$$

   and therefore at least one of n_1 or n_2 is nonzero.

(a) $n_1 \neq 0$. Then $j = 1$.
(b) $n_2 \neq 0$. Then $j = 2$.

Case (2): $i = 1$. We have to prove that if $n_1 \neq 0$, then there exists j, $2 \leq j \leq 5$ such that $n_j \neq 0$. From Eq. (9.2) we have

$$n_3 + n_4 + n_5 > 0$$

and therefore at least one of n_3, n_4, or n_5 is nonzero.

1. $n_3 \neq 0$. Then $j = 3$.
2. $n_4 \neq 0$. Then $j = 4$.
3. $n_5 \neq 0$. Then $j = 5$.

Similarly we can prove that the lemma holds for the cases when $i = 2, 3, 4, 5, 6, 7$.

Theorem A.1. A B-code of the form $A = \langle (i, j) \mid (\mathbf{d}_0 : n_0)(\mathbf{d}_1 : n_1)\cdots(\mathbf{d}_7 : n_7)\rangle$ is a restricted domain or a line at 45° or 135°.

Proof. We have to show that the B-code represents a discretely convex shape. This we will do by finding the interior angles at each of the vertices and showing that the are concave (less than 180°). In the case when it is exactly 180° it becomes a line but is still discretely convex. Only the lines at 45° and 135° are 8-connected. All the rest of the lines and restricted domains are 4-connected. Details of this method can be found in the book by Preparata and Shamos [8].

Since A is a B-code, either none, two, or more than two of the lengths are nonzero. Hence, we will consider the following three cases:

Case (1): All lengths are zero. Since $n_i = 0$, for all i, $0 \leq i \leq 7$, A reduces to the point (i, j). Thus, A is a restricted domain.

Case (2): Two lengths are nonzero. Let i and j, $0 \leq i < j \leq 7$, such that $n_i \neq 0$ and $n_j \neq 0$. From Lemma A.2 we see that that $j = i + 4$, and A represents a line. Thus A is discretely convex.

Case (3): Three or more lengths are nonzero. We will proceed by first ordering the vertices of A in an increasing order of counterclockwise angular displacement around the starting point. Then we take three consecutive vertices at a time and check if the angle between them is concave. If all such consecutive triples have angles less than 180°, the binary shape represented by the B-code is discretely convex. Let $v_0, v_1, \ldots, v_7, v_8 = v_0$ be the eight vertices of A. If n_i is zero, vertices v_i and v_{i+1} will coincide. Only the distinct vertices of the shape will contribute to the edges of the convex hull. Let $v_{k+1}, v_{l+1}, v_{m+1}$, with $0 \leq k + 1 < l + 1 < m + 1 \leq 7$ be three distinct consecutive vertices of A. A detailed discussion of the technique can be found in [8].

The vertices $v_{k+1}, v_{l+1}, v_{m+1}$ form a concave angle if

$$\begin{vmatrix} \cos\theta_l & \sin\theta_l & 0 \\ \cos\theta_m & \sin\theta_m & 0 \\ 0 & 0 & 1 \end{vmatrix} = \sin(\theta_m - \theta_l) \geq 0 \tag{9.68}$$

where θ_i is the angle formed by the direction d_i with the positive x-axis. Since $\theta_i = i\pi/4$, Eq. (9.68) reduces to

$$\sin(\theta_m - \theta_l) = \sin(m - l)\frac{\pi}{4} \tag{9.69}$$

Consider the following two cases:

1. $0 \le k + 1 < l + 1 \le 4$. Since the vertices $v_{k+1}, v_{l+1}, v_{m+1}$ are distinct and consecutive we have

   $n_k \ne 0$

   $n_l \ne 0$

   $n_m \ne 0$

   Since $l \le 3$, from Lemma A.2, there exists i, $l + 1 \le i \le l + 4$, such that $n_i \ne 0$. Since v_{m+1} is consecutive to v_{l+1}, we have

   $$l + 1 \le m \le i \le l + 4 \tag{9.70}$$

   Therefore,

   $$1 \le m - l \le 4 \tag{9.71}$$

   and $\sin(\theta_m - \theta_l) \ge 0$.
2. $l \ge 4$. Since v_{l+1} and v_{m+1} are consecutive, we have

   $$l + 1 \le m \le 8 \tag{9.72}$$

   Since $l \ge 4$, we have $m - l \le 8 - 4 = 4$,

   $$1 \le m - l \le 4 \tag{9.73}$$

   and $\sin(\theta_m - \theta_l) \ge 0$.

Hence, A is a discretely convex shape. Since the edges of A are in the directions d_i, they are oriented at angles that are multiples of 45° with respect to the positive x-axis. Thus, A is a restricted domain or a line at 45° or 135°.

Chapter 10

On a Distance Function Approach for Gray-Level Mathematical Morphology

Françoise Preteux

Département Images, Télécom Paris, Paris, France

I. INTRODUCTION

In its set approach, pattern recognition by mathematical morphology (MM) consists of analyzing the relationships between an object (subset of $\mathbf{R}^n$) and its environment, using structuring elements, i.e., predefined geometrical sets.

From a practical point of view, the morphological operators depend on the choice of these structuring elements, the relationships between them, and the image to be transformed. However, the underlying concepts are those of local neighborhood configuration and distance function. Thus, on binary images, the Golay operators are obtained by neighborhood operators [1], while erosion, dilation, skeleton, and ultimate erosion [2] are performed from the distance function to a set or to its complement [3]. The main advantage of such an approach is that it reduces the complexity of the algorithm, which becomes independent of the size of the structuring element. Unfortunately, such a convenient interpretation of basic operators in binary MM in terms of distance function cannot be directly transposed in gray-level MM, where previous work dealt only with neighborhood operators [4].

In this chapter dealing with gray-level MM, we propose a novel approach based on the notion of distance function and define the topographical and differential distances derived, respectively, from the concepts of connection cost and deviation cost.

The principle of the connection cost is to associate with two different points the smallest altitude for which the two points belong to the same connected component with respect to the series of thresholds of the initial gray-level image. The underlying mathematical notion is the geodesic distance.

The concept of deviation cost translates the energy required to link, in a monotonic way, two different points by a path that should not be the path of greatest slope.

This theoretical framework is powerful for characterizing the topographical information of gray-level images by introducing a topographical MM, developing new propagation primitives relative to these distances, and establishing the equivalence between skeleton by influence zone (SKIZ) and watershed (WS)—two basic and fundamental operators in segmentation, respectively in binary and gray-level images. This leads to an identification of binary versus gray-level images as well as new fast algorithms for computing the SKIZ and WS.

The concepts mentioned above were applied to the segmentation of various anatomical features in computed tomography (CT), establishing efficiency and robustness of the corresponding algorithms.

In Section II we briefly describe the basic operators of binary MM in terms of Euclidean and geodesic distances and present two specific examples of MM segmentation exploiting the information generated by a distance function. In Section III we introduce the connection cost concept, study its main properties, define the topographical distance function, and mention some topological properties. We develop a topographical MM (specifically a topographical dilation and closing) and show a first relationship between SKIZ and WS. This is followed by Section IV presenting the differential distance based on the deviation cost concept and establishing the complete equivalence between SKIZ and WS. Section V deals with the algorithmic aspects of the developed concepts. Specifically, a fast algorithm for computing the connection cost leads to a new algorithm of SKIZ and WS applied to segmentation in medical imaging.

II. ON DISTANCE FUNCTIONS IN BINARY MATHEMATICAL MORPHOLOGY

A. Euclidean Distance

In the following, the Euclidean distance from point x to point y in $\mathbf{R}^n$ (respectively to a nonempty subset X in $\mathbf{R}^n$) is referred to as $d(x,y)$ (respectively $d(x,X) = \inf\{d(x,y),\ y \in X\}$). Let us notice that X and its closure, denoted by $\bar{X}$, are undistinguishable with respect to d, that is, $d(.,X) = d(.,\bar{X})$.

If we consider as an isotropic structuring element a ball of radius ρ centered at the origin, denoted by $B(0,\rho)$, the eroded set of X with respect to $B(0,\rho)$ is given by

$$X \ominus B(0,\rho) = \{x \in \mathbf{R}^n,\ B(x,\rho) \subseteq X\} = \{x \in \mathbf{R}^n,\ d(x,X^c) > \rho\}$$

where $(.)^c$ denotes the complement.

More generally, the isolevel curves of $d(.,X^c)$ correspond to the boundaries of the successive eroded sets with respect to increasing balls, and the maxima of $d(.,X^c)$ correspond to the ultimate eroded sets of X.

According to the usual duality principle with respect to the complement, the dilation is expressed by

$$X \oplus B(0,\rho) = (X^c \ominus B(0,\rho))^c = \{x \in \mathbf{R}^n, B(x,\rho) \cap X \neq \emptyset\}$$

It may be interpreted in terms of distance via the mapping $d(.,X)$. We then have the following properties:

Proposition 1

1. $\forall \rho > 0, \forall x \in \mathbf{R}^n, d(x,X) < \rho \Leftrightarrow x \in X \oplus \mathring{B}(0,\rho)$ where $\mathring{B}(0,\rho)$ denotes the interior of $B(0,\rho)$.
2. $\forall \rho \geq 0, \forall x \in \mathbf{R}^n, d(x,X) > \rho \Leftrightarrow x \notin \bar{X} \oplus B(0,\rho)$.
3. $\forall \rho \geq 0, \forall x \in \mathbf{R}^n, d(x, X \oplus B(0,\rho)) = \max\{0, d(x,X) - \rho\}$.

By combination, opening and closing with respect to balls may also be interpreted in terms of distance.

Moreover, the concept of SKIZ introduced in [5] and derived from the notion of influence zone is directly based on the Euclidean distance. Let us briefly recall its principle.

Let $X = \cup_{p \in \mathbf{N}} X_p$ be a countable union of disjoint compact subsets of $\mathbf{R}^n$. The *influence zone* of X_p, denoted by $\mathrm{IZ}(X_p)$, is the set of points x of $\mathbf{R}^n$ that are closer to X_p than to any other set X_q:

$$\mathrm{IZ}(X_p) = \{x \in \mathbf{R}^n, \forall q \in \mathbf{N}, q \neq p\ d(x,X_p) < d(x,X_q)\}$$

Here, $\mathrm{IZ}(X_p)$ is an open set of $\mathbf{R}^n$ [5].

The SKIZ of X, denoted by SKIZ(X), is the set of points that belong to no influence zone. It is defined by

$$\mathrm{SKIZ}(X) = \left[\bigcup_{p \subset \mathbf{N}} \mathrm{IZ}(X_p)\right]^c$$

It is also expressed by

$$\mathrm{SKIZ}(X) = \partial\left[\bigcup_{p \in \mathbf{N}} \mathrm{IZ}(X_p)\right]$$

where $\partial(.)$ denotes the boundary [5].

From a topological point of view, SKIZ(X) separates the compact sets that define X, has an empty interior, and satisfies interesting properties of connectivity and continuity [5].

From a practical point of view, the SKIZ concept has been used in binary imaging for analyzing the distribution of particle population [5] and for image representation and compression [6,7]. Let us mention that in this last application the usual terminology is the *Voronoï tesselation*, by reference to graph theory and computational geometry.

B. Geodesic Distance

Geodesic MM consists of considering a strict subspace of $\mathbf{R}^n$ and performing the morphological operators relative to it. The fundamental concept is that of a path, taking into account connectivity properties.

Let us briefly recall the main notions of path, geometric arc, and arc length in the framework of a normed vector space E. For a more complete presentation see [8].

Let I be an interval in $\overline{\mathbf{R}}$. Any continuous function γ from I to E is called a path, which is said to be compact if and only if (iff) I is a compact set. Two paths γ and γ' defined, respectively, from I to E and J to E are equivalent if there exists a bijection β from I to J such that $\gamma = \gamma' \bigcirc \beta$. The fact that two equivalent paths have the same image in E (the reciprocity being false) leads to the notion of geometric arc, which is defined by the class of equivalent paths.

In order to define the arc length, we use an approximation of the arc γ by means of polygonal lines. Let us denote the norm in E by $\|\ \|$. With any subdivision $\sigma = (s_0, s_2, \ldots, s_p)$ of I ($s_0 = \inf I$, $s_p = \sup I$), let us associate the polygonal line $(M_0, M_2, \ldots, M_p)$ in γ of vertices $M_i = \gamma(t_i)$. The length of the polygonal line is given by

$$\mathcal{L}(\gamma,\sigma) = \sum_{i=0}^{p-1} \|M_i M_{i+1}\| = \sum_{i=0}^{p-1} \|\gamma(t_{i+1}) - \gamma(t_i)\|$$

Let $\mathcal{S}$ be the set of subdivisions of I; the length of the arc γ is defined by

$$\mathcal{L}(\gamma) = \sup\{\mathcal{L}(\gamma,\sigma),\ \sigma \in \mathcal{S}\}$$

By definition, an arc γ is said to be rectifiable iff $\mathcal{S}(\gamma)$ is finite.

In the following, in order to simplify, we will use the generic term "path" to refer to both the path and the arc. More precisely, a path linking x and y in a subset X in $\mathbf{R}^n$ will be a continuous mapping γ from $[0,1]$ to X such that $\gamma(0) = x$ and $\gamma(1) = y$. We will denote by $\mathcal{P}_X(x,y)$, the set of paths in X linking x and y.

Definition 1. Let X be a nonempty subset of $\mathbf{R}^n$. Given two arbitrary points x and y in X, let us define the number $\delta_X(x,y)$ as follows: $\delta_X(x,y)$ is the lower bound (i.e., the infimum) of the lengths of the paths γ of $\mathcal{P}_X(x,y)$ if such paths exist, and $+\infty$ if not. Mathematically, it can be expressed by

$$\forall\ (x,y) \in X^2, \qquad \delta_X(x,y) = \begin{cases} \inf\{\mathcal{L}(\gamma),\ \gamma \in \mathcal{P}_X(x,y)\} & \text{iff } \mathcal{P}_X(x,y) \neq \emptyset \\ +\infty & \text{otherwise} \end{cases}$$

where $\mathcal{L}(\gamma)$ is the length of the path. Here, $\delta_X(x,y)$ is called the *geodesic distance* from x to y with respect to X (Figure 1).

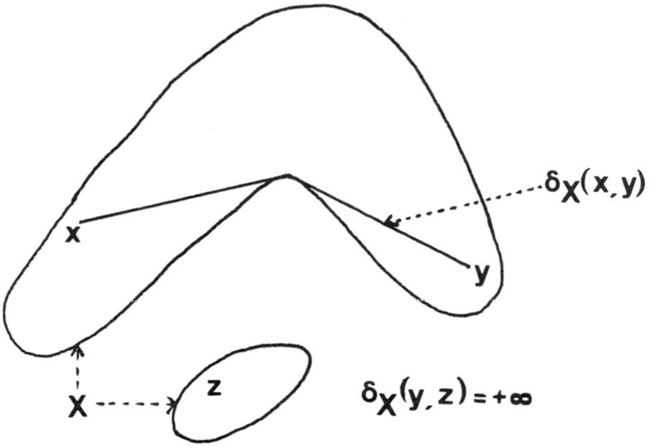

Figure 1. Example of a geodesic distance relative to the set X.

It can easily be shown that the mapping δ_X from $\mathbf{R}^n \times \mathbf{R}^n$ to $\mathbf{R}^+ \cup \{+\infty\}$ satisfies the three properties of a distance function (separability, symmetry, triangular inequality). Nevertheless, δ_X is not a distance function since it can take infinite values, especially if x and y belong to two different connected components of X. Here, δ_X is called the *generalized distance function* from $\mathbf{R}^n \times \mathbf{R}^n$ to $\mathbf{R}^+ \cup \{+\infty\}$.

Let us remark that the metric defined on X by δ_X is not always equivalent to the metric induced on X from the Euclidean distance d.

Definition 2. Let X be a nonempty subset of $\mathbf{R}^n$. The *extended geodesic distance* δ_X with respect to X is defined as follows:

$$\forall\ (x,y) \in X^2, \qquad \tilde{\delta}_X(x,y) = \delta_X(x,y)$$

$$\forall\ (x,y) \in (\mathbf{R}^n)^2 \backslash X^2, \qquad \tilde{\delta}_X(x,y) = \begin{cases} +\infty & \text{iff } x \neq y \\ 0 & \text{otherwise} \end{cases}$$

Obviously, $\tilde{\delta}_x$ is a generalized distance function on $\mathbf{R}^n \times \mathbf{R}^n$.

The concept of geodesic ball of radius ρ centered at y, with respect to X, is given by $B_X(y,\rho) = \{x \in X,\ \tilde{\delta}_X(x,y) \leq \rho\}$.

Let Y be a subset of X. By definition, the geodesic erosion (resp. dilation) of Y with respect to B_X with radius ρ, denoted by $\mathrm{Er}_X(Y,\rho)$ (resp. $\mathrm{Di}_X(Y,\rho)$), is expressed by

$$\mathrm{Er}_X(Y,\rho) = \{x \in X,\ B_X(x,\rho) \subseteq Y\}$$

$$(\text{resp. } \mathrm{Di}_X(Y,\rho) = \{x \in X,\ B_X(x,\rho) \cap Y \neq \emptyset\})$$

Apart from the usual geodesic opening, closing, IZ, and SKIZ, let us define an MM operator that makes sense only in the geodesic context: the particle reconstruction. For $Y \subseteq X = \cup_{p \in N} X_p$, the reconstruction of the set X from the marker Y, denoted by $R_X(Y)$, is defined as follows:

$$R_X(Y) = \lim_{\rho \to +\infty} \mathrm{Di}_X(Y,\rho)$$

In general, the differences between the geodesic and Euclidean frameworks lie in a less rich range of possible structuring elements and a truncation effect due to the considered region of interest.

From an algorithmic point of view, the Euclidean distance is usually approximated by discrete distance functions according to a given connectivity (4-, 6-, or 12-connectivity, respectively, for square, hexagonal, or dodecagonal distance) and computed using more or less fast algorithms according to their type: parallel [9], sequential [10], or based on loops [11]. More recently, an algorithm of a new type based on chains [12] has been developed which yields an exact estimate of the Euclidean distance in a very fast way. It makes it possible to perform exact Euclidean MM operators.

C. On the Use of Distance Functions in Pattern Recognition

If MM operators are fundamentally based on the distance concept, the distance function as an intrinsic tool in pattern recognition is of a great interest. Specifically, it allows us to solve difficult problems of automatic parameter adaptivity and feature segmentation. These two aspects will be illustrated for the complex problem of the quantitative analysis of the pulmonary parenchyma in CT [13–15].

1. Recognition of the Right and Left Lungs in Computed Tomography: Autoadaptive Erosion

We do not provide the details of the clinical aspects of our research [13,16,17] but only briefly describe the segmentation problem in hand.

Our data set consists of the CT examinations of both healthy subjects and ill patients suffering from diffuse infiltrative lung diseases. For each patient we have a series of about 30 slices, each 1.5 mm in thickness, with 10 mm between two consecutive slices and in 512×512 reconstruction matrices. Radiologically, the two normal lungs appear as pulmonary interstitium with low densities containing vascular elements with middle and high densities and elements with very low densities. They are surrounded by ribs with very high densities (Figure 2).

In the case of healthy right and left lungs, the morphological procedure developed to recognize and separate them [13] is based on density and connectivity criteria. This very simple procedure may not be robust in the case of overinflated lungs (Figure 3a–c) and may yield a global pulmonary mask in which the two lungs are connected (Figure 3d–f). The quantitative analysis for each of them becomes impossible. How do we perform the separation of the two lungs?

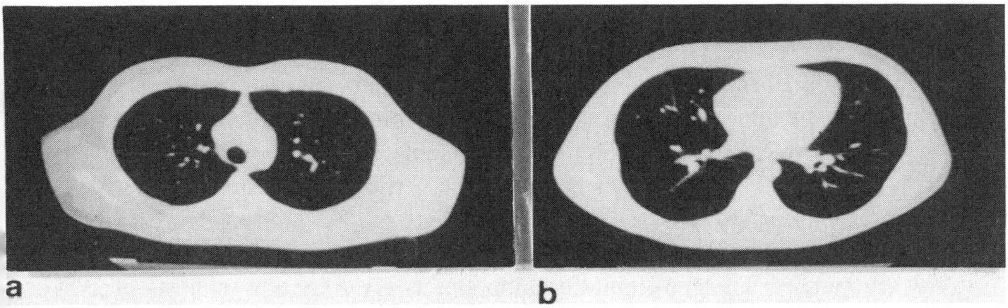

Figure 2. Two CT images with normal lung parenchyma.

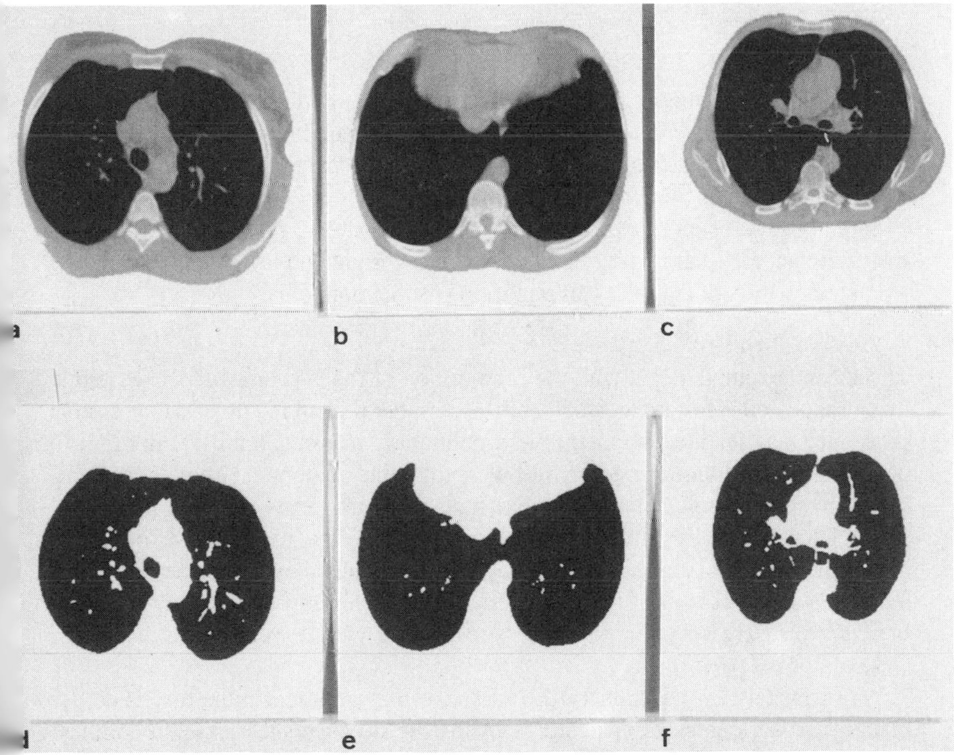

Figure 3. Three CT images with overinflated lungs (a–c) Associated global pulmonary mask (d–f) in which the two lungs are connected. Notice that the number, location, and size of connections vary.

Experience shows that the number, the location, and the size of connections vary. However, connections always appear as strictures with variable thickness of the global pulmonary mask. This leads us to use an erosion in order to realize the separation. The difficulty lies in the way to determine automatically the minimal size necessary for the separation. The principle of the developed procedure [18,14] exploits the information given by the vertical Euclidean distance function to the pulmonary mask considered only through a region of interest (ROI) defined by geometrical criteria. For each connection, let us consider its smallest vertical diameter. The convenient erosion size is given by the greatest value of these diameters considered through the ROI. The operational procedure is given by the following steps (Figure 4a–i):

$$f_{/\mathrm{ROI}} \longrightarrow \max v(d_v(f_{/\mathrm{ROI}})) \times \mathrm{Fr}(f_{/\mathrm{ROI}}) = g$$
$$g \longrightarrow \min c(g) = h$$
$$\mathrm{Size} = \max(h)$$

where $f_{/\mathrm{ROI}}$ denotes the global pulmonary mask through the ROI, d_v the vertical Euclidean distance, Fr the boundary, max v the mapping that associates with each vertical segment its maximum value, and min c the mapping that associates with each connected component its minimum value.

In practice, erosion sizes vary from 6 to 24 pixels. This justifies fully an adaptive approach for each image. Finally, the correct right and left pulmonary masks are obtained by topographical propagation (see Section V.C).

2. Rib Segmentation

The second example deals with the recognition of the two lungs in the case that hyperattenuated areas are located near the peripheral pleura. The previous simple procedure may lead to two incomplete pulmonary masks (Figure 5). In order to obtain the missing lung parenchyma, we exploit the anatomical knowledge dealing with the proximity of the external lung contours to the ribs: the peripheral pleura leans on the internal side of the ribs [18,14]. The idea consists of propagating the pulmonary masks previously obtained until they reach the ribs. This requires a segmentation of the ribs. Because CT gives absolute density information and ribs have very high density, thresholding with respect to the density of cortical bone is performed.

Unfortunately, in addition to the ribs, we may segment humerus, shoulder blades, and various calcified features inside the rib cage such as micronodules, nodules, calcified aorta, etc. (Figure 6a–c). The question is, how do we discriminate the ribs from the other calcified elements?

In order to guarantee a robust and parameter-independent procedure, we propose a procedure based on the propagation of the outside of the anatomical slice relative to the distance function to the set of the calcified elements. Let us present in detail the different steps of this procedure in the most complicated config-

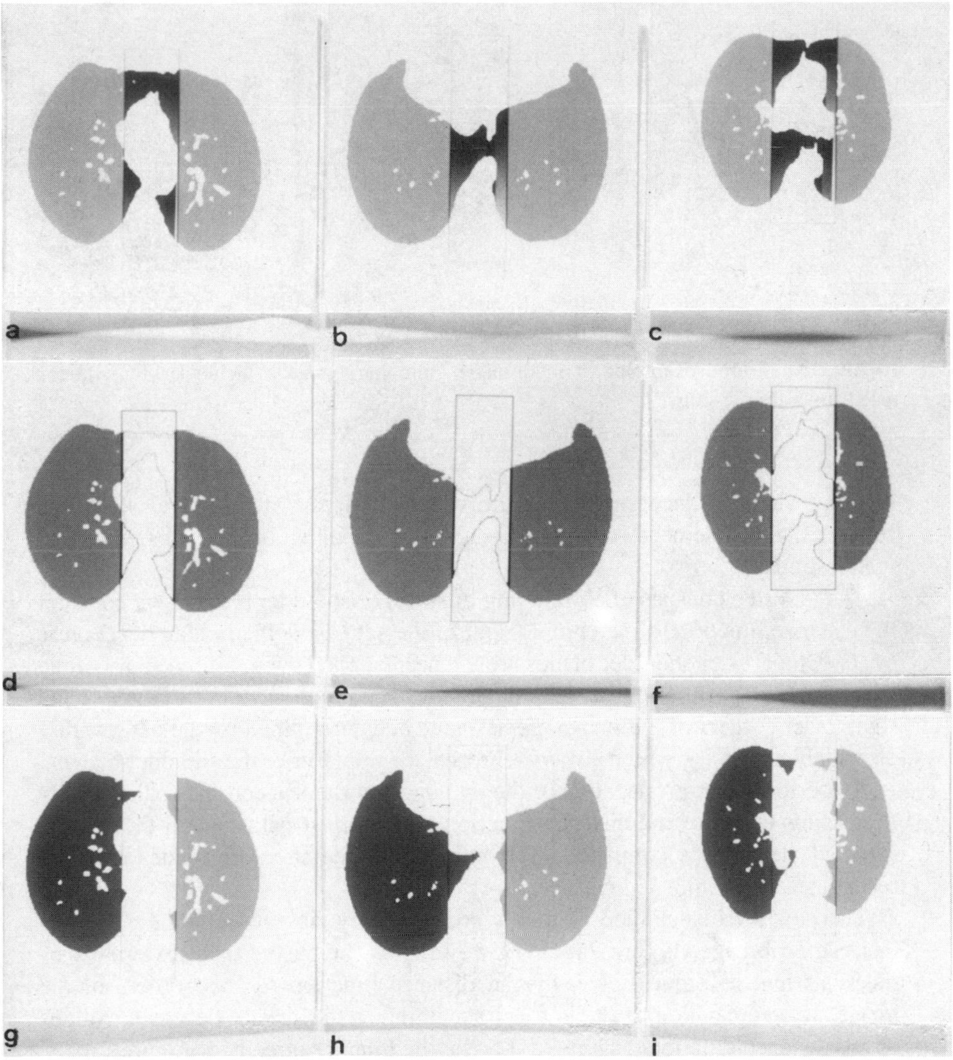

Figure 4. Main steps of the autoadaptive erosion for separating and recognizing the left and right lungs. (a–c) The vertical Euclidean distance of figure 3d–f computed in a ROI defined by geometric criteria. (d–f) The boundaries of the global pulmonary masks restricted to the ROI assigned to the maximum of the distance values of (a)–(c). (g–i) The separation of the two lungs of figure 3d–f obtained by autoadaptive erosion.

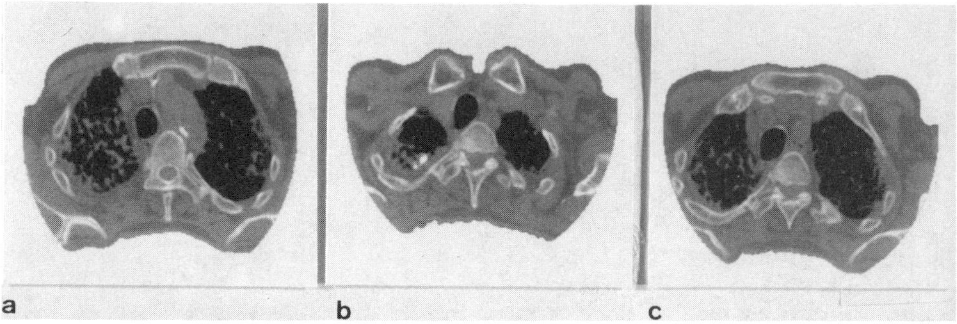

Figure 5. Three examples of incomplete pulmonary masks (in black); hyperattenuated areas are missing.

uration, which is encountered for the superior slices. At this level slice, the set of calcified elements contains humerus or shoulder blades, ribs, calcified features, etc.

The first step consists of eliminating humerus or shoulder blades. We consider the distance function to the complement of the set of calcified elements (Figure 6d–f). Humerus or shoulder blades are the nearest pixels with high density of the left and right lower corners of the image. Distance function values of the left and right lower corners of the image are assigned and determined by the nearest pixels with high density, which always represent the humerus or the shoulder blades.

By performing a propagation of the two lower corners according to the strictly decreasing values of the distance function, the shortest path toward the nearest calcified elements is computed. The humerus or the shoulder blades are then recognized and eliminated (Figure 6g–i).

The new set of calcified elements contains only ribs and calcified features inside the rib cage (Figure 7a–c). However, ribs, being the most external particles, assign the values taken by the distance function on the borders of the image. By propagation of the borders according to the strictly deceasing values of this distance function (Figure 7d–f), ribs are finally segmented (Figure 7g–i).

Let us note that the underlying notion of these propagations is that of SKIZ. It would be equivalent to perform the SKIZ of the set of calcified elements (Figure 8b), to eliminate the IZ connected to the two lower corners, to perform again the SKIZ of the remaining calcified elements, and to keep the IZ connected to the image borders (Figure 8c). In comparison, the use of the distance function yields much better computing performance: robust results are obtained in four screenings of the complete image.

In the mentioned examples, the Euclidean distance may be considered directional or not. Let us notice that the notion of distance function has been used

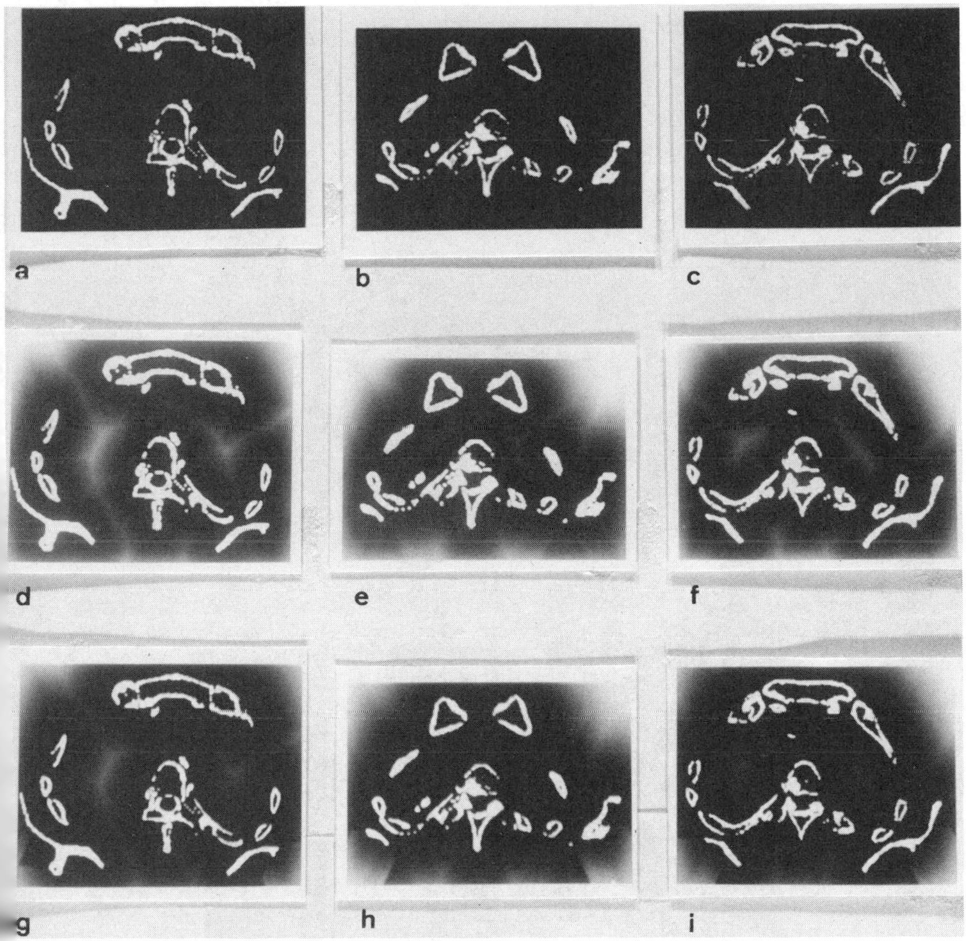

Figure 6. Rib segmentation: first step corresponding to the detection of shoulder blades. (a–c) The set of calcified elements: shoulder blades, ribs nodules, calcified aorta. (d–f) The distance function to the complement of (a)–(c). (g–i) The propagation of the two lower corners (in black) according to the decreasing values of (d)–(f): the shoulder blades are recognized.

more recently in order to define and control the spatial dependence of the external energy field in active contour models [19,20].

Whereas MM dealing with binary images exploits widely the information on distance to a set or to its complement relative to the whole space or to a subset of it, gray-level MM is mainly based on the analysis of neighborhood configurations [4]. In the sequel, we tackle gray-level MM from a distance point of view

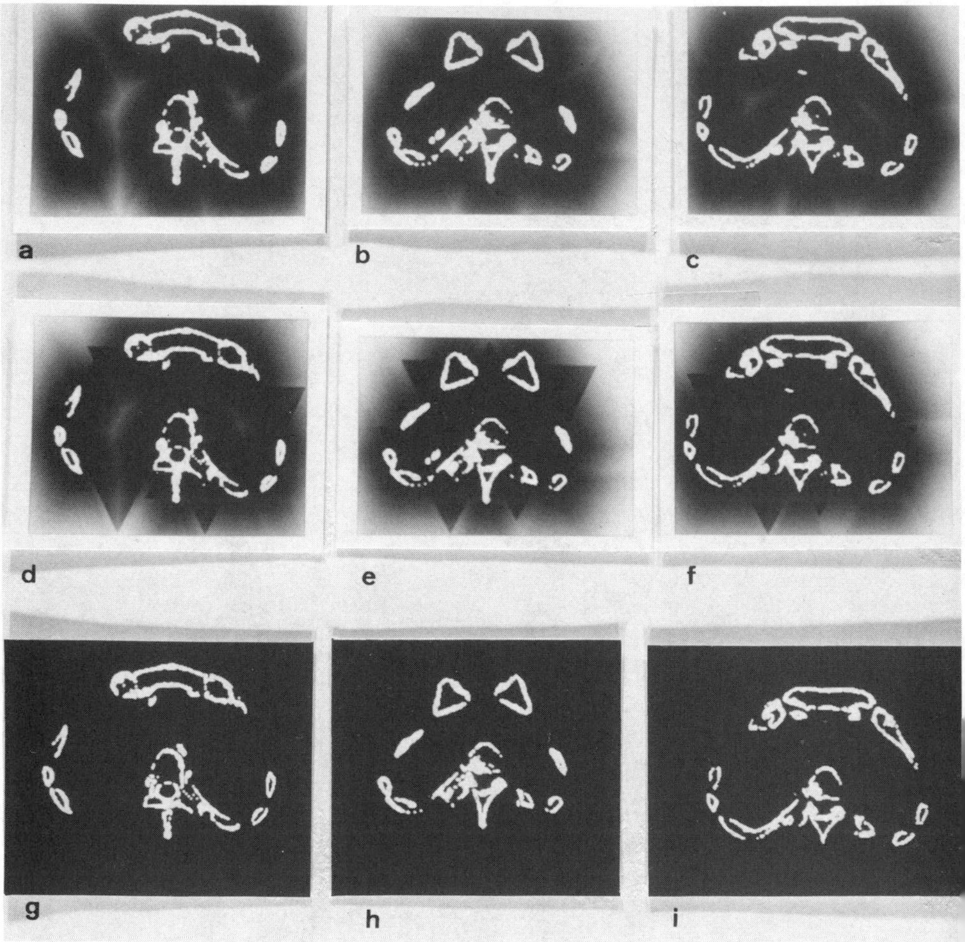

Figure 7. (a–c) The distance function to the complement of the set defined by the ribs and the calcified elements in the rib cage. (d–f) The propagation of the image borders (in black) relative to the decreasing values of (a)–(c). (g–i) The segmented ribs.

and introduce the topographical and differential distances based on connection and deviation cost, respectively.

III. TOPOGRAPHICAL DISTANCE FUNCTION

The set of upper semicontinuous functions from $\mathbf{R}^n$ to $\mathbf{R}$, denoted by $\mathcal{F}_u(\mathbf{R}^n,\mathbf{R})$, equipped with the hit-or-miss topology [21], forms the most general and appropriate mathematical framework for gray-tone image analysis. For a function f,

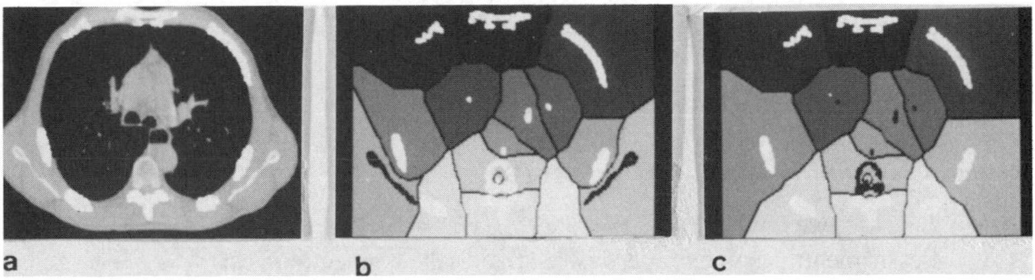

Figure 8. Ribs segmentation based on SKIZ. (a) The initial CT image with the set of calcified elements (in white). (b) The SKIZ relative to the calcified elements and the detection of the shoulder blades (in black) by eliminating the IZ connected to the two lower corners of the image. (c) The SKIZ relative to the remaining calcified elements and the detection of the ribs (in white) by eliminating the IZ connected to the image borders.

supp(f) denotes its support. For the sake of brevity, proofs are not given here. They can be found in [22].

A. Connection Cost: Definition and Properties

In the following, we consider the subset of $\mathcal{F}_u(\mathbf{R}^n,\mathbf{R})$, denoted by $\mathcal{F}_{u,c}(\mathbf{R}^n,\mathbf{R})$, of functions f with a connected support and upper bounded on any bounded subset in supp(f).

The mapping $\mathcal{T} : \mathcal{F}_u(\mathbf{R}^n,\mathbf{R}) \times \mathbf{R} \longrightarrow \mathcal{P}(\mathbf{R}^n)$ which associates (f,λ) with $X_{f,\lambda} = \{x \in \text{supp}(f),\ f(x) \leq \lambda\}$ defines the *threshold* $X_{f,\lambda}$ of f at level λ. Obviously, $\mathcal{T}(f,.)$ is an increasing mapping and $\mathcal{T}(.,\lambda)$ is a decreasing one.

To simplify the notation, we use $\delta_{f,\lambda}$ and $\tilde{\delta}_{f,\lambda}$ in place of $\delta_{X_{f,\lambda}}$ and $\tilde{\delta}_{X_{f,\lambda}}$, respectively. The mapping $\tilde{\delta}_f$ that associates λ with $\tilde{\delta}_{f,\lambda}$ is decreasing.

Proposition-Definition 1. Let f be a function in $\mathcal{F}_{u,c}(\mathbf{R}^n,\mathbf{R})$. Let us consider the mapping $\mathcal{C}_f$ from $\mathbf{R}^n \times \mathbf{R}^n$ to $\mathbf{R}$ such that

$$\forall (x,y) \in \mathbf{R}^n \times \mathbf{R}^n,\ x \neq y, \qquad \mathcal{C}_f(x,y) = \inf\{\lambda \in \mathbf{R},\ \tilde{\delta}_{f,\lambda}(x,y) < +\infty\}$$

By convention, we set $\mathcal{C}_f(x,x) = -\infty$. We refer to $\mathcal{C}_f(x,y)$ as the *connection cost* of x and y with respect to f. The inf is well defined because the set $\{\lambda \in \mathbf{R},\ \tilde{\delta}_{f,\lambda}(x,y) < +\infty\}$ is a nonempty and lower-bounded subset of $\mathbf{R}$ for $x \neq y$.

Intuitively, the connection cost is a concept familiar to the mountain hiker: it corresponds to the altitude of the lowest neck of the mountain linking two valleys on either side. Obviously, taking the neck of the mountain may increase the length of the trip, but it guarantees a minimum ascent. The higher the ridge between two valleys, the higher will be the connection cost.

Nevertheless, it is important to notice that, in general, the connection cost is an infimum and not a minimum. Effectively, we may have $\tilde{\delta}_{f,\mathcal{C}_f(x,y)}(x,y) = +\infty$, as in the following example. Let x and y be two arbitrary points of $\mathbf{R}^2$. Let D be the median of the segment $[x,y]$. Let us define the function f from $\mathbf{R}^2$ into $\mathbf{R}^+$ such that

$$f(z) = \begin{cases} \exp(-d(x,z)) & \forall z \in D \\ 0 & \forall z \in \mathbf{R}^2 \setminus D \end{cases}$$

In this case, we have $\mathscr{C}_f(x,y) = 0$ and $\tilde{\delta}_{f,0}(x,y) = +\infty$.

Let us mention some properties relative to the connection cost.

Proposition 2

1. $\forall \lambda = \in \mathbf{R}, \lambda < \mathscr{C}_f(x,y), \tilde{\delta}_{f,\lambda}(x,y) = +\infty.$
2. $\forall \lambda \in \mathbf{R}, \lambda > \mathscr{C}_f(x,y), \tilde{\delta}_{f,\lambda}(x,y) \in \mathbf{R}\ \tilde{\delta}_{f,\lambda}(x,y) \leq \tilde{\delta}_{f,\mathscr{C}_f(x,y)}(x,y).$
3. $\forall (x,y) \in (\mathrm{supp}(f))^2, x \neq y,$

$$\mathscr{C}_f(x,y) = \inf\left\{\max\{f(h),\ h \in \gamma\},\ \gamma \in \mathscr{P}_{\mathrm{supp}(f)}(x,y)\right\}$$

Proposition 3

1. $\forall\ (x,y) \in (\mathrm{supp}(f))^2, \mathscr{C}_f(x,y) = \mathscr{C}_f(y,x).$
2. $\forall\ (x,y,z) \in (\mathrm{supp}(f))^3, \mathscr{C}_f(x,z) \leq \max(\mathscr{C}_f(x,y), \mathscr{C}_f(y,z)).$
3. $\forall\ (x,y) \in (\mathrm{supp}(f))^2, x \neq y, \mathscr{C}_f(x,y) \geq \max(f(x), f(y)).$

In order to study the morphological properties of the connection cost, we introduce the notion of connection cost relative to a subset. In the same way that we define the distance function from a point x in $\mathbf{R}$ to a subset Y of $\mathbf{R}^n$, we set the following definition.

Definition 3. Let f be a function of $\mathscr{F}_{u,c}(\mathbf{R}^n,\mathbf{R})$. The *connection cost* of a point x of $\mathbf{R}^n$ to a nonempty subset Y of $\mathbf{R}^n$, denoted by $\mathscr{C}_f(x,Y)$, is given by

$$\mathscr{C}_f(x,Y) = \begin{cases} \inf\{\lambda \in \mathbf{R},\ \tilde{\delta}_{f,\lambda}(x,Y) < +\infty\} & \text{iff } x \notin Y \\ -\infty & \text{otherwise} \end{cases}$$

where $\tilde{\delta}_{f,\lambda}(x,Y)$ is the extended geodesic distance from x to Y with respect to $X_{f,\lambda}$.

This definition is consistent with the case $Y = \{y\}$. Moreover, we have the following interesting property:

Proposition 4

$$\mathscr{C}_f(x,Y) = \inf\{\mathscr{C}_f(x,y)\ /\ y \in Y\}.$$

We specialize the set Y as the set of the regional minima of f.

Definition 4. The set M_h is a *regional minimum with altitude h* of the function $f \in \mathscr{F}^1_{l,c}(\mathbf{R}^n,\mathbf{R})$ iff it is a connected plateau of altitude h from which it is impossible to reach a point with lower altitude without need to ascend.

$$\forall x \in M_h,\ \forall y \notin M_h,\ f(y) \leq f(x),\ \forall \gamma \in \mathscr{P}_{\mathrm{supp}(f)}(x,y),$$
$$\exists s \in [0,1[,\ f(\gamma(s)) > f(x) = h$$

In the following, $M = (\cup M_h)_{h\in\mathbf{R}}$ will denote the set of regional minima of f.

Proposition 5. Let M be the set of all the regional minima of $f \in \mathcal{F}_{u,c}(\mathbf{R}^n,\mathbf{R})$. The connection cost from a point x to the set M ($x \notin M$) with respect to f is given by $\forall\, x \in \text{supp}(f)\backslash M\ \mathcal{C}_f(x,M) = f(x)$.

It results from the foregoing that for any point x not belonging to Y, the connection cost from x to Y is greater than or equal to $f(x)$. Inside Y, the connection cost is uniformaly equal to $-\infty$, which is not convenient for a morphological study. This leads us to a new definition:

Proposition-Definition 2. Let Y be a subset in $\mathbf{R}^n$. Let ϕ_Y be the mapping from $\mathcal{F}_{u,c}(\mathbf{R}^n,\mathbf{R})$ into itself defined by

$$\forall f \in \mathcal{F}_{u,c}(\mathbf{R}^n,\mathbf{R}),\ \forall\, x \in \mathbf{R}^n, \qquad \phi_Y(f)(x) = \max(\mathcal{C}_f(x,Y), f(x))$$

Here, ϕ_Y is an extensive morphological filter and is called *topographical closing*.

B. Topographical Distance

Let us remark that the connection cost with respect to f is not a distance function since the separability condition ($\mathcal{C}_f(x,y) = 0 \Leftrightarrow x = y$) is not satisfied in general. To overcome this problem and to get only finite positive values, we combine the connection cost with the exponential function. Let us note that any strictly increasing and symmetric mapping ψ such that $\psi(-\infty) = 0$ is adequate.

Proposition 6. Let f be a function of $\mathcal{F}_{u,c}(\mathbf{R}^n,\mathbf{R})$. The mapping Δ_f from $\mathbf{R}^n \times \mathbf{R}^n$ to $\mathbf{R}$ defined by

$$\Delta_f : (x,y) \longrightarrow \Delta_f(x,y) = \exp\{\mathcal{C}_f(x,y)\}$$

is a distance function.

Definition 5. Δ_f is called the *topographical distance* and the *topographical distance to a subset* Y in $\mathbf{R}^n$ is expressed by $\Delta_f(x,Y) = \inf\{\Delta_f(x,y),\ y \in Y\}$.

Obviously, we have the following property:

$$\forall\, x \in \mathbf{R}^n,\ \forall\, Y \subseteq \mathbf{R}^n, \qquad \Delta_f(x,Y) = \exp\{\mathcal{C}_f(x,Y)\}$$

Proposition 7. Δ_f is an ultrametric distance on $\mathbf{R}^n \times \mathbf{R}^n$, that is,

$$\forall\, (x,y,z) \in (\mathbf{R}^n)^3 \qquad \Delta_f(x,z) \le \max(\Delta_f(x,y),\ \Delta_f(y,z))$$

This is directly derived from the property of the connection cost (Proposition 3). Let us notice that for any nonnegative function f in $\mathcal{F}_{u,c}(\mathbf{R}^n,\mathbf{R})$, the open ball centered at x with a radius $\rho < 1$ is such that

$$\forall\, \rho < 1 \qquad B(x,\rho) = \{y \in \mathbf{R}^n,\ \Delta_f(x,y) < \rho\} = \{x\}$$

More generally, upon letting $\rho < \exp(\inf(f))$, we have $B(x,\rho) = \{x\}$. This implies that each point is open. Here, $(\mathbf{R}^n,\Delta_f)$ is a discrete space. Moreover, $(\mathbf{R}^n,\Delta_f)$ is a noncompact space but is locally compact. It is not separable because the minimum basis is the set of all points. Let us mention that Δ_f cannot be associated with a norm.

From the topographical distance, we can define a *topographical dilation* of size r, denoted by Θ_r, as follows: $\Theta_r(f) = \{x \in \mathbf{R}^n, \Delta_f(x,Y) < r\}$, which characterizes the zones of $\mathbf{R}^n$ reachable from Y without "climbing too much." By duality, iteration, and combination, we build a topographical MM.

One advantage of the topographical distance is that it enables us to characterize the topography of gray-level images by introducing the concepts of topographical graph and topographical canonical form [23]. This leads to defining an appropriate mathematical framework for a solution to the problem of isomorphism, a fundamental issue of pattern recognition [24,25].

Here, we will show that the topographical distance allows us to establish a relationship between the two fundamental segmentation operators in MM: the SKIZ and the WS.

C. Topographical Distance, SKIZ and WS

While the SKIZ is defined on binary images, the WS is fundamentally a gray-level operator introduced in [26]. Its principle is close to the one presented in [27] but without using any fixed parameters and with a more global point of view.

What is a WS? Intuitively, the notion of WS, well known in geography, is based on the following physical principle: a drop of water placed on a relief will go down following the greatest slope until it reaches an outlet. The set of points leading to the same outlet is called a catchment basin. The watershed corresponds to the line with a null flow separating at least two different catchment basins. It is defined as the locus of apparent points at the origin of divergent streamlines on the piezometric surface [28]. In such a hydrologic context, because nature does not like ambiguity, there exists at each point a greatest slope, except in zones corresponding to deltas.

Despite the apparent correspondance between gray-level images and topographical surfaces, if this concept is directly applied to image analysis, it can result in certain ambiguities. Specifically, the questions that arise are what happens to plateaus and what is the topological status of a catchment basin?

All these difficulties form the origin of a rich literature dealing with algorithms for the detection of WS: algorithms based on flooding [26], arrowing [29], and more recently queues [12], yielding a particular successful technical implementation. The WS concept has been used for segmenting regions or finding edges from gray-level images [30], for classifying Landsat images [31], for analyzing medical images [20], and so on. Let us note that another interesting approach consists of detecting the WS of the gradient [12,20]. In spite of the fact that numerous algorithms exist, a lack of theoretical developments can still be felt.

Here we propose a rigorous formulation of the WS concept in order to establish relationships between SKIZ and WS. Let us consider the space of lower semicomplete (regional minima are the only plateaus) and continuously differentiable functions from $\mathbf{R}^n$ to $\mathbf{R}$ with connected compact support, denoted by $\mathcal{F}^1_{l,c}(\mathbf{R}^n,\mathbf{R})$. Here, the notion of regional minimum of a function $f \in \mathcal{F}^1_{l,c}(\mathbf{R}^n,\mathbf{R})$ represents the role of an outlet.

Let f be a function in $\mathcal{F}^1_{l,c}(\mathbf{R}^n,\mathbf{R})$. Let us introduce $\Gamma_{x,y}$, the set of monotonic paths from x to y ($x \neq y$):

$$\Gamma_{x,y} = \left\{ \begin{array}{l} \gamma \in \mathcal{P}_{\mathrm{supp}(f)}, \ \forall (s_1,s_2) \in [0,1]^2, \\ s_1 < s_2 \Rightarrow f(\gamma(s_1)) \leq f(\gamma(s_2)) \text{ or } f(\gamma(s_1)) \geq f(\gamma(s_2)) \end{array} \right\}$$

We will refer to the descending path from x to y as a monotonic decreasing path.

Definition 6. Let f be a function in $\mathcal{F}^1_{l,c}(\mathbf{R}^n,\mathbf{R})$ and let $M = (M_h)_{h\in\mathbf{R}}$ be its regional minima. The *catchment basin associated with* M_h, denoted by $\mathrm{CB}_f(M_h)$, is the interior of the set of points $x \in \mathbf{R}^n$ for which all the descending paths of greatest slope originating from x lead exclusively to M_h.

Definition 7. The *watershed* relative to $M = (\cup M_h)_{h\in\mathbf{R}}$, the set of the regional minima of $f \in \mathcal{F}^1_{l,c}(\mathbf{R}^n,\mathbf{R})$, denoted by $\mathrm{WS}(f)$, is the complement of the union of all CBs:

$$\mathrm{WS}(f) = \left[\bigcup_{h\in\mathbf{R}} \mathrm{CB}_f(M_h) \right]^c$$

By definition, WS(f) is a closed set. Nevertheless, complex configurations [12] show that the interior of a watershed may be nonempty. As a consequence, the notion of watershed cannot be reinterpreted in terms of boundary as we do for the SKIZ.

Although the mathematical framework of lower complete functions is particularly convenient for defining the notion of WS by eliminating points without a greatest slope, it may be quite constraining in image analysis practice. In order to preserve the generality of the class of images to be analyzed, we consider, in the following, the more general space of continuously differentiable functions with connected compact support, denoted by $\mathcal{F}^1_c(\mathbf{R}^n,\mathbf{R})$.

In such a general framework, what happens if a drop of water is placed on a flat surface? Where would it end up? The answer generally given, at least from the algorithmic point of view [26], is to cut the plateau into influence zones: a SKIZ of the plateau is performed relative to its lower neighbors; each point is associated with the closest lower neighbor.

In fact, we face two different configurations: either the lower neighbors all lead to the same minimum (the plateau is in a way included in a basin), or they lead to at least two different minima (the edges of the plateau are situated at the border of different CBs). The answer to the first possibility is elementary: we associate the points of the plateau with the corresponding CB. Dealing with the second configuration is more complicated and leads to distinguishing two types of CB:

1. The *wide catchment basin* (WCB), which is the interior of the set defined by the union of the initialy obtained CB and the plateau
2. The *narrow catchment basin* (NCB), which is the interior of the set defined by the union of the initial CB, the IZ, and the SKIZ of edges in the corresponding part of the considered CB

Thick (resp. *thin*) *watershed*, denote by TWS (resp. WS), is defined from the notion of WCB (resp. NCB). By default, we will use the terms watershed and catchment basin in place of the thin watershed and the narrow catchment basin.

Proposition 8

1. Let $X = \bigcup_{p \in \mathbf{N}} X_p$ be the union of compact subsets of $\mathbf{R}^n$. The WS of the Euclidean distance $d(.,X)$ is the SKIZ of X with respect to d.
2. Let f be an element in $\mathcal{F}_c^1(\mathbf{R}^n,\mathbf{R})$ and let M be the set of its regional minima. The WS f is a subset of the SKIZ of M relative to the topographical distance: $\mathrm{TWS}(f) \subseteq \mathrm{SKIZ}_{\Delta_f}(M)$ and $\mathrm{WS}(f) \subseteq \mathrm{SKIZ}_{\Delta_f}(M)$.

However, this last inclusion, in general, does not hold with equality because the topographical distance does not take into account information relative to the slope (see the one-dimensional counterexample in Figure 9). To handle this situation, it is necessary to introduce the differential distance.

IV. DIFFERENTIAL DISTANCE FUNCTION

In the same way as we introduced the connection cost concept, exploiting the altitude information according to a connectivity criterion, let us define the notion of deviation cost in order to control information relative to the greatest slope on a topographical surface.

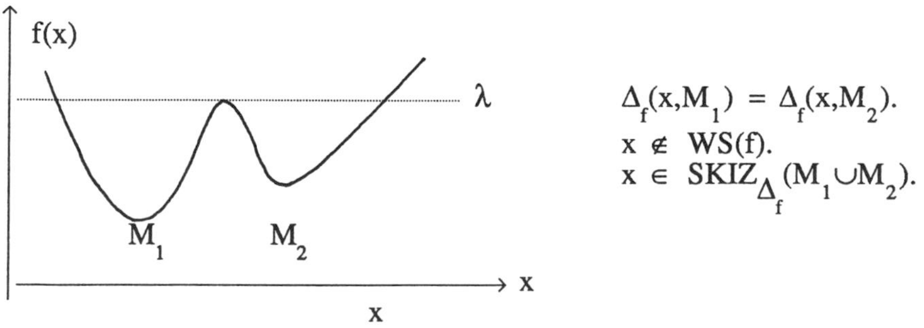

Figure 9. One-dimensional counterexample establishing $\mathrm{WS}(f) \neq \mathrm{SKIZ}\ \Delta_f(M)$.

A. Deviation Cost: Definition and Properties

Proposition-Definition 3. Let $\mathcal{D}_f$ be the mapping from $\mathbf{R}^n \times \mathbf{R}^n$ to $\mathbf{R}^+ \cup \{-\ln(2), +\infty\}$ such that

$$\forall x \in \mathbf{R}^n, \qquad \mathcal{D}_f(x,x) = -\log(2)$$

$$\forall x \in \mathbf{R}^n, \forall y \in \mathbf{R}^n, x \neq y,$$

$$\mathcal{D}_f(x,y) = \begin{cases} \inf \left\{ \int_\gamma \left[f'_{\gamma_m}(h) - f'_\gamma(h) \right] dh, \ \gamma \in \Gamma_{x,y} \right\} & \text{iff } \Gamma_{x,y} \neq \emptyset \\ +\infty & \text{otherwise} \end{cases}$$

where $f_{\gamma_m}(h) = \tau_{\gamma_m}(h).k$ (resp. $f_\gamma(h) = \tau_\gamma(h).k$) is the scalar product of the tangent vector at h along the path of greatest slope γ_m (resp. along the path γ) with the vector k, orthogonal to $\mathbf{R}^n$. Here $\mathcal{D}_f(x,y)$ is called the *deviation cost* of x and y with respect to the function f.

The deviation cost corresponds to the cost to be paid (in terms of energy, constraint, time, etc.) in order to take a path deviating from the path of greatest slope. The deviation cost is so much higher that the deviation will be significant. It is minimum on the path of greatest slope. In particular, the deviation cost from a point belonging to a CB to the minimum associated with this CB is null.

Proposition 9

1. $\forall (x,y) \in \text{supp}(f)^2, \mathcal{D}_f(x,y) = \mathcal{D}_f(y,x)$.
2. $\forall (x,y) \in \text{supp}(f)^2, x \neq y,$

$$\mathcal{D}_f(x,y) = 0 \quad \Leftrightarrow x \text{ and } y \text{ belong to a path of greatest slope}$$

$$\mathcal{D}_f(x,y) < +\infty \Leftrightarrow \text{there is a monotonic path from } x \text{ to } y$$

$$\mathcal{D}_f(x,y) = +\infty \Leftrightarrow \text{there is no monotonic path from } x \text{ to } y$$

As usual, the deviation cost of the point x to the nonempty subset Y of $\mathbf{R}^n$ is defined by $\mathcal{D}_f(x,Y) = \inf\{\mathcal{D}_f(x,y), y \in Y\}$.

Thus the deviation cost summarizes useful properties about the slope between x and y. However, its transformation into a distance function is more complicated than in the case of the connection cost, since the triangular inequality is not verified by the deviation cost.

B. Differential Distance and Equivalence of SKIZ and WS

Proposition 10. The mapping D_f from $\mathbf{R}^n \times \mathbf{R}^n$ to $\mathbf{R}^+$ such that

$$\forall x \in \mathbf{R}^n, \forall y \in \mathbf{R}^n, \qquad D_f(x,y) = 2 - \exp(-\mathcal{D}_f(x,y))$$

is a distance function.

Definition 9. D_f is called the *differential distance function*.

Let us notice that the space $(\mathbf{R}^n, D_f)$ is a discrete space.

Properties of D_f derive from those of the deviation cost:

1. $\forall (x,y) \in \mathrm{supp}(f)^2,\ D_f(x,y) = 0 \Leftrightarrow x = y.$
2. $\forall (x,y) \in \mathrm{supp}(f)^2,\ x \neq y,$

$$D_f(x,y) = 1 \quad \Leftrightarrow x \text{ and } y \text{ belong to a path of greatest slope}$$

$$1 \leq D_f(x,y) < 2 \Leftrightarrow \text{there is a monotonic path from } x \text{ to } y$$

$$D_f(x,y) = 2 \quad \Leftrightarrow \text{there is no monotonic path from } x \text{ to } y$$

We deduce from the above developments the following proposition dealing with the TWS.

Proposition 11. Let f be an element of $\mathcal{F}_c^1(\mathbf{R}^n,\mathbf{R})$. Let M be the set of the regional minima associated with f.

The SKIZ of M relative to the differential distance is the TWS of f: $\mathrm{SKIZ}_{D_f}(M) = \mathrm{TWS}(f)$.

In order to establish a similar relation dealing with the (thin) WS, considering the slope information is not enough. We need to add to it geometric information by introducing the notion of plateau depth.

Let x and y be points in $\mathbf{R}^n$ such that $x \neq y$ and $\Gamma_{x,y} \neq \emptyset$. Let us refer to a path linking x to y by $\gamma_{x,y}$. Let us consider the set

$$\Lambda_{x,y} = \{z \in \gamma_{x,y},\ \forall h \in \gamma_{x,z}, f(h) = f(x)\}$$

We then define the mapping $\psi_{x,y}$ from $\Gamma_{x,y}$ to $\mathbf{R}^n$ such that:

1. $f(x) > f(y) \Rightarrow \psi_{x,y}(\gamma_{x,y}) \in \Lambda_{x,y}$ and $\forall h \in \gamma_{\psi_{x,y}(\gamma_{x,y}),y},\ h \in \Lambda_{x,y}$
2. $f(x) < f(y) \Rightarrow \psi_{x,y}(\gamma_{x,y}) \in \Lambda_{x,y}$ and $\forall h \in \gamma_{x,\psi_{x,y}(\gamma_{x,y})},\ h \notin \Lambda_{y,x}$
3. $f(x) = f(y) \Rightarrow \psi_{x,y}(\gamma_{x,y}) = x.$

The plateau depth, denoted by Π_f, is defined as follows:

If $x \neq y$ and $\Gamma_{x,y} \neq \emptyset$,

if $f(x) > f(y)$, $\Pi_f(x,y) = \inf\{\delta_{f,f(x)}(x,\psi_{x,y}(\gamma)),\ \gamma \in \Gamma_{x,y}\}$,

else if $f(x) < f(y)$, $\Pi_f(x,y) = \inf\{\delta_{f,f(y)}(y,\psi_{x,y}(\gamma),\ \gamma \in \Gamma_{x,y}\}$,

else $\Pi_f(x,y) = \delta_{f,f(x)}(x,y)$,

else $\Pi_f(x,y) = 0$.

Despite the asymmetry of the definition, we have

$$\forall (x,y) \in \mathrm{supp}(f)^2,\ \Pi_f(x,y) = \Pi_f(y,x)$$

Let us notice that this information is useful only in the cases of plateaus situated on the border of a CB and for $D_f(x,y) = 1$.

Definition 9. Let us introduce the mapping $\check{D}_f$ from $\mathbf{R}^n \times \mathbf{R}^n$ to $\mathbf{R}$ defined as follows:

$$\forall\ (x,y) \in (\mathbf{R}^n)^2,$$
$$D_f(x,y) = 0 \text{ then } \check{D}_f(x,y) = 0,$$
$$\text{else if } D_f(x,y) = 1 \text{ then } \check{D}_f(x,y) = (D_f(x,y) + 2) - \exp(-\Pi_f(x,y)),$$
$$\text{else } D_f(x,y) > 1 \text{ then } \check{D}_f(x,y) = D_f(x,y) + 2$$

Here, $\check{D}_f$ is a distance function. For practical reasons, we will also call it *differential distance*. Obviously, we have the following properties:

$$\forall\ (x,y) \in \text{supp}(f)^2,$$
$$\check{D}_f(x,y) = 0 \quad \Leftrightarrow x = y.$$
$$2 \leq \check{D}_f(x,y) < 3 \Leftrightarrow x \text{ and } y \text{ belong to a path of greatest slope},$$
$$3 \leq \check{D}_f(x,y) < 4 \Leftrightarrow \text{there is a monotonic path from } x \text{ to } y,$$
$$\check{D}_f(x,y) = 4 \quad \Leftrightarrow \text{there is no monotonic path from } x \text{ to } y.$$

So, $\check{D}_f$ is a distance which characterizes perfectly the slope between x and y, taking into account the plateau information. We can now establish the complete equivalence of SKIZ and WS:

Proposition 12. Let f be an element of $\mathcal{F}_c^1(\mathbf{R}^n,\mathbf{R})$. Let M be the set of the regional minima associated with f. The SKIZ of M relative to the differential distance $\check{D}_f$ is the (thin) WS of f: $\text{SKIZ}_{\check{D}_f}(M) = \text{WS}(f)$.

In summary, it should not be surprising that we used two different distances (the Euclidean and differential ones) in order to identify the two concepts of SKIZ and WS; their different nature, binary versus gray-level, explains the asymmetry of the approach.

The advantage of the unifying concepts developed here is not only theoretical, because they lead to very fast algorithms for computing the connection cost, the plateau depth, the SKIZ, and the WS.

V. SOME NEW ALGORITHMS

A. Computation of the Connection Cost

We propose the following algorithm for computing the connection cost relative to a subset ($Y \subset \text{supp}(\text{fl})$). In order to initialize the process, let us define the following images:

IM1 is the image binary or labeled associated with Y.
IM2 is the gray-level image associated with f.
IM3 is defined by

$$\text{IM3}(x) = \begin{cases} \text{IM2}(x) & \text{if IM1}(x) \neq 0 \\ +\infty & \text{else} \end{cases}$$

Then the algorithm performs recursively, back and forth and until convergence (i.e., one scanning with no change), the following operation:

$$\text{if IM1}(x) = 0,\ \text{IM3}(x) \longleftarrow \max(\text{IM2}(x),\ \min_{v \in V(x)} (\text{IM3})(v)))$$

where $V_u(x)$ is the elementary upstream neighborhood of x (with x excluded) with respect to the current scanning. Image borders are processed according to the geodesic framework.

The final step is obtained by performing: if $\text{IM1}(x) > 0$, $\text{IM3}(x) \longleftarrow -\infty$. The result is in IM3.

In practice, convergence occurs in few scannings.

For obvious reasons, the same algorithm is used to comptue the topographical distance.

B. Computation of the WS Based on the Connection Cost

Let us recall that the proposed algorithm for the WS (based on the connection cost) yields an approximation of the WS and deals with a possible strict subset of the minima of f.

Let us define the following images:

IM1 is the image of the labeled selected minima of f.
IM2 is the connection cost relative to the selected minima.

The initialization step is the following: IM3-IM1

$$\text{if} \begin{Bmatrix} \text{IM1 } (x) = 0 \\ v \in V(x)\ \text{IM1}(v) > 0 \\ \text{IM2 } (v) \leq \text{IM2}(x) \end{Bmatrix} \text{then IM1}(x) \longleftarrow \text{IM1}(v)$$

The final result (in IM1) is obtained in one scanning:

$$\text{if IM1}(x) < \text{IM1}(v),\ \text{IM1}(v) \longleftarrow 0$$

By using the connection cost, the rejected minima are filled up and the propagation from the selected minima can always go upward.

C. Computation of the SKIZ and Conditional Propagation

This algorithm for computing the SKIZ is based on Proposition 8 and on the previous algorithm for computing the WS. Considering the simple shape of the Euclidean distance function, the proposed procedure gives an exact result.

The initialization step is the following:

IM1 corresponds to the labeled connection components of X.
IM2 is the gray-level image defined by $d(.,X^c)$.

We perform, until convergence, the following operations (label propagation):

$$\text{if} \begin{Bmatrix} \text{IM1}\ (x) = 0 \\ v \in V(x)\ \text{IM1}(v) > 0 \\ \text{IM2}\ (v) < \text{IM2}\ (x) \end{Bmatrix} \text{ then IM1}(x) \longleftarrow \text{IM1}(v)$$

The final step is performed in one scanning:

$$\text{if} \begin{Bmatrix} \text{IM1}(x) \neq \text{IM1}(v) \\ \text{IM2}(v) \leq \text{IM2}(x) \end{Bmatrix}, \qquad \text{IM1}(v) \longleftarrow 0$$

Let us notice that this propagating approach (according to increasing values of the distance function) may be generalized to the other MM operators. We define in this way dilation, erosion, thickening, thinning, convex hull, etc. relative to the increasing/decreasing values of one/several distance functions to different features.

Whereas geodesic operators yield drastic effects, these conditioned operators make it possible to obtain finer effects.

Let us now illustrate this approach with the specific example of the segmentation of the two lungs in CT. Let us consider the markers of the right and left lungs defined by the autoadaptive erosion (Figure 4g and h) (see Section II.C.1). We compute the connection cost (Figure 10a and b) to these markers relative to the initial CT image (Figure 3a and b). One may notice that the trachea has been filled up even if there were no markers inside it. The result looks like a gray-level hole filling.

The exact lung contours are finally obtained (Figure 10e and f) by performing a topographical propagation according to the algorithm presented above. Figure 10c and d illustrate the paths followed by the propagation.

D. Computation of the WS Based on the Differential Distance

The algorithm for computing the WS includes two steps. The first one deals with the computation of the plateau depth (more precisely $\delta_{f,f(x)}(x, \psi_{x,M_h})$). The second one directly concerns the WS.

1. Plateau Depth Algorithm

Let us denote by $V^*(x)$ the set of the neighbors of x (not including x) to which a value has already been assigned in the current scanning.

The initialization step is given by:

IM2, the image of the connection cost relative to the selected minima.
IM3, the gray-level hexagonal erosion of size one of IM2.
IM4 is uniformly equal to $+\infty$; it will contain the result.

Let us set $A(x) = \{v \in V^*(x), \text{IM2}(v) = \text{IM3}(x)\}$. Let us introduce an intermediate variable, "val." We perform back and forth, until convergence, the following operation:

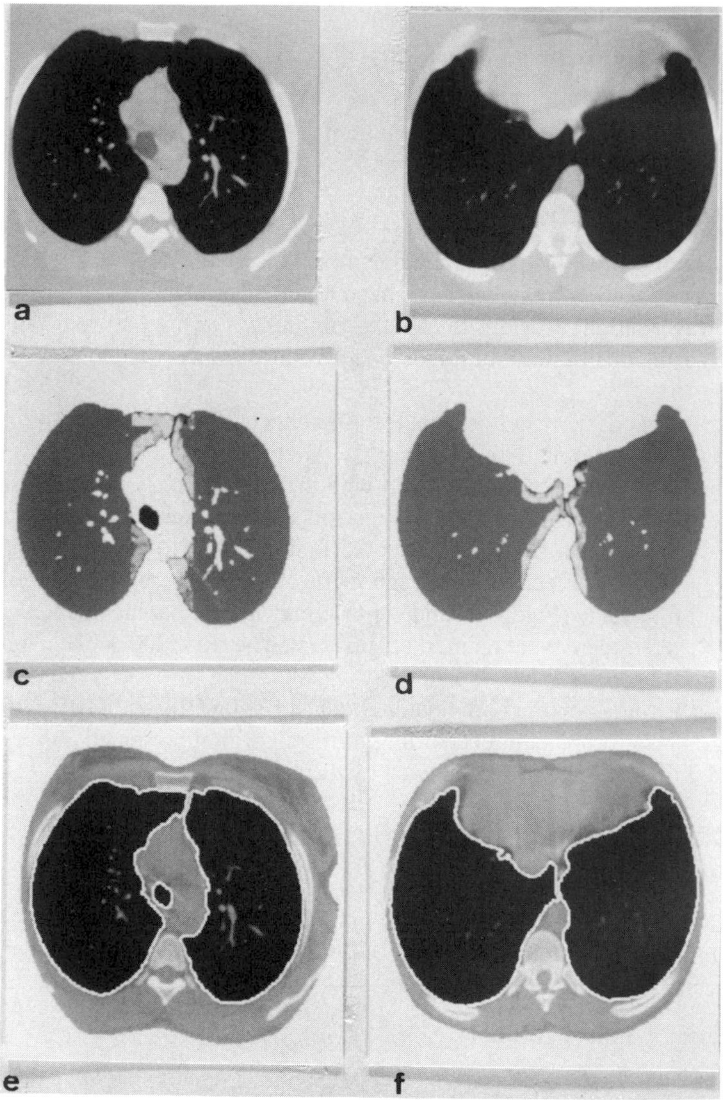

Figure 10. Segmentation of the right and left lung based on topographical propagation. (a and b) The connection cost relative to Figure 4g and h. Notice that the trachea has been filled up. (c and d) The paths followed by the propagation. (e and f) The lung contours obtained (in white) superimposed on the initial CT image.

If $\mathrm{IM3}(x) = \mathrm{IM2}(x)$,

if $A(x) \neq \emptyset$, val $\longleftarrow \min\{\mathrm{IM4}(v),\ v \in V^*(x),\ \mathrm{IM2}(v) = \mathrm{IM3}(x)\} + 1$

else val $\longleftarrow +\infty$

if val $< \mathrm{IM4}(x)$, $\mathrm{IM4}(x) \longleftarrow$ val

else $\mathrm{IM4}(x) \longleftarrow 1$.

We perform in one scanning:

If $\mathrm{IM4}(x) = +\infty$, $\mathrm{IM4}(x) \longleftarrow 1$

2. WS Algorithm

For reasons of simplicity, the presented algorithm is based on the deviation cost taking into account only the maximum of the deviation to the path of greatest slope in the place of the summation of the different deviations.

Let us consider:

IM1, the image of the labeled selection minima.
IM2, the image of the connection cost relative to the selected minima.
IM3, the gray-level hexagonal erosion of size one of IM2.
IM4 is the plateau depth of IM2.

The initialization step is given by $\mathrm{IMS}(x) = 0$:

$$\left.\begin{cases}\mathrm{IM1}(x) = 0\\ v \in V(x)\\ \mathrm{IM1}(v) > 0\\ \mathrm{IM2}(v) = \mathrm{IM3}(x)\\ \mathrm{IM2}(x) \neq \mathrm{IM2}(v) \text{ or } \mathrm{IM4}(v) \leq \mathrm{IM4}(x)\end{cases}\right\}, \qquad \mathrm{IM1}(x) \longleftarrow \mathrm{IM1}(v)$$

The final step is obtained in one scanning:

$$\text{if}\left.\begin{cases}\mathrm{IM1}(x) \neq \mathrm{IM1}(v)\\ \mathrm{IM2}(v) < \mathrm{IM2}(x) \text{ or } \begin{cases}\mathrm{IM2}(v) = \mathrm{IM2}(x)\\ \mathrm{IM4}(v) < \mathrm{IM4}(x)\end{cases}\end{cases}\right\}, \qquad \mathrm{IM1}(x) \longleftarrow 0$$

The result is in IM1.

Let us notice that the exact catchment basins are in IM1.

VI. CONCLUSION

We have presented here a novel theory for gray-tone MM by introducing the topographical and differential distances based, respectively, on the concepts of connection cost and deviation cost. The theoretical framework developed is powerful for exploiting and characterizing the topographical information contained in gray-tone images and yielding a perfect duality between SKIZ and WS. The concepts mentioned above were applied to the segmentation of various anatomical features in computed tomography, establishing the efficiency and robustness of the corresponding algorithms.

The range of application of the segmentation algorithms introduced is not limited to medical imaging. It encompasses a wide range of other application areas such as robotics, industrial radiography, and nondestructive imaging.

ACKNOWLEDGMENTS

This research was supported by l'Assistance Publique (Service de Radiologie, Prof. Ph. Grenier, Hôpital Salpétrière, Paris, France).

The author wishes to express her gratitude to Prof. N. Farvardin for making constructive comments on the English structure of the text.

REFERENCES

1. Golay, M., Hexagonal parallel pattern transformation, *IEEE Trans. Comput.*, *C18*, 733–740 (1969).
2. Serra, J., *Image Analysis and Mathematical Morphology,* Academic Press, London, 1982.
3. Coster, M., and Chermant, J. L. *Précis d'Analyse d'Images*, CNRS, 1989.
4. Sternberg, S. R., Grayscale morphology. CVGIP, *35*, 333–355 (1986).
5. Lantuejoul, Ch., La squelettisation et son application aux mesures topologiques des mosaïques polycristallines, Thèse Dr Ing., Ecole des Mines de Paris, 1978.
6. Ahuja, N., An, B., and Schachter, B., Image representation using Voronoï tesselation. *CVGIP*, *29*, 286–295 (1985).
7. Rom, H., and Peleg, S., Image representation using Voronoï tesselation: adaptive and secure. *CVPR*, 282–285 (1988).
8. Lelong-Ferrand, J., and Arnaudies, J. M., Géométrie et cinématique, Dunod Université, 1975.
9. Borgefors, G., Distance transformations in digital images, *CVGIP*, *34*, 334–371 (1986).
10. Rosenfeld, A., and Pfaltz, J. L., Sequential operations in digital picture processing, *J. ACM*, *13*(4), 471–494 (1966).
11. Schmitt, M., Des algorithmes morphologiques à l'intelligence artificielle, Thèse de Doctorat, Ecole des Mines de Paris, 1989.
12. Vincent, L., Algorithmes morphologiques à base de files d'attente et de lacets. Extension aux graphes, Thèse de Doctorat, Ecole des Mines de Paris, 1990.
13. Preteux, F. Description et interprétation des images par la morphologie mathématique. Application à l'imagerie médicale, Thèse d'Etat, Université Paris VI, 1987.
14. Merlet, N., Preteux, F., and Grenier, Ph., Maladies bronchopulmonaires: une analyse quantitative des lésions par tomodensitométrie, 37èmes Journées Françaises de Radiologie, Paris, France, Nov. 7–9, 1988.

15. Preteux, F., Mathématique et imagerie médicale. Ed. Tech., Encycl. Méd. Chir., Paris, France, 1990.
16. Preteux, F., Merlet, N., and Grenier, Ph., Algorithms for automated evaluation of pulmonary lesions with high-resolution CT via image analysis, Radiol. Society of North America, Chicago, Nov. 26–Dec. 1, 1989.
17. Preteux, F., Hel-Or, Y., and Remy, M., Pattern recognition of micronodular profusion in pulmonary CT, Radiol. Society of North America, Chicago, Nov. 24–30, 1990.
18. Merlet, N., Caractérisation et quantification de lésions pulmonaires en TDM, Rapport de stage CMM-Ecole des Mines de Paris, June 1988.
19. Preteux, F., and Hel-Or, Y., Mathematical morphology and active contour model: a cooperative approach for segmentation of lung contours in CT, in *Proceedings CAR'91*, Berlin, July 1991.
20. Rougon, N., and Preteux, F., Deformable markers: mathematical morphology for active contour models control, in *Image Algebra and Morphological Image Processing*, SPIE, San Diego, July 21–26, 1991.
21. Matheron, G., *Random Sets and Integral Geometry*, Wiley, New York, 1975.
22. Preteux, F., Some new primitives in gray-level mathematical morphology: theoretical and computational aspects, Internal Report, Telecom Paris, Department IMA, 1990.
23. Preteux, F., and Merlet, N., On the topographical distance function, in *7th SCIA*, Aalborg, Denmark, Aug. 1991.
24. Simon, J. C., *La reconnaissance de formes par algorithmes*, Masson, Paris, 1985.
25. Pavel, M., *Fundamentals of Pattern Recognition*, Marcel Dekker, New York, 1989.
26. Digabel, H., and Lantuejoul, Ch., Iterative algorithms, in *Proceedings, Second European Symposium of Quantitative Analysis of Microstructures*, Chermant Ed., R. Verlag, Stuttgart, 1978, pp. 85–99.
27. Haralick, R. M., Ridges and valleys on digital images, *CVGIP*, *22*, 28–38 (1983).
28. Casrany, G., and Margat, J., *Dictionnaire français d'hydrogéologie*, Ed. BRGM, 1977.
29. Beucher, S., Watersheds of functions and picture segmentation, *Proceedings, IEEE, International Conference on Acoustics, Speech and Signal Processing*, Paris, May 1982, pp. 1928–1931.
30. Beucher, S., and Lantuejoul, Ch., Use of watersheds in contour detection, in *International Workshop on Image Processing, Real Time Edge and Motion Detection*, CCETT, Rennes, France, Sept. 1989.
31. Watson, A. I., A new method of classification for Landsat data using the "watershed" algorithm, *PRL*, *6*, 15–19 (1987).

Chapter 11

Invariant Characterizations and Pseudocharacterizations of Finite Multidimensional Sets Based on Mathematical Morphology

Divyendu Sinha

College of Staten Island
City University of New York
Staten Island, New York

Hanjin Lee

Daewoo Telecom Co., Ltd.
Siheung City, Kyungki Do, South Korea

I. INTRODUCTION

Consider a dynamically growing collection $\mathfrak{C}$ of D-dimensional discrete sets that constitute the so-called *pattern database* and an observed D-dimensional set $\mathcal{A}$. The problem addressed in this chapter is to find an element $\mathcal{B}$ of the collection $\mathfrak{C}$ such that $\mathcal{A}$ can be obtained from $\mathcal{B}$ via some a priori chosen geometric transformations. This process of selection is known as *pattern matching*.

In pattern recognition, one usually requires that a set of relevant features be precomputed for all elements of the set $\mathfrak{C}$. These features are computed for the observed set $\mathcal{A}$, and the matching process consists of comparison of feature vectors of $\mathcal{A}$ with that of an element of $\mathfrak{C}$. The underlying philosophy is that the computation of feature vectors is $\mathcal{A}$ and the corresponding comparisons will be less time consuming.

Thus, by definition, we require definition and extraction of features that are invariant (or predictably variant) under permissible geometric operations. Features of an object generally include any information that can be deduced from the shape of an object. Since our primary interest is in object recognition, we need only concern ourselves with the features that are essential to the discrimination process. Examples of features include area, perimeter, moments, skeletons, compactness, eccentricity, convex deficiency, the Fourier descriptors of the contour, curvature, and signature of the contour.

Usually, one is interested in recognizing objects irrespective of their position, orientation, and scaling. We hasten to add that some or most of these invariance properties may not be desirable in certain applications. For example, we should not allow 180° rotation in the recognition of English alphabets. The pairs of English alphabets {'b','q'}, {'d','p'}, {'m','w'}, and {'n','u'} would be recognized as same characters if we allowed 180° rotation. If dealing with the oriental character set, we must allow neither 90° nor 180° rotations. For integrated circuit (IC) chip and printed circuit board inspection, we should permit only translation invariance.

Numerous geometric pattern recognition algorithms have been proposed in the literature, and some of these have been based on mathematical morphology. But a practical implementation of most (if not all) of these algorithms is not able to characterize images uniquely. In the literature, the uniqueness question is restricted to a fixed and usually very small collection of objects of interest as opposed to the entire class of finite objects). Since we are interested in characterizations of all finite sets, we will survey only techniques that strive to achieve this. We must add that, by the very definition of the problem, these algorithms will be extremely sensitive to noise and be, therefore, less "practical."

In the area of computational geometry, several algorithms have been proposed for exact as well as approximate congruences; for example, refer to [1–4]. In [3] it has been shown that to compare the equality of two finite sets $\mathcal{A}$ and $\mathcal{B}$, one needs at least $\Omega(n \log n)$ time with $n = \max(\text{card } \mathcal{A}, \text{card } \mathcal{B})$. Since the detection of congruence between $\mathcal{A}$ and $\mathcal{B}$ is a superset of the "set equality" problem, it will also require at least $\Omega(n \log n)$ time. In fact, $O(n \log n)$ time algorithms exist for detecting congruences between sets $\mathcal{A}$ and $\mathcal{B}$ if $\mathcal{A}, \mathcal{B} \subseteq \Re^k$, where $k \leq 3$ and $\Re$ is the set of reals [1].

It is important to notice that there is a significant difference between the computational geometry problems and the problem we address in this chapter. This difference is imposed by the domain of application that we seek for our algorithms. Employing traditional computational geometry schemes in matching leads to computationally intensive matching. Perhaps this has led and continues to lead researchers in pattern recognition to seek heuristic-based approaches to pattern matching. Another reason is the desire to handle approximate congruences.

New algebraically derived features of images were proposed in [5,6] in an attempt to recognize a two-dimensional binary image directly in its discrete domain. These features were defined using morphological operations. These features can be also obtained from an image by inspecting the "distances" between two elements of the set. These methods provided only pseudocharacterization of objects; that is, there exist two geometrically dissimilar objects that will be lumped together by these methods. An algebraic and geometric characterization of the induced equivalence class of approximately congruent objects was also investigated. The second method proposed in [5,7] was based on the notion of morphological annular openings [8] and is related to the previous scheme. Even the most general form of this scheme is unable to classify objects uniquely, and, moreover, it is computationally intensive.

In this chapter, we present a new algebraically derived feature for geometric pattern recognition whose performance is better than that of the schemes presented in [5–7]. We provide an intuitive description of this approach in Section III. In Sections IV and V, we propose a feature that can be employed to characterize one- and two-dimensional objects. The set of two-dimensional images, called irregular images, which cannot be characterized uniquely by the proposed scheme, have been shown to possess very special geometric properties. In particular, in such images, the existence (absence) of a pixel determines the existence (absence) of a host of other pixels! We have provided complete characterization of the irregular images. In Section VII, we show that a straightforward generalization of these results holds for d-dimensional images, $d \geq 3$.

The proposed method has applications in areas where exact matching is required. Examples of such domains include printed international character recognition, automated visual inspection (for example, IC chip and printed circuit board Inspection). When the set $\mathfrak{C}$ contains only one object, our algorithm is inefficient compared to the known schemes in computational geometry. Also, we are investigating the problem of noiseless pattern matching. As can be guessed, algorithms that strive for exact congruences are extremely sensitive to noise. Finally, it must also be stated that the proposed algorithm lends itself to an efficient implementation on parallel machines.

A Note on the Notation

In this chapter, we use different notation for morphological operations than in the other chapters. In what follows, we assume functional (or prefix) notation to define all morphological operations. We feel that this slight change makes the various formula more "readable." We assume that the reader is familiar with the operations of dilation (**DILATE**), erosion (**ERODE**), and opening (**OPEN**). The translation or shift operation is denoted as **T**. Similarly, we denote the 90° rotation and reflection operations by **N** and **R**, respectively.

II. PREVIOUS MATHEMATICAL MORPHOLOGY–BASED APPROACHES

Let us first consider the morphological features studied in references [5–7]. Let **IMAGE** denote the collection of all finite subsets of $\mathfrak{R}^2$ where $\mathfrak{R}$ represents the set of integers. Let $\mathfrak{N}^+$ denote the set of positive integers. As mentioned earlier, a two-dimensional discrete binary image will be modeled as an element of **IMAGE**.

Let us consider three parameterized family of structuring elements:

$$h_n = \{\langle 0,0\rangle, \langle n,0\rangle\}$$
$$v_n = \{\langle 0,0\rangle, \langle 0,\mathrm{n}\rangle\}$$
$$b_{n,m} = \{\langle -n,0\rangle, \langle n,0\rangle, \langle 0,-m\rangle, \langle 0,m\rangle\}$$

where n and m are nonnegative integers. We define the following three sets of functions for any given image $\mathcal{A}$:

Set 1: $\mathbf{HOR}\ (\mathcal{A},n) = \text{card}\ \mathbf{OPEN}\mathcal{A},h_n)$

$\mathbf{VER}(\mathcal{A},n) = \text{card}\ \mathbf{OPEN}\ (\mathcal{A},v_n)$

Set 2: $\mathbf{HCOV}\ (\mathcal{A},n) = \text{card}\ \mathbf{ERODE}(\mathcal{A},\mathbf{N}^2(h_n))$

$\mathbf{VCOV}\ (\mathcal{A},n) = \text{card}\ \mathbf{ERODE}\ (\mathcal{A},\mathbf{N}^2(v_n))$

Set 3: $\mathbf{ANNULAR}(\mathcal{A},n,m) = \text{card}\ \mathcal{A} \cap \mathbf{DILATE}(\mathcal{A},b_{n,m})$

Since the image $\mathcal{A}$ is fixed, we view these quantities as functions of the variable n (and m in case of **ANNULAR** function). The **HCOV** and **VCOV** functions have been employed in granulometrics and are usually called the *horizontal and vertical covariance functions*, respectively. The **ANNULAR** function was introduced in [8]. These three sets of functions deal with related sets:

$$\mathbf{OPEN}(\mathcal{A},h_n) = \mathcal{A} \cap [\mathbf{T}\ (\mathcal{A};\langle -n,0\rangle) \cup \mathbf{T}(\mathcal{A};\langle -n,0\rangle)]$$
$$\mathbf{ERODE}\ (\mathcal{A},\mathbf{N}^2(h_n)) = \mathcal{A} \cap \mathbf{T}\ (\mathcal{A};\langle n,0\rangle)$$
$$\mathcal{A} \cap \mathbf{DILATE}\ (\mathcal{A},b_{n,m}) = \mathbf{OPEN}\ (\mathcal{A},h_n) \cup \mathbf{OPEN}\ (\mathcal{A},v_m)$$

The preceding formulas give rise to efficient implementations of these functions. One also obtains

$$\mathbf{HOR}(\mathcal{A},n) = \text{card}\{\langle x,y\rangle \in \mathcal{A} : \textit{either}\ \langle x + n, y\rangle \in \mathcal{A}\ \textit{or}\ \langle x - n, y\rangle \in \mathcal{A}\} \tag{11.1}$$
$$\mathbf{HCOV}\ (\mathcal{A},n) = \text{card}\{\langle x,y\rangle \in \mathcal{A} : \langle x + n, y\rangle \in \mathcal{A} \tag{11.2}$$

The formulas for **VER** and **VCOV** are analogous.

In references [5–7] it has been shown that the functions **HOR** and **VER** (or **HCOV** and **VCOV**) must be considered in order to arrive at a decent pseudo-characterization of two-dimensional objects. Furthermore, it has been shown that the characterization qualities of the functions of sets 1 and 2 are identical. In order to pursue this in depth, let us define the following notion of *structure of an image* $\mathcal{A}$, denoted as $\nabla(\mathcal{A})$:

$$\nabla(\mathcal{A}) = \{\langle x,y\rangle \in \mathcal{A} : \exists a,b \ni a \neq x,\ b \neq y,\ \langle a,y\rangle \in \mathcal{A},\ \text{and } \langle x,b\rangle \in \mathcal{A}\}$$

Intuitively speaking, a pixel $\langle x,y\rangle$ of $\mathcal{A}$ forms a part of its structure if and only if there exists another pixel of $\mathcal{A}$ in the same row and another pixel of $\mathcal{A}$ in the same column as the pixel $\langle x,y\rangle$. The structure can be defined via morphological operations:

$$\nabla(\mathcal{A}) = \bigcup_{n=-\infty}^{\infty} \bigcup_{m=-\infty}^{\infty} \mathbf{ERODE}(\mathcal{A}, \psi_{n,m}) \tag{11.3}$$

where the structuring element $\psi_{n,m}$ is defined as

$$\psi_{n,m} = \{\langle 0,0\rangle, \langle n,0\rangle, \langle 0,m\rangle\}$$

First, note that $\nabla(\mathcal{A}) \subseteq \mathcal{A}$. Now, since the image $\mathcal{A}$ is finite, only a finite number of erosions in Eq. (11.3) will be nonempty. In reference [5] it was shown that if two images $\mathcal{A}$ and $\mathcal{B}$ have identical **HOR** and **VER** functions, that is, if **HOR** $(\mathcal{A},\cdot)$ = **HOR**$(\mathcal{B},\cdot)$ and **VER** $(\mathcal{A},\cdot)$ = **VER**$(\mathcal{B},\cdot)$, then $\nabla(\mathcal{A}) = \nabla(\mathcal{B})$ ignoring translation and 180° rotation. The converse is not true. The two images of Figure 1 have the same structure, but their **VER** functions are different. In reference [5], an operation called *generalized mirror reflection* was defined so as to characterize geometrically all images that have identical **HOR** and **VER** functions.

If the set $\mathcal{A}$ is large (in terms of cardinality), then in most situations of practical interest the cardinality of set $\mathcal{A} \setminus \nabla(\mathcal{A})$ will be very small compared to the cardinality of $\mathcal{A}$. (We are mindful of the fact that $\nabla(\mathcal{A})$ can be empty: consider any straight line of thickness one pixel. Also, see Figure 2.) Thus, if **HOR** and **VER** functions of two images $\mathcal{A}$ and $\mathcal{B}$ are equal, then $\mathcal{A}$ and $\mathcal{B}$ can be considered more or less identical; that is, these two functions, taken together, provide a pseudocharacterization of objects.

In reference [5] a complete characterization has also been provided. In order to discuss these characterizations, we must define two new terms: Consider any

Figure 1. Two images that have the same structure and **HOR** function but their **VER** functions are different.

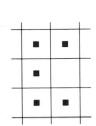

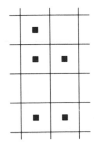

Figure 2. Two images that have the same **ANNULAR** function. For these two images, the structure is empty set. These images also have same $\mathbf{H}^+$, $\mathbf{H}^-$, $\mathbf{V}^+$, and $\mathbf{V}^-$ functions. Later, in section 5, we will call these images 3 × 3 basic irregular images.

image $\mathcal{A}$, and an integer v. The *support of* $\mathcal{A}$ *in row* v, denoted as $\mathcal{A}_{[v]}$, is defined as a one-dimensional set

$$\mathcal{A}_{[v]} = \{i : \langle i,v \rangle \in \mathcal{A}\} \tag{11.4}$$

Similarly, we can define the *support of* $\mathcal{A}$ *in column* v as

$$\mathcal{A}^{[v]} = \{j : \langle v,j \rangle \in \mathcal{A}\} \tag{11.5}$$

If we think of $\mathcal{A}$ as a matrix, then $\mathcal{A}_{[\alpha]}$ and $\mathcal{A}^{[\beta]}$ denote the αth row vector and βth column vector of $\mathcal{A}$, respectively. If we let $\mathfrak{N}$ denote the set of integers, then $\mathcal{A}_{[v]} \subseteq \mathfrak{N}$ and $\mathcal{A}^{[v]} \subseteq \mathfrak{N}$. The *rows of support* for $\mathcal{A}$ are defined as the set

$$\mathfrak{R}_{\mathcal{A}} \equiv \{v \in \mathfrak{N} : \mathcal{A}_{[v]} \neq \emptyset\} = \{\mathrm{j} \in \mathfrak{N} : \exists i \in \mathfrak{N} \ni \langle i,j \rangle \in \mathcal{A}\} \tag{11.6}$$

Similarly, one can define the *columns of support* $\mathfrak{C}_{\mathcal{A}}$ as

$$\mathfrak{R}_{\mathcal{A}} \equiv \{v \in \mathfrak{N} : \mathcal{A}^{[v]} \neq \emptyset\} = \{i \in \mathfrak{N} : \exists j \in \mathfrak{N} \ni \langle i,j \rangle \in \mathcal{A}\} \tag{11.7}$$

Now we are in a position to state the characterization mentioned in [5]:

Theorem 2.1. Consider any two images $\mathcal{A}$ and $\mathcal{B}$. Then the following two conditions are equivalent:

(a) **HOR** functions of two images $\mathcal{A}$ and $\mathcal{B}$ are identical.

(b) There exists a bijective mapping $\zeta : \mathfrak{R}_{\mathcal{A}} \to \mathfrak{R}_{\mathcal{B}}$ such that $\mathcal{A}_{[v]}$ can be obtained from $\mathcal{B}_{[\zeta(v)]}$ by translation and/or reflection.

Theorem 2.2. Consider any two images $\mathcal{A}$ and $\mathcal{B}$. Then the following two conditions are equivalent:

(a) **VER** functions of two images $\mathcal{A}$ and $\mathcal{B}$ are identical.

(b) There exists a bijective mapping $\xi : \mathfrak{C}_{\mathcal{A}} \to \mathfrak{C}_{\mathcal{B}}$ such that $\mathcal{A}^{[v]}$ can be obtained from $\mathcal{B}^{[\xi(v)]}$ by translation and/or reflection.

Identical statements hold for the covariance functions. In fact, in [5] it was shown that:

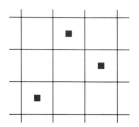

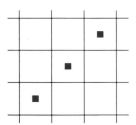

Theorem 2.3. Consider any two images $\mathcal{A}$ and $\mathcal{B}$. The following two equivalencies hold:

$$\text{(a) } \mathbf{HOR}(\mathcal{A},\cdot) = \mathbf{HOR}(\mathcal{B},\cdot) \text{ if and only if (iff) } \mathbf{HCOV}(\mathcal{A},\cdot) = \mathbf{HCOV}(\mathcal{B},\cdot).$$

$$\text{(b) } \mathbf{VER}(\mathcal{A},\cdot) = \mathbf{VER}(\mathcal{B},\cdot) \text{ iff } \mathbf{VCOV}(\mathcal{A},\cdot) = \mathbf{VCOV}(\mathcal{B},\cdot).$$

The function **ANNULAR**, as stated, is extremely time intensive. In [5] it was wrongly claimed that this function provides a complete characterization of two-dimensional objects. The images of Figure 2 have the same **ANNULAR** function. Readers interested in the analysis of **ANNULAR** function when $n = m$ should refer to [5].

In reference [6] a slight variation of the above scheme was also studied. Say the **HOR** and **VER** functions of two images $\mathcal{A}$ and $\mathcal{B}$ were identical. One can then investigate the characterization of sets $\mathcal{A}^c$ and $\mathcal{B}^c$. Since $\mathcal{A}$ and $\mathcal{B}$ are assumed to be finite, their complements will be infinite sets. Hence, we must define a window within which the complement will be restricted. For simplicity, let the window for $\mathcal{A}^c$ ($\mathcal{B}^c$) be the smallest rectangle that encloses $\mathcal{A}$ ($\mathcal{B}$). In Figure 3 we show two Chinese characters that have identical **HOR** and **VER** functions but their complements have different **VER** functions. This success is also very short-lived: The complements of images of Figure 2 also have identical **HOR** and **VER** functions (see Figure 4). In any case, if we incorporate this idea we arrive at a better pseudocharacterization of objects.

Failure of these schemes, it seems to us, is due to the fact they look at only the foreground (set) or the background (complement of set) at a time. An object, as the famous illusions tell us, can be defined only by an interplay between the background and the foreground. This has led us to believe that a complete and efficient morphological characterization, if there exists one, must investigate the given set $\mathcal{A}$ and its complement concurrently.

There are a few problems that we must address. First, how should we define the complement of a finite set, that is, what should be the window? This question is important as there is no straightforward mechanism for handling windows in mathematical morphology. In particular, should the window be dependent on the object (as we had in our earlier discussion)? If the answer is yes, then we cannot employ the "wrap-around" morphological algebra proposed in [9], as the window changes with the image under investigation. Second, how can we incorporate the notion of window within mathematical morphology? Third, do these new functions also have a simpler formulation (in terms of distances) like the functions of sets 1 and 2 (see Eqs. (11.1) and (11.2)? The last question is relevant because almost all of the proofs stated in [5] were based on Eqs. (11.1) and (11.2).

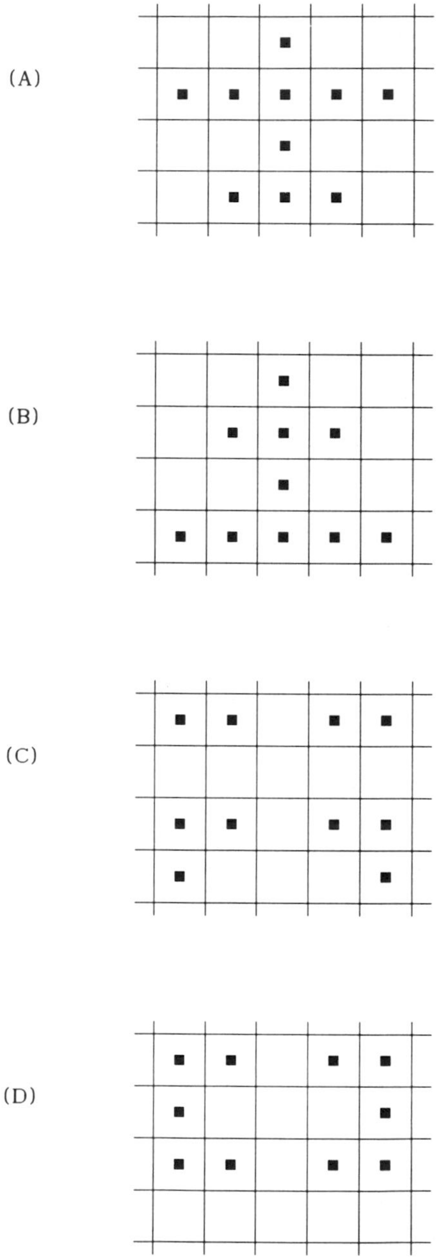

Figure 3. The two Chinese characters that have identical **HOR** and **VER** functions are shown in (A) and (B). Their complements are shown in (C) and (D) respectively. The images (C) and (D) have different **VER** functions.

Figure 4. The complement of two images of Figure 11.2. These complements also have identical **HOR** and **VER** functions.

We must also decide on the definition of scaling that we will use. In the literature, authors have used the Euclidean definition of scaling, that is, a transformation of the type

$$\begin{bmatrix} a & 0 \\ 0 & b \end{bmatrix}$$

This definition is not at all suitable in the discrete case as (1) it does not magnify the area at all and (2) it creates "holes" in the object. In particular, it transforms a horizontal straight line into a collection of pixels that are separated by $a - 1$ units from the nearest one.

We use the definition given in [5]. Let $\mathbf{MAG}_x(\mathcal{A},\alpha)$ and $\mathbf{MAG}_y(\mathcal{A},\beta)$ denote the magnifications in the horizontal and vertical directions by amounts α and β, respectively. We will assume that the magnification factors are positive integers. In reference [5] it was suggested that we define

$$\mathbf{MAG}_x\,(\mathcal{A},\alpha) = \left\{ \langle i,j \rangle : \left\langle \left\lfloor \frac{i}{\alpha} \right\rfloor, j \right\rangle \in \mathcal{A} \right\} \tag{11.8}$$

$$\mathbf{MAG}_y\,(\mathcal{A},\beta) = \left\{ \langle i, j \rangle : \left\langle i, \left\lfloor \frac{j}{\beta} \right\rfloor \right\rangle \in \mathcal{A} \right\} \tag{11.9}$$

It is possible to relate this definition to the usual Euclidean definition of scaling by using the dilation operation: Let $\hat{\mathcal{A}}$ denote the image obtained from Euclidean scaling by amount α in the horizontal direction:

$$\hat{\mathcal{A}} = \{\langle \alpha i,j \rangle : \langle i,j \rangle \in \mathcal{A}\} \tag{11.10}$$

and define the set $\psi_\alpha = \{\langle 0,0 \rangle, \langle 1,0 \rangle, \ldots, \langle \alpha - 1, 0 \rangle\}$. It is easy to show that

$$\mathbf{MAG}_x(\mathcal{A},\alpha) = \mathbf{DILATE}(\hat{\mathcal{A}},\psi_\alpha) \tag{11.11}$$

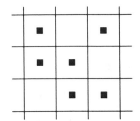

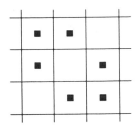

The set ψ_α can be stated in terms of repeated dilations of the structuring element h_1 (defined at the beginning of this section):

$$\psi_\alpha = \begin{cases} h_1 & \text{if } \alpha = 1 \\ \mathbf{DILATE}(\psi_{\alpha-1}, h_1) & \text{if } \alpha > 1 \end{cases} \tag{11.12}$$

Expression (11.12) is not efficient; it only serves the purpose of illustrating that the magnification can be defined as an *n-fold dilation*. The vertical magnification can be defined accordingly. To get an idea of what magnification entails, let us state another result from [5]:

Theorem 2.4. Consider two images $\mathcal{A}$ and $\mathcal{B}$ and positive integers α and β such that

$$\mathcal{B} = \mathbf{MAX}_x\left[\mathbf{MAG}_y(\mathcal{A},\beta),\ \alpha\right] = \mathbf{MAX}_y\left[\mathbf{MAG}_x(\mathcal{A},\alpha),\ \beta\right]$$

Then, for all nonnegative integers n, we obtain

$$\mathbf{HOR}\ (\mathcal{B}\alpha n) = \alpha\beta\ \mathbf{HOR}(\mathcal{A}, n) \qquad \text{and}$$

$$\mathbf{VER}\ (\mathcal{A},n) \qquad \mathbf{VER}(\mathcal{B},\beta n) = \alpha\beta$$

The preceding theorem essentially states that **HOR** and **VER** are homogeneous functions of order 1. Analogous results hold for **HCOV** and **VCOV**.

The notion of magnification brings out another problem that we must address: When we scale an image, what happens to the associated window? Does it also get scaled? We present the proposed answers to all these questions in the next section.

III. AN OUTLINE OF THE PRESENT APPROACH

The *pseudocomplement* $\hat{\mathcal{A}}$ of an image $\mathcal{A}$ will be defined as $\hat{\mathcal{A}} = \mathcal{A}^c \cap \mathcal{W}_{\mathcal{A}}$ where $\mathcal{A}^c$ is the set-theoretic complement of $\mathcal{A}$ and $\mathcal{W}_{\mathcal{A}}$ is a rectangular region that contains $\mathcal{A}$:

$$\mathcal{W}_{\mathcal{A}} = \left\{\langle x,y\rangle \in \mathfrak{N}^2 : x \in \mathfrak{X},\ y \in \mathfrak{Y}\right\} \tag{11.13}$$

where the sets $\mathfrak{X}$ and $\mathfrak{Y}$ are subsets of $\mathfrak{N}$:

$$\mathfrak{X} = \{\mathrm{i} \in \mathfrak{N} : \mathfrak{C}_{\mathcal{A}} - \mathfrak{w}_1 \le \mathrm{i} \le \mathfrak{C}_{\mathcal{A}} + \mathfrak{w}_2\} \tag{11.14}$$

$$\mathfrak{Y} = \{\mathrm{i} \in \mathfrak{N} : \mathfrak{R}_{\mathcal{A}} - \mathfrak{h}_1 \le \mathrm{i} \le \mathfrak{R}_{\mathcal{A}} + \mathfrak{h}_2\} \tag{11.15}$$

where $\mathfrak{w}_1$, $\mathfrak{w}_2$, $\mathfrak{h}_1$, and $\mathfrak{h}_2$ are some positive constants, called *windowing coefficients*. Usually, we let $\mathfrak{w}_1 = \mathfrak{w}_2 = \mathfrak{h}_1 = \mathfrak{h}_2 = 1$. Note that if $\mathcal{C} \subseteq \mathfrak{N}$, then we let min $\mathcal{C}$ (max $\mathcal{C}$) denote the smallest (largest) element of $\mathcal{C}$. The reason for not defining $\mathcal{B}_{\mathcal{A}}$ to be the smallest rectangle that encloses $\mathcal{A}$ will be apparent in Section IV. We now define four functions on the set **IMAGE**.

Since we have decided to work within a windowing environment, we must now deal with the following two situations:

(a) pixels within the window

(a.1) foreground pixels

(a.2) background pixels

(b) pixels that are outside the window

Let us now define a valuation space **L** for our binary images. Serra [8] has shown that the natural framework for defining morphological operations is to assume that **L** is a complete lattice. The lattice **L** induces a lattice structure onto the set $\mathbf{L}^{\mathfrak{R}\times\mathfrak{R}}$ as outlined in [10]. To this end, define two binary operations **MAX** and **MIN** on $\mathbf{L}^{\mathfrak{R}\times\mathfrak{R}}$ as follows: For all $v \in \mathfrak{R}\times\mathfrak{R}$, let

$$\mathbf{MAX}(f,g)(v) = f(v) \bigvee g(v)$$

$$\mathbf{MIN}(f,g)(v) = f(v) \bigwedge g(v)$$

We will follow the standard terminology by writing "$f \leq \psi$" whenever $\mathbf{MAX}(f,\psi) = \psi$. Both of these operations can be extended to handle three or more arguments. Now we are in a position to define a morphological operation that is of interest to us. The operation

$$\mathbf{OPEN} : \mathbf{L}^{\mathfrak{R}\times\mathfrak{R}} \times \mathbf{L}^{\mathfrak{R}\times\mathfrak{R}} \to \mathbf{L}^{\mathfrak{R}\times\mathfrak{R}} \tag{11.16}$$

is defined as follows:

$$\mathbf{OPEN}(f,\psi) = \mathbf{MAX}\left\{\mathbf{T}(\psi;v) : \mathbf{T}(\psi;v) \leq f\right\}$$

L must contain at least three elements: 0 (case a.2), 1 (case a.1), and α (case b). We assume **L** to be the boolean algebra of Figure 5. When an image is observed, its gray values are either 0 or 1. We then transform this image into a window environment by making use of the transformation $\mathfrak{L} : \mathbf{IMAGE} \to \mathbf{L}^{\mathfrak{R}\times\mathfrak{R}}$ such that

$$\left[\mathfrak{L}(\mathcal{A})\right](v) = \begin{cases} 1 & \text{if } v \in \mathcal{A} \\ 0 & \text{if } v \in \mathcal{W}_{\mathcal{A}} \setminus \mathcal{A} \\ \alpha & \text{otherwise} \end{cases}$$

where $\mathcal{W}_{\mathcal{A}}$ is the window defined via Eq. (11.13) etc. If we assume that the gray value α states that the image at this pixel is undefined, then the transform $\mathfrak{L}$ converts an observed image $\mathcal{A}$ into an image $\mathfrak{L}(\mathcal{A})$ that is defined only within a window. Of course, this window is itself dependent on the image $\mathcal{A}$. The gray value β can be assumed to define ambiguity and can arise only from the **MAX** operation. Even though β never arises in our discussion, it is needed to make **L** a complete lattice (otherwise, the **MAX** operation is not well defined). It is also of interest to note that one could not model the window-type environment using either the wrap-around algebra of [9] or the pointed fuzzy set of [11].

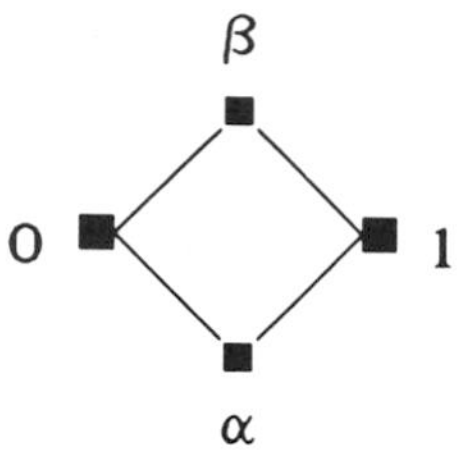

Figure 5. The lattice **L**.

It is of interest to note that $\mathfrak{L}^{-1}$, which is defined as

$$\mathfrak{L}^{-1}(\mathfrak{f}) = \{v \in \mathfrak{N}^2 : \mathfrak{f}(v) = 1\}$$

is a one-to-many mapping. Let us define a cardinality-type operation

$$\delta : \mathbf{L}^{\mathfrak{N}\times\mathfrak{N}} \rightarrow \mathfrak{N}^+ \cup \{0\}$$

such that $\delta(\mathfrak{f}) = \text{card}\{\text{v} \in \mathfrak{N}\times\mathfrak{N} : \mathfrak{f}(v) = 1\}$.

Let us now define four parameterized family of structuring elements

$$\left\{\mathfrak{h}_n^+\right\}_{n\in\mathfrak{N}+}, \quad \left\{\mathfrak{h}_n^-\right\}_{n\in\mathfrak{N}+}, \quad \left\{\mathfrak{v}_n^+\right\}_{n\in\mathfrak{N}+}, \quad \left\{\mathfrak{v}_n^-\right\}_{n\in\mathfrak{N}+}$$

as follows: for $v \in \mathfrak{N}^2$

$$\mathfrak{h}_n^+(v) = \begin{cases} 1 & \text{if } v = \langle 0,0\rangle \\ 0 & \text{if } v = \langle n,0\rangle \\ \alpha & \text{otherwise} \end{cases}$$

$$\mathfrak{v}_n^+(v) = \begin{cases} 1 & \text{if } v = \langle 0,0\rangle \\ 0 & \text{if } v = \langle 0,\text{n}\rangle \\ \alpha & \text{otherwise} \end{cases}$$

$$\mathfrak{h}_n^-(v) = \begin{cases} 1 & \text{if } v = \langle 0,0\rangle \\ 0 & \text{if } v = \langle -\text{n},0\rangle \\ \alpha & \text{otherwise} \end{cases}$$

$$\mathfrak{v}_n^-(v) = \begin{cases} 1 & \text{if } v = \langle 0,0\rangle \\ 0 & \text{if } v = \langle 0,-\text{n}\rangle \\ \alpha & \text{otherwise} \end{cases}$$

For any observed image $\mathcal{A}$ and a positive integer n, we define the following four functions:

$$\mathbf{H}^+(\mathcal{A},n) = \delta[\mathbf{OPEN}(\mathfrak{L}(\mathcal{A}),\mathfrak{h}_n^+)]$$

$$\mathbf{H}^-[\mathcal{A},n) = \delta]\mathbf{OPEN}(\mathfrak{L}(\mathcal{A}),\mathfrak{h}_n^-)]$$

$$\mathbf{V}^+(\mathcal{A},n) = \delta[\mathbf{OPEN}(\mathfrak{L}(\mathcal{A}), \mathfrak{v}_n^-)]$$

$$\mathbf{V}^-(\mathcal{A},n) = \delta[\mathbf{OPEN}(\mathfrak{L}(\mathcal{A}), \mathfrak{v}_n^+)]$$

Example 3.1. As a simple example, let us consider the image $\mathcal{A}$ in Figure 6. We transform the image $\mathcal{A}$ into $\mathfrak{L}(\mathcal{A})$ as shown in Figure 7a. Figure 7b represents the structuring element $\mathfrak{h}_2^+$. If we perform opening on image $\mathcal{A}$ by structuring element $\mathfrak{h}_2^+$, we obtain $\mathbf{H}^+(\mathcal{A},2) = 2$. The various steps involved in the process are shown in Figure 8.

From the preceding example and the definition, it may not be clear that the functions $\mathbf{H}^+$, $\mathbf{H}^-$, $\mathbf{V}^+$, and $\mathbf{V}^-$ look into the sets $\mathcal{A}$ and $\mathcal{A}^c$ concurrently. To this end, we have the following result:

Theorem 3.1. For any $\mathcal{A} \in \mathbf{IMAGE}$ and $n \in \mathfrak{N}^+$, we have

$$\mathbf{H}^+(\mathcal{A},n) = \text{card}\{\langle\alpha,\beta\rangle \in \mathcal{A} : \exists\langle\gamma,\beta\rangle \in \hat{\mathcal{A}} \ni \gamma = \alpha + n\}$$

$$\mathbf{V}^+(\mathcal{A},n) = \text{card}\{\langle\alpha,\beta\rangle \in \mathcal{A} : \exists\langle\alpha,\gamma\rangle \in \hat{\mathcal{A}} \ni \gamma = \beta + n\}$$

$$\mathbf{H}^-(\mathcal{A},n) = \text{card}\{\langle\alpha,\beta\rangle \in \mathcal{A} : \exists\langle\gamma,\beta\rangle \in \hat{\mathcal{A}} \ni \gamma = \alpha - n\}$$

$$\mathbf{V}^-(\mathcal{A},n) = \text{card}\{\langle\alpha,\beta\rangle \in \mathcal{A} : \exists\langle\alpha,\gamma\rangle \in \hat{\mathcal{A}} \ni \gamma = \beta - n\}$$

Proof. We shall establish the result only for $\mathbf{H}^+$.

$$\begin{aligned}\mathbf{H}^+(\mathcal{A},n) &= \text{card}\{\langle i,j\rangle \in \mathcal{A} : \langle i+n, j\rangle \in \hat{\mathcal{A}}\} \\ &= \text{card}\{\langle i,j\rangle \in \mathfrak{N}^2 : \mathfrak{L}(\mathcal{A})(i,j) \\ &= 1 \textit{ and } \mathfrak{L}(\mathcal{A})(i+n,j) = 0\} \\ &= \text{card}\{v \in \mathfrak{N}^2 : \mathbf{T}(\mathfrak{h}_n^+;v) \le \mathfrak{L}(\mathcal{A})\}\end{aligned} \tag{11.17}$$

Finally, if $\psi \equiv \mathbf{OPEN}(\mathfrak{L}(\mathcal{A}), \mathfrak{h}_n^+)$ then we have

$$\psi(v) = 1 \quad \text{iff} \quad \mathbf{T}(\mathfrak{h}_n^+;v) \le \mathfrak{L}(\mathcal{A})$$

Hence,

$$\begin{aligned}\delta(\psi) &= \text{card}\{v \in \mathfrak{N}^2 : \mathbf{T}(\mathfrak{h}_n^+;v) \le \mathfrak{L}(\mathcal{A})\} \\ &= \mathbf{H}^+(\mathcal{A},n) \qquad \text{from Eq. (11.17).}\end{aligned}$$

The preceding equations are thus counterparts of Eqs. (11.1) and (11.2) (which were the equivalent formulations of the **HOR** and **VER** functions). From Theorem 3.1 it is clear that the proposed four functions investigate the set $\mathcal{A}$ and its complement simultaneously. From now on, *we will prefer to work with the equivalent formulations (as stated in Theorem 3.1) of the proposed functions.* Now, the matching can be performed with respect to one of the following four pairs:

$$\{\mathbf{H}^+, \mathbf{V}^+\}, \{\mathbf{H}^+, \mathbf{V}^-\}, \{\mathbf{H}^-, \mathbf{V}^+\} \text{ and } \{\mathbf{H}^-, \mathbf{V}^-\} \tag{11.18}$$

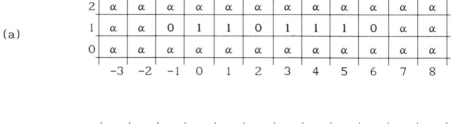

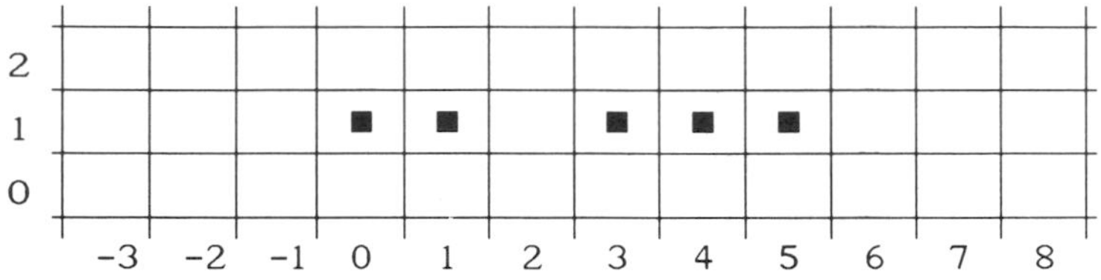

Figure 6. Image $\mathscr{A}$ of Example 3·1.

(a)

	-3	-2	-1	0	1	2	3	4	5	6	7	8
2	α	α	α	α	α	α	α	α	α	α	α	α
1	α	α	0	1	1	0	1	1	1	0	α	α
0	α	α	α	α	α	α	α	α	α	α	α	α

(b)

	-3	-2	-1	0	1	2	3	4	5	6	7	8
1	α	α	α	α	α	α	α	α	α	α	α	α
0	α	α	α	1	α	0	α	α	α	α	α	α
1	α	α	α	α	α	α	α	α	α	α	α	α

Figure 7. (a) The function $\mathfrak{L}(\mathscr{A})$ corresponding to the image $\mathscr{A}$ of Figure 6; (b) the structuring element $\mathfrak{h}_2^+$.

Example 3.2. Consider two Chinese characters F and G of Figure 3. The computation of the function $\mathbf{H}^+(F,\cdot)$ proceeds as follows:

$$\mathbf{H}^+(F,1) = \text{card}\,\{\langle 5,1\rangle,\langle 3,2\rangle,\langle 4,3\rangle,\langle 3,4\rangle\} = 4$$

$$\mathbf{H}^+(F,2) = \text{card}\,\{\langle 4,1\rangle,\langle 3,2\rangle,\langle 3,3\rangle,\langle 4,3\rangle,\langle 3,4\rangle\} = 5$$

$$\mathbf{H}^+(F,3) = \text{card}\{\langle 3,1\rangle,\langle 3,2\rangle,\langle 2,3\rangle,\langle 3,3\rangle,\langle 3,4\rangle\} = 5$$

$$\mathbf{H}^+(F,4) = \text{card}\{\langle 2,1\rangle,\langle 2,3\rangle\} = 2$$

$$\mathbf{H}^+(F,5) = \text{card}\{\langle 1,1\rangle\} = 1$$

$$\mathbf{H}^+(F,6) = \text{card}\,\emptyset = 0$$

$$\mathbf{H}^+(F,7) = \text{card}\,\emptyset = 0$$

etc. Similarly, $\mathbf{H}^+(G,\cdot)$ can be computed and it can easily be checked that $\mathbf{H}^+(F,\cdot) = \mathbf{H}^+(G,\cdot)$. Note that the $\mathbf{V}^+$ functions of the images F and G are

(a)

2	α	α	α	α	α	α	α	α	α	α	α	α
1	α	α	α	1	α	0	α	α	α	α	α	α
0	α	α	α	α	α	α	α	α	α	α	α	α
	−3	−2	−1	0	1	2	3	4	5	6	7	8

(b)

2	α	α	α	α	α	α	α	α	α	α	α	α
1	α	α	α	α	α	α	α	1	α	0	α	α
0	α	α	α	α	α	α	α	α	α	α	α	α
	−3	−2	−1	0	1	2	3	4	5	6	7	8

(c)

2	α	α	α	α	α	α	α	α	α	α	α	α
1	α	α	α	1	α	0	α	1	α	0	α	α
0	α	α	α	α	α	α	α	α	α	α	α	α
	−3	−2	−1	0	1	2	3	4	5	6	7	8

Figure 8. The various steps involved in computing **OPEN**($\mathfrak{L}(\mathcal{A}),\mathfrak{h}_2^+$). The only translations of $\mathfrak{h}_2^+$ that fit inside $\mathfrak{L}(\mathcal{A})$ are **TRAN**($\mathfrak{h}_2^+;\langle 0,1\rangle$) and **TRAN**($\mathfrak{h}_2^+;\langle 4,1\rangle$) and are shown in (a) and (b). In (c) we show the result of **MAX** operation on the images of (a) and (b).

different: $\mathbf{V}^+(F,3) = 5 \neq 3 = \mathbf{V}^+(G,3)$. Hence, the pair $\{\mathbf{H}^+, \mathbf{V}^+\}$ can distinguish between the two Chinese characters F and G.

Now, does one of the pairs listed in (11.18) characterize finite two-dimensional discrete images uniquely? The answer is no. The images F and G of Figure 2 have same $\mathbf{H}^+$, $\mathbf{H}^-$, $\mathbf{V}^+$, and $\mathbf{V}^-$ functions, and so do the images of Figure 9. These examples were illustrative in many ways. First, all the known morphology-based functions (of one parameter) [5–7] as well as the proposed functions fail to distinguish between these two images. Even more remarkably, the joint central moments all taken together cannot characterize all these images (see Section V.E). Second, these images exhibit a very peculiar property that may be extremely difficult to satisfy in nonsynthetic images. We will look into this property in Section V. In particular, we will specify all images that have the same characterization based on the proposed pairs of functions $\{\mathbf{H}^+, \mathbf{V}^+\}$, $\{\mathbf{H}^+, \mathbf{V}^-\}$,

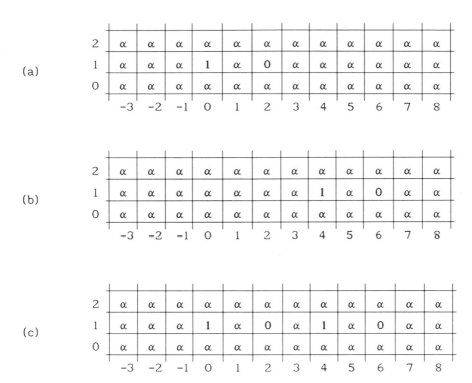

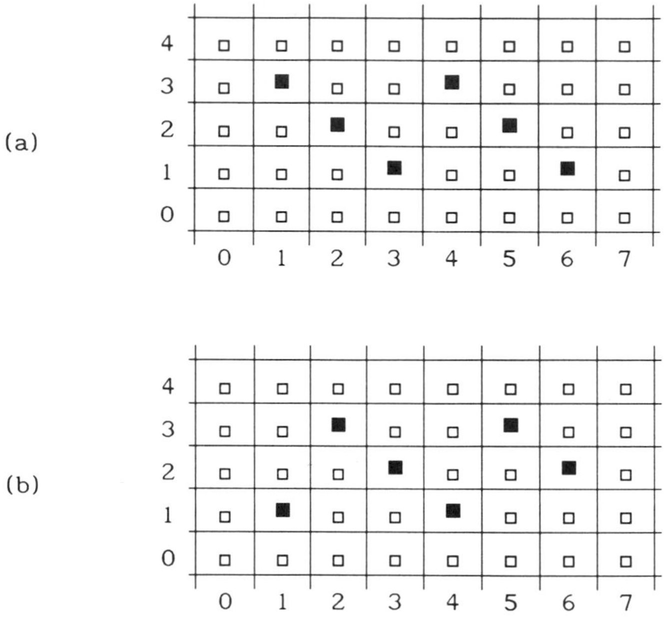

Figure 9. The images H and I that have identical $\mathbf{H}^+$, $\mathbf{H}^-$, $\mathbf{V}^+$, and $\mathbf{V}^-$ functions. As will be seen later in section 5, these images are 6×6 basic irregular images. It is also of interest to note that the image H (I) can be constructed by forming a union of two translated copies of the image F (G) of Figure 2. Note that □ denotes pixel of pseudo-complement.

$\{\mathbf{H}^-,\mathbf{V}^+\}$, or $\{\mathbf{H}^-, \mathbf{V}^-\}$. For the time being, we simply illustrate the ideas that will be presented later on.

Example 3.3. The two images of Figure 2 can be constructed using the following generic formula (see Theorem 5.5):

$$\mathcal{A} = \{\langle i,j\rangle : i + j - \nu - \tau \equiv 0 \pmod{\kappa}\}$$

where $\kappa = \gcd(\text{card } \mathfrak{C}\mathcal{A},\ \text{card } \mathfrak{R}\mathcal{A})$ and $\langle \nu,\tau\rangle$ is any pixel in image $\mathcal{A}$. Consider the construction of image $\mathcal{A}$ and let $\nu = \tau = 2$. Since $\kappa = 3$, we obtain $\mathcal{A} = \{\langle 1,3\rangle, \langle 2,2\rangle, \langle 3,1\rangle\}$. Similarly, we can show that the image of Figure 2b is $\{\langle 1,1\rangle, \langle 2,3\rangle, \langle 3,2\rangle\}$.

But what happened to our claim that if we investigate the set and its complement simultaneously, we can arrive at a unique characterization? To this end, we define the following parameterized family of structuring elements

$$\left\{\mathfrak{hv}^{++}_{n,m}\right\}_{n,m\in\mathfrak{R}^*}$$

where

$$\mathfrak{N}^* = \{\langle i,j\rangle : i \geq 0, j \geq 0, \text{ and } i + j \geq 1\}$$

and

$$\mathfrak{hv}^{++}_{n,m}(v) = \begin{cases} 1 & \text{if } v = \langle 0,0\rangle \\ 0 & \text{if } v = \langle n,m\rangle \\ \alpha & \text{otherwise} \end{cases}$$

Of course, $\mathfrak{hv}^{++}_{n,0} = \mathfrak{h}^{+}_{n}$ and $\mathfrak{hv}^{++}_{0,n} = \mathfrak{v}^{+}_{n}$. For every image $\mathcal{A}$, and $\langle n,m\rangle \in \mathfrak{N}^*$, we define a function **B** as follows:

$$\mathbf{B}(\mathcal{A};n,m) = \delta\left[\mathbf{OPEN}\left(\mathfrak{L}(\mathcal{A}),\ \mathfrak{hv}^{++}_{n,m}\right)\right]$$

We can again show that

$$\mathbf{B}(\mathcal{A};n,m) = \text{card}\{\langle\alpha,\beta\rangle \in \mathcal{A} : \langle\alpha + n, \beta + m\rangle \in \hat{\mathcal{A}}\}$$

In particular, we have $\mathbf{B}(\mathcal{A};n,0) = \mathbf{H}^{+}(\mathcal{A},n)$ and $\mathbf{B}(\mathcal{A};0,m) = \mathbf{V}^{+}(\mathcal{A},n)$. Hence, the characterization qualities of **B** cannot be worse than those of the $\mathbf{H}^{+}$ and $\mathbf{V}^{+}$ pair. As will be shown in Section VI, the function **B** characterizes the set $\mathcal{A}$ uniquely.

Example 3.4. Consider the image $\mathcal{A}$ of Figure 10, whose transformation $\mathfrak{L}(\mathcal{A})$ is shown in Figure 11. Consider the computation of $\mathbf{B}(\mathcal{A};1,2)$, whose value is 8, within the morphological framework. We need to perform the opening on $\mathfrak{L}(\mathcal{A})$ by the structuring element $\mathfrak{h}^{++}_{1,2}$ (shown in Figure 12). The image $\mathbf{OPEN}(\mathfrak{L}(\mathcal{A}), \mathfrak{h}^{++}_{1,2})$ is shown in Figure 13.

Example 3.5. Consider the two images of Figure 2. The **B** function is able to distinguish between these images. Let us call the images (a) and (b) $\mathcal{A}$ and $\mathcal{B}$, respectively. We find that $\mathbf{B}(\mathcal{A};3,3) = 0 \neq 1 = \mathbf{B}(\mathcal{B};3,3)$.

An obvious question that arises now is: How can we extend these functions to handle multidimensional objects? The answer is via various types of projections of $\mathfrak{N}^{\mathbf{D}}$ onto $\mathfrak{N}^{\mathbf{d}}$ with $\mathbf{d} \leq \mathbf{D}$, and using a recursive definition. This process is carried out in Section VII.

IV. CHARACTERIZATION OF FINITE ONE-DIMENSIONAL DISCRETE SETS

Before we discuss the higher-dimensional images, we will investigate the one-dimensional images first. Let **X** denote the collection of all finite subsets of $\mathfrak{N}$. We define the pseudocomplement $\hat{\mathcal{A}}$ of image $\mathcal{A} \in \mathbf{X}$ and two functions on finite one-dimensional discrete images as follows.

Definition 4.1. Consider any one-dimensional image $\mathcal{A}$. The *pseudocomplement* $\hat{\mathcal{A}}$ of image $\mathcal{A}$ is defined as $\hat{\mathcal{A}} = \mathcal{A}^c \cap \mathcal{W}_{\mathcal{A}}$, where $\mathcal{A}^c$ is the set-theoretic complement of $\mathcal{A}$ and

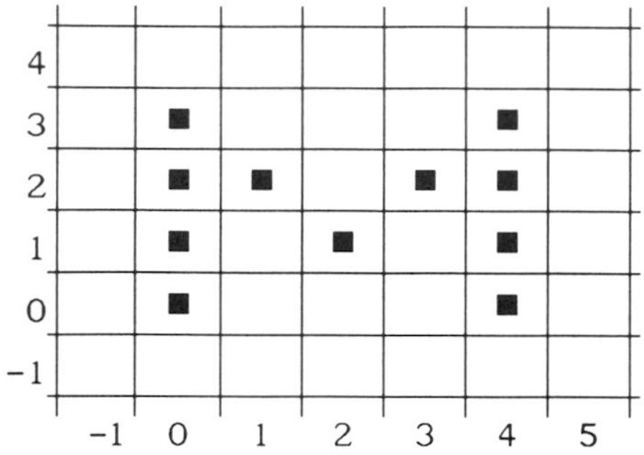

Figure 10. Image $\mathscr{A}$ of Example 3·4.

	−2	−1	0	1	2	3	4	5	6
5	α	α	α	α	α	α	α	α	α
4	α	0	0	0	0	0	0	0	α
3	α	0	1	0	0	0	1	0	α
2	α	0	1	1	0	1	1	0	α
1	α	0	1	0	1	0	1	0	α
0	α	0	1	0	0	0	1	0	α
−1	α	0	0	0	0	0	0	0	α
−2	α	α	α	α	α	α	α	α	α

Figure 11. The signal $\mathscr{L}(A)$ corresponding to image $\mathscr{A}$ of Figure 10.

	−1	0	1	2
3	α	α	α	α
2	α	α	0	α
1	α	α	α	α
0	α	1	α	α
−1	α	α	α	α

Figure 12. The structuring element $\mathfrak{hv}_{12}^{++}$.

5	α	α	α	α	α	α	α	α	α
4	α	α	α	0	0	α	0	0	α
3	α	α	α	0	α	0	α	0	α
2	α	α	1	1	α	1	1	0	α
1	α	α	1	α	1	α	1	α	α
0	α	α	α	α	α	α	1	α	α
−1	α	α	α	α	α	α	α	α	α
−2	α	α	α	α	α	α	α	α	α
	−2	−1	0	1	2	3	4	5	6

Figure 13. The result of opening signal $\mathfrak{L}(\mathcal{A})$ of Figure 11 by the structuring element $\mathfrak{hv}_{12}^{++}$ of Figure 3·8.

$$\mathcal{W}_{\mathcal{A}} = \{x \in \Re : \min \mathcal{A} - w_1 \leq x \leq \max \mathcal{A} + w_2\}$$

where w_1 and w_2 are fixed positive integers called *widow coefficients*. Clearly, $\hat{\mathcal{A}}$ is also a one-dimensional image. Also, define the *width* of $\mathcal{A}$ by

$$\max \mathcal{A} - \min \mathcal{A} + 1.$$

Definition 4.2. Define two functions $\lambda^+, \lambda^- : \mathbf{X}^2 \times \Re^+ \rightarrow \Re^+ \cup \{0\}$ such that

$$\lambda^+(\mathcal{A},\hat{\mathcal{A}};\text{n}) = \text{card}\{\alpha \in \mathcal{A} : \alpha + n \in \hat{\mathcal{A}}\}$$
$$\lambda^-(\mathcal{A},\hat{\mathcal{A}};\text{n}) = \text{card}\{\alpha \in \mathcal{A} : \alpha - n \in \hat{\mathcal{A}}\}$$

If we assume that window coefficients are fixed for all images, the explicit specification of $\hat{\mathcal{A}}$ in $\lambda^+(\mathcal{A},\hat{\mathcal{A}};n)$ and $\lambda^-(\mathcal{A},\hat{\mathcal{A}};n)$ is redundant. However, see page 420.

For an image $\mathcal{A} \in \mathbf{X}$, the only valid geometric operations are

$$\mathbf{T}(\mathcal{A};\alpha) = \{x + \alpha : \text{x} \in \mathcal{A}\}$$
$$\mathbf{N}^2(\mathcal{A}) = \{-x : \text{x} \in \mathcal{A}\}$$

We define the following sets to simplify the subsequent proof:

$$\chi^+(\mathcal{A},\hat{\mathcal{A}};n) = \{\alpha \in \mathcal{A} : \alpha + n \in \hat{\mathcal{A}}\}$$
$$\chi^-(\mathcal{A},\hat{\mathcal{A}};n) = \{\alpha \in \mathcal{A} : \alpha - n \in \hat{\mathcal{A}}\}$$
$$\hat{\mathcal{A}}_+ = \{\alpha \in \hat{\mathcal{A}} : \alpha > \min \mathcal{A}\}$$
$$\hat{\mathcal{A}}_- = \{\alpha \in \hat{\mathcal{A}} : \alpha < \max \mathcal{A}\}$$

Clearly, $\hat{\mathcal{A}} = \hat{\mathcal{A}}_+ \cup \hat{\mathcal{A}}_-$. Furthermore, for all $n \in \mathfrak{N}^+$,

$$\chi^+(\mathcal{A},\hat{\mathcal{A}};n) = \chi^+(\mathcal{A},\hat{\mathcal{A}}_+;n)$$

$$\chi^-(\mathcal{A},\hat{\mathcal{A}};n) = \chi^-(\mathcal{A}, \hat{\mathcal{A}}_+;n)$$

$$\lambda^+(\mathcal{A},\hat{\mathcal{A}};n) = \lambda^+(\mathcal{A},\hat{\mathcal{A}}_+;n)$$

$$\lambda^-(\mathcal{A},\hat{\mathcal{A}};n) = \lambda^-(\mathcal{A}, \hat{\mathcal{A}}_-;n)$$

Example 4.1. Consider the image $\mathcal{A}$ of Figure 14. Assume that $w_1 = 3$ and $w_2 = 2$. Then, we have $\mathcal{A} = \{3,5,7,8\}$, $\hat{\mathcal{A}} = \{0,1,2,4,6,9,10\}$, $\hat{\mathcal{A}}_- = \{4,6,9,10\}$, and $\hat{\mathcal{A}}_- = \{0,1,2,4,6\}$. Furthermore,

$$\chi^+(\mathcal{A},\hat{\mathcal{A}}_+;2) = \chi^+(\mathcal{A},\hat{\mathcal{A}};2) = \text{card}\{7,8\} = 2$$

$$\chi^+(\mathcal{A},\hat{\mathcal{A}}_+;3) = \chi^+(\mathcal{A},\hat{\mathcal{A}};3) = \text{card}\{3,7\} = 2$$

$$\chi^-(\mathcal{A},\hat{\mathcal{A}}_-;2) = \chi^-(\mathcal{A},\hat{\mathcal{A}};2) = \text{card}\{3,8\} = 2$$

$$\chi^-(\mathcal{A},\hat{\mathcal{A}}_-;4) = \chi^-(\mathcal{A},\hat{\mathcal{A}};4) = \text{card}\{5,8\} = 2$$

Our first result states that the finite one-dimensional discrete images can be uniquely characterized (up to translation) by their λ^+ or λ^- functions.

Theorem 4.1. For any two image $\mathcal{A}$ and $\mathcal{B}$ in **X**, the following two relationships hold for some $v \in \mathfrak{N}$:

(i) $\lambda^+(\mathcal{A}\hat{\mathcal{A}};\cdot) = \lambda^+(\mathcal{B},\hat{\mathcal{B}};\cdot) \Leftrightarrow \mathcal{A} = \mathbf{T}(\mathcal{B};v)$ *and* $\max \hat{\mathcal{A}} - \max \hat{\mathcal{B}}$
$$= \max \mathcal{A} - \max \mathcal{B}$$

(ii) $\lambda^-(\mathcal{A},\hat{\mathcal{A}};\cdot) = \lambda^-(\mathcal{B},\hat{\mathcal{B}};\cdot) \Leftrightarrow \mathcal{A} = \mathbf{T}(\mathcal{B};v)$ *and* $\min \hat{\mathcal{A}} - \min \hat{\mathcal{B}}$
$$= \min \mathcal{A} - \min \mathcal{B}$$

Proof. We will show (i).
(*Sufficiency*) Directly follows from Definition 4.2.
(*Necessity*) If $\lambda^+(\mathcal{A},\hat{\mathcal{A}};\cdot) = \lambda^+(\mathcal{B},\hat{\mathcal{B}};\cdot)$, then we must have

$$\forall n \geq 1, \quad \text{card } \chi^+(\mathcal{A},\hat{\mathcal{A}};n) = \text{card } \chi^+(\mathcal{B},\hat{\mathcal{B}};n)$$

For every pair $\alpha \in \chi^+(\mathcal{A},\hat{\mathcal{A}};n)$ and $\beta \in \hat{\mathcal{A}}_+$ such that $\beta - \alpha = n$, there exists a unique pair $\kappa \in \chi^+(\mathcal{B},\hat{\mathcal{B}};n)$ and $v \in \hat{\mathcal{B}}_+$ such that $v - \kappa = n$. Let us define two mappings

$$\theta_1 ; \mathcal{A} \rightarrow \mathcal{B} \quad \text{and} \quad \theta_2 : \hat{\mathcal{A}}_+ \rightarrow \hat{\mathcal{B}}_+$$

such that

$$\forall i \in \mathcal{A}, \forall j \in \hat{\mathcal{A}}_+, \quad j - i = \theta_2(j) - \theta_1(i) \tag{11.19}$$

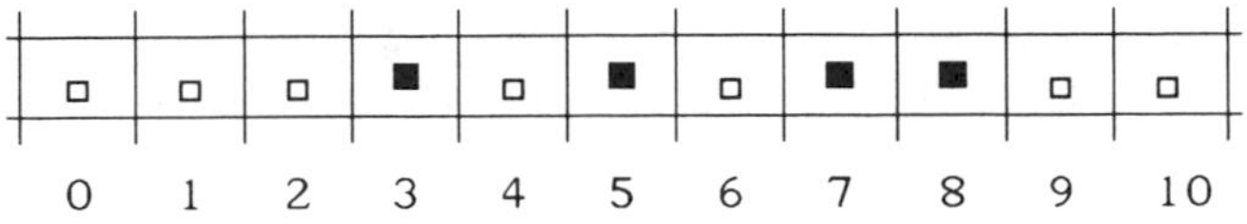

Figure 14. Example of a 1-dimensional image.

Consider any n. We have

$$x \in \chi^+(\mathcal{A}, \hat{\mathcal{A}}; n) \quad \textit{iff}\ x \in \mathcal{A}\ x + n \in \hat{\mathcal{A}}_+$$
$$\textit{iff}\ \theta_2(x + n) - \theta_1(x) = n$$
$$\textit{iff}\ \theta_1(x) \in \chi^+(\mathcal{B}, \hat{\mathcal{B}}; n)$$

Since $\lambda^+(\mathcal{A},\hat{\mathcal{A}};\cdot) = \lambda^+(\mathcal{B},\hat{\mathcal{B}};\cdot)$ and sincc

$$\forall x \in \mathcal{A} \qquad \exists n \in \mathfrak{N}^+ \ni x \in \chi^+(\mathcal{A}\hat{\mathcal{A}};n)$$

θ_1 must be a bijective mapping. Consequently, the constraint (11.19) implies that θ_2 is also a bijective mapping. Hence card $\mathcal{A}$ = card $\mathcal{B}$ and card $\hat{\mathcal{A}}_+$ = card $\hat{\mathcal{B}}_+$. Furthermore, the constraint (11.19) implies that

$$\forall i,j \in \mathcal{A}, \qquad i - j = \theta_1(i) - \theta_1(j) \tag{11.20}$$

$$\forall i,j \in \hat{\mathcal{A}}_+, \qquad i - j = \theta_2(i) - \theta_2(j) \tag{11.21}$$

Now, since card $\mathcal{A}$ is finite, we can assume, without any loss of generality, that θ_1 and θ_2 are finite-degree polynomials.

$$\theta_1(x) = \sum_{i=0}^{n} \alpha_i x^i \qquad \text{and} \qquad \theta_2(x) = \sum_{i=0}^{m} \beta_i x^i$$

with $\alpha_n \neq 0$ and $\beta_m \neq 0$. In order to satisfy the constraints (11.20) and (11.21), we must have $n = m = 1$ and $\alpha_1 = \beta_1 = 1$. Finally, by substituting the values of α_1 and β_1 in Eq. (11.19), we obtain $\alpha_0 = \beta_0$. Therefore, if $\lambda^+(\mathcal{A},\hat{\mathcal{A}};\cdot) = \lambda^+(\mathcal{B},\hat{\mathcal{B}};\cdot)$ then $\mathcal{A} = \mathbf{T}(\mathcal{B};v)$.

To complete the proof, note that card $\hat{\mathcal{A}}_+$ = card $\hat{\mathcal{B}}_+$ and card $\mathcal{A}$ = card $\mathcal{B}$, a fortiori determine that

$$\max \hat{\mathcal{A}}_+ - \max \mathcal{A} = \max \hat{\mathcal{B}}_+ - \max \mathcal{B}.$$

Similarly, one can show (ii).

With rcspcct to one-dimensional images, the only other (valid) geometric operation is mirror reflection. In view of Theorem 4.1 one may wonder about the relationship between the characterizations of an image $\mathcal{A}$ and its mirror reflection

$\mathbf{N}^2(\mathcal{A})$. At the same time, we do not know of any relationship between the functions λ^+ and λ^-. Our second result states that the answers to these two inquiries are closely coupled:

Theorem 4.2. Consider any two images $\mathcal{A}$ and $\mathcal{B}$ in $\mathbf{X}$. Then the following two conditions are equivalent:

(i) $\lambda^+(\mathcal{A},\hat{\mathcal{A}};\cdot) = \lambda^-(\mathcal{B},\hat{\mathcal{B}};\cdot)$

(ii) $\mathbf{N}^2(\mathcal{A}) = \mathcal{B}$ *and* $\max \hat{\mathcal{A}} - \max \mathcal{A} = \min \mathcal{B} - \min \hat{\mathcal{B}}$ ignoring translation.

Proof. Note that $\mathcal{B} = \mathbf{N}^2(\mathcal{A}) = \{-x : x \in \mathcal{A}\}$.

$$x \in \chi^+(\mathcal{A},\hat{\mathcal{A}}; n) \Leftrightarrow x \in \mathcal{A} \text{ and } x + n \in \hat{\mathcal{A}}_+$$

$$\Leftrightarrow -x \in \mathcal{B} \text{ and } -x - n \in \hat{\mathcal{B}}_-$$

$$\Leftrightarrow -x \in \chi^-(\mathcal{B},\hat{\mathcal{B}};n)$$

V. PSEUDOCHARACTERIZATION OF FINITE TWO-DIMENSIONAL DISCRETE SETS

In Section IV we showed that all finite one-dimensional images can be characterized uniquely by λ^+ and λ^- functions. In this section, we extend λ^+ and λ^- functions to two-dimensional images and show the consequences.

In Section V.A, we give the formal definition of $\mathbf{H}^+$, $\mathbf{H}^-$, $\mathbf{V}^+$, $\mathbf{V}^-$ functions and show that they can be computed in term of the summation of λ^+ and λ^- functions for each rows. The process can easily be visualized as follows: Consider a two-dimensional image as a fixed set of lines kept next to each other. We characterize each of these lines uniquely via the λ^+/λ^- function. However, while summing the values, we lose track of the relative position of these lines, which is very valuable information. But, at least, we can say that if two images F and G have the same $\mathbf{H}^+/\mathbf{H}^-$ ($\mathbf{V}^+/\mathbf{V}^-$) function then the rows (columns) of F can be obtained from the rows (columns) of G.

The natural question that arises next is: what if we considered the feature pair $\{\mathbf{H}^+,\mathbf{V}^+\}$, $\{\mathbf{H}^+,\mathbf{V}^-\}$, $\{\mathbf{H}^-,\mathbf{V}^+\}$, or $\{\mathbf{H}^-,\mathbf{V}^-\}$? Unfortunately, even then the two-dimensional images cannot be characterized uniquely. In Section V.B we show that the set of images that cannot be characterized uniquely possess very special properties such that there exist interdependencies between the elements of such images. We call those images *irregular images*.

In Section V.C we introduce basic irregular images that will be used as basic building blocks of irregular images. The basic irregular images can be constructed very easily. We propose a representation scheme that relates their construction to the solutions of certain two-variate Diophantine equations.

In Section V.D, we outline the construction of irregular images. Irregular images can be constructed by taking the union of disjoint basic irregular images.

In Section V.E, we investigate a few geometric and closure properties of the irregular images.

In Sections V.F and V.G we show that $\mathbf{H}^+$, $\mathbf{H}^-$, $\mathbf{V}^+$, $\mathbf{V}^-$ functions are invariant under rotation and magnification. The rotation invariance can be achieved by considering the combinations of the functions. Also, the magnification invariance can be achieved only for a certain window.

All the missing proofs can be found in reference [12].

A. H⁺, H⁻, V⁺, and V⁻ Functions

Let **IMAGE** denote the set of all finite subsets of $\Re^2$. Consider any image $\mathfrak{f}$, and assume that $\mathfrak{f}$ has support only in the rows (columns) indexed by $\mathfrak{R}_\mathfrak{f}$ ($\mathfrak{C}_\mathfrak{f}$) as stated in Eq. 3.6 (3.7). We now define the pseudocomplement of a two-dimensional image and the $\mathbf{H}^+$, $\mathbf{H}^-$, $\mathbf{V}^+$, and $\mathbf{V}^-$ functions. We follow the definitions given in section III, especially, Theorem 3.1. *Except* for *Section* V.G, *we will assume that* $\mathfrak{w}_1 = \mathfrak{w}_2 = \mathfrak{h}_1 = \mathfrak{h}_2 = 1$. We also define the *width* and *height* of f as max $\mathfrak{C}_\mathfrak{f}$ − min $\mathfrak{C}_\mathfrak{f}$ + 1 and max $\mathfrak{R}_\mathfrak{f}$ − min $\mathfrak{R}_\mathfrak{f}$ + 1, respectively.

Since the functions $\{\mathbf{H}^+,\mathbf{H}^-,\mathbf{V}^+,\mathbf{V}^-\}$ are alike, we will concentrate only on the functions $\mathbf{H}^+$ and $\mathbf{H}^-$, allowing the reader to develop analogous arguments for $\mathbf{V}^+$ and $\mathbf{V}^-$. Also, we will concentrate only on the $\{\mathbf{H}^+,\mathbf{V}^+\}$ pair. Our first result relates the computation of $\mathbf{H}^+$ and $\mathbf{H}^-$ functions to that of λ^+ and λ^- functions. Essentially, it states that the computation of the $\mathbf{H}^+$ function of two-dimensional images can be decomposed into the computation of the λ^+ function of line images. A line image is essentially a one-dimensional image in the sense that it has support either in a single row or in a single column.

To simplify the subsequent discussion, for each image f, we define six sets as follows.

$$\mathfrak{A}^+(\mathfrak{f}, n) = \{\langle\alpha,\beta\rangle \in \mathfrak{f} : \exists\langle\gamma,\beta\rangle \in \mathfrak{f} \ni \gamma = \alpha + n\}$$
$$\mathfrak{B}^+(\mathfrak{f}, n) = \{\langle\alpha,\beta\rangle \in \mathfrak{f} : \exists\langle\alpha,\gamma\rangle \in \mathfrak{f} \ni \gamma = \beta + n\}$$
$$\mathfrak{A}^-(\mathfrak{f}, n) = \{\langle\alpha,\beta\rangle \in \mathfrak{f} : \exists\langle\gamma,\beta\rangle \in \mathfrak{f} \ni \gamma = \alpha - n\}$$
$$\mathfrak{B}^-(\mathfrak{f}, n) = \{\langle\alpha,\beta\rangle \in \mathfrak{f} : \exists\langle\alpha,\gamma\rangle \in \mathfrak{f} \ni \gamma = \beta - n\}$$
$$\hat{\mathfrak{f}}^+ = \{\langle x, y\rangle \in \hat{\mathfrak{f}} : \text{card } \mathfrak{f}_{[y]} > 0 \text{ and } x \in \hat{\mathfrak{f}}_{[y]+}\}$$
$$\hat{\mathfrak{f}}^- = \{\langle x, y\rangle \in \hat{\mathfrak{f}} : \text{card } \mathfrak{f}_{[y]} > 0 \text{ and } x \in \hat{\mathfrak{f}}_{[y]-}\}$$

Theorem 5.1. Consider any image $\mathfrak{f}$. Then for all $n \in \Re^+$,

$$\mathbf{H}^+(\mathfrak{f}, n) = \sum_{v\in\mathfrak{R}_\mathfrak{f}} \lambda^+\left(\mathfrak{f}_{[v]}, \hat{\mathfrak{f}}_{[v]}; n\right)$$

$$\mathbf{H}^-(f, n) = \sum_{v \in \mathfrak{R}_f} \lambda^-\left(f_{[v]}, \hat{f}_{[v]}; n\right).$$

Proof. Fix n. Consider a pixel $\langle x, y\rangle \in f$.

$$\langle x, y\rangle \in \mathfrak{A}^+(f, n) \quad \text{iff } \exists\, x \in f_{[y]} \text{ and } x + n \in \hat{f}_{[y]}$$
$$\text{iff } x \in \chi^+(f_{[y]}, \hat{f}_{[y]}; n)$$

Hence,

$$\mathbf{H}^+(f, n) = \sum_{v \in \mathfrak{R}_f} \text{card}\ \chi^+(f_{[v]}, \hat{f}_{[v]}; n) = \sum_{v \in \mathfrak{R}_f} \lambda^+(f_{[v]}, \hat{f}_{[v]}; n).$$

Now, it is easy to check that $\mathbf{H}^+$ and $\mathbf{V}^+$ functions provide information about the width and height of an image f.

Corollary 5.1. Consider any image f and let its height and width be η and ω, respectively. Then η and ω are the smallest integers such that

$$n \geq \eta + 1 \Rightarrow \mathbf{V}^+(f, n) = 0 \qquad \text{and} \qquad n \geq \omega + 1 \Rightarrow \mathbf{H}^+(f, n) = 0.$$

Proof. Consider any image f, and let ω_v be the width of row $v \in \mathfrak{R}_f$ and $\omega c_v = \max \hat{f}_{[v]} - \max f_{[v]}$. Then we have, due to Proposition 3.1(b) and Theorem 5.1,

$$\forall n \geq \max\{\omega_v + \omega c_v : v \in \mathfrak{R}_f\}, \qquad \mathbf{H}^+(f, n) = 0$$

Now, note that

$$\max_{v \in \mathfrak{R}_f}(\omega_v + \omega c_v) = \max_{v \in \mathfrak{R}_f}\left[\max \hat{f}_{[v]} - \min f_{[v]} + 1\right]$$
$$= \max_{v \in \mathfrak{R}_f} \max \hat{f}_{[v]} - \min_{v \in \mathfrak{R}_f} \min f_{[v]} + 1$$
$$= (\max \mathfrak{C}_f + 1) - (\min \mathfrak{C}_f) + 1 = \omega + 1$$

where ω is the so-called width of image f. Similarly, $\mathbf{V}^+(f, n) = 0$ for all $n \geq \eta + 1$.

In the proof of Theorem 3.1, we showed that if $\lambda^+(f, \hat{f}; \cdot) = \lambda^+(\psi, \hat{\psi}; \cdot)$, then card f = card ψ and card $\hat{f}_+$ = card $\hat{\psi}_+$. Lemma 5.1 states that this observation is also true for the $\mathbf{H}^+$ function. Lemma 5.2 states that the $\mathbf{H}^+$ function for any line image is unique. That is, for a line image, there does not exist a two-dimensional image having the same $\mathbf{H}^+$ function.

Lemma 5.1. If $\mathbf{H}^+(f, \cdot) = \mathbf{H}^+(\psi, \cdot)$, then card f = card ψ and card $\hat{f}^+$ = card $\hat{\psi}^+$.

Proof. If $\mathbf{H}^+(f, \cdot) = \mathbf{H}^+(\psi, \cdot)$, then we must have for all $n \geq 1$ (see Section III and Theorem 5.1):

$$\sum_{\nu \in \Re_{f}} \text{card } \chi^{+}(f_{[\nu]}, \hat{f}_{[\nu]}; n) = \sum_{\mu \in \Re_{\psi}} \text{card } \chi^{+}(\psi_{[\mu]}, \hat{\psi}_{[\mu]}; n) \tag{11.22}$$

Consider any $\nu \in \Re_{f}$. Then, for every pair $\alpha \in \chi^{+}(f_{[\nu]}, \hat{f}_{[\nu]}; n)$ and $\beta \in \hat{f}_{[\nu]+}$ such that $\beta = \alpha + n$, there exists a unique pair $\kappa \in \chi^{+}(\psi_{[\mu]}, \hat{\psi}_{[\mu]}; n)$ and $\varepsilon \in \hat{\psi}_{[\mu]+}$ such that $\varepsilon = \kappa + n$. Denote this correspondence as

$$\langle\langle \beta, \nu; \rangle, \langle \alpha, \nu \rangle\rangle \; \Xi_{n} \; \langle\langle \varepsilon, \mu \rangle, \langle \kappa, \mu \rangle\rangle$$

Now, define two mappings, $\theta_1 : f \rightarrow \psi$ and $\theta_2 : \hat{f}^{+} \rightarrow \hat{\psi}^{+}$ such that

$$[\exists n \ni \langle\langle \beta, \nu \rangle, \langle \alpha, \nu \rangle\rangle \; \Xi_{n} \; \langle\langle \varepsilon, \mu \rangle, \langle \kappa, \mu \rangle\rangle] \tag{11.23}$$

$$\textit{iff } \theta_2(\beta, \nu) = \langle \varepsilon, \mu \rangle \quad \text{and} \quad \theta_1(\alpha, \nu) = \langle \kappa, \mu \rangle$$

Note that condition (11.23) implies that θ_1 and θ_2 should map any two pixels in the same row of f into two pixels in a single row of ψ.

Consider any $n \geq 1$. We have

$$\begin{aligned} x \in \chi^{+}(f_{[\nu]}, \hat{f}_{[\nu]}; n) &\Leftrightarrow \langle x, \nu \rangle \in f \textit{ and } \langle x + n, \nu \rangle \in \hat{f}^{+} \\ &\Leftrightarrow \theta_2(x + n, \nu) = \theta_1(x, \nu) + \langle n, 0 \rangle \\ &\Leftrightarrow \exists \mu \in \Re_{\psi}, \exists \kappa \in \mathfrak{C}_{\psi} \ni \kappa \in \chi^{+}(\psi_{[\mu]}, \hat{\psi}_{[\mu]}; n) \end{aligned}$$

with $\langle \kappa, \mu \rangle \equiv \theta_1(x, \nu)$

Now, since (11.22) holds and since

$$\forall \nu \in \Re_{f}, \forall x \in f_{[\nu]}, \exists n \geq 1 \ni x \in \chi^{+}(f_{[\nu]}, \hat{f}_{[\nu]}; n)$$

θ_1 must be a bijective mapping. Consequently, (11.23) implies that θ_2 is also a bijective mapping. Hence the results.

Lemma 5.2. If f is a one-dimensional image, then there does *not* exist a two-dimensional image ψ such that

$$\text{either } \mathbf{H}^{+}(\psi, \cdot) = \lambda^{+}(f, \hat{f}; \cdot) \text{ or } \mathbf{H}^{-}(\psi, \cdot) = \lambda^{-}(f, \hat{f}; \cdot)$$

except for the degenerate case where ψ has support only in one row (say κ), and $\psi_{[\kappa]}$ is the same as f (ignoring translation).

Proof. To the contrary, assume that there exists a ψ and let ψ have support in rows indexed by $\Re_{\psi}$. Clearly, card $\Re_{\psi} \geq 2$. By Lemma 5.1, card f = card ψ and

$$\text{card } \hat{f}_{+} = \text{card } \hat{\psi} f^{+} \tag{11.24}$$

Moreover, by Corollary 5.1, we must have width f = width ψ. Let $\alpha \in \Re_{\psi}$ be such that $\min \psi_{[\alpha]} \leq \min_{\nu \in \Re_{f}} (\min \psi_{[\nu]})$, that is, row α of ψ has the leftmost activated pixel. Then card $f >$ card $\psi_{[\alpha]}$, and consequently card $\hat{f}_{+} <$ card $\hat{\psi}_{[\alpha]+}$. This contradicts (11.24).

As a consequence of Theorem 5.1 and Lemmas 5.1 and 5.2, one can deduce Theorem 5.2.

Theorem 5.2. For any two images $\mathfrak{f}$ and ψ, the following three conditions are equivalent:

(i) $\mathbf{H}^{+}(\mathfrak{f},\cdot) = \mathbf{H}^{+}(\psi,\cdot)$.

(ii) $\mathbf{H}^{-}(\mathfrak{f},\cdot) = \mathbf{H}^{-}(\psi,\cdot)$.

(iii) There exists a bijective mapping $\omega : \mathfrak{R}_{\mathfrak{f}} \to \mathfrak{R}_{\psi}$ such that for fixed $\tau \in \mathfrak{N}$, we have for all $v \in \mathfrak{R}_{\mathfrak{f}}$, $\mathfrak{f}_{[v]} = \mathbf{T}\left(\psi_{[\omega(v)]}; \tau\right)$.

Proof. We will show that (i) iff (iii). (ii) iff (iii) can be proved in similar fashion.

(*Sufficiency*) Directly follows from Definition 5.2.

(*Necessity*) From Lemma 5.2, we know that the cardinalities of $\mathfrak{R}_{\mathfrak{f}}$ and $\mathfrak{R}_{\psi}$ must be the same. Therefore, there must be a mapping $\omega : \mathfrak{R}_{\mathfrak{f}} \to \mathfrak{R}_{\psi}$ such that the λ^{+} functions of $\mathfrak{f}_{[\alpha]}$ and $\psi_{[\omega(\alpha)]}$ are identical. Now, from Theorem 3.1, we know that $\mathfrak{f}_{[\alpha]}$ and $\psi_{[\omega(\alpha)]}$ are at most translated versions of each other. Hence, the outcome of Theorem 3.1 and Lemma 5.2 is

$$\exists\omega : \mathfrak{R}_{\mathfrak{f}} \to \mathfrak{R}_{\psi} \ni \forall v \in \mathfrak{R}_{\mathfrak{f}}, \qquad \mathfrak{f}_{[v]} = \mathbf{T}\left(\psi_{[\omega(v)]}; \tau_v\right)$$

where τ_v may be dependent on v. Now, if we impose the results of Lemma 5.1, we conclude that τ_v must be independent of v; otherwise the relationship between the cardinalities of $\hat{\mathfrak{f}}^{+}$ and $\hat{\psi}^{+}$ will be violated (or the result of Corollary 5.1 will be violated).

In exactly similar fashion one can show the corresponding result for $\mathbf{V}^{+}$ and $\mathbf{V}^{-}$ functions.

Theorem 5.3. For any two images $\mathfrak{f}$ and ψ, the following three conditions are equivalent:

(i) $\mathbf{V}^{+}(\mathfrak{f},\cdot) = \mathbf{V}^{+}(\psi,\cdot)$.

(ii) $\mathbf{V}^{-}(\mathfrak{f},\cdot) = \mathbf{V}^{-}(\psi,\cdot)$.

(iii) There exists a bijective mapping $\omega : \mathfrak{C}_{\mathfrak{f}} \to \mathfrak{C}_{\psi}$ such that for a fixed

$\tau \in \mathfrak{N}$, we have for all $v \in \mathfrak{C}_{\mathfrak{f}}$, $\mathfrak{f}^{[v]} = \mathbf{T}(\psi^{[\sigma(v)]}; \tau)$.

Proof. Same as the proof of Theorem 5.2.

Example 5.1. Via this example, we illustrate the difference between the implications of Theorem 5.2 and Theorem 3.1, thereby establishing the superiority of the $\mathbf{H}^{+}$ function over **HOR**.

Consider the two images F and G of Figure 15. They have the same **HOR** functions but their $\mathbf{H}^{+}$ functions are different. Row 3 of F and row 2 of G have

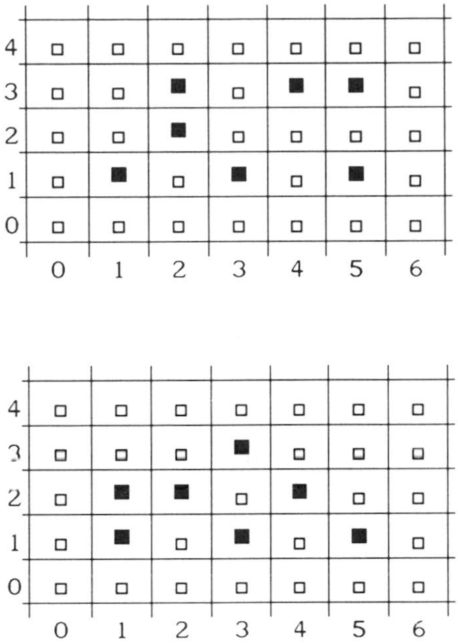

Figure 15. The two images F and G that have different $\mathbf{H}^+$, but, have the same **HOR** function.

the same **HOR** function but their $\mathbf{H}^+$ functions are different. The same discussion applies to the second row of F and third row of G.

Next, if we consider two images H and I of Figure 16, they have same **HOR** functions as well as $\mathbf{H}^+$ functions. The mapping between rows, ω, is defined as follows:

1st row of $H \overset{\omega}{\rightarrow}$ 3rd row of I
3rd row of $H \overset{\omega}{\rightarrow}$ 1st row of I
4th row of $H \overset{\omega}{\rightarrow}$ 2nd row of I.

Note that $\mathbf{H}^+$ functions are the same if there exists a bijective mapping between the rows of the two images. Therefore in two-dimensional images, the $\mathbf{H}^+$ ($\mathbf{H}^-$) function does not characterize an image uniquely, even though the λ^+ (λ^-) function characterizes a one-dimensional image uniquely.

By Theorems 5.2 and 5.3, the proposed functions are *invariant under translation*. Now, we wish to deal with the consequences of Theorems 5.2 and 5.3. Consider two images f and ψ such that

$$\mathbf{H}^+(f,\cdot) = \mathbf{H}^+(\psi,\cdot) \qquad \text{and} \qquad \mathbf{V}^+(f,\cdot) = \mathbf{V}^+(\psi,\cdot)$$

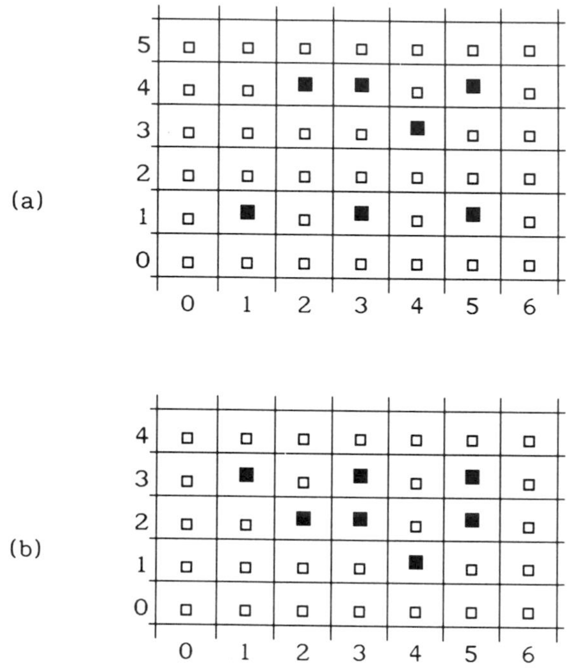

Figure 16. Images H and I have same $\mathbf{H}^+$ and **HOR** function.

If we ignore the translation, that is, if we let

$$\min \mathfrak{C}_f = \min \mathfrak{C}_\psi \qquad \text{and} \qquad \min \mathfrak{R}_f = \min \mathfrak{R}_\psi$$

then for some $n,m \geq 1$, we can assume that

$$\mathfrak{R}_f = \mathfrak{R}_\psi = \{\mathfrak{r}_1, \mathfrak{r}_2, \ldots, \mathfrak{r}_n\} \qquad \text{and} \qquad \mathfrak{C}_f = \mathfrak{C}_\psi = \{\mathfrak{c}_1, \mathfrak{c}_2, \ldots, \mathfrak{c}_m\}$$

Now, consider the two mappings

$$\omega : \mathfrak{R}_f \to \mathfrak{R}_\psi \qquad \text{and} \qquad \sigma : \mathfrak{C}_f \to \mathfrak{C}_\psi$$

as defined in Theorems 5.2 and 5.3, respectively. *If there exists a identity mapping for rows (columns), then* ω (σ) *will be assumed to be an identity mapping*. Now, it follows that our algorithm works only if ω and σ are identity mappings.

What does "there exist σ and ω which are identity mappings" mean? Consider the two images F and G of Figure 17. We can define the mappings σ and ω as follows (employing the standard terminology for representing permutations):

$$\omega : \begin{pmatrix} 1 & 2 & 3 \\ 1 & 3 & 2 \end{pmatrix} \qquad \text{and} \qquad \sigma : \begin{pmatrix} 1 & 2 & 3 \\ 3 & 2 & 1 \end{pmatrix}$$

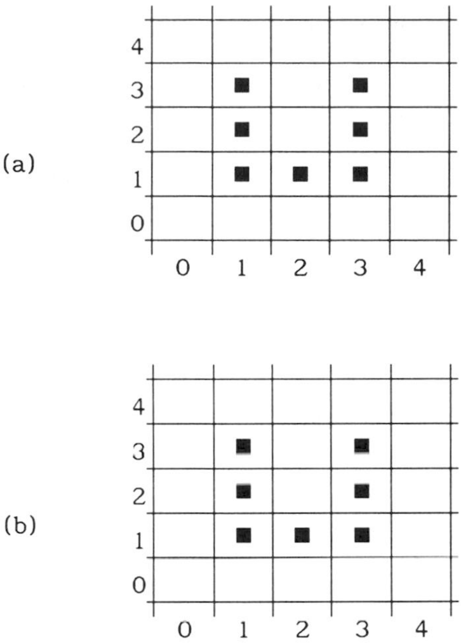

Figure 17. Example of images which have identical σ and ω mappings.

Also, σ and ω can be defined as identity mappings such that

$$\omega : \begin{pmatrix} 1 & 2 & 3 \\ 1 & 2 & 3 \end{pmatrix} \quad \text{and} \quad \sigma : \begin{pmatrix} 1 & 2 & 3 \\ 1 & 2 & 3 \end{pmatrix}$$

In this case, since there exist identity mappings, σ and ω are assumed to be identity mappings.

The uniqueness question has thus been reduced to: Is it true that

$$\mathbf{H}^+(f,\cdot) = \mathbf{H}^+(\psi,\cdot) \qquad \text{and} \qquad \mathbf{V}^+(f,\cdot) = \mathbf{V}^+(\psi,\cdot)$$

iff there exist identity mappings ω and σ?

The two images of Figure 2 show that ω and σ need not be identity mappings. The mappings ω and σ are defined as:

$$\omega : \begin{pmatrix} 1 & 2 & 3 \\ 2 & 3 & 1 \end{pmatrix} \quad \text{and} \quad \sigma : \begin{pmatrix} 1 & 2 & 3 \\ 2 & 3 & 1 \end{pmatrix}$$

B. Irregular Images

Before we proceed further, we pause to state one of the most important consequences of Theorems 5.2 and 5.3.

*Proposition 5.1**. If $\mathbf{H}^+(\mathfrak{f},\cdot) = \mathbf{H}^+(\psi,\cdot)$ and $\mathbf{V}^+(\mathfrak{f},\cdot) = \mathbf{V}^+(\psi,\cdot)$, then we have for all $k \neq 0$, $\langle i,j\rangle \in \mathfrak{f}$ *iff* $\langle\sigma^k(i),\omega^{-k}(j)\rangle \in \mathfrak{f}$.

Proof. Consider any pixel $\langle i,j\rangle$ in $\mathfrak{f}$. From Theorem 5.3 we know that the ith columns of $\mathfrak{f}$ is identical to the $\sigma(i)$th column of ψ. Hence,

$$\langle i,j\rangle \in \mathfrak{f} \text{ iff } \langle\sigma(i),j\rangle \in \psi \tag{11.25}$$

From Theorem 5.2 we know that the jth row of $\mathfrak{f}$ is identical to the $\omega(j)$th row of ψ. Therefore, from Eq. (11.25) we obtain

$$\langle i,j\rangle \in \mathfrak{f} \text{ iff } \langle\sigma(i),j\rangle \in \psi \text{ iff } \langle\sigma(i),\omega^{-1}(j)\rangle \in \mathfrak{f} \tag{11.26}$$

Repeated application of Eq. (11.26) gives the desired result.

Example 5.2. Consider two image F and G of Figure 2. They have the same $\mathbf{H}^+$ and $\mathbf{V}^+$ functions. The ω and σ mappings are as follows:

$$\omega : \begin{pmatrix} 1 & 2 & 3 \\ 2 & 3 & 1 \end{pmatrix} \quad \text{and} \quad \sigma : \begin{pmatrix} 1 & 2 & 3 \\ 2 & 3 & 1 \end{pmatrix}$$

Now we show the interrelationship between the pixels in image F:

$$\begin{aligned} \langle 1,3\rangle \in \mathfrak{f} &\Leftrightarrow \langle\sigma(1),\omega^{-1}(3)\rangle = \langle 2,2\rangle \in \mathfrak{f} \\ &\Leftrightarrow \langle\sigma(2),\omega^{-1}(2)\rangle = \langle 3,1\rangle \in \mathfrak{f} \\ &\Leftrightarrow \langle\sigma(3),\omega^{-1}(1)\rangle = \langle 1,3\rangle \in \mathfrak{f} \end{aligned}$$

That is, the mappings σ and ω between rows and columns of F and G determine the pixels in image F. The same arguments hold for G.

This result is remarkable in the sense that it states that the existence (absence) of a pixel in $\mathfrak{f}$ determines the existence (absence) of a host of other pixels in $\mathfrak{f}$, unless, of course, ω and σ are identity mappings. This type of relationship normally exists in symmetric images only, and the irregular images are not necessarily symmetric.

Proposition 5.1 focuses our attention on the two mappings ω and σ. Consider the following two conditions:

(†1) ω is an identity mapping: $\forall\alpha \in \mathfrak{R}_{\mathfrak{f}}$, $\omega(\alpha) = \alpha$.

(†2) σ is an identity mapping: $\forall\alpha \in \mathfrak{C}_{\mathfrak{f}}$, $\sigma(\alpha) = \alpha$.

*Let $\lambda : \mathfrak{A} \rightarrow \mathfrak{A}$ be a bijective mapping. A standard convention in algebra is to define the mappings $\lambda^k,\lambda^{-k} : \mathfrak{A} \rightarrow \mathfrak{A}$, $k \geq 1$, recursively as

$$\lambda^k(\alpha) = \lambda\left[\lambda^{k-1}(\alpha)\right] \quad \text{and} \quad \lambda^{-k}(\alpha) = \lambda^{-1}\left[\lambda^{-k+1}(\alpha)\right]$$

Also, $\lambda^0(\alpha) = \alpha$.

These give rise to the following four situations:

(a) (†1) and (†2) are both true.

(b) (†1) is true but, (†2) is false.

(c) (†2) is true but, (†1) is false.

(d) (†1) and (†2) are both false.

We intend to show that situations (b) and (c) cannot be true. If images $\mathfrak{f}$ and ψ are identical, then we obtain situation (a). The images of Figure 2 come under situation (d). Hence, from the perspective of exact geometric characterization, the $\{\mathbf{H}^+, \mathbf{V}^+\}$ pair fails for the images that give rise to situation (d). We will show that this class contains a very special type of images (called irregular images).

Proposition 5.2. Consider situation (b). There exists an identity mapping $\hat{\sigma} : \mathfrak{C} \rightarrow \mathfrak{C}_\psi$ that satisfies Theorem 5.3.

Proof. Our construction of $\hat{\sigma}$ will be based on σ. Assume that for some α, $\sigma(\alpha) \neq \alpha$, and define $\tau \in \mathfrak{R}$ so that $\sigma^\tau(\alpha) = \alpha$. Let $\sigma^{\tau-1}(\alpha) = \beta$. Since ω is an identity mapping, we obtain from Proposition 5.1:

$$\forall j \in \mathfrak{R}, \qquad \langle \alpha, j\rangle \in \mathfrak{f} \Leftrightarrow \langle \sigma^{\tau-1}(\alpha), j\rangle \equiv \langle \beta, j\rangle \in \mathfrak{f}$$

Hence, columns α and β are identical. Now, define a new mapping $\hat{\sigma} : \mathfrak{C}_\mathfrak{f} \rightarrow \mathfrak{C}_\psi$ such that $\bar{\sigma}(\alpha) = \sigma(\beta) = \sigma(\alpha)$, and

$$\forall x \in \mathfrak{C}_\mathfrak{f} \setminus \{\alpha, \beta\}, \qquad \hat{\sigma}(x) = \sigma(x)$$

In $\hat{\sigma}$, we have altered the columns of ψ that get mapped into from columns α and β of $\mathfrak{f}$; others have remained the same. In particular, we have

$$\textit{either } \operatorname{card}\{x \in \mathfrak{C}_\mathfrak{f} : \sigma(x) \neq x\} = \operatorname{card}\{x \in \mathfrak{C}_\mathfrak{f} : \hat{\sigma}(x) \neq x\} + 1$$
$$\textit{or } \operatorname{card}\{x \in \mathfrak{C}_\mathfrak{f} : \sigma(x) \neq x\} = \operatorname{card}\{x \in \mathfrak{C}_\mathfrak{f} : \hat{\sigma}(x) \neq x\} + 2$$

Thus $\hat{\sigma}$ has brought us one step closer to situation (a). Repeating the above process for $\hat{\sigma}$, we can make $\operatorname{card}\{x \in \mathfrak{C}_\mathfrak{f} : \hat{\sigma}(x) \neq x\} = 0$. That is, $\hat{\sigma}$ is an identity mapping, which leads to the contradiction of assumption (b).

Proposition 5.3. Consider situation (c). There exists an identity mapping $\hat{\omega} : \mathfrak{R}_\mathfrak{f} \rightarrow \mathfrak{R}_\psi$ that satisfies Theorem 5.2.

Proof. Same as the proof of Proposition 5.2.

Next, we turn our attention to the most complicated situation, (d). Before we can analyze this situation, we must introduce some new notation.

Definition 5.4. A two-dimensional image $\mathfrak{f}$ will be called an *irregular image* iff there exist two permutations $\zeta : \mathfrak{C}_\mathfrak{f} \rightarrow \mathfrak{C}_\mathfrak{f}$ and $\xi : \mathfrak{R}_\mathfrak{f} \rightarrow \mathfrak{R}_\mathfrak{f}$ such that

(i) $\langle i,j\rangle \in \mathfrak{f}$ *iff* $\langle \zeta(i), \xi^{-1}(j)\rangle \in \mathfrak{f}$.

(ii) $\exists \alpha \in \mathfrak{C}_\mathfrak{f} \ni \zeta(\alpha) \neq \alpha$ *and* $\mathfrak{f}^{[\alpha]} \neq \mathfrak{f}^{[\zeta(\alpha)]}$.

(iii) $\exists \alpha \in \mathfrak{R}_\mathfrak{f} \ni \xi(\alpha) \neq \alpha$ *and* $\mathfrak{f}_{[\alpha]} \neq \mathfrak{f}_{[\xi(\alpha)]}$.

Remarks

1. Note that conditions (ii) and (iii) ensure that ζ and ξ cannot be converted into identity mappings. In (ii) and (iii) we have implicitly assumed that card $\mathfrak{C}_{\mathfrak{f}}$, card $\mathfrak{R}_{\mathfrak{f}} \geq 2$. This is justified from the fact that line images can be identified uniquely and hence can never be classified as irregular images.
2. An image $\mathfrak{f}$ is said to be a filled rectangle *iff* $\mathfrak{f} = \mathfrak{C}_{\mathfrak{f}} \times \mathfrak{R}_{\mathfrak{f}}$. Note that we do not require the contiguity of rows and/or columns. Even though a filled rectangle is not an irregular image, we will assume that *all filled rectangular images are irregular*, as an exception. This exception is necessary when we discuss the construction of irregular images in Section V.D.

Example 5.3. Examples of irregular images are *F* and *G* of Figure 2, *H* and *I* of Figure 9, and *J* and *K* of Figure 18. For image *H* of Figure 9, the mappings ζ and ξ are

$$\zeta : \begin{pmatrix} 1 & 2 & 3 & 4 & 5 & 6 \\ 2 & 3 & 1 & 5 & 6 & 4 \end{pmatrix} \quad \text{and} \quad \xi : \begin{pmatrix} 1 & 2 & 3 \\ 2 & 3 & 1 \end{pmatrix}$$

It is easy to check that the mappings ζ and ξ satisfy requirements (ii) and (iii) of Definition 5.4. Similarly, for image *J* of Figure 18 we have

$$\zeta : \begin{pmatrix} 1 & 2 & 3 & 4 & 5 & 6 \\ 2 & 3 & 1 & 5 & 6 & 4 \end{pmatrix} \quad \text{and} \quad \xi : \begin{pmatrix} 1 & 2 & 3 & 4 & 5 & 6 \\ 2 & 3 & 1 & 5 & 6 & 4 \end{pmatrix}$$

and they satisfy the requirements.

Examples of images that are not irregular are an empty image and the images shown in Figures 1, 3, and 15 through 17.

The following two results state that the class of irregular images has a very unique closure property:

Proposition 5.4. For every irregular image $\mathfrak{f}$, there exists an irregular image ψ such that we obtain situation (d).

Proof. Since $\mathfrak{f}$ is an irregular image, there exist two bijective mappings $\zeta : \mathfrak{C}_{\mathfrak{f}} \rightarrow \mathfrak{C}_{\mathfrak{f}}$ and $\xi : \mathfrak{R}_{\mathfrak{f}} \rightarrow \mathfrak{R}_{\mathfrak{f}}$ such that

$$\langle i,j \rangle \in \mathfrak{f} \text{ iff } \langle \zeta(i), \xi^{-1}(j) \rangle \in \mathfrak{f}$$

With the help of mappings ζ and ξ, we define another image ψ as follows:

$$\langle i,j \rangle \in \mathfrak{f} \text{ iff } \langle \zeta(i), j \rangle \in \psi$$

Clearly, $\mathfrak{C}_{\mathfrak{f}} = \mathfrak{C}_{\psi}$ and $\mathfrak{R}_{\mathfrak{f}} = \mathfrak{R}_{\psi}$. Now, the mapping $\zeta : \mathfrak{C}_{\mathfrak{f}} \rightarrow \mathfrak{C}_{\psi}$ has the following property:

$$\forall \alpha \in \mathfrak{C}_{\mathfrak{f}}, \qquad \mathfrak{f}^{[\alpha]} = \psi^{[\zeta(\alpha)]}$$

Hence the $\mathbf{V}^+$ functions of $\mathfrak{f}$ and ψ are identical (see Theorem 5.3). All we need to shown now is that the $\mathbf{H}^+$ functions of $\mathfrak{f}$ and ψ are identical. To this end, we first note that

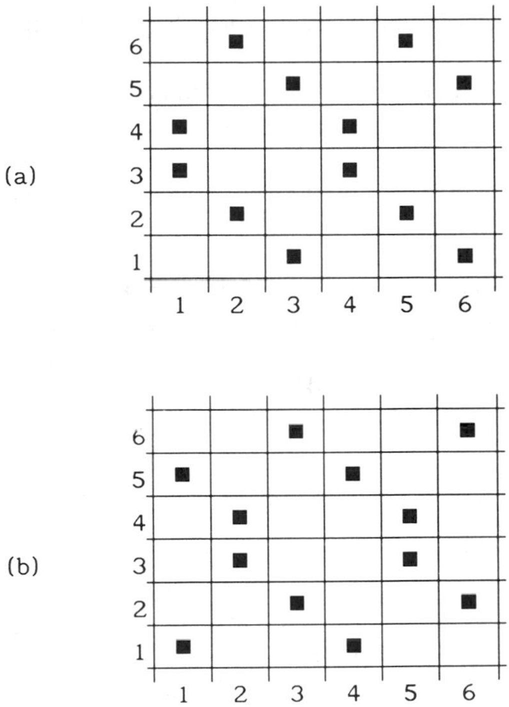

Figure 18. Examples of irregular images; (a) and (b) are refered to as images J and K respectively in the text.

$$\langle i,j\rangle \in f \text{ iff } \langle \zeta(i),\xi^{-1}(j)\rangle \in f \text{ iff } \langle \zeta^{-1}(i),\xi(j)\rangle \in f$$
$$\text{iff } \langle i,\xi(j)\rangle \in \psi$$

Hence the mapping $\xi : \Re_f \rightarrow \Re_\psi$ has the following property:

$$\forall \alpha \in \Re_f, \qquad f_{[\alpha]} = \psi_{[\xi(\alpha)]}$$

From Theorem 5.2, the $\mathbf{H}^+$ functions are identical.

To complete the proof, we first state that

$$\langle i,j\rangle \in \psi \text{ iff } \langle \zeta^{-1}(i),j\rangle \in f \text{ iff } \langle \zeta^{-1}(i),\xi(j)\rangle \in \psi$$

Let $\sigma : \mathfrak{C}_\psi \rightarrow \mathfrak{C}_\psi$ and $\omega : \Re_\psi \rightarrow \Re_\psi$ be such that $\sigma \equiv \zeta^{-1}$ and $\omega \equiv \xi^{-1}$. Then σ and ω satisfy all the requirements for ψ to be classified as an irregular image; for example, if $\zeta(\alpha) \neq \alpha$, then $\sigma(\alpha) = \zeta^{-1}(\alpha) \neq \alpha$.

Proposition 5.1. Consider any two images f and ψ that give rise to situation (d). Then f and ψ must both be irregular.

From the above propositions, we can define the equivalence class for irregular images as follows.

Definition 5.5. If two-dimensional images $\mathfrak{f}$ and ψ satisfy the constraints of Proposition 5.1 or 5.4, then we say that the images $\mathfrak{f}$ and ψ belong to the same *equivalence irregular class*.

It easily follows that this is an equivalence relation; that is, this relationship splits the class of all irregular images into mutually disjoint collections. We summarize our findings in the following result.

Theorem 5.4. Consider any two images $\mathfrak{f}$ and ψ. Then $\mathbf{H}^+(\mathfrak{f},\cdot) = \mathbf{H}^+(\psi,\cdot)$ and $\mathbf{V}^+(\mathfrak{f},\cdot) = \mathbf{V}^+(\psi,\cdot)$ iff either $\mathfrak{f} = \mathbf{T}(\psi,v)$ with $v \in \mathfrak{R}^2$ or $\mathfrak{f}$ and ψ are equivalent irregular images.

Proof. From Propositions 5.1, 5.2, 5.3, and 5.4.

C. Basic Irregular Images

From the unique closure property of irregular images discussed in the last section, one may wish to arrive at a scheme for constructing new irregular images from the given ones. To this end, we first define the basic building blocks of such constructions.

Consider an irregular image $\mathfrak{f}$ with card $\mathfrak{C}_\mathfrak{f} = m$ and card $\mathfrak{R}_\mathfrak{f} = n$. If there exist ζ and ξ mappings defined in Definition 5.4 containing only *one cycle** and the image $\mathfrak{f}$ is the smallest possible such image among the images with n rows and m columns, then the irregular image $\mathfrak{f}$ is called an $(n \times m)$ basic irregular image. More formally, we have

Definition 5.6. An irregular image $\mathfrak{f}$ is called a *basic irregular image* iff there exist mappings ζ and ξ such that

(i) $\forall \alpha \in \mathfrak{R}_\mathfrak{f}$, $\xi^k(\alpha) = \alpha$ iff $k \equiv 0 \pmod{n}$.

(ii) $\forall \beta \in \mathfrak{C}_\mathfrak{f}$, $\zeta^k(\beta) = \beta$ iff $k \equiv 0 \pmod{m}$.

(iii) If $\psi \subset \mathfrak{f}$ with $\mathfrak{R}_\mathfrak{f} = \mathfrak{R}_\psi$ and $\mathfrak{C}_\mathfrak{f} = \mathfrak{C}_\psi$, then ψ is not a basic irregular image.

Recall that n = card $\mathfrak{R}_\mathfrak{f}$, m = card $\mathfrak{C}_\mathfrak{f}$, and the mappings ζ and ξ are as defined in Definition 5.4.

Example 5.4. The examples of basic irregular images are the images F and G of Figure 2 (see Example 5.2) and the images H and I of Figure 9 (see Example 5.3). Note that, in Example 5.3, the mappings ζ and ξ for the image H were defined as

*Consider any permutation ω of a set $\mathfrak{A}$. Then, for every choice of $\alpha \in \mathfrak{A}$, the set $\{\omega^k(\alpha) : k \geq 0\}$ is called a *cycle* in ω.

$$\zeta : \begin{pmatrix} 1 & 2 & 3 & 4 & 5 & 6 \\ 2 & 3 & 1 & 5 & 6 & 4 \end{pmatrix} \quad \text{and} \quad \xi ; \begin{pmatrix} 1 & 2 & 3 \\ 2 & 3 & 1 \end{pmatrix}$$

But also, the mappings ζ and ξ can be defined as

$$\zeta : \begin{pmatrix} 1 & 2 & 3 & 4 & 5 & 6 \\ 2 & 3 & 4 & 5 & 6 & 1 \end{pmatrix} \quad \text{and} \quad \xi : \begin{pmatrix} 1 & 2 & 3 \\ 2 & 3 & 1 \end{pmatrix}$$

That is, there exist ζ and ξ mappings containing only *one cycle* each. Furthermore, there does not exist an image $\psi \subset H$ such that $\mathfrak{R}_\psi = \mathfrak{R}_f$, $\mathfrak{C}_\psi = \mathfrak{C}_f$, and ζ and ξ have only one cycle each. Therefore, the image H of Figure 9 is a basic irregular image.

The examples of nonbasic irregular images are the images F and G of Figure 19 and the images H and I of Figure 20. The mappings ζ and ξ for the image F of Figure 19 are defined as

$$\zeta : \begin{pmatrix} 1 & 2 & 3 & 4 \\ 4 & 3 & 2 & 1 \end{pmatrix} \quad \text{and} \quad \xi : \begin{pmatrix} 1 & 2 & 3 & 4 \\ 4 & 3 & 2 & 1 \end{pmatrix}$$

The mappings ζ and ξ contain two cycles and are the only possible mappings. Therefore, the irregular image F is not a basic irregular image. The same arguments hold for the images of Figure 20.

(a)

(b)

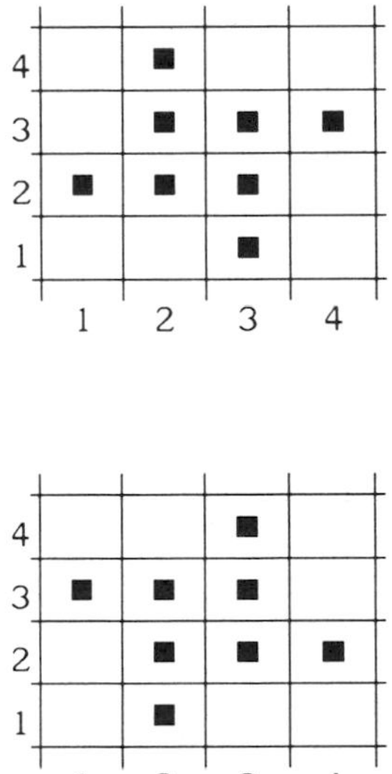

Figure 19. Examples of non-basic irregular images; (a) and (b) are refered to as images F and G respectively in the text.

(a)

(b)

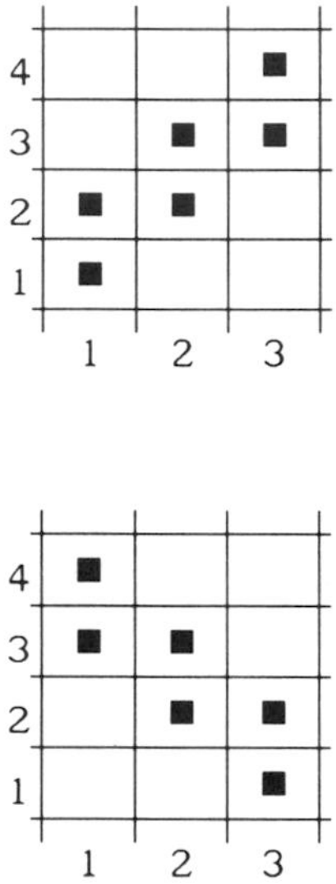

Figure 20. Examples of non-basic irregular images; (a) and (b) represent characters '/' and '\' respectively, and are refered to as images H and I respectively in the text.

Finally, if we look at the images J and K of Figure 18 they are not basic irregular images. Even though the images satisfy requirements (i) and (ii), they do not satisfy requirement (iii). If we consider an image $\mathfrak{f}$ such that there exists only one activated pixel in each six rows/columns like the images of Figure 2, then the image $\mathfrak{f} \subset J\ (K)$ and the mappings ζ and ξ have only one cycle each.

Since the mappings ζ and ξ contain only one cycle each for the basic irregular images, *we establish the following convention*:* Let

$$\mathfrak{C}_{\mathfrak{f}} = \{c_1, c_2, \ldots, c_m\} \qquad \text{and} \qquad \mathfrak{R}_{\mathfrak{f}} = \{r_1, r_2, \ldots, r_n\}$$

such that

$$\zeta(c_i) = c_{(i \bmod m\, +\, 1)} \qquad \text{and} \qquad \xi(r_j) = r_{(j \bmod n\, +\, 1)} \tag{11.27}$$

In Figure 21 we list all of the six 3×3 basic irregular images. Note that the rows and columns need not be adjacent. Furthermore, we can also relabel the axes as (rows: bottom → top; columns: left → right):

Images 1, 4, 6: rows are r_1, r_3, and r_2; columns are c_1, c_3, and c_2.
Images 2, 3, 5: rows are r_1, r_2, and r_3; columns are c_1, c_3, and c_2.

* This will remove the need to specify the mappings ζ and ξ for basic irregular images; all one needs to do is label the axes appropriately.

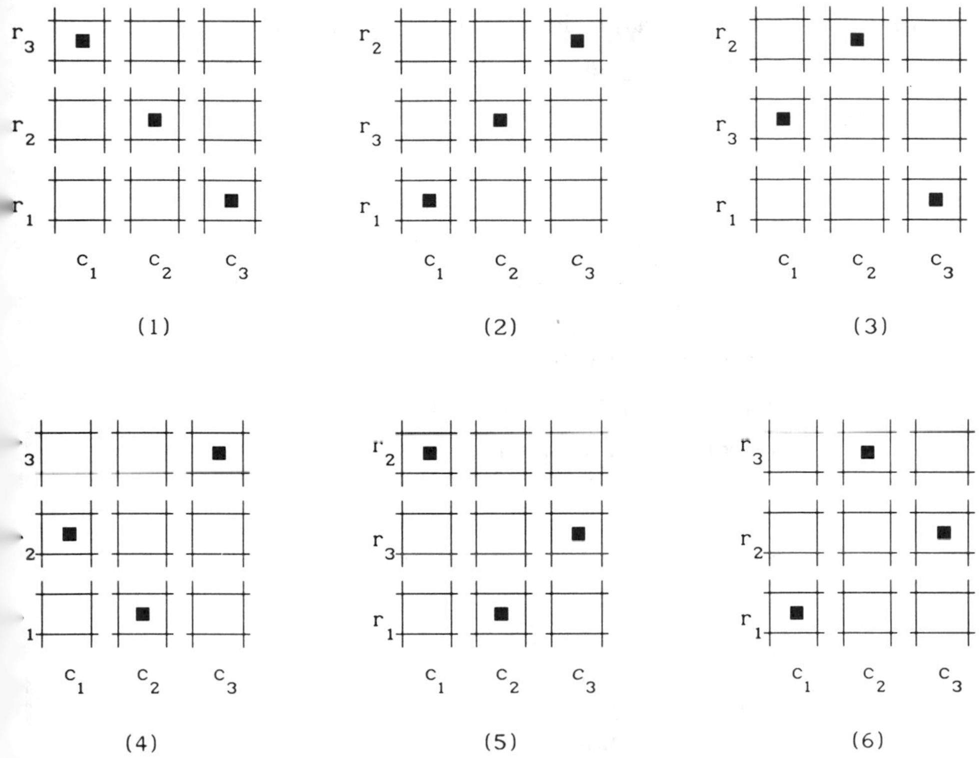

Figure 21. Six different 3 × 3 basic irregular images.

Since the basic irregular images are the simplest of the irregular images, one may wonder if we can specify exactly what they are. Our next results state how to construct all nontrivial basic irregular images. *Note that the mappings ζ and ξ appear implicitly in the construction* as they were used for labeling the columns and rows, respectively [see Eq. (11.27)]. Before we can establish Theorem 5.5, we need the following two simple results on congruences; see, for example, [13]. We follow the standard convention by letting $\gcd(\alpha,0) = \alpha$, where gcd stands for greatest common divisor; also, *the least common multiple* will be denoted as lcm.

Lemma 5.3. (Bezout's Identity). Consider any three fixed integers n, m, and κ. Then a solution(s) to the equation

$$\exists x,y \in \mathfrak{N} \ni nx + my = \kappa$$

exists *iff* κ is divisible by $\gcd(n,m)$. Let $\langle x_0,y_0\rangle$ be any one of the solutions to the equation. Then all the solutions are of the following form: $x = x_0 - m\alpha$ and $y = y_0 + n\alpha$ where $\alpha \in \mathfrak{N}$.

Theorem 5.5. Consider any basic irregular image $\mathfrak{f}$ such that

$$\text{card } \mathfrak{R}_{\mathfrak{f}} = n, \qquad \text{card } \mathfrak{C}_{\mathfrak{f}} = m, \qquad \langle c_v, r_\tau \rangle \in \mathfrak{f}$$

Depending on the values of n and m, we can uniquely define the image $\mathfrak{f}$ as follows:

$$\mathfrak{f} = \{\langle c_i, r_j \rangle : i + j - v - \tau \equiv 0 \pmod{\kappa}\} \qquad \text{where } \kappa = \gcd(n,m)$$

In particular, if $\kappa = 1$, then $\mathfrak{f}$ is a filled rectangle.

Proof. Since $\mathfrak{f}$ is an irregular image, there exist two mappings ζ and ξ such that

$$\langle c_v, r_\tau \rangle \in \mathfrak{f} \Leftrightarrow \langle \zeta^j(c_v), \xi^{-j}(r_\tau) \rangle \in \mathfrak{f}$$

Furthermore, for any $\alpha \in \{1,2,\ldots,n\}$ and $\beta \in \{1,2,\ldots,m\}$,

$$\xi^{-j}(r_\tau) = r_\alpha \text{ iff } \xi^j(r_\alpha) = r_\tau \text{ iff } \exists p \in \mathfrak{N} \ni j = pn + \tau - \alpha$$

$$\zeta^j(c_v) = c_\beta \text{ iff } \exists q \in \mathfrak{N} \ni j = qm + \beta - v$$

Combining the two equivalences, we conclude that

$$\langle c_\beta, r_\alpha \rangle \in \mathfrak{f} \text{ iff } \exists p,q \in \mathfrak{N} \ni pn - qm = \alpha + \beta - v - \tau \tag{11.28}$$

Hence, from Bezout's identity (Lemma 5.3), we know that the solution exists iff $\alpha + \beta - v - \tau \equiv 0 \ (\text{mod}(\gcd(n,m)))$. To complete the proof, note that if $\gcd(n,m) = 1$, then the solution to Eq. (11.28) always exists. Since this is true for all α and β, image $\mathfrak{f}$ will be an $n \times m$ filled rectangle.

Example 5.5. Let us construct image (2) of Figure 21 and call it $\mathfrak{f}$. Assume that $v = \tau = 1$. Since $\kappa = 3$, we obtain

$$\langle c_i, r_j \rangle \in \mathfrak{f} \text{ iff } \langle i,j \rangle \in \{\langle 1,1 \rangle, \langle 2,3 \rangle, \langle 3,2 \rangle\}$$

Hence, the image has following pixels: $\langle c_1, r_1 \rangle$, $\langle c_2, r_3 \rangle$, and $\langle c_3, r_2 \rangle$. Similarly, we can construct the remaining images of Figure 21.

Next, we show the construction of image H of Figure 9. Assume that $\langle 1,3 \rangle \in H$. Since $\kappa = 3$, we obtain

$$H = \{\langle 1,3 \rangle, \langle 2,2 \rangle, \langle 3,1 \rangle, \langle 4,3 \rangle, \langle 5,2 \rangle, \langle 6,1 \rangle\}$$

From Theorem 5.5, we can obtain several useful corollaries that will be used later on. Their proofs are fairly straightforward:

Corollary 5.2. The area of any $n \times m$ basic irregular image is $\text{lcm}(n,m)$.

Corollary 5.3. One can construct exactly $\gcd(n,m)$ different basic irregular images on n rows and m columns for fixed ζ and ξ mappings.

Corollary 5.4. There are $n!$ different $n \times n$ basic irregular images.

Corollary 5.5. Any $n \times m$ basic irregular image $\mathfrak{f}$ can be represented as the union of exactly nm/κ^2 many translated versions of a single $\kappa \times \kappa$ basic irregular image. $\kappa = \gcd(n,m)$.

Proof. Let us partition the image f into nm/κ^2 different regions

$$\{ f_{pq} : 0 \leq p < m/\kappa \text{ and } 0 \leq q < n/\kappa\}$$

where

$$f_{pq} = \{\langle c_i, r_j\rangle \in f : p\kappa < i \leq (p+1)\kappa;\ q\kappa < j \leq (q+1)\kappa\}$$

In order to prove Corollary 5.2, we must have essentially shown that

$$f_{[r_v]} = f_{[r_{\alpha(v)}]} \qquad \text{and} \qquad f^{[c_\tau]} = f^{[c_{\beta(\tau)}]}$$

where

$$\alpha(v) \equiv v \pmod{\kappa} \qquad \text{and} \qquad \beta(\tau) \equiv \tau \pmod{\kappa}$$

Now it follows trivially that for all $0 \leq p,r < m/\kappa$ and $0 \leq q,s \leq n/\kappa$, $f_{pq} = f_{rs}$. To complete the proof, all we need to show is that f_{pq} is a $\kappa \times \kappa$ basic irregular image. Using standard notation (introduced on page 386), we can rewrite f_{pq} as

$$f_{pq} = \{\langle c_i^*, r_j^*\rangle : 1 \leq i \leq \kappa;\ 1 \leq j \leq \kappa\}$$

where $c_i^* = c_{p\kappa - i}$ and $r_j^* = r_{q\kappa - j}$. Now, for some $2 \leq \gamma \leq n + m$,

$$\langle c_i, r_j\rangle \in f \text{ iff } i + j \equiv \gamma \pmod{\kappa}$$

Therefore,

$$\begin{aligned} \langle c_i^*, r_j^*\rangle \in f_{pq} &\text{ iff } \langle c_{p\kappa - i}, r_{q\kappa - j}\rangle \in f \\ &\text{ iff } (p + q)\kappa - (i + j) \equiv \gamma \pmod{\kappa} \\ &\text{ iff } i + j \equiv \gamma \pmod{\kappa} \end{aligned}$$

Hence f_{pq} is basic irregular (see Theorem 5.5).

Corollary 5.6. One can construct at most $\gcd(n,m)!$ different basic irregular images on n rows and m columns.

Example 5.6. Consider the construction of 4×6 ($n = 4$ and $m = 6$) basic irregular images. Let $v = \tau = 1$. We obtain

$$\langle c_i, r_j\rangle \in f \text{ iff } i + j \equiv 0 \pmod{2}$$

Therefore, we obtain image F of Figure 22. There exists only one more choice for such a basic irregular image: G of Figure 22. To obtain this image, one can let $v = 1$ and $\tau = 2$ (among others) in Theorem 5.5. We can get the same result with Corollaries 5.5 and 5.6.

D. Construction of Irregular Images

In the previous section, we showed how to construct basic irregular images. In this section, we show how to construct irregular images. In the construction of irregular images, basic irregular images are used as basic building blocks.

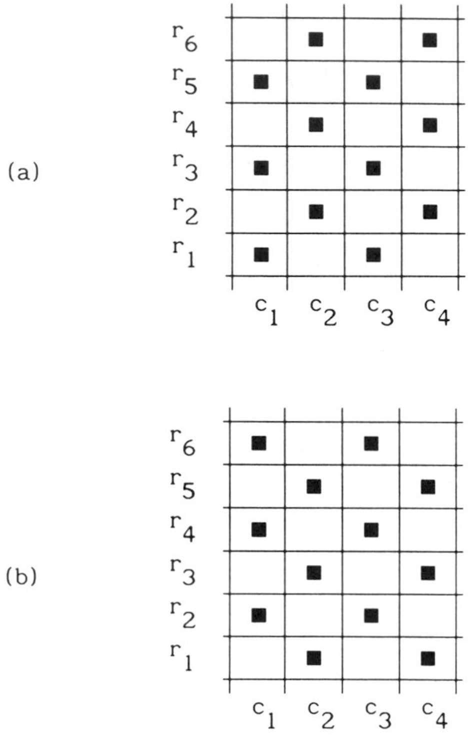

Figure 22. Two different 6 × 4 basic irregular images; (a) and (b) are refered to as images F and G respectively in the text.

Going back to Figure 18, we showed that the image J is not a basic irregular image in Example 5.4. However, it can be thought of as the union of two basic irregular images H and I of Figure 9. Furthermore, it can be represented as the union of images (1) and (6) of Figure 21. Our next two results state that, under less general restrictions, these observations hold.

Proposition 5.5. Consider any two irregular images $\mathcal{A}$ and $\mathcal{B}$. Let the two mappings for $\theta \in \{\mathcal{A}, \mathcal{B}\}$, as defined in Definition 5.4, be denoted as ζ_θ and ξ_θ. Then $\mathcal{A} \cup \mathcal{B}$ is an irregular image if*

(i) $\mathcal{A} \cap \mathcal{B} = \emptyset$

(ii) *either* $\mathfrak{R}_{\mathcal{A}} \cap \mathfrak{R}_{\mathcal{B}} = \mathfrak{C}_{\mathcal{A}} \cap \mathfrak{C}_{\mathcal{B}} = \emptyset$

or $\mathfrak{R}_{\mathcal{A}} \subseteq \mathfrak{R}_{\mathcal{B}}$, $\xi_{\mathcal{B}}|_{\mathfrak{R}_{\mathcal{A}}} = \xi_{\mathcal{A}}$, and $\mathfrak{C}_{\mathcal{A}} \cap \mathfrak{C}_{\mathcal{B}} = \emptyset$

*Consider any function $\omega : \mathbf{A} \to \mathbf{B}$, and let $\mathbf{C} \subseteq \mathbf{A}$. Consider the function

$$\sigma : \mathbf{C} \to \mathbf{B} \quad \text{such that } \forall \alpha \in \mathbf{C}, \sigma(\alpha) = \omega(\alpha)$$

Then σ is said to be a *restriction* of ω, and σ is often denoted as $\omega|_{\mathbf{C}}$.

or $\mathfrak{C}_{\mathcal{A}} \subseteq \mathfrak{C}_{\mathcal{B}}$, $\zeta_{\mathcal{B}}|_{\mathfrak{C}_{\mathcal{A}}} = \zeta_{\mathcal{A}}$, and $\mathfrak{R}_{\mathcal{A}} \cap \mathfrak{R}_{\mathcal{B}} = \emptyset$

or $\mathfrak{R}_{\mathcal{A}} = \mathfrak{R}_{\mathcal{B}}$, $\mathfrak{C}_{\mathcal{A}} = \mathfrak{C}_{\mathcal{B}}$, $\xi_{\mathcal{A}} = \xi_{\mathcal{B}}$, and $\zeta_{\mathcal{A}} = \zeta_{\mathcal{B}}$

Proof. If $\mathcal{A} \cap \mathcal{B} = \emptyset$, $\mathfrak{R}_{\mathcal{A}} \subseteq \mathfrak{R}_{\mathcal{B}}$, $\xi_{\mathcal{B}}|\,\mathfrak{R}_{\mathcal{A}} = \xi_{\mathcal{A}}$, and $\mathfrak{C}_{\mathcal{A}} \cap \mathfrak{C}_{\mathcal{B}} = \emptyset$, then we need to show that $\phi \equiv \mathcal{A} \cup \mathcal{B}$ is an irregular image. We will construct two mappings $\zeta_\phi : \mathfrak{C}_\phi \to \mathfrak{C}_\phi$ and $\xi_\phi : \mathfrak{R}_\phi \to \mathfrak{R}_\phi$ which satisfy the criteria of Definition 5.4. Note that $\mathfrak{C}_\phi = \mathfrak{C}_{\mathcal{A}} \cup \mathfrak{C}_{\mathcal{B}}$ and $\mathfrak{R}_\phi = \mathfrak{R}_{\mathcal{A}} \cup \mathfrak{R}_{\mathcal{B}} = \mathfrak{R}_{\mathcal{B}}$. Now, we let $\xi_\phi = \xi_{\mathcal{B}}$ and construct ζ_ϕ as follows:

Consider any $\langle i,j \rangle \in \mathfrak{C}_\phi \times \mathfrak{R}_\phi$. Two situations can arise:

Case 1: $i \in \mathfrak{C}_{\mathcal{A}}$. Then for all k $\in \mathfrak{R}$, $\langle i,k \rangle \notin \mathcal{B}$. Hence, $\phi^{[i]} = \mathcal{A}^{[i]}$.

Let $\zeta_\phi(i) = \zeta_{\mathcal{A}}(i)$

Case 2: $i \in \mathfrak{C}_{\mathcal{B}}$. Then for all $k \in \mathfrak{R}$, $\langle i,k \rangle \notin \mathcal{A}$. Hence, $\phi^{[i]} = \mathcal{B}^{[i]}$.

Let $\zeta_\phi(i) = \zeta_{\mathcal{B}}(i)$

Now, we claim that $\langle i,j \rangle \in \phi$ *iff* $\langle \zeta_\phi(i), \xi_\phi^{-1}(j) \rangle \in \phi$. This follows directly from the construction itself. Since neither $\zeta_{\mathcal{A}}$ nor $\zeta_{\mathcal{B}}$ is an identity mapping, ζ_ϕ cannot be an identity mapping. Via similar arguments, we conclude that ξ_ϕ is not an identity mapping. Hence ϕ must be irregular.

Similarly, one can show that $\mathcal{A} \cup \mathcal{B}$ is an irregular image in other situations.

Remarks. To illustrate the necessity of condition (i), let $\mathcal{A}$ and $\mathcal{B}$ be defined as the basic irregular images (1) and (2) of Figure 21. Then the image $P \equiv \mathcal{A} \cup \mathcal{B}$ is not irregular (see Figure 23).

We will now investigate the converse of Proposition 5.5.

Proposition 5.6. Given any nonbasic irregular image $\mathcal{A}$, one can find two irregular images $\mathcal{B}$ and $\mathcal{C}$ such that $\mathcal{B} \cup \mathcal{C} = \mathcal{A}$ as specified in Proposition 5.5.

Proof. Let the mappings, as specified in Definition 5.4, for columns and rows of $\mathcal{A}$ be $\zeta_{\mathcal{A}}$ and $\xi_{\mathcal{A}}$, respectively. Since $\mathcal{A}$ is a nonbasic irregular image, either one of the following three situations must be true:

(i) $\exists\ \alpha \in \mathfrak{R}_{\mathcal{A}} \ni \xi^k(\alpha) = \alpha$ and $k \not\equiv 0 \pmod{n}$.

(ii) $\exists\ \alpha \in \mathfrak{C}_{\mathcal{A}} \ni \zeta^k(\alpha) = \alpha$ and $k \not\equiv 0 \pmod{m}$.

(iii) There is a basic irregular image $\mathcal{C} \in \mathcal{A}$ with

$$\mathfrak{R}_{\mathcal{A}} = \mathfrak{R}_{\mathcal{C}} \qquad \text{and} \qquad \mathfrak{C}_{\mathcal{A}} = \mathfrak{C}_{\mathcal{C}}$$

In situation (i), Let $\alpha \in \mathfrak{R}_{\mathcal{A}}$ be such that $\xi^k(\alpha) = \alpha$ and $k \not\equiv 0 \pmod{n}$. We define a set

$$\mathfrak{R}_{\mathcal{C}} = \left\{\xi^k(\alpha) : k \in \mathfrak{N}\right\} \qquad \text{and} \qquad \mathfrak{R}_{\mathcal{B}} = \mathfrak{R}_{\mathcal{A}} \setminus \mathfrak{R}_{\mathcal{C}}$$

Let $\mathfrak{C}_{\mathcal{A}} = \mathfrak{C}_{\mathcal{C}} = \mathfrak{C}_{\mathcal{B}}$, and define two images $\mathcal{C}$ and $\mathcal{B}$ as follows:

$$\mathcal{C} = \mathcal{A} \cap \mathfrak{R}_{\mathcal{C}} \times \mathfrak{C}_{\mathcal{C}} \qquad \text{and} \qquad \mathcal{B} = \mathcal{A} \cap \mathfrak{R}_{\mathcal{B}} \times \mathfrak{C}_{\mathcal{B}}$$

Clearly, $\mathcal{C}$ and $\mathcal{B}$ satisfy the requirements of Proposition 5.5. Situation (ii) can be handled similarly.

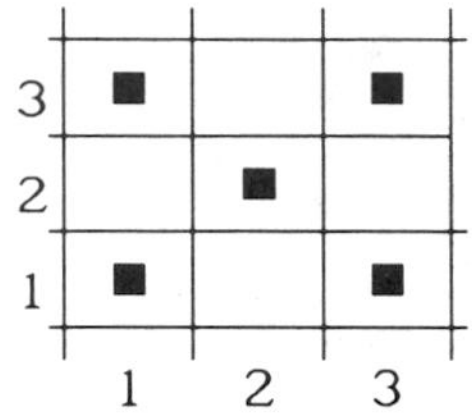

Figure 23. Example of a nonirregular image.

In situation (iii), consider any pixel $\langle i,j \rangle \in \mathcal{A}$ and define the images $\mathcal{C}$ and $\mathcal{B}$ as

$$\mathcal{C} = \left\{ \langle \zeta^k(i), \xi^{-k}(j) \rangle : k \in \mathfrak{N} \right\} \qquad \text{and} \qquad \mathcal{B} = \mathcal{A} \setminus \mathcal{C}$$

Clearly, $\mathcal{C}$ and $\mathcal{B}$ satisfy the requirements of Proposition 5.5.

As an immediate consequence of Proposition 5.6 we obtain an important characterization of irregular images.

Theorem 5.6. Any irregular image can be expressed as a finite disjoint union of some basic irregular images.

Example 5.7. We illustrate the constructions specified in Proposition 5.5 by a slightly complicated example. All images referred to in this example are shown in Figure 24. The image F is clearly an irregular image with

$$n = 4, \qquad m = 3, \qquad \zeta_F : \begin{pmatrix} 1 & 2 & 3 \\ 3 & 2 & 1 \end{pmatrix}, \qquad \xi_F : \begin{pmatrix} 1 & 2 & 3 & 4 \\ 4 & 3 & 2 & 1 \end{pmatrix}$$

The permutations ζ_F and ξ_F contain two cycles each. Let us define the following quantities:

$$\mathfrak{C}_1 = \{1,3\}, \qquad \mathfrak{C}_2 = \{2\}, \qquad \mathfrak{R}_1 = \{1,4\}, \qquad \mathfrak{R}_2 = \{2,3\}$$

$$\zeta_1 : \begin{pmatrix} 1 & 3 \\ 3 & 1 \end{pmatrix}, \qquad \zeta_2 : \begin{pmatrix} 2 \\ 2 \end{pmatrix}, \qquad \xi_1 : \begin{pmatrix} 1 & 4 \\ 4 & 1 \end{pmatrix}, \qquad \xi_2 : \begin{pmatrix} 2 & 3 \\ 3 & 2 \end{pmatrix}$$

We can construct six basic irregular images with these row-column pairs. Of course, one has to employ the appropriate ζ_1/ζ_2 and ξ_1/ξ_2 mappings so as to adhere to the construction specified in Proposition 5.5. We show three of these basic irregular images: A, B, and C. Note that $\zeta_A = \zeta_2$, $\zeta_B = \zeta_C = \zeta_1$, $\xi_A = \xi_C = \xi_2$, and $\xi_B = \xi_1$. The images B and C satisfy the constraint of Proposition 5.5 and can be combined to obtain the image P. P is an irregular image with $\zeta_P = \zeta_1$ and $\xi_P = \xi_F$. Now, note that P and A satisfy the constraints of Proposition 5.5, implying that $P \cup A \equiv F$ is an irregular image.

Alternatively, one can decompose F into P and A and then P into B and C. We thus obtain $F = A \cup B \cup C$ with A as a 1×2 basic irregular image and both B

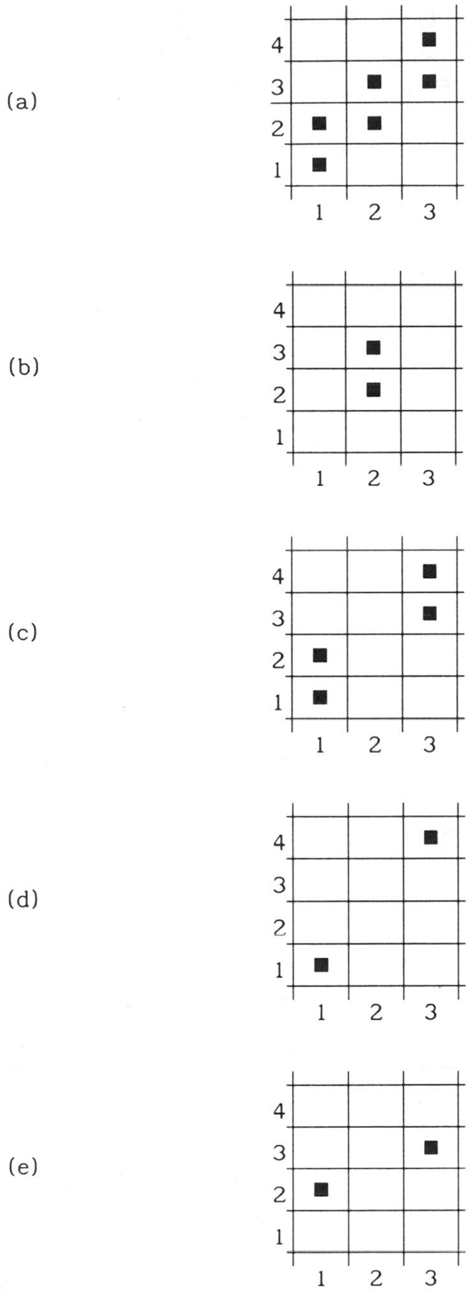

Figure 24. Construction of the irregular image F (shown in (a)). The images A, P, B, and C are shown in (b), (c), (d), and (e) respectively.

and C as 2×2 basic irregular images. For the first decomposition, we isolated the cycle ζ_2 in F and defined $\mathfrak{C}_A = \{2\}$, $\mathfrak{C}_P = \{1,3\}$. Also, the rows of A and P were defined so as to satisfy $A \cup P = F$. For the second decomposition, we isolated the cycle ξ_1 in P and defined two images B and C with $\mathfrak{R}_B = \{1,4\}$, $\mathfrak{R}_C = \{2,3\}$. The columns of B and C were defined so as to maintain $B \cup C = P$. In Corollary 5.6 we show that this observation is true in general.

It is easy to verify that if an irregular image $\mathcal{A}$ is such that the ζ and ξ mappings have exactly one cycle, then $\mathcal{A}$ can be represented (see Proposition 5.5) as a union of at most gcd(n,m) basic irregular images (see Corollary 5.2). Note that n and m are numbers of rows and columns in $\mathcal{A}$, respectively. We now generalize this observation.

Corollary 5.7. Consider any irregular image $\mathcal{A}$ such that the ζ and ξ mappings have $\alpha \geq 1$ and $\beta \geq 1$ cycles, respectively. Let the length of cycles in ζ be $\{\kappa_i : 1 \leq i \leq \alpha\}$ and in ξ be $\{\tau_i : 1 \leq i \leq \beta\}$. Then $\mathcal{A}$ can be expressed as a union of at most

$$\sum_{i=1}^{\alpha} \sum_{j=1}^{\beta} \gcd[\kappa_i,\tau_j]$$

disjoint basic irregular images.

E. Geometric and Closure Properties of Irregular Images

At this juncture, we wish to state and prove certain geometric and closure properties of the irregular images. We first deal with the closure properties, as they are easier to state:

Theorem 5.7. The following four conditions are equivalent:

(i) $\mathcal{A}$ is an irregular image.

(ii) $\hat{\mathcal{A}}$ is an irregular image.

(iii) $\mathbf{T}(\mathcal{A};v)$ is an irregular images for all $v \in \mathfrak{R}^2$.

(iv) $\mathbf{N}^j \bigcirc \mathbf{R}^i(\mathcal{A})$ is an irregular image for all integers i, j.

Proof. We will prove the result only for $\mathcal{B} \equiv \mathbf{N}(\mathcal{A})$. Others can be proved in a similar manner. First note that

$$\langle i,j \rangle \in \mathcal{A} \text{ iff } \langle -j,i \rangle \in \mathcal{B} \tag{11.29}$$

Clearly, $\mathfrak{C}_{\mathcal{A}} = \mathfrak{R}_{\mathcal{B}}$ and $\mathfrak{R}_{\mathcal{A}} = \{-\alpha : \alpha \in \mathfrak{C}_{\mathcal{B}}\}$. If $\mathcal{A}$ is an irregular image, then there exist bijective mappings $\omega : \mathfrak{R}_{\mathcal{A}} \to \mathfrak{R}_{\mathcal{A}}$ and $\sigma : \mathfrak{C}_{\mathcal{A}} \to \mathfrak{C}_{\mathcal{A}}$ such that

$$\langle i,j \rangle \in \mathcal{A} \text{ iff } \langle \sigma(i),\omega^{-1}(j) \rangle \in \mathcal{A} \tag{11.30}$$

From Eqs. (11.29) and (11.30), we obtain

$$\langle i,j \rangle \in \mathcal{A} \text{ iff } \langle \sigma(i),\omega^{-1}(j) \rangle \in \mathcal{A} \textit{ iff } \langle -\omega^{-1}(j),\sigma(i) \rangle \in \mathcal{B}$$

Therefore,

$$\langle -j,i\rangle \in \mathcal{B} \text{ iff } \langle -\omega^{-1}(j),\sigma(i)\rangle \in \mathcal{B}$$

that is,

$$\langle \alpha,\beta\rangle \in \mathcal{B} \text{ iff } \langle \omega^{-1}(\alpha),\sigma(\beta)\rangle \in \mathcal{B}$$

Hence, the mappings ω^{-1} and σ^{-1} satisfy the constraints of Definition 5.4 for columns and rows of $\mathcal{B}$, respectively; that is, $\mathcal{B}$ is an irregular image.

For $\mathcal{B} = \mathbf{N}^{\alpha} \bigcirc \mathbf{R}^{\beta}(\mathcal{A})$, we have

$$\alpha + \beta \text{ is odd:} \quad \langle i,j\rangle \in \mathcal{A} \text{ iff } \langle \kappa j,\tau i\rangle \in \mathcal{B}$$
$$\alpha + \beta \text{ is even:} \quad \langle i,j\rangle \in \mathcal{A} \text{ iff } \langle \kappa i,\tau j\rangle \in \mathcal{B}$$

where $\kappa, \tau \in \{-1,1\}$. One can show that $\mathcal{B}$ is an irregular image. In particular, if $\alpha + \beta$ is odd, the mappings for rows and columns of $\mathcal{B}$ are ω^{-1} and σ^{-1}, respectively; else there mappings are σ and ω.

The results for other operations trivially follow.

Next, we turn our attention to geometric properties of irregular images. Before we can analyze this situation, we must introduce some new notation.

Definition 5.7. Let $\varphi : \mathbf{IMAGE} \to \mathfrak{R}^4$ such that for any image $\mathcal{A}$, if we let

$$\mathfrak{A} = \{i \in \mathfrak{R} : \exists j \in \mathfrak{R} \ni \langle i,j\rangle \in \mathcal{A}\}$$
$$\mathfrak{B} = \{j \in \mathfrak{R} : \exists i \in \mathfrak{R} \ni \langle i,j\rangle \in \mathcal{A}\}$$

then $\varphi(\mathcal{A}) = \langle \min \mathfrak{A}, \max \mathfrak{A}, \min \mathfrak{B}, \max \mathfrak{B}\rangle \equiv \langle \mathcal{A}^{-\infty}, \mathcal{A}^{\infty}, \mathcal{A}_{-\infty}, \mathcal{A}_{\infty}\rangle$. We will let $\max \emptyset = 0$ and $\min \emptyset = 0$. For any image $\mathcal{A}$, $\varphi(\hat{\mathcal{A}}) - \varphi(\mathcal{A}) = \langle -\mathfrak{w},\mathfrak{w},-\mathfrak{h},\mathfrak{h}\rangle$, where $\mathfrak{w}$ and $\mathfrak{h}$ are window coefficients introduced in Definition 5.1.

Definition 5.8. Consider any image $\mathcal{A}$. For any $v \in \mathfrak{R}_{\mathcal{A}}$ and $\kappa \in \mathfrak{C}_{\mathcal{A}}$, we define the one-dimensional images $-\mathcal{A}_{[v]}$ and $-\mathcal{A}^{[\kappa]}$ as follows:

$$-\mathcal{A}_{[v]} = \mathbf{T}[\mathbf{N}^2(\mathcal{A}); \langle \mathcal{A}^{-\infty} + \mathcal{A}^{\infty}, 0\rangle]_{[-v]}$$
$$-\mathcal{A}^{[\kappa]} = \mathbf{T}[\mathbf{N}^2(\mathcal{A}); \langle 0, \mathcal{A}_{-\infty} + \mathcal{A}_{\infty}\rangle]^{[-\kappa]}$$

Example 5.8. For image F of Figure 2, we have $F_{[1]} = -F_{[2]}$, $F_{[2]} = F_{[1]}$, $F_{[3]} = -F_{[3]}$, $F^{[1]} = -F^{[2]}$, $F^{[2]} = -F^{[1]}$, and $F^{[3]} = -F^{[3]}$.

It is of interest to note that the "$-$" operator on images is, like the $\mathbf{N}^2$ operator, involutory. This was evident in the images of Figure 2.

Proposition 5.7. Consider any image $\mathcal{A}$. For any $v \in \mathfrak{R}_{\mathcal{A}}$ and $\kappa \in \mathfrak{C}_{\mathcal{A}}$, we have

$$-[-\mathcal{A}_{[v]}] = \mathcal{A}_{[v]} \quad \text{and} \quad -[-\mathcal{A}^{[\kappa]}] = \mathcal{A}^{[\kappa]}$$

Definition 5.9. We will call a two-dimensional image $\mathcal{A}$ *reflective* iff there exist two bijective mappings $\sigma : \mathfrak{C}_{\mathcal{A}} \to \mathfrak{C}_{\mathcal{A}}$ and $\omega : \mathfrak{R}_{\mathcal{A}} \to \mathfrak{R}_{\mathcal{A}}$ such that for every row α and column β, we have

$$\mathcal{A}_{[\alpha]} = -\mathcal{A}_{[\omega(\alpha)]} \quad \text{and} \quad \mathcal{A}^{[\beta]} = -\mathcal{A}^{[\sigma(\beta)]}$$

Example 5.9. An empty image trivially satisfies this property. Other examples of such trivial images are filled rectangles $\mathcal{A}$. All the images of Figure 2 are reflective. The following result states that this observation is true in general.

Theorem 5.8. Consider a basic irregular image $\mathcal{A}$ such that $\mathfrak{R}_{\mathcal{A}}$ and $\mathfrak{C}_{\mathcal{A}}$ are contiguous sets.* Then $\mathcal{A}$ is reflective.

Proof. Consider the terminology of Theorem 5.5 with the assumption that $c_x = x$ and $r_x = x$. We know that

$$\langle v,\tau\rangle \in \mathcal{A} \text{ iff } \langle\alpha,\beta\rangle \in \mathcal{A} \text{ with } \alpha + \beta - v - \tau \equiv 0 \pmod{\kappa}$$

Let $\langle v,\tau\rangle \in \mathcal{A}$. We wish to show that for every column i of $\mathcal{A}$, there exists another column I of $\mathcal{A}$ such that $\mathcal{A}^{[i]} = -\mathcal{A}^{[I]}$.

Consider any $j \in \mathfrak{R}_{\mathcal{A}}$. If $\langle i,j\rangle \in \mathcal{A}$ then

$$\exists p \in \mathfrak{N} \qquad \text{such that } i + j - v - \tau = p\kappa \tag{11.31}$$

If column I has the stated property, then we must have $\langle I, n + 1 - j\rangle \in \mathcal{A}$, that is

$$\exists q \in \mathfrak{N} \qquad \text{such that } I + n + 1 - j - v - \tau = q\kappa \tag{11.32}$$

Eliminating the variable j from Eqs. (11.31) and (11.32), we obtain the condition as

$$\exists r \in \mathfrak{N} \qquad \text{such that } r\kappa - I = i + n - 2(v + \tau) + 1 \tag{11.33}$$

Equation (11.33) has two variables r and I, the coefficients of which are κ and -1, respectively. Since $\gcd(\kappa, -1) = 1$, from Bezout's identity we know that a solution to (11.33) always exists. Hence the result.

Similarly, one can show that for every row j of $\mathcal{A}$, there exists a row J of $\mathcal{A}$ such that $\mathcal{A}_{[j]} = -\mathcal{A}_{[J]}$ and J satisfies the equation

$$J + j + m - 2(v + \tau) + 1 \equiv 0 \pmod{\kappa}$$

Hence the result.

Remarks. The converse of Theorem 4.8 is not true in general: the image P of Figure 23 is reflective but is not even irregular. The same is true for an empty image.

*That is, $\mathcal{r}_i = \mathcal{r}_{i-1} + 1$ and $\mathcal{c}_i = \mathcal{c}_{i-1} + 1$. Since we are not interested in the effects of translation, we will let $\mathcal{c}_1 = \mathcal{r}_1 = 1$.

F. How Difficult Is It to Recognize Irregular Images?

In this section we wish to show that even the joint central moments [14] fail to distinguish between the basic irregular images. We denote the joint central moments as μ_{pq} where p and q are positive integers.

From Corollary 5.4 we know that the equivalence class of $N \times N$ basic irregular images contains $N!$ different images. We consider $N \times N$ (with $N = 2, 3, 4, 5$) basic irregular images whose rows and columns of supports are continuous sets (see Theorem 5.8). We first compute the moment μ_{11} for all these images. The numbers of images whose moments μ_{11} are same are listed in Table 1.

It is interesting to note that all of the equivalent $N \times N$ images which have the same moment $\mu_{11} \neq 0$ could be distinguished by μ_{12}. However, for images having $\mu_{11} = 0$ we must consider other higher moments.

The optimism is quickly quenched by the images of Figures 25 and 26. For these four images, we have the following equalities:

$$\textit{either } p \textit{ or } q \textit{ is odd:} \quad \mu_{pq}(F) = \mu_{pq}(G) = \mu_{pq}(H) = \mu_{pq}(I) = 0$$

$$p \textit{ and } q \textit{ both are even:} \quad \mu_{pq}(F) = \mu_{pq}(G),\ \mu_{pq}(H) = \mu_{pq}(I)$$

These observations also hold for the two irregular images J and K in Figure 27. These examples demonstrate the limitation of joint central moments for invariant and unique characterization of digitized images.

G. Rotation Invariance

Until now, we have discussed the translation invariance and uniqueness of $\mathbf{H}^+$ and $\mathbf{V}^+$ functions. Now, we analyze of the effects of $\mathbf{N}$ and $\mathbf{R}$ operations as defined in Definition 3.3.

Theorem 5.9. Consider any image $\mathcal{A}$, and let $\mathcal{B} = \mathbf{N}^i \bigcirc \mathbf{R}^j(\mathcal{A})$ with $j \in \{0,1\}$ and $i \in \{0,1,2,3\}$. Then $\mathbf{H}^+(\mathcal{A},\cdot) = \theta_1(\mathcal{B},\cdot)$ and $\mathbf{V}^+(\mathcal{A},\cdot) = \theta_2(\mathcal{B},\cdot)$ where $\theta_1,\theta_2 \in \{\mathbf{H}^+,\mathbf{H}^-,\mathbf{V}^+,\mathbf{V}^-\}$ are shown in Table 2.

Proof. We will prove the result only for $\mathcal{B} = \mathbf{N} \bigcirc \mathbf{R}(\mathcal{A})$. Others can be proved in a similar manner. First note that $\mathbf{N} \bigcirc \mathbf{R}(\mathcal{A}) = \{\langle x,-y\rangle : \langle x,y\rangle \in \mathcal{A}\}$. Now

$$\begin{aligned}\langle x,y\rangle \in \mathfrak{A}^+(\mathcal{A},n) &\Leftrightarrow \langle x,y\rangle \in \mathcal{A} \text{ and } \langle x+n, y\rangle \in \hat{\mathcal{A}}\\ &\Leftrightarrow \langle x,-y\rangle \in \mathcal{B} \text{ and } \langle x+n, -y\rangle \in \hat{\mathcal{A}}\\ &\Leftrightarrow \langle x,-y\rangle \in \mathfrak{A}^+(\mathcal{B},n)\end{aligned}$$

$$\begin{aligned}\langle x,y\rangle \in \mathfrak{B}^+(\mathcal{A},n) &\Leftrightarrow \langle x,y\rangle \in \mathcal{A} \text{ and } \langle x, y+n\rangle \in \hat{\mathcal{A}}\\ &\Leftrightarrow \langle x,-y\rangle \in \mathcal{B} \text{ and } \langle x, -y-n\rangle \in \hat{\mathcal{A}}\\ &\Leftrightarrow \langle x,-y\rangle \in \mathfrak{B}^-(\mathcal{B},n)\end{aligned}$$

Table 1. Number of $N \times N$ Basic Irregular Images that Have the Same Moment μ_{11}

μ_{11}	−10	−9	−8	−7	−6	−5	−4	−3	−2	−1	0	1	2	3	4	5	6	7	8	9	10
2										1		1									
3									1	2		2	1								
4						1	3	1	4	2	2	2	4	1	3	1					
5	1	4	3	6	7	6	4	10	6	10	6	10	6	10	4	6	7	6	3	4	1

The rows represent the values of N, and the columns represent the values of μ_{11}. The square in row n and column m denotes the number of $n \times n$ basic irregular images that have $\mu_{11} = m$.

(a)

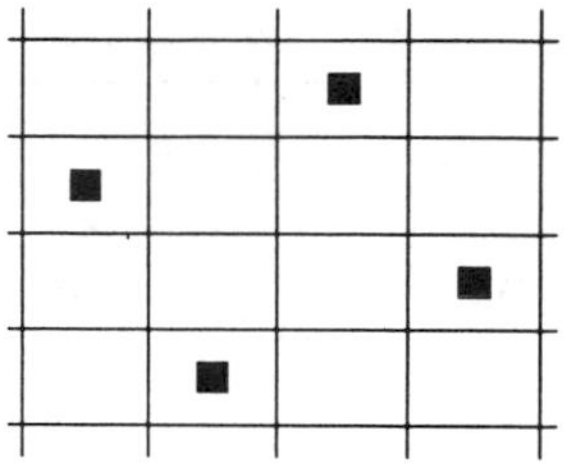

(b)

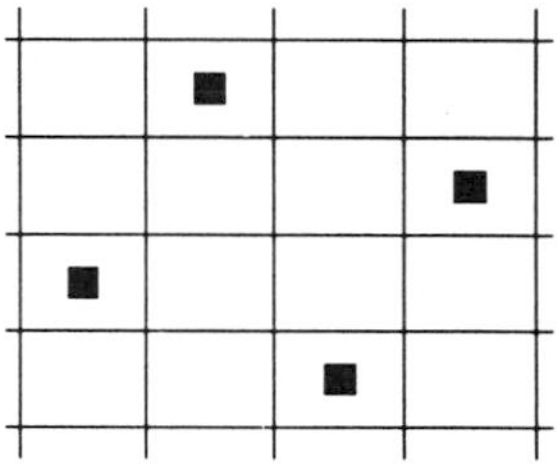

Figure 25. The two 4 × 4 basic irregular images that have same central moments μ_{pq} for all p and q; (a) and (b) are called images F and G in the text.

(a)

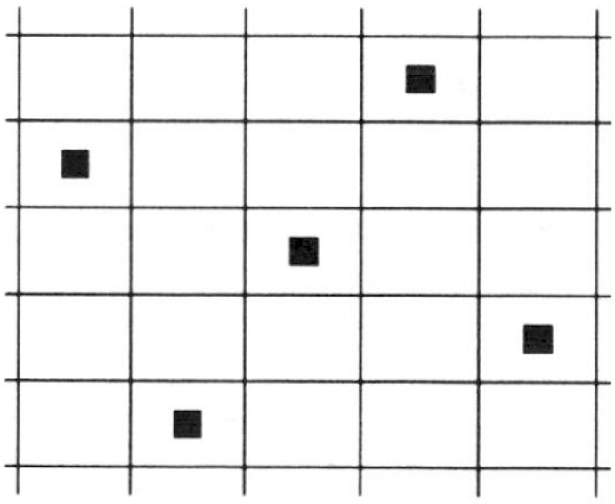

(b)

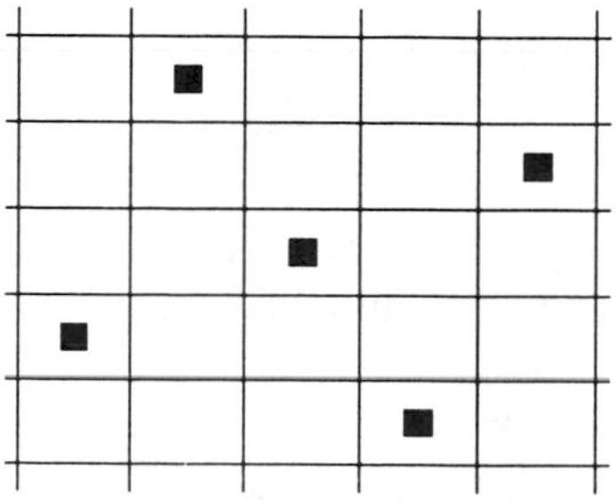

Figure 26. The two 5 × 5 basic irregular images that have same central moments μ_{pq} for all p and q; (a) and (b) are called images H and I in the text.

Table 2. Effect of **NINETY** and **FLIP** on the $\mathbf{H}^+$, $\mathbf{H}^-$, $\mathbf{V}^+$, and $\mathbf{V}^-$ Functions

	⟨0,0⟩	⟨1,0⟩	⟨2,0⟩	⟨3,0⟩	⟨0,1⟩	⟨1,1⟩	⟨2,1⟩	⟨3,1⟩
θ_1	$\mathbf{H}^+$	$\mathbf{V}^-$	$\mathbf{H}^-$	$\mathbf{V}^+$	$\mathbf{V}^-$	$\mathbf{H}^+$	$\mathbf{V}^+$	$\mathbf{H}^-$
θ_2	$\mathbf{V}^+$	$\mathbf{H}^+$	$\mathbf{V}^-$	$\mathbf{H}^-$	$\mathbf{H}^-$	$\mathbf{V}^-$	$\mathbf{H}^+$	$\mathbf{V}^+$

(a)

(b)

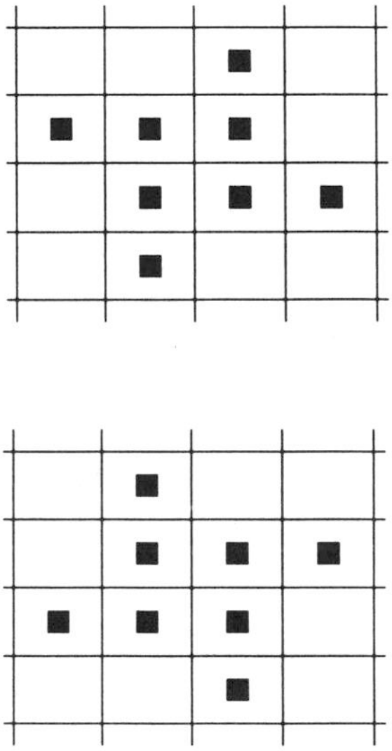

Figure 27. The two irregular images that have same central moments μ_{pq} for all p and q; (a) and (b) are called images J and K in the text.

In order to achieve rotational invariance, we must do something extra. The process will be based on the notion of the *theory of group averaging*, which is discussed later in Section VI. For the time being, let us simply note that the function

$$\mathbf{S}(\mathscr{A},n) = \mathbf{H}^+(\mathscr{A},n) + \mathbf{H}^-(\mathscr{A},n) + \mathbf{V}^+(\mathscr{A},n) + \mathbf{V}^-(\mathscr{A},n)$$

is rotationally invariant (see Theorem 5.9). The performance of the function **S** cannot be worse than that of the functions $\mathbf{H}^+$, $\mathbf{H}^-$, $\mathbf{V}^+$, and $\mathbf{V}^-$ combined. In fact, it is the same: two objects can be distinguished by the $\{\mathbf{H}^+,\mathbf{H}^-,\mathbf{V}^+,\mathbf{V}^-\}$ functions if and only if they can be distinguished by the **S** function.

H. Magnification Invariance

Now we discuss the effect of magnification. First, we borrow the concept of discrete magnification from [5]; see Eqs. 11.8 and 11.9. Assume that the window coefficients $w_1 = w_2 = h_1 = h_2 = 1$. Since

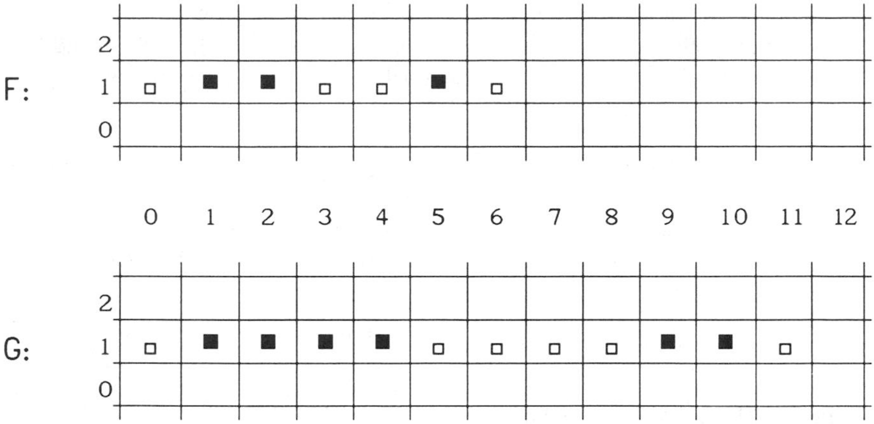

Figure 28. Image G has been obtained from F by performing a magnification in the horizontal direction by a factor of 2.

$$\mathcal{B} = \mathbf{MAG}_x(\mathcal{A},\Delta) \Rightarrow \hat{\mathcal{B}} \subseteq \mathbf{MAG}_x(\hat{\mathcal{A}},\Delta) \tag{11.34}$$

holds, one can easily show that

$$\begin{aligned} &\mathbf{H}^+\left(\mathbf{MAG}_x(\mathcal{A},\Delta),\ \Delta n\right) \leq \Delta\mathbf{H}^+(\mathcal{A},n) \\ &\mathbf{V}^+\left(\mathbf{MAG}_x(\mathcal{A},\Delta),\Delta n\right) \leq \Delta\mathbf{V}^+(\mathcal{A},n) \end{aligned} \tag{11.35}$$

Example 5.11. Consider the two images F and G of Figure 28. Image G is the magnified version of image F: $G = \mathbf{MAG}_x(F,2)$. We obtain

$$\begin{aligned} &\mathbf{H}^+(G,2) = 3 \leq 4 = 2 \cdot \mathbf{H}^+(F,1) \\ &\mathbf{H}^+(G,4) = 4 = 2 \cdot \mathbf{H}^+(F,2) \\ &\mathbf{H}^+(G,6) = 2 = 2 \cdot \mathbf{H}^+(F,3) \\ &\mathbf{H}^+(G,8) = 2 = 2 \cdot \mathbf{H}^+(F,4) \\ &\mathbf{H}^+(G,10) = 1 \leq 2 = 2 \cdot \mathbf{H}^+(F,5) \\ &\mathbf{H}^+(G,2n) = 0 = 2 \cdot \mathbf{H}^+(F,n) \qquad \text{for } n \geq 6 \end{aligned}$$

Let denote the activated pixel of F in position 5 as α. Also denote the actiated pixel of G in position 10 (which is the pixel obtained from α via the magnification) as β. Note that α contributes 1 to $\mathbf{H}^+(F,1)$. But β does not contribute to the value of $\mathbf{H}^+(G,2)$. This is why $\mathbf{H}^+(G,2) < 2{\cdot}\mathbf{H}^+(F,1)$. However, if $\hat{G}$ is also defined in position 12, then

$$\mathbf{H}^+(G,2) = 4 = 4 = 2{\cdot}\mathbf{H}^+(F,1)$$

Therefore, we should change the window $\mathfrak{B}_{\mathcal{A}}$ defined in Definition 5.1 in order to achieve magnification invariance.

Now, in order to achieve magnification invariance (that is, equality in both the above expressions of (11.34)), we must have the window coefficients w and h as the width and height of image $\mathcal{A}$, respectively. With this bigger window $\mathfrak{B}_{\mathcal{A}}$, we have

$$\mathcal{B} = \mathbf{MAG}_x(\mathcal{A}\,\Delta) \text{ iff } \hat{\mathcal{B}} = \mathbf{MAG}_x(\hat{\mathcal{A}},\Delta) \tag{11.36}$$

This is based on the fact that the width of $\hat{\mathcal{B}}$ is Δ times the width of $\hat{\mathcal{A}}$. Now, one can show that the $\mathbf{H}^+$ and $\mathbf{V}^+$ functions display the characteristics of a multivariate homogeneous (of order 1) function.

Theorem 5.10. Consider any $\mathcal{A}$ in **IMAGE** and $\Delta > 1$. We have for all $n \in \mathfrak{N}^+$

$$\begin{aligned} \mathbf{H}^+\left(\mathbf{MAG}_x(\mathcal{A},\Delta),\Delta n\right) &= \Delta\mathbf{H}^+(\mathcal{A},n) \\ &= \Delta\mathbf{V}^+(\mathcal{A},n) \quad \text{and} \quad \mathbf{V}^+\left(\mathbf{MAG}_x(\mathcal{A},\Delta),n\right) \\ \mathbf{V}^+\left(\mathbf{MAG}_y(\mathcal{A},\Delta),\Delta n\right) &= \Delta\mathbf{V}^+(\mathcal{A},n) \\ &= \Delta\mathbf{H}^+(\mathcal{A},n) \quad \text{and} \quad \mathbf{H}^+\left(\mathbf{MAG}_y(\mathcal{A},\Delta),n\right) \end{aligned}$$

Exactly same results hold for $\mathbf{H}^-$ and $\mathbf{V}^-$ functions.

Proof. Assume that $\mathcal{B} = \mathbf{MAG}_x(\mathcal{A},\Delta)$. We shall show that

$$\begin{aligned} \text{card } \mathbf{H}^+(\mathcal{B},\Delta n) &= \Delta \text{ card } \mathbf{H}^+(\mathcal{A},n) \quad \text{and} \quad \mathbf{V}^+(\mathcal{B},n) \\ &= \Delta \text{ card } \mathbf{V}^+(\mathcal{A},n) \end{aligned}$$

Note that, for the $\mathbf{H}^+$ function, it suffices to show the result with respect to $\mathcal{B}_{[v]} = \varphi$ and $\mathcal{A}_{[v]} = \mathcal{C}$ for any $v \in \mathfrak{R}_{\mathcal{A}}$. Consider any $x \in \mathcal{C}$, and let $\gamma_x = \{\alpha : x = \lfloor \alpha/\Delta \rfloor\}$. Clearly, card $\gamma_x = \Delta$ and $\gamma_x \subseteq \varphi$.

Similarly, for any $y \in \hat{\mathcal{C}}$, let $\delta_y = \{\beta : y = \lfloor \beta/\Delta \rfloor\}$. Then

$$\text{card } \delta_y = \Delta \quad \text{and} \quad \delta_y \subseteq \tilde{\varphi}$$

If $x \in \mathcal{C}$ and $y \in \hat{\mathcal{C}}$ are such that $y - x = n$, then there exist $\gamma \in \gamma_x$ and $\tau \in \delta_y$ such that $\tau - \gamma = \Delta n$. In particular, let $\gamma = \Delta x + \kappa$ and $\tau = \Delta y + \kappa$ with

$$\kappa \in \{0,1,2, \ldots ,\Delta - 1\}$$

Hence, card $\mathbf{H}^+(\varphi,\Delta n) \geq \Delta$ card $\mathbf{H}^+(\mathcal{C},n)$. Similarly, one can show that

$$\text{card } \mathbf{H}^+(\varphi,\Delta n) \leq \Delta \text{ card } \mathbf{H}^+(\mathcal{C},n)$$

To complete the proof, note that each column in $\mathcal{A}$ has been duplicated Δ times to obtain $\mathcal{B}$. Hence, card $\mathbf{V}^+(\mathcal{B},n) = \Delta$ card $\mathbf{V}^+(\mathcal{A},n)$.

Finally, note that all the results listed, except for Corollary 5.1, hold in the framework. The modifications needed in Corollary 5.1 are fairly straightforward.

VI. CHARACTERIZATION OF FINITE TWO-DIMENSIONAL DISCRETE SETS

In Section V, it has been shown that if $\mathbf{H}^+(\mathcal{A},\cdot) = \mathbf{H}^+(\mathcal{B},\cdot)$ and $\mathbf{V}^+(\mathcal{A},\cdot) = \mathbf{V}^+(\mathcal{B},\cdot)$, then $\mathcal{A} = \mathcal{B}$ (ignoring translation) or $\mathcal{A}$ and $\mathcal{B}$ are equivalent irregular images (see Theorem 5.4). After careful inspection of an equivalent irregular image, one can realize that the nature of the equivalence classes is due to the independent computation of horizontal and vertical distance measures.

It has been shown that there exists a bijective mapping between rows and columns of equivalent irregular images $\mathcal{A}$ and $\mathcal{B}$ (see Theorems 5.2 and 5.3). Therefore, a uniqueness algorithm must be able to identify the mappings between rows or columns. The equivalent irregular images can be obtained from each other by permuting rows and columns. Hence, if the algorithm can distinguish the permutation of rows and columns, then the characterization of irregular images can be achieved.

To complete the theoretical analysis of the algorithms which are based on the distance measure between activated and nonactivated pixels, we will consider an algorithm which is more space and time intensive. In [5] and [7] it has been noted that the **ANNULAR** function which has $O(H \cdot W)$ space complexity provides better characterization. It is interesting to note that a similar function, which we call $\mathbf{C}_2$, provides complete characterization of finite two-dimensional discrete sets.

A. Characterization Without Invariance Property

Now, define a function $\mathbf{C}_2 : \mathbf{IMAGE}^2 \times \mathfrak{N}^* \rightarrow \mathfrak{N}^+ \cup \{0\}$ such that for any $\mathcal{A} \in \mathbf{IMAGE}$ and $\langle n,m \rangle \in \mathfrak{N}^*$,

$$\mathbf{C}_2(\mathcal{A},\hat{\mathcal{A}};n,m) = \text{card}\{\langle \alpha,\beta \rangle \in \mathcal{A} : \langle \alpha + n, \beta + m \rangle \in \hat{\mathcal{A}}\}$$

Note that the set $\mathfrak{N}^*$ was defined in Section III. If a fixed pseudocomplement $\hat{\mathcal{A}}$ is applied to $\mathbf{C}_2$, $\mathbf{H}^+$, and $\mathbf{V}^+$ functions, it is easy to check that

$$\mathbf{C}_2(\mathcal{A},\hat{\mathcal{A}};n,0) = \mathbf{H}^+(\mathcal{A},n) \qquad \text{and} \qquad \mathbf{C}_2(\mathcal{A},\hat{\mathcal{A}};0,m) = \mathbf{V}^+(\mathcal{A},m)$$

Hence the characterization performance of the $\mathbf{C}_2$ function cannot be worse than that of the $\{\mathbf{H}^+,\mathbf{V}^+\}$ pair. The reason for including the pseudocomplement in the definition of the $\mathbf{C}_2$ function will become apparent when we deal with the higher-dimensional images in the next section (see page 412).

The following theorem states that the unique characterization is achieved by the $\mathbf{C}_2$ function.

Theorem 6.1. $\mathbf{C}_2(\mathcal{A},\hat{\mathcal{A}};\cdot) = \mathbf{C}_2(\mathcal{B},\hat{\mathcal{B}};\cdot)$ *iff* the following three conditions are met

(i) $\mathcal{A} = \mathcal{B}$ ignoring translation.

(ii) max $\mathfrak{C}_{\hat{\mathcal{A}}}$ − max $\mathfrak{C}_{\mathcal{A}}$ = max $\mathfrak{C}_{\hat{\mathcal{B}}}$ − max $\mathfrak{C}_{\mathcal{B}}$

(iii) max $\mathfrak{R}_{\hat{\mathcal{A}}}$ − max $\mathfrak{R}_{\mathcal{A}}$ = max $\mathfrak{R}_{\hat{\mathcal{B}}}$ − max $\mathfrak{R}_{\mathcal{B}}$

Proof. (*Sufficiency*) Directly follows from the definition.
(*Necessity*) Since

$$\mathbf{C}_2(\mathcal{A},\hat{\mathcal{A}};n,0) = \mathbf{C}_2(\mathcal{B},\hat{\mathcal{B}};n,0) \quad \text{and} \quad \mathbf{C}_2(\mathcal{A},\hat{\mathcal{A}};0,m) = \mathbf{C}_2(\mathcal{B},\hat{\mathcal{B}};0,m)$$

we conclude (see Corollary 5.1 and Lemma 5.1):

(i) width $\mathcal{A}$ = width $\mathcal{B}$ and max $\mathfrak{C}_{\hat{\mathcal{A}}}$ − max $\mathfrak{C}_{\mathcal{A}}$ = max $\mathfrak{C}_{\hat{\mathcal{B}}}$ − max $\mathfrak{C}_{\mathcal{B}}$

(ii) height $\mathcal{A}$ = height $\mathcal{B}$ and max $\mathfrak{R}_{\hat{\mathcal{A}}}$ − max $\mathfrak{R}_{\mathcal{A}}$
= max $\mathfrak{R}_{\hat{\mathcal{B}}}$ − max $\mathfrak{R}_{\mathcal{B}}$.

(iii) card $\mathcal{A}$ = card $\mathcal{B}$.

Assume, without any loss of generality, that $\mathfrak{R}_{\mathcal{A}} = \mathfrak{R}_{\mathcal{B}}$ and $\mathfrak{C}_{\mathcal{A}} = \mathfrak{C}_{\mathcal{B}}$.

Let $\theta : \mathcal{A} \rightarrow \mathcal{B}$ be such that $\langle\alpha,\beta\rangle$ contributes 1 to $\mathbf{C}_2(\mathcal{A},\hat{\mathcal{A}};n,m)$ iff $\theta(\alpha,\beta)$ contributes 1 to $\mathbf{C}_2(\mathcal{B},\hat{\mathcal{B}};n,m)$. More formally,

$$\langle\alpha,\beta\rangle \in \mathcal{A} \quad \text{and} \quad \langle\alpha + n, \beta + m\rangle \in \hat{\mathcal{A}}$$
$$\text{iff } \theta(\alpha,\beta) \in \mathcal{B} \text{ and } \theta(\alpha,\beta) + \langle n,m\rangle \in \hat{\mathcal{B}}$$

All we need to show is that θ is an identity mapping. Proceeding as in [5], we can show that θ can be written as

$$\theta(p,q) = \langle\omega(p), \sigma(q)\rangle$$

for some appropriate choice of functions $\omega : \mathfrak{C}_{\mathcal{A}} \rightarrow \mathfrak{C}_{\mathcal{B}}$ and $\sigma : \mathfrak{R}_{\mathcal{A}} \rightarrow \mathfrak{R}_{\mathcal{B}}$.

Now, consider any $\langle x,y\rangle \in \mathcal{A}$, and fix $n,m \in \mathfrak{N}^+$ such that

$$x + n = \max \mathfrak{C}_{\hat{\mathcal{A}}} \quad \text{and} \quad y + m = \max \mathfrak{R}_{\hat{\mathcal{A}}}$$

If $x < \omega(x)$, then $\omega(x) + n \notin \mathfrak{C}_{\mathcal{B}}$ and therefore $\theta(\alpha,\beta) + \langle n,m\rangle \notin \hat{\mathcal{B}}$—a contradiction. Similarly, if $y < \sigma(y)$, then $\theta(\alpha,\beta) + \langle n,m\rangle \notin \hat{\mathcal{B}}$. Hence, we must have $x \geq \omega(x)$ and $y \geq \sigma(y)$.

Therefore, σ and ω must be nonincreasing functions. Since $\mathfrak{R}_{\mathcal{A}} = \mathfrak{R}_{\mathcal{B}}$ and $\mathfrak{C}_{\mathcal{A}} = \mathfrak{C}_{\mathcal{B}}$, the only way to accomplish this is by having σ and ω both as identity mappings.

Next, we state the effect of rotation and reflection operations on the $\mathbf{C}_2$ function. To show the effect of all operations in the transformation group, we need to define the following three variants of the $\mathbf{C}_2$ function: For $\mathcal{A} \in$ **IMAGE** and $\langle n,m\rangle \in \mathfrak{N}^*$ we let

$$\mathbf{C}_2^{+-}(\mathcal{A},\hat{\mathcal{A}};n,m) = \text{card}\{\langle\alpha,\beta\rangle \in \mathcal{A} : \langle\alpha + n, \beta - m\rangle \in \hat{\mathcal{A}}\}$$

$$\mathbf{C}_2^{-+}(\mathcal{A},\hat{\mathcal{A}};n,m) = \text{card}\{\langle\alpha,\beta\rangle \in \mathcal{A} : \langle\alpha - n, \beta + m\rangle \in \hat{\mathcal{A}}\}$$

$$\mathbf{C}_2^{--}(\mathcal{A},\hat{\mathcal{A}};n,m) = \text{card}\{\langle\alpha,\beta\rangle \in \mathcal{A} : \langle\alpha - n, \beta - m\rangle \in \hat{\mathcal{A}}\}$$

To maintain this terminology, we will let $\mathbf{C}_2^{++} \equiv \mathbf{C}_2$.

Theorem 6.2. Assume that the windowing coefficients are equal: $w_1 = w_2 = h_1 = h_2$. Consider any image $\mathcal{A}$ and integers $0 \leq i \leq 3$ and $0 \leq j \leq 1$. Let $\mathcal{B} = \mathbf{N}^i \bigcirc \mathbf{R}^j(\mathcal{A})$. Then

$$\mathbf{C}_2(\mathcal{A},\hat{\mathcal{A}};n,m) = \begin{cases} \mathbf{C}_2^{pq}(\mathcal{B},\hat{\mathcal{B}};n,m) & \text{if } i + j \text{ is even} \\ \mathbf{C}_2^{pq}(\mathcal{B},\hat{\mathcal{B}};m,n) & \text{otherwise} \end{cases}$$

where

$$pq = \begin{cases} ++ & \text{if } \langle i,j\rangle \in \{\langle 0,0\rangle,\langle 2,1\rangle\} \\ +- & \text{if } \langle i,j\rangle \in \{\langle 1,0\rangle,\langle 1,1\rangle\} \\ -+ & \text{if } \langle i,j\rangle \in \{\langle 3,0\rangle,\langle 3,1\rangle\} \\ -- & \text{otherwise} \end{cases}$$

Finally, we state the result for magnification. As was done with the functions introduced in Section V, we will assume that

$$w_1 = w_2 = \text{width } \mathcal{A} \qquad \text{and} \qquad h_1 = h_2 = \text{height } \mathcal{A}$$

Theorem 6.3. Consider any image $\mathcal{A}$ and two integers $\Delta_x,\Delta_y \in \mathfrak{N}^+$. Let

$$\mathcal{B} = \mathbf{MAG}_{xy}(\mathcal{A};\Delta_x,\Delta_y) \qquad \text{and} \qquad \mathfrak{C} = \mathbf{MAG}_{xy}(\hat{\mathcal{A}};\Delta_x,\Delta_y)$$

Then for all $\langle n,m\rangle \in \mathfrak{N}^*$ and all $p,q \in \{+,-\}$, we have

$$\mathbf{C}_2^{pq}\left[\mathcal{B},\mathfrak{C};\ \Delta_x n,\Delta_y m\right] = \Delta_x\Delta_y\ \mathbf{C}_2^{pq}(\mathcal{A},\hat{\mathcal{A}};\ n,\ m)$$

B. Characterization with Invariance Property: Theory of Group Averaging

To make the $\mathbf{C}_2$ function an invariant measure, we outline a scheme based on the *theory of group averaging*. The development is essentially due to [15]. Let $\mathfrak{G}$ be the associated finite transformation group [10]. For example, if the underlying two-dimensional image tessellation is based on squares, then

$$\mathfrak{G} = \{1, \mathbf{N}, \mathbf{N}^2, \mathbf{N}^3, \mathbf{R}, \mathbf{N} \bigcirc \mathbf{R}, \mathbf{N}^2 \bigcirc \mathbf{R}, \mathbf{N}^3 \bigcirc \mathbf{R}\}$$

We will assume that $\mathfrak{G}$ is finite.

Let $\mathfrak{F} : \mathbf{IMAGE} \rightarrow \mathfrak{R}$ be any (image) functional such that $\langle\mathfrak{R};+,\cdot,0,1\rangle$ is a ring and $\mathfrak{F}$ is translation invariant.

Define a measure $\mu : \mathfrak{X} \rightarrow \mathfrak{R}$ such that for any $X \in \mathfrak{X}$,

$$\mu(X) = \kappa \cdot \left[\sum_{g \in \mathfrak{G}} \mathfrak{F}[(g)X]\right]$$

where $\kappa \in \mathfrak{R}$ is fixed and may be dependent on the cardinality of set $\mathfrak{G}$. Then $\mu(X)$ can be thought of as an "average" of the values assumed by functional $\mathfrak{F}$ on all the set-theoretic images of $\mathfrak{X}$ under transformations in $\mathfrak{G}$.

Theorem 6.4 (Group Averaging Theorem). The measure μ, as defined above, is invariant under all transformations in $\mathfrak{G}$ and is also translation invariant.

Proof. Consider any $\psi \in \mathfrak{G}$ and any $X \in \mathfrak{X}$. We have

$$\sum_{g \in \mathfrak{G}} \mathfrak{F}[g(X)] = \sum_{f \in \mathfrak{G}} \mathfrak{F}[f(\psi(X))]$$

where, since $\mathfrak{G}$ is a group, $f = g \bigcirc \psi^{-1}$ is well defined. Hence, $\mu(X) = \mu[\psi(X)]$. That is, μ is invariant under the transformation ψ. The translation invariance is directly from the definition of $\mathfrak{F}$.

The preceding result is remarkable in the sense that a simple group averaging provides all of the desired invariants. If the rotation invariance under the multiple of **N** and **R** operations is required, then averaging of the values of $\mathbf{C}_2^{++}$, $\mathbf{C}^{+-2}$, $\mathbf{C}_2^{-+}$, $\mathbf{C}_2^{--}$ functions provides an invariant measure. We will denote the resultant function by $\mathbf{C}_2^*$.

Definition 6.1. For any $\mathcal{A} \in$ **IMAGE** and $v \in \mathfrak{R}^*$, define a function $\mathbf{C}_2^*$ such that

$$\mathbf{C}_2^*(\mathcal{A},\hat{\mathcal{A}};v) = \mathbf{C}_2^{++}(\mathcal{A},\hat{\mathcal{A}};v) + \mathbf{C}_2^{+-}(\mathcal{A},\hat{\mathcal{A}};v) + \mathbf{C}_2^{-+}(\mathcal{A},\hat{\mathcal{A}};v) + \mathbf{C}_2^{--}(\mathcal{A},\hat{\mathcal{A}};v)$$

Corollary 6.1. Consider any two images $\mathcal{A}$ and $\mathcal{B}$, and assume that the sets $\mathcal{W}_{\mathcal{A}}$ and $\mathcal{W}_{\mathcal{B}}$ are computed with a fixed value of $w_1 = w_2 = h_1 = h_2$. Then the following two conditions are equivalent:

(i) $\mathcal{A} = \mathbf{T}[\mathbf{N}^i \bigcirc \mathbf{R}^j(\mathcal{B});v]$ for some $i \in \{0,1,2,3\}$, $j \in \{0,1\}$, and $v \in \mathfrak{R}^2$.

(ii) $\mathbf{C}_2^*(\mathcal{A},\hat{\mathcal{A}};\cdot) = \mathbf{C}_2^*(\mathcal{B},\hat{\mathcal{B}};\cdot)$.

Proof. Follows directly from Theorems 2.1, 6.2, and 6.3.

To conclude, we would like to mention that the rotation and magnification invariants cannot be incorporated at the same time in the $\mathbf{C}_2^*$ function. However, the $\mathbf{C}_2^*$ function is invariant under magnification and operations of the form $\mathbf{N}^i \bigcirc \mathbf{R}^j$ with $i + j$ as nonnegative even integers. To this end, simply let $w_1 = w_2 =$ width $\mathcal{A}$ and $h_1 = h_2 =$ height $\mathcal{A}$.

C. Morphological Definitions

Proceeding along the lines of Section III, we define the following four parameterized families of structuring elements

$$\{\mathfrak{hv}^{++}_{n,m}\}_{n,m\in\mathfrak{N}^*}, \qquad \{\mathfrak{hv}^{+-}_{n,m}\}_{n,m\in\mathfrak{N}^*},$$

$$\{\mathfrak{hv}^{-+}_{n,m}\}_{n,m\in\mathfrak{N}^*}, \qquad \{\mathfrak{hv}^{--}_{n,m}\}_{n,m\in\mathfrak{N}^*}$$

as follows, for $v \in \mathfrak{N}^2$:

$$\mathfrak{hv}^{++}_{n,m}(v) = \begin{cases} 1 & \text{if } v = \langle 0,0\rangle \\ 0 & \text{if } v = \langle n,m\rangle \\ \alpha & \text{otherwise} \end{cases}$$

$$\mathfrak{hv}^{+-}_{n,m}(V) = \begin{cases} 1 & \text{if } v = \langle 0,0\rangle \\ 0 & \text{if } v = \langle n,-m\rangle \\ \alpha & \text{otherwise} \end{cases}$$

$$\mathfrak{hv}^{-+}_{n,m}(v) = \begin{cases} 1 & \text{if } v = \langle 0,0\rangle \\ 0 & \text{if } v = \langle -n,m\rangle \\ \alpha & \text{otherwise} \end{cases}$$

$$\mathfrak{hv}^{--}_{n,m}(v) = \begin{cases} 1 & \text{if } v = \langle 0,0\rangle \\ 0 & \text{if } v = \langle -n,-m\rangle \\ \alpha & \text{otherwise} \end{cases}$$

Of course,

$$\mathfrak{hv}^{++}_{n,0} = \mathfrak{hv}^{+-}_{n,0} = \mathfrak{h}^{+}_{n}$$

$$\mathfrak{hv}^{-+}_{n,0} = \mathfrak{hv}^{--}_{n,0} = \mathfrak{h}^{-}_{n}$$

$$\mathfrak{hv}^{++}_{0,n} = \mathfrak{hv}^{-+}_{0,n} = \mathfrak{v}^{+}_{n}$$

$$\mathfrak{hv}^{+-}_{0,n} = \mathfrak{hv}^{--}_{0,n} = \mathfrak{v}^{-}_{n}$$

Proceeding along the lines of Theorem 3.1, we can show that:

Theorem 6.5. Consider any image $\mathcal{A}$, a tuple N in $\mathfrak{N}^*$, and $p,q \in \{+,-\}$. We have

$$\mathbf{C}^{pq}_{2}(\mathcal{A};N) = \delta\left[\mathbf{OPEN}\left(\mathfrak{L}(\mathcal{A}),\ \mathfrak{hv}^{pq}_{N}\right)\right]$$

VII. GENERALIZATIONS TO FINITE MULTIDIMENSIONAL SETS

In this section, we shall generalize the results of Sections V and VII to handle finite subsets of $\mathfrak{N}^{\mathbf{D}}$ with $\mathbf{D} \geq 3$. As can be guessed, in this general setting the notation gets very complicated. However, all the results are, intuitively speaking, obvious based on the discussion of the previous sections. We will state the representative results for the general case.

For **D**-dimensional images, we consider **D** distance measures, one for each dimension. As was done in the case of two-dimensional images, we define the *i*th dimensional distance, $1 \leq i \leq \mathbf{D}$, between two points to be the Euclidean distance between them provided the line joining these two points is parallel to the *i*th axis; otherwise the distance is undefined.

We present three different versions of the generalization of results presented in Sections V and VI. Before we discuss these schemes, it may be helpful to look back at the generalization carried out from one-dimensional images to two-dimensional images. In Section V we looked at each dimensional component of the image separately and employed the characterization of one-dimensional images to these individual components. The outcome of this process was only a pseudocharacterization of the two-dimensional images. In Section VI we considered both distance measures simultaneously to arrive at a unique characterization.

For the complete characterization, we present the **D**-dimensional analog of the $\mathbf{C}_2$ function that was introduced in Section VI. We consider all the **D** distance measures simultaneously to arrive at a single feature for each image.

For the pseudocharacterizations, there are two possible "good" approaches, each being justifiable in some sense.

The first approach is based on the assumption that the $\mathbf{D} - 1$ dimension sets can be uniquely characterized by the $\mathbf{D} - 1$ dimensional analog of the $\mathbf{C}_2$ function. Essentially, very much like viewing two-dimensional images as lines placed on top of each other, we view a **D**-dimensional image as $\mathbf{D} - 1$ dimensional hyperplanes stacked on each other (like a deck of cards). These hyperplanes are obtained by slicing the image at various coordinates on a fixed axis. We thus obtain **D** features—one for each axis—and the subsequent analysis follows that of the Section V very closely. For example, here we will observe that each of these features is not able to distinguish between the permutations of a fixed set of hyperplanes very much like the $\mathbf{H}^+$ $(\mathbf{V}^+)$ function, which failed to differentiate between the permutations of a fixed set of horizontal (vertical) lines. In fact, we obtain the **D**-dimensional analogy of all results of Section V.

In the second approach, we project the given image on a two-dimensional plane in **D** different ways, one for each dimension. For the projection corresponding to *i*th dimension, we derive a feature based on the *i*th distance measure. The characterization of the original image based on these **D** features is then investigated. This analysis is also related to that presented in Section V; however, we were not able to obtain the analogs of all results that were presented in Section V for the **D**-dimensional images.

A. Projection Functions

For convenience, we will borrow the following shorthand notation (which is widely used in combinatorics): for any $N \in \mathfrak{R}^+$, let $[N] = \{1,2, \ldots ,N\}$. Now, we define (information-preserving) projection functions on $\mathfrak{R}^{\mathbf{D}}$, columns and rows of support, and the line images of **D**-dimensional images. These concepts will be used throughout the section.

Definition 7.1. For each dimension $i \in [\mathbf{D}]$, define two *projection functions*

$$\Pi_i : \mathfrak{R}^{\mathbf{D}} \rightarrow \mathfrak{R} \qquad \text{and} \qquad \Omega_i : \mathfrak{R}^{\mathbf{D}} \rightarrow \mathfrak{R}^{\mathbf{D}-1}$$

as follows:

$$\Pi_i\,(\alpha_1, \ldots ,\alpha_{\mathbf{D}}) = \alpha_i$$

$$\Omega_i(\alpha_1, \ldots \alpha_{\mathbf{D}}) = (\alpha_1, \ldots ,\alpha_{i-1},\alpha_{i+1}, \ldots ,\alpha_{\mathbf{D}})$$

Definition 7.2. For any **D**-dimensional image $\mathcal{A}$, define the *columns of support* and *rows of support* respectively as

$$\{C_{\mathcal{A},i} : i \in [\mathbf{D}]\} \qquad \text{and} \qquad \{R_{\mathcal{A},i} : i \in [\mathbf{D}]\}$$

such that

$$C_{\mathcal{A},i} = \{\Pi_i(x) : x \in \mathcal{A}\} \qquad \text{and} \qquad R_{\mathcal{A},i} = \{\Omega_i(x) : x \in \mathcal{A}\}$$

It is easy to observe that for any image $\mathcal{A}$, the following inequality holds:

$$\mathcal{A} \subseteq C_{\mathcal{A},1} \times C_{\mathcal{A},2} \times \cdots \times C_{\mathcal{A},\mathbf{D}}$$

If the equality holds in the above expression, we will call $\mathcal{A}$ a *filled parallelepiped*. Also, for any $i \in [\mathbf{D}]$, we have

$$\mathcal{A} \subseteq \left\{x \in \mathfrak{R}^{\mathbf{D}} : \Pi_i(x) \in C_{\mathcal{A},i} \text{ and } \Omega_i(x) \in R_{\mathcal{A},i}\right\}$$

The equality holds for all $i \in [\mathbf{D}]$ only when $\mathcal{A}$ is a filled parallelepiped.

The *width of image $\mathcal{A}$ in the ith dimension*, $W_i(\mathcal{A})$, is defined as

$$W_i(\mathcal{A}) = \max C_{\mathcal{A},i} - \min C_{\mathcal{A},i} + 1$$

If card $C_{\mathcal{A},i} = 1$ (or, equivalently, $W_i(\mathcal{A}) = 1$) for some $i \in [\mathbf{D}]$, then $\mathcal{A}$ is said to be a *line image in the ith dimension*.

Example 7.1. Consider the following three-dimensional image $\mathcal{A}$:

$$\mathcal{A} = \{\langle 0,1,0\rangle, \langle 0,1,1\rangle, \langle 0,1,2\rangle, \langle 0,3,0\rangle\ \langle 0,3,3\rangle\}$$

We have $C_{\mathcal{A},1} = \{0\}$, $C_{\mathcal{A},2} = \{1,3\}$, and $C_{\mathcal{A},3} = \{0,1,2,3\}$. Also,

$$R_{\mathcal{A},1} = \{\langle 1,0\rangle, \langle 1,1\rangle, \langle 1,2\rangle, \langle 3,0\rangle, \langle 3,3\rangle\}$$
$$R_{\mathcal{A},2} = \{\langle 0,0\rangle, \langle 0,1\rangle, \langle 0,2\rangle, \langle 0,3\rangle\}$$
$$R_{\mathcal{A},3} = \{\langle 0,1\rangle, \langle 0,3\rangle\}$$

The widths of $\mathcal{A}$ are 1, 3, and 4. Finally, $\mathcal{A}$ is a line image in the first dimension.

Definition 7.3. For any **D**-dimensional image $\mathcal{A}$, define its pseudocomplement $\hat{\mathcal{A}}$ as $\mathcal{A}^c \cap W_{\mathcal{A}}$ where $\mathcal{A}^c$ is the set-theoretic complement of $\mathcal{A}$ and*

$$W_{\mathcal{A}} = \{x \in \Re^{\mathbf{D}} : \forall i \in [\mathbf{D}], \alpha_i \leq \Pi_i(x) \leq \beta_i + w_i\}$$

with for all $i \in [\mathbf{D}]$, $w_i \in \Re^+$, $\alpha_i = \min C_{\mathcal{A},i}$, and $\beta_i = \max C_{\mathcal{A},i}$.

For convenience, we introduce a special subset of $\Re^{\mathbf{D}}$.

Definition 7.4. The set $\mathfrak{T}_{\mathbf{D}}$ contains all those tuples $\tau \in (\Re^+ \cup \{0\})^{\mathbf{D}}$ for which $\Sigma_{i\in[\mathbf{D}]} \Pi_i(\tau) > 0$. It is easy to verify that the set $\mathfrak{T}_2$ is the same as the set $\Re^*$.

B. Unique Characterization: $\mathbf{C_D}$ Function

Exactly similar to the $\mathbf{C}_2$ function, we have the function $\mathbf{C_D}$ for the **D**-dimensional images.

Definition 7.5. For any **D**-dimensional image $\mathcal{A}$ and a tuple $\mathbf{n} \in \mathfrak{T}_{\mathbf{D}}$, define

$$\mathbf{C_D}(\mathcal{A},\hat{\mathcal{A}};n) = \text{card}\{x \in \mathcal{A} : x + n \in \hat{\mathcal{A}}\}$$

where + denotes the "vector addition" in $\Re^{\mathbf{D}}$. Recall that $\mathfrak{T}_{\mathbf{D}} \subseteq \Re^{\mathbf{D}}$.

Example 7.2. Consider the following three-dimensional image $\mathcal{A}$:

$$\mathcal{A} = \{\langle 1,1,1\rangle, \langle 2,2,2\rangle, \langle 3,3,3\rangle\}$$

Assume that $w_1 = w_2 = w_3 = 1$. First note that $W_1(\mathcal{A}) = W_2(\mathcal{A}) = W_3(\mathcal{A}) = 3$. Therefore, we must compute the values of $\mathbf{C}_3(\mathcal{A},\hat{\mathcal{A}};p,q,r)$ for $0 \leq p,q,r \leq 3$; for other values of p, q, and r, $\mathbf{C}_3(\mathcal{A},\hat{\mathcal{A}};p,q,r) = 0$. Note that $\mathbf{C}_3(\mathcal{A},\hat{\mathcal{A}};0,0,0)$ is not defined. We now list the nonzero values of the function $\mathbf{C}_3$.

(i) $\mathbf{C}_3(\mathcal{A},\hat{\mathcal{A}};p,q,r) = 3$ for the following triplets $\langle p,q,r\rangle$: $\langle 0,0,1\rangle$, $\langle 0,1,0\rangle$, $\langle 0,1,1\rangle$, $\langle 1,0,0\rangle$, $\langle 1,0,1\rangle$, and $\langle 1,1,0\rangle$

(ii) $\mathbf{C}_3(\mathcal{A},\hat{\mathcal{A}};p,q,r) = 1$ for the following triplets $\langle p,q,r\rangle$: $\langle 3,*,*\rangle$, $\langle *,3,*\rangle$, $\langle *,*,3\rangle$, $\langle 2,2,2\rangle$, and $\langle 1,1,1\rangle$ where * stands for any value in $\{0,1,2,3\}$.

(iii) $\mathbf{C}_3(\mathcal{A},\hat{\mathcal{A}};p,q,r) = 2$ for the remaining triplets $\langle p,q,r\rangle$ in $\{0,1,2,3\}^3$.

Hence, $\mathbf{C}_3(\mathcal{A},\hat{\mathcal{A}};p,q,r)$ assumes nonzero values for all triplets $\langle p,q,r\rangle$ in the set $\{0,1,2,3\}^3$, except $\langle 0,0,0\rangle$.

For the present, it suffices to mention that, analogous to the special cases of the $\mathbf{C}_2$ function (as stated in Section VI), here also we can specify relationships

*Since we wish to generalize only $\mathbf{H}^+$, $\mathbf{V}^+$, and $\mathbf{C}_2$ functions, the nonactivated pixels in $\hat{f}$ only need to be the right of α_i in each dimension i.

of $\mathbf{C_D}$ with the functions that will be introduced in the next two sections. Now, we mention the analog of Theorem 6.2.

Theorem 7.1. Consider any two $\mathbf{D}$-dimensional images $\mathcal{A}$ and $\mathcal{B}$. Then $\mathbf{C_D}(\mathcal{A},\hat{\mathcal{A}};\cdot) = \mathbf{C_D}(\mathcal{A},\hat{\mathcal{B}};\cdot)$ iff $\mathcal{A} = \mathbf{T}(\mathcal{A};v)$ with $v \in \Re^{\mathbf{D}}$ and for all $\in [\mathbf{D}]$,

$$\max C_{\mathcal{A},i} - \max C_{\hat{\mathcal{A}},i} - \max C_{\mathcal{B},i} - \max C_{\hat{\mathcal{B}},i}$$

Proof. The proof, which is based on induction on $\mathbf{D}$, will be postponed till the end of next section. However, for the purpose of next section, we will assume that this result holds for $\mathbf{D} - 1$ dimensions (in this sense, the entire Section VII.C is part of this proof).

Example 7.3. consider the following image $\mathcal{B}$:

$$\mathcal{B} = \{\langle 3,1,1\rangle, \langle 2,2,2\rangle, \langle 1,3,3\rangle\}$$

and the image $\mathcal{A}$ of Example 7.2. Assume that $w_1 = w_2 = w_3 = 1$. Then their $\mathbf{C}_3$ functions are different: $\mathbf{C}_3(\mathcal{A},\hat{\mathcal{A}};1,1,1) = 1 \neq 3 = \mathbf{C}_3(\mathcal{B},\hat{\mathcal{B}};1,1,1)$.

C. A Pseudocharacterization: $\mathbf{F}_i$ Functions

In Section V, we viewed a two-dimensional image as a collection of lines placed next to each other and found that the changes in the relative position of these lines were not detected by the introduced features. Now, we wish to a view a $\mathbf{D}$-dimensional image as $\mathbf{D} - 1$ dimensional hyperplanes stacked on each other (very much like a deck of cards). The spirit of this generalization a fortiori assumes that the $\mathbf{D} - 1$ dimension sets can be uniquely characterized (we will assume that the $\mathbf{C_{D-}}$ function can be employed for this purpose). We will observe that the subsequent analysis follows that of Section V very closely.

Note that, as was the case in Section V, we will assume that $w_i = 1$, for all $i \in [\mathbf{D}]$, unless otherwise specified (see Definition 7.3).

Definition 7.6. For any $\mathbf{D}$-dimensional image $\mathcal{A}$, integers $i \in [\mathbf{D}]$ and $v \in C_{\mathcal{A},i}$ define the hyperplanes $P_i(\mathcal{A};v)$ as

$$P_i(\mathcal{A};v) = \left\{\Omega_i(x) : x \in \mathcal{A} \text{ and } \Pi_i(x) = v\right\}$$

Also define the pseudocomplement $\hat{P}_i(\mathcal{A};v)$ of $P_i(\mathcal{A};v)$ as

$$\hat{P}_i(\mathcal{A};v) = \left\{\Omega_i(x) : x \in \hat{\mathcal{A}} \text{ and } \Pi_i(x) = v\right\} = P_i(\hat{\mathcal{A}};v)$$

For all $i \in [\mathbf{D}]$ and $v \in C_{\mathcal{A},i}$, the hyperplanes $P_i(\mathcal{A};v)$ and $\hat{P}_i(\mathcal{A};v)$ are $\mathbf{D} - 1$ dimensional images. Furthermore, for all $i \in [\mathbf{D}]$,

$$\mathcal{A} = \left\{x \in \Re^{\mathbf{D}} : \Omega_i(x) \in P_i\left(\mathcal{A};\Pi_i(x)\right)\right\}$$

$$\hat{\mathscr{A}} = \left\{x \in \Re^{\mathbf{D}} : \Omega_i(x) \in \hat{P}_i\left(\mathscr{A};\Pi_i(x)\right)\right\}$$

Example 7.4. Consider the image $\mathscr{A}$ of Example 7.1. We can define the following seven hyperplanes:

$$P_1(\mathscr{A};0) = \{\langle 1,0\rangle, \langle 1,1\rangle, \langle 1,2\rangle, \langle 3,0\rangle, \langle 3,3\rangle\}$$
$$P_2(\mathscr{A};1) = \{\langle 0,0\rangle, \langle 0,1\rangle, \langle 0,2\rangle\}$$
$$P_2(\mathscr{A};3) = \{\langle 0,0\rangle, \langle 0,3\rangle\}$$
$$P_3(\mathscr{A};0) = \{\langle 0,1\rangle, \langle 0,3\rangle\}$$
$$P_3(\mathscr{A};1) = P_3(\mathscr{A};2) = \{\langle 0,1\rangle\}$$
$$P_3(\mathscr{A};3) = \{\langle 0,3\rangle\}$$

Similarly, we can define $\hat{P}_i(\mathscr{A};v)$, $v \in C_{\mathscr{A},i}$ and $i \in [3]$. For example, the sets $\hat{P}_2(\mathscr{A};1)$ and $\hat{P}_2(\mathscr{A};3)$ are shown in Figure 29.

We now introduce **D** features based on the projections for each dimension as follows (it is worth comparing this definition to Theorem 6.1):

Definition 7.7. For any **D**-dimensional image $\mathscr{A}$ and an integer $i \in [\mathbf{D}]$, define the function $\mathbf{F}_i(\mathscr{A};n)$ with $n \in \mathfrak{T}_{\mathbf{D}-1}$ as

$$\mathbf{F}_i(\mathscr{A};n) = \sum_{v \in \mathbf{C}_{\mathscr{A},i}} \mathbf{C}_{\mathbf{D}-1}\left[P_i(\mathscr{A};v), \hat{P}_i(\mathscr{A};v);\ n\right]$$

Remarks. Consider any image $\mathscr{A}$, and assume that $\Pi_i(n) = 0$ for some $n \in \mathfrak{T}_{\mathbf{D}}$ and $i \in [\mathbf{D}]$. It is easy to verify that the following equality holds:

$$\mathrm{C}_{\mathbf{D}}(\mathscr{A},\hat{\mathscr{A}};n) = \mathbf{F}_i\left[\mathscr{A};\ \Omega_i(n)\right]$$

It is also of interest to note that the pseudocomplement of $P_i(\mathscr{A};v)$ is *not* the set $\hat{P}_i(\mathscr{A};v)$. Hence, if we did not explicitly specify the value of $\hat{P}_i(\mathscr{A};v)$ in the definition of the $\mathbf{C}_{\mathbf{D}-1}$ function, there would be no means of determining this set.

Example 7.5. Consider the image $\mathscr{A}$ of Example 7.1:

$$\mathscr{A} = \{\langle 0,1,0\rangle, \langle 0,1,1\rangle, \langle 0,1,2\rangle, \langle 0,3,0\rangle, \langle 0,3,3\rangle\}$$

First, consider the computation of the $\mathbf{F}_2$ function. We need to compute the $\mathbf{C}_2$ functions of the images $P_2(\mathscr{A};1)$ and $P_2(\mathscr{A};3)$ specified in Example 7.4 (see Figure 29). The *nonzero* values of $\mathbf{F}_2$ are

$$\mathbf{F}_2(\mathscr{A};0,1) = \mathbf{F}_2(\mathscr{A};1,3) = 3$$
$$\mathbf{F}_2(\mathscr{A};0,2) = 4$$
$$\mathbf{F}_2(\mathscr{A};0,3) = \mathbf{F}_2(\mathscr{A};0,4) = \mathbf{F}_2(\mathscr{A};1,4) = 2$$
$$\mathbf{F}_2(\mathscr{A};1,0) = \mathbf{F}_2(\mathscr{A};1,1) = \mathbf{F}_2(\mathscr{A};1,2) = 5$$

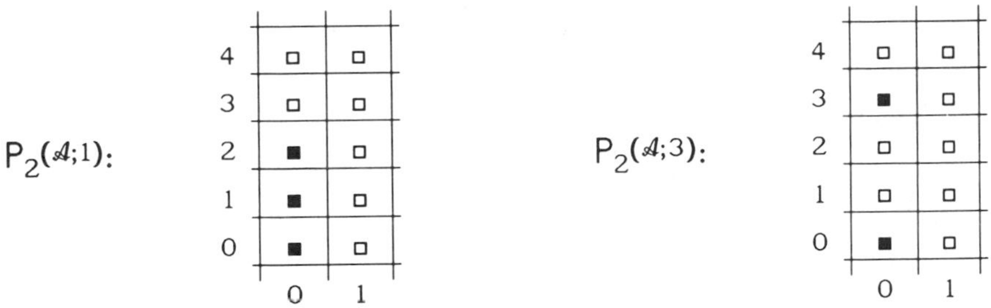

Figure 29. The two hyperplanes $P_2(\mathcal{A},1)$ and $P_2(\mathcal{A},3)$ used in Example 7.4 and 7.5.

Similarly, we can compute $\mathbf{F}_1$ and $\mathbf{F}_3$.

Our investigation of functions $\mathbf{F}_i$ proceeds along the lines of Section V.

Lemma 7.1. Consider two images $\mathcal{A}$ and $\mathcal{B}$ such that for some $i \in [\mathbf{D}]$, $\mathbf{F}_i(\mathcal{A};\cdot) = \mathbf{F}_i(\mathcal{B};\cdot)$. Then card $\mathcal{A}$ = card $\mathcal{B}$. Furthermore, if the rows of $\mathcal{A}$ and $\mathcal{B}$ are contiguous, then card $\hat{\mathcal{A}}$ = card $\hat{\mathcal{B}}$.

Proof. If $\mathbf{F}_i(\mathcal{A};\cdot) = \mathbf{F}_i(\mathcal{B};\cdot)$, then we must have for all $n \in \mathfrak{T}_{\mathbf{D}-1}$:

$$\sum_{v \in \mathbf{C}_{\mathcal{A},i}} \mathbf{C}_{\mathbf{D}-1}\left[P_i(\mathcal{A};v),\hat{P}_i(\mathcal{A};v);\, n\right] = \sum_{\tau \in \mathbf{C}_{\mathcal{B},i}} \mathbf{C}_{\mathbf{D}-1}\left[P_i(\mathcal{B};\tau),\hat{P}_i(\mathcal{B};\tau);\, n\right]$$

Consider any pair $x \in P_i(\mathcal{A};v)$ and $x + n \in \hat{P}_i(\mathcal{A};v)$. Then there exists a pair $y \in P_i(\mathcal{B};\tau)$ and $y + n \in \mathrm{P}_i(\mathcal{B};v)$. Denote this correspondence as

$$\langle X',X''\rangle \,\Xi_n\, \langle Y',Y''\rangle$$

where

$$\begin{array}{ll} \Omega_i(X') = x & \Omega_i(X'') = x + n \\ \Omega_i(Y') = y & \Omega_i(Y'') = y + n \\ \Pi_i(X') = \Pi_i(X'') = v & \Pi_i(Y') = \Pi_i(Y'') = \tau \end{array}$$

Define two mappings $\theta_1 : \mathcal{A} \rightarrow \mathcal{B}$ and $\theta_2 : \hat{\mathcal{A}} \rightarrow \hat{\mathcal{B}}$ such that $\alpha \in \mathcal{A}$ contributes 1 to $\mathbf{F}_i(\mathcal{A};n)$ iff $\theta_1(\alpha)$ contributes 1 to $\mathbf{F}_i(\mathcal{B};n)$. More formally, we have

$$\exists n \ni \langle X',X''\rangle \,\Xi_n\, \langle Y',Y''\rangle \text{ iff } \theta_1(X') = Y' \text{ and } \theta_2(X'') = Y''$$

Not that the above condition implies that θ should map any two voxels in $P_i(\mathcal{A};v)$ into two voxels in a single $P_i(\mathcal{B};\tau)$.

Proceeding as in the proof of Lemma 5.1, one can show that θ_1 and θ_2 must be bijective mappings.

Lemma 7.2. If $\mathcal{A}$ is a line image in the ith dimension, then there does *not* exist a nonline image $\mathcal{B}$ such that $\mathbf{F}_i(\mathcal{B};\cdot) = \mathbf{F}_i(\mathcal{A};\cdot)$.

Theorem 7.2. Consider any $i \in [\mathbf{D}]$. For any two images $\mathcal{A}$ and $\mathcal{B}$, the following two conditions are equivalent:

(i) $\mathbf{F}_i(\mathcal{A};\cdot) = \mathbf{F}_i(\mathcal{B};\cdot)$.

(ii) There exists a bijective mapping $\omega_i : C_{\mathcal{A},i} \to C_{\mathcal{B},i}$ such that for a fixed $\tau \in \Re^{\mathbf{D}-1}$ we have for all $v \in C_{\mathcal{A},i}$,

$P_i(\mathcal{A};v) = \mathbf{T}[P_i(\mathcal{A};\, \omega_i(v));\, \tau]$

Proof. (*Sufficiency*) Directly follows from the definition.

(*Necessity*) From Lemma 7.2, we know that the cardinalities of $C_{\mathcal{A},i}$ and $C_{\mathcal{B},i}$ must be the same. Therefore, there exists a $\omega_i : C_{\mathcal{A},i} \to C_{\mathcal{B},i}$ such that the $\mathbf{C_D} - 1$ functions of $P_i(\mathcal{A};v)$ and $P_i(\mathcal{A};\omega_i(v))$ are identical. Now, from theorem 7.1, we know that $P_i(\mathcal{A};v)$ and $P_i(\mathcal{A};\omega_i(v))$ are translated versions of each other. Hence,

$$\forall v \in C_{\mathcal{A},i},\ \exists \tau_v \in \Re^{\mathbf{D}-1} \ni P_l(\mathcal{A};v) = \mathbf{T}\left[P_i\left(\mathcal{A};\ \omega_i(v)\right);\ \tau_v\right]$$

Now, if we impose the results of Lemma 7.1, we conclude that τ_v must be independent of v; otherwise the relationship between the cardinalities of the pseudo-complements of $P_i(\mathcal{A};v)$ and $P_i(\mathcal{A};\omega_i(v))$ will be violated.

Consider two images $\mathcal{A}$ and $\mathcal{B}$ such that

$$\forall i \in [\mathbf{D}], \qquad \mathbf{F}_i(\mathcal{A};\cdot) = \mathbf{F}_i(\mathcal{B};\cdot)$$

Lew $\omega_i : C_{\mathcal{B},i} \to C_{\mathcal{B},i}$ be as defined in Theorem 7.2. As was done in Section V, we will ignore translations, that is, we will let $C_{\mathcal{A},i}$. Now *if there exists an identity mapping for the columns of support in* i*th dimension, then* ω_i *will be assumed to be an identity mapping*. Then, it follows that our algorithm works only if all the ω_i's are identity mappings. The counterpart of Proposition 5.1 is also true for the **D**-dimensional images.

Proposition 7.1. If for all $i \in [\mathbf{D}]$, $\mathbf{F}_i(\mathcal{A};\cdot) = \mathbf{F}_i(\mathcal{B};\cdot)$, then the voxels in $\mathcal{A}$ satisfy the following condition:

$$\langle x_1, \ldots, x_{\mathbf{D}}\rangle \in \mathcal{A} \text{ iff } \langle \omega_1^{p_1}(x_1), \omega_2^{p_2}(x_1), \ldots \omega_{\mathbf{D}}^{p_{\mathbf{D}}}(x_1)\rangle \in \mathcal{A}$$

where the mappings ω_i are defined in Theorem 7.2 and the integers p_i are such that $p_1 + p_2 + \cdots + p_{\mathbf{D}} = 0$.

Proof. Let $\mathbf{x} \equiv \langle x_1, x_2, \ldots, x_{\mathbf{D}}\rangle$. From Theorem 7.2 we know that

$$\mathbf{x} \in \mathcal{A} \text{ iff } \langle \omega_1(x_1), x_2, \ldots, x_{\mathbf{D}}\rangle \in \mathcal{B} \tag{11.37.1}$$

$$\mathbf{x} \in \mathcal{A} \text{ iff } \langle x_1, \omega_2(x_2), \ldots, x_{\mathbf{D}}\rangle \in \mathcal{B} \tag{11.37.2}$$

$$\vdots$$

$$\mathbf{x} \in \mathcal{A} \text{ iff } \langle x_1, x_2, \ldots, \omega_{\mathbf{D}}(x_{\mathbf{D}})\rangle \in \mathcal{B} \tag{11.37.D}$$

Following the proof of Proposition 5.1, we combine the equivalencies (11.37.i), $1 \le i \le \mathbf{D} - 1$, and (11.37.$\mathbf{D}$) to obtain

$$\mathbf{x} \in \mathcal{A} \text{ iff } \langle \omega_1^{p1}(x_1), x_2, \ldots, x_{\mathbf{D}-1}, \omega_{\mathbf{D}}^{-p1}(x_{\mathbf{D}})\rangle \in \mathcal{A} \tag{11.38.1}$$

$$\mathbf{x} \in \mathcal{A} \text{ iff } \langle x_1, \omega_1^{p2}(x_2), x_3, \ldots, x_{\mathbf{D}-1}, \omega_{\mathbf{D}}^{-p2}(x_{\mathbf{D}})\rangle \in \mathcal{A} \tag{11.38.2}$$

$$\vdots$$

$$\mathbf{x} \in \mathcal{A} \text{ iff } \langle x_1, \ldots, x_{\mathbf{D}-2}, \omega_{\mathbf{D}-1}{}^{p}{}_{\mathbf{D}-1}(x_{\mathbf{D}-1}), \omega_{\mathbf{D}}{}^{-p}{}_{\mathbf{D}-1}(x_{\mathbf{D}})\rangle \in \mathcal{A} \tag{11.38.D$-$1}$$

Combining the above $\mathbf{D} - 1$ equivalences, we obtain $\mathbf{x} \in \mathcal{A}$

$$\text{iff } \langle \omega_1^{p1}(x_1), x_2, x_3, \ldots, x^{\mathbf{D}-1}, \omega_{\mathbf{D}}^{-p1}(x^{\mathbf{D}})\rangle \in \mathcal{A} \qquad \text{[from (11.38.1)]}$$

$$\text{iff } \langle \omega_1^{p1}(x_1), \omega_1^{p2}(x_2), x_3, \ldots, x_{\mathbf{D}-1}, \omega_{\mathbf{D}}^{-p1-pl2}(x_{\mathbf{D}})\rangle \in \mathcal{A} \qquad \text{[from (11.38.2)]}$$

$$\vdots$$

$$\text{iff } \langle \omega_1^{p1}(x_1), \omega_2^{p2}(x_2), \ldots, \omega_{\mathbf{D}}^{p\mathbf{D}}(x_{\mathbf{D}})\rangle \in \mathcal{A} \qquad \text{[from (11.38.D} - 1)]$$

Let us now illustratc the above result.

Example 7.6. Consider the following two 3-dimensional images $\mathcal{A}$ and $\mathcal{B}$:

$$\mathcal{A} = \{\langle 1,1,1\rangle, \langle 1,2,2\rangle, \langle 2,2,1\rangle, \langle 2,1,2\rangle\}$$
$$\mathcal{B} = \{\langle 1,1,2\rangle, \langle 1,2,1\rangle, \langle 2,1,1\rangle, \langle 2,2,2\rangle\}$$

It can be easily verified that their $\mathbf{F}_i$, $\mathbf{F}_2$, and $\mathbf{F}_3$ functions are identical. Furthermore, the corresponding column mappings ω_1, ω_2, and ω_3 are given as follows:

$$\omega_1 : \begin{pmatrix} 1 & 2 \\ 2 & 1 \end{pmatrix}, \qquad \omega_2 : \begin{pmatrix} 1 & 2 \\ 2 & 1 \end{pmatrix}, \qquad \omega_3 : \begin{pmatrix} 1 & 2 \\ 2 & 1 \end{pmatrix}$$

Consider $\langle 1,1,1\rangle \in \mathcal{A}$. Then we have

$$\langle \omega_1^2(1), \omega_2(1), \omega_3^{-3}(1)\rangle = \langle 1,2,2\rangle$$
$$\langle \omega_1^3(1), \omega_2(1), \omega_3^{-4}(1)\rangle = \langle 2,2,1\rangle$$
$$\langle \omega_1^3(1), \omega_2^2(1), \omega_3^{-5}(1)\rangle = \langle 2,1,2\rangle$$

Hence, the voxel $\langle 1,1,1\rangle$ determines all the other voxels in $\mathcal{A}$.

Definition 7.8. A $\mathbf{D}$-dimensional image $\mathcal{A}$ will be called an *irregular image* iff there exist $\mathbf{D}$ permutations $\zeta_i : C_{\mathcal{A},i} \to C_{\mathcal{A},i}$ such that

(i) $\langle x_1, \ldots, x_{\mathbf{D}}\rangle \in \mathcal{A}$ iff $\langle \zeta_1{}^{p1}(x_1), \zeta_2{}^{p2}(x_1), \ldots, \zeta_{\mathbf{D}}{}^{p\,\mathbf{D}}(x_1)\rangle \in \mathcal{A}$ for $p_1 + p_2 + \cdots + p_{\mathbf{D}} = 0$.

(ii) $\forall i \in [\mathbf{D}]$, if card $C_{\mathcal{A},i} \geq 2$,

then $\exists\ \alpha \in C_{\mathcal{A},i} \ni \zeta_i(\alpha) \neq \alpha$ and $P_i(\mathcal{A};\alpha) \neq P_i[\mathcal{A};\zeta_i(\alpha)]$.

(iii) $C_{\mathcal{A},i} \geq 2$ for at least two different $i \in [\mathbf{D}]$.

Remarks

1. Note that condition (ii) ensures that ζ_i's cannot be converted into identity mappings. The condition "card $C_{\mathcal{A},i} \geq 2$" is needed to handle degenerate **D**-dimensional irregular images—a problem that cannot arise in Section V.
2. As in the two-dimensional case, even though a filled parallelepiped is not an irregular image, we will assume that *all filled parallelepiped images are irregular*, as an exception. As was the case in Section V, this exception is necessary when we discuss the construction of irregular images.
3. As was the case in Section V (see Propositions 5.1 and 5.4), we can show that if $\mathbf{F}_i(\mathcal{A};\cdot) = \mathbf{F}_i(\mathcal{B};\cdot)$ for all $i \in [\mathbf{D}]$, then $\mathcal{A}$ is an irregular image iff $\mathcal{B}$ is an irregular image. $\mathcal{A}$ and $\mathcal{B}$ will be called an *equivalent irregular pair*.

Example 7.7. The two images $\mathcal{A}$ and $\mathcal{B}$ of Example 7.6 are irregular images. As another example, consider the following two (degenerate) three-dimensional images:

$$P = \{\langle 1,1,1\rangle,\langle 1,2,2\rangle,\langle 1,3,3\rangle\}$$
$$Q = \{\langle 1,1,3\rangle,\langle 1,2,2\rangle,\langle 1,3,1\rangle\}$$

It can easily be checked that P and Q have the same $\mathbf{F}_1$, $\mathbf{F}_2$, and $\mathbf{F}_3$ functions; that is, P and Q are irregular images. To complete this example, note that

$$\omega_1 : \begin{pmatrix} 1 \\ 1 \end{pmatrix}, \qquad \omega_2 : \begin{pmatrix} 1 & 2 & 3 \\ 3 & 2 & 1 \end{pmatrix}, \qquad \omega_3 : \begin{pmatrix} 1 & 2 & 3 \\ 3 & 2 & 1 \end{pmatrix}$$

We now wish to show that even the construction of irregular images can be carried out in a manner similar to that of Section V. Let us assume that for all $i \in [\mathbf{D}]$,

$$C_{\mathcal{A},i} = \{c_{i,1}, c_{i,2}, \ldots, c_{i,m_i}\}$$

with $m_i \geq 1$. If an irregular image $\mathcal{A}$ is such that each ζ_i contains exactly one cycle, and if $\mathcal{A}$ is the smallest possible such image among the images with same number of columns in each dimension, then the irregular image $\mathcal{A}$ is called a basic irregular image. More formally,

Definition 7.9. An irregular image $\mathcal{A}$ is called a *basic irregular image* iff the mappings ζ_i, $i \in [\mathbf{D}]$, also satisfy the following properties:

(i) $\forall i \in [\mathbf{D}]$, $\forall \alpha \in C_{\mathcal{A},i}$, $\xi_i^k(\alpha) = \alpha$ iff $k \equiv 0$ (modi).

(ii) If $\mathcal{B} \subset \mathcal{A}$ with $C_{\mathcal{A},i} = C_{\mathcal{B},i}$ for all $i \in [\mathbf{D}]$, then $\mathcal{B}$ is not a basic irregular image.

Since the mappings ζ_i contain only one cycle each for the basic irregular images, *we establish the following convention*. Let

$$C_{\mathcal{A},i} = \{c_{i,1}, c_{i,2}, \ldots, c_{i,m_i}\}$$

such that $m_i \in \mathfrak{N}^+$, and for all $j \in [m_i]$, $\zeta_i(c_{i,j}) = c_{i,(j \bmod m_i + 1)}$.

Theorem 7.3. Consider any basic irregular image $\mathcal{A}$ such that

$$\langle c_{1,\nu_1}, c_{2,\nu_2}, \ldots, c_{\mathbf{D},\nu_\mathbf{D}} \rangle \in \mathcal{A}$$

We can uniquely define the image $\mathcal{A}$ as follows:

$$\mathcal{A} = \{\langle c_{1,\alpha_1}, \ldots, c_{\mathbf{D},\alpha_\mathbf{D}} \rangle : \alpha_1 + \alpha_2 + \cdots$$
$$+ \alpha_\mathbf{D} - \nu_1 - \nu_2 - \cdots \nu_\mathbf{D} \equiv 0 \pmod{\kappa}\}$$

where $\kappa = \gcd(m_1, m_2, \ldots, m_\mathbf{D})$. In particular, if $\kappa = 1$, then $\mathcal{A}$ is a filled parallelepiped.

Proof. We need to employ the generalized version of Bezout's identity:

> Consider any $\mathbf{D} + 1$ fixed integers $n_1, n_2, \ldots, n_\mathbf{D}$ and κ. Then the solution to the equation $n_1x_1 + n_2x_2 + \cdots + n_\mathbf{D}x_\mathbf{D} = \kappa$ exists iff κ is divisible by $\gcd(n_1, n_2, \ldots, n_\mathbf{D})$.

Now, the proof is exactly similar to that of Theorem 5.5.

Example 7.8. The two images $\mathcal{A}$ and $\mathcal{B}$ of Example 7.6 are $2 \times 2 \times 2$ basic irregular images. Let us construct the image $\mathcal{A}$. Assume that $\nu_1 = \nu_2 = \nu_3 = 1$. Since $\gcd(2,2,2) = 2$, we obtain

$$\mathcal{A} = \{\langle c_{1,\alpha_1}, c_{2,\alpha_2}, c_{3,\alpha_3} \rangle : \alpha_1 + \alpha_2 + \alpha_3 \equiv 1 \pmod{2}\}$$
$$\Rightarrow \langle \alpha_1, \alpha_2, \alpha_3 \rangle \in \{\langle 1,1,1 \rangle, \langle 1, 2, 2 \rangle, \langle 2, 2, 1 \rangle, \langle 2, 1, 2 \rangle\}$$

The rest of the analysis can be also carried out; however, we will not belabor this point. So far we have concentrated on the construction of a **D**-dimensional irregular image from simpler or smaller **D**-dimensional irregular images. Now we wish to investigate a recursive scheme to construct **D**-dimensional irregular images. In particular, given a $\mathbf{D} - 1$ dimensional irregular images, can we construct a **D**-dimensional irregular image and vice versa? Before plunging into this analysis, it may be helpful to review two examples.

Example 7.9. Consider the two images P and Q that were introduced in Example 7.7. We have

$$P_1(P;1) = \{\langle 1,1 \rangle, \langle 2,2 \rangle, \langle 3,3 \rangle\}$$
$$P_1(Q;1) = \{\langle 1,3 \rangle, \langle 2,2 \rangle, \langle 3,1 \rangle\}$$

$P_1(P;1)$ and $P_1(Q;1)$ are 3×3 two-dimensional basic irregular images (see Example 5.5). The rest of the projections are 1×1 basic irregular images.

Example 7.10. Consider the following irregular image pair:

$$\mathcal{A} = \{\langle 1,1,1\rangle,\langle 1,3,3\rangle,\langle 2,2,2\rangle,\langle 3,1,3\rangle,\langle 3,3,1\rangle\}$$
$$\mathcal{B} = \{\langle 1,1,3\rangle,\langle 1,3,1\rangle,\langle 2,2,2\rangle,\langle 3,1,1\rangle,\langle 3,3,3\rangle\}$$

We have

$$P_1(\mathcal{A};1) = \{\langle 1,1\rangle,\langle 3,3\rangle\} = P_1(\mathcal{B};3)$$
$$P_1(\mathcal{A};2) = \{\langle 2,2\rangle\} = P_1(\mathcal{B};2)$$
$$P_1(\mathcal{A};3) = \{\langle 1,3\rangle,\langle 3,1\rangle\} = P_1(\mathcal{B};1)$$

etc.

Clearly, all the above projections of $\mathcal{A}$ and $\mathcal{B}$ are two-dimensional irregular images. Furthermore, $P_1(\mathcal{A};1)$ and $P_1(\mathcal{B};3)$ are equivalent irregular images. So are $P_1(\mathcal{A};2)$ and $P_1(\mathcal{B};2)$, etc.

Proposition 7.2. Consider a $\mathbf{D} - 1$ dimensional irregular image $\mathcal{A}$ and fix an integer $i \in [\mathbf{D}]$. Define a $\mathbf{D}$-dimensional image F as

$$x \in F \text{ iff } \Omega_i(x) \in \mathcal{A}$$

Furthermore, we require that card $C_{F,i} = 1$. Then F is a $\mathbf{D}$-dimensional irregular image.

Proof. Since $\mathcal{A}$ is a $\mathbf{D} - 1$ dimensional irregular image, there exist $\mathbf{D} - 1$ permutations ζ_j of $C_{\mathcal{A},j}$, $j \in [\mathbf{D} - 1]$, which satisfy the requirements of Definition 7.8 (for $\mathbf{D} - 1$ dimensional images). Now,

$$C_{F,j} = C_{\mathcal{A},j} \qquad \text{for } j = 1,2, \ldots ,i - 1$$
$$C_{F,j+1} = C_{\mathcal{A},j} \qquad \text{for } j = i, i + 1, \ldots ,\mathbf{D} - 1$$
$$C_{F,i} = \{v\} \qquad \text{for some fixed } \varphi \in \mathfrak{N}$$

Let ξ_j be permutations of $C_{F,j}$, $j \in [\mathbf{D}]$, such that

$$\xi_j \equiv \zeta_j \qquad \text{for } j = 1,2, \ldots ,i - 1$$
$$\xi_{j+1} \equiv \zeta_j \qquad \text{for } j = i, i + 1, \ldots ,\mathbf{D} - 1$$
$$\xi_i : \begin{pmatrix} v \\ v \end{pmatrix}$$

Clearly, $\{\xi_i\}$ satisfies the requirements of Definition 7.8 for $\mathbf{D}$-dimensional irregular images.

Proposition 7.3. Let $\mathcal{A}$ be any $\mathbf{D}$-dimensional irregular image. For all $i \in [\mathbf{D}]$ and $v \in C_{\mathcal{A},i}$, $P_i(\mathcal{A};v)$ is a $\mathbf{D} - 1$ dimensional irregular image.

Even though the previous two propositions specify a hierarchical construction mechanism, they fail to provide us with a recursive mechanism to check for irregularity. As of yet, we have not been able to arrive at a recursive irregularity verification algorithm. However, Example 7.10 suggests that such a scheme might exist.

Finally, we are in a position to provide the proof of Theorem 7.1.

Proof of Theorem 7.1. The proof is done by induction. The result clearly holds for one- and two-dimensional images. Next, assume that the result holds for $\mathbf{D} - 1$ dimensional images. Then the entire analysis of the present section is true. Now the arguments needed to complete the proof are exactly similar to those of Theorem 6.2.

(*Sufficiency*) Directly follows from Definition 7.5.

(*Necessity*) Since $\mathbf{C}_{\mathbf{D}-1}$ characterize $\mathbf{D} - 1$ dimensional images uniquely, we conclude that (see Lemmas 7.1 and 7.2 and Theorem 7.2) for all $i \in [\mathbf{D}]$.

(i) $W_i(\mathcal{A}) = W_i(\mathcal{B})$.

(ii) $\max C_{\mathcal{A},i} - \max C_{\hat{\mathcal{A}},i} = \max C_{\mathcal{B},i} - \max C_{\hat{\mathcal{B}},i}$.

(iii) card $\mathcal{A}$ = card $\mathcal{B}$.

Assume, without any loss of generality, that $C_{\mathcal{A},i} = C_{\mathcal{B},i}$, for all $i \in [\mathbf{D}]$. Let $\theta : \mathcal{A} \to \mathcal{B}$ be such that for every $n \in \mathcal{T}_{\mathbf{D}}$,

$$\alpha \text{ contributes 1 to } \mathbf{C}_{\mathbf{D}}(\mathcal{A},\hat{\mathcal{A}};n) \text{ iff } \theta(\alpha) \text{ contributes 1 to } \mathbf{C}_{\mathbf{D}}(\mathcal{B},\hat{\mathcal{B}};n)$$

Now all we need to show is that θ is an identity mapping. As was the case in Section V, proceeding along the lines in [46], we can show that θ can be written as

$$\theta\left(\langle\alpha_1,\alpha_2, \ldots ,\alpha_{\mathbf{D}}\rangle\right) = \left\langle\omega_1(\alpha_1),\omega_2(\alpha_2), \ldots ,\omega_{\mathbf{D}}(\alpha_{\mathbf{D}})\right\rangle$$

for some appropriate choice of functions $\left\{\omega_i : C_{\mathcal{A},i} \to C_{\mathcal{B},i}\right\}_{i\in[\mathbf{D}]}$.

Now, consider any $\langle\delta_1,\delta_2, \ldots ;\delta_{\mathbf{D}}\rangle \in \mathcal{A}$, and fix $\eta_1,\eta_2, \ldots ,\eta_{\mathbf{D}} \in \Re^+$ such that for all $i \in [\mathbf{D}]$, $\delta_i + \eta_i = \max C_{\hat{\mathcal{A}},i}$. Now,

$$\forall i \in [\mathbf{D}], \qquad \delta_i < \omega_i(\delta_i) \Rightarrow \omega_i(\delta_i) + \eta_i \notin C_{\mathcal{B},i}$$

Therefore, if $\delta_i < \omega_i(\delta_i)$, then

$$\theta\left(\langle\alpha_1\alpha_2, \ldots ,\alpha_{\mathbf{D}}\rangle\right) + \langle\eta_1,\eta_2, \ldots ,\eta_{\mathbf{D}}\rangle \notin \hat{\mathcal{B}}$$

Hence, we must have $\delta_i \geq \omega_i(\delta_i)$ for all $i \in [\mathbf{D}]$. This implies that the ω_i's are identity mappings. Hence the result.

D. Another Pseudocharacterization: H_i Functions

In Section VII.C we viewed a **D**-dimensional image as $\mathbf{D} - 1$ dimensional hyperplanes stacked on each other. In this section, we project the given image on a two-dimensional plane in **D** different ways, one for each dimension. For the projection corresponding to ith dimension, we derive a feature based on the ith distance measure. The characterization of the original image based on these $\mathbf{D}$ features is then investigated.

Again, note that as was the case in Section V, we will assume that $w_i = 1$ for all $i \in [\mathbf{D}]$, in Definition 7.3, unless otherwise specified.

Definition 7.10. For any $i \in [\mathbf{D}]$ define the *line images*

$$\{\mathcal{A}_{i;\alpha} : \alpha \in R_{\mathcal{A},i}\} \quad \text{and} \quad \{\hat{\mathcal{A}}_{i;\alpha} : \alpha \in R_{\mathcal{A},i}\}$$

in the ith dimension as

$$\mathcal{A}_{i;\alpha} = \{\Pi_i(x) : x \in \mathcal{A} \text{ and } \Omega_i(x) = \alpha\}$$

$$\hat{\mathcal{A}}_{i;\alpha} = \{\Pi_i(x) : x \in \hat{\mathcal{A}} \text{ and } \Omega_i(x) = \alpha\}$$

In Example 7.1, if we let $i = 3$, then the two line images are

$$\mathcal{A}_{3;\langle 0,1\rangle} = \{0,1,2\} \quad \text{and} \quad \mathcal{A}_{3;\langle 0,3;\rangle} = \{0,3\}$$

Also, one can easily derive the following relations for all $i \in [\mathbf{D}]$:

$$\begin{aligned} C_{\mathcal{A},i} &= \cup \{\mathcal{A}_{i;\alpha} : \alpha \in R_{\mathcal{A},i}\} \\ \mathcal{A} &= \{x \in \Re^{\mathbf{D}} : \Pi_i(x) \in \mathcal{A}_{i;\alpha} \text{ and } \Omega_i(x) = \alpha\} \qquad (11.39) \\ \hat{\mathcal{A}} &= \{x \in \Re^{\mathbf{D}} : \Pi_i(x) \in \hat{\mathcal{A}}_{i;\alpha} \text{ and } \Omega_i(x) = \alpha\} \end{aligned}$$

As was the case with two-dimensional images, we define $\mathbf{D}$ functions on the set of $\mathbf{D}$-dimensional images as follows:

$$\{\mathbf{H}_i : \mathbf{P} \times \Re^{+} \to \Re^{+} \cup \{0\}\}_{i \in [\mathbf{D}]}$$

Definition 7.11. Consider any $i \in [\mathbf{D}]$. For any $\mathbf{D}$-dimensional image $\mathcal{A}$ and $n \in \Re^{+}$, define

$$\begin{aligned} \mathbf{H}_i(\mathcal{A},n) = \operatorname{card}\{x \in \mathcal{A} : \exists y \in \hat{\mathcal{A}} \ni \Pi_i(y) - \Pi_i(x) = n \\ \text{and } \Omega_i(y) = \Omega_i(x)\} \end{aligned}$$

Analogous to the Theorem 5.1, we have for every $i \in [\mathbf{D}]$,

$$\mathbf{H}_i(\mathcal{A},n) = \sum_{\alpha \in R_{\mathcal{A},i}} \lambda^{+}\left[\mathcal{A}_{i;\alpha}, \hat{\mathcal{A}}_{i;\alpha}; n\right] \qquad (11.40)$$

Equation (11.40) suggests a representation scheme by which the manual computation of $\mathbf{H}_i$ for $\mathbf{D}$-dimensional images is identical to the computation of the $\mathbf{H}^{+}$ function for two-dimensional images. Furthermore, since $\hat{\mathcal{A}}_{i;\alpha}$ is not the pseudo-complement of $\mathcal{A}_{i;\alpha}$, it justifies why we chose to include it explicitly in the definition of the function λs.

Definition 7.12. For any image $\mathcal{A}$, define the sets $R_{\mathcal{A},i}$, $i \in [\mathbf{D}]$, as

$$R_{\mathcal{A},i} = \left\{ r_{i,1}, r_{i,2}, \ldots, r_{i,\eta_i} \right\}$$

Consider any $i \in [\mathbf{D}]$, and define the meta image $\mathfrak{M}_i(\mathcal{A})$ of $\mathcal{A}$ to be a two-dimensional image such that

$$\forall v \in [\eta_i], \qquad \mathfrak{M}_i(\mathcal{A})_{[v]} = \mathcal{A}_{i;r_{i,v}}$$

Define the pseudocomplement of $\mathfrak{M}_i(\mathcal{A})$ as $\mathfrak{M}_i(\hat{\mathcal{A}})$ (otherwise, as stated earlier, this relation is not true).

Relation (11.40) can now be restated as

$$\forall i \in [\mathbf{D}], \qquad \mathbf{H}_i(\mathcal{A},\cdot) = \mathbf{H}^+\left(\mathfrak{M}_i(\mathcal{A}),\cdot\right) \tag{11.41}$$

Hence, it should come as no surprise that the analog of Theorem 5.2 also holds. Before we state the result, it may be helpful first to consider an example.

Example 7.11. Consider the following two 3-dimensional images:

$$f = \{\langle 1,1,1\rangle, \langle 1,1,2\rangle, \langle 1,2,2\rangle, \langle 1,2,3\rangle\}$$
$$g = \{\langle 1,1,2\rangle, \langle 1,1,3\rangle, \langle 2,2,1\rangle, \langle 2,2,2\rangle\}$$

Consider the computation of $\mathbf{H}_3(f,\cdot)$ and $\mathbf{H}_3(g,\cdot)$. For this purpose, it may be helpful to visualize the images f and g as the two-dimensional meta images F and G of Figure 30 respectively. The rows and columns of the meta image F are

$$R_{f,3} = \{\langle 1,1\rangle, \langle 1,2\rangle\} \qquad \text{and} \qquad C_{f,3} = \{1,2,3\}$$

respectively. Similarly, the meta image G can be constructed from g. From this construction and Eq. (11.40), it follows that the computation of $\mathbf{H}_3(f,\cdot)$ is exactly same as the computation of $\mathbf{H}^+(F,\cdot)$. Now, from Theorem 5.2, we conclude that f and g have identical $\mathbf{H}_3$ functions. The row mapping is as follows:

$$\text{meta row mapping:} \qquad \begin{pmatrix} \langle 1,1\rangle & \langle 1,2\rangle \\ \langle 2,2\rangle & \langle 1,1\rangle \end{pmatrix}$$

Similarly, one can show that $\mathbf{H}_1$ and $\mathbf{H}_2$ functions of $\mathcal{A}$ and $\mathcal{B}$ are different. (See Figures 31 and 32.)

Theorem 7.4. For any $i \in [\mathbf{D}]$, $\mathbf{H}_i(\mathcal{A},\cdot) = \mathbf{H}_i(\mathcal{B},\cdot)$ iff there exist a bijective mapping $\varphi_i : R_{\mathcal{A},i} \rightarrow R_{\mathcal{B},i}$ such that for some *fixed* $v \in \mathfrak{R}$ we have

$$\forall \alpha \in R_{\mathcal{A},i}, \qquad \mathcal{A}_{i;\alpha} = \mathbf{T}\left[\mathcal{B}_{i;\varphi_i(\alpha)}; v\right]$$

Proof. Directly follows from Definition 7.12 and Theorem 5.2.

Proceeding along the lines of Section V, we investigate the implications of $\mathbf{H}_i(\mathcal{A},\cdot) = \mathbf{H}_i(\mathcal{B},\cdot)$ for all $i \in [\mathbf{D}]$. From Theorem 7.4, we know the existence

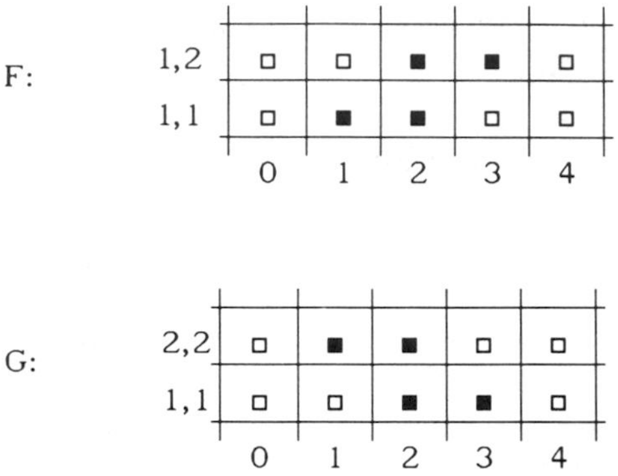

Figure 30. Representation of two 3-dimensional images of Example 7.11 as 2-dimensional "meta images" for the purpose of computing their $\mathbf{H}_3$ function.

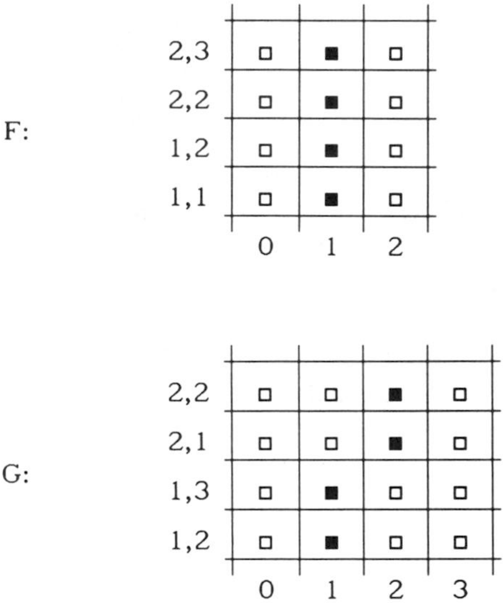

Figure 31. Representation of two 3-dimensional images of Example 7.11 as 2-dimensional "meta images" for the purpose of computing their $\mathbf{H}_1$ function.

of a collection $\{\varphi_i : R_{\mathcal{A},i} \to R_{\mathcal{B},i}\}_{i\in[\mathbf{D}]}$ of bijective mappings. In Section V, the irregular images are defined for two-dimensional images. The image $\mathcal{A}$ is irregular iff there exist two permutations ζ and ξ such that

$$\langle i,j\rangle \in \mathcal{A} \text{ iff } \langle \zeta(i),\xi^{-1}(j)\rangle \in \mathcal{A} \tag{11.42}$$

and ζ and ξ are not an identity mapping.

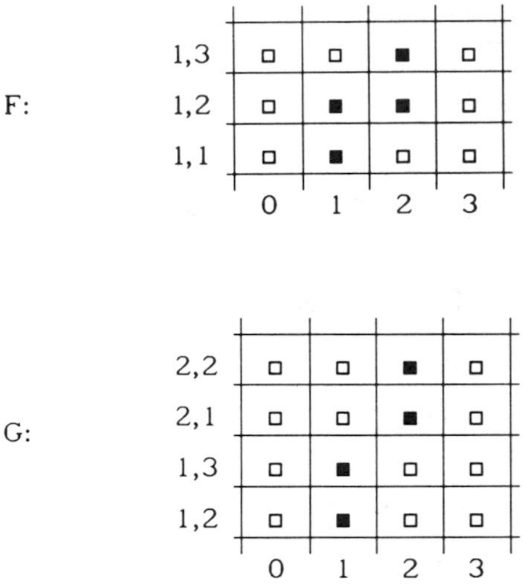

Figure 32. Representation of two 3-dimensional images of Example 7.11 as 2-dimensional "meta images" for the purpose of computing their $\mathbf{H}_2$ function.

Consider three-dimensional image $\mathcal{B}$ and let $\mathfrak{M}_3(\mathcal{B})$ be the corresponding meta image for the z direction. If we apply Eq. (11.42) to the image $\mathfrak{M}_3(\mathcal{B})$,

$$\langle r_{3,\nu}, z\rangle \in \mathcal{A} \text{ iff } \langle \varphi_3(r_{3,\nu}), \xi^{-1}(z)\rangle \in \mathcal{A} \tag{11.43}$$

where φ_3 is a row mapping of meta images defined in Theorem 7.1. Therefore, the definition of three-dimensional irregular images looks pretty much like that of two-dimensional irregular images. However, the mapping $\varphi_3(r_{3,\nu})$ in Eq. (11.43) is not convenient for our purpose. In Example 7.11, the mappings φ_3 is defined as follows:

$$\varphi_3(1,1) = \langle 2,2\rangle \qquad \text{and} \qquad \varphi_3(1,2) = \langle 1,1\rangle$$

As $\varphi_3(1,\cdot) = \{\langle 1,\cdot\rangle, \langle 2,\cdot\rangle\}$, φ_3 is not even a function in the first variable. This fact complicates the analysis of images which are of dimension greater than 2. In order to obtain results similar to those of Section V, we must define the expression "ignoring translation." By this, we will imply

$$C_{\mathcal{B},i} = C_{\mathcal{A},i} \qquad \text{for all } i \in [\mathbf{D}]$$

Since the domains $R_{\mathcal{A},i}$ of different φ_i are different, it is convenient for our purpose to extend their domain of definition as follows.

Definition 7.13. Define a collection of mappings

$$\{\Gamma_i : \mathcal{A} \rightarrow \mathcal{B}\}_{i\in[}\mathbf{D}]$$

such that for all $i \in [\mathrm{D}]$, and all $x \in \mathcal{A}$,

Example 7.12. Consider the two images f and g defined in Example 7.11. Then the mapping Γ_3 can be defined as follows:

$$\Gamma_3(1,1,1) = \langle 2,2,1\rangle \qquad \Gamma_3(1,1,2) = \langle 2,2,2\rangle$$
$$\Gamma_3(1,2,2) = \langle 1,1,2\rangle \qquad \Gamma_3(1,2,3) = \langle 1,1,3\rangle$$

Since φ_i's are objective mappings, one can also define the collection of mappings $\{\Gamma_i^{-1} : ,\mathcal{B} \rightarrow \mathcal{A}\}_{i\in[\mathbf{D}]}$ such that

$$\forall i \in [\mathbf{D}], \forall x \in \mathcal{A}, \qquad \Pi_i[\Gamma_i^{-1}(x)] = \Pi_i(x) \qquad \text{and}$$
$$\Omega_i[\Gamma_i^{-1}(x)] = \varphi_i^{-1}(x)$$

Clearly, the mapping Γ_i^{-1} is the inverse of the mapping Γ_i, that is,

$$\forall i \in [\mathbf{D}], \forall x \in \mathcal{A}, \qquad \Gamma_i[\Gamma_i^{-1}(x)] = x = \Gamma_i^{-1}[\Gamma_i(x)]$$

Before we go any further, it must be noted that the mapping Γ_i will be assumed to be an identity mapping if there exists a identity mapping that satisfies the requirements of Theorem 7.4.

Example 7.13. Consider the following two 3-dimensional images f and g.

$$f = \{\langle 1,1,1\rangle, \langle 3,1,1\rangle, \langle 2,2,2\rangle, \langle 1,3,3\rangle, \langle 3,3,3\rangle\}$$
$$g = \{\langle 1,1,1\rangle, \langle 3,1,1\rangle, \langle 2,2,2\rangle, \langle 1,3,3\rangle, \langle 3,3,3\rangle\}$$

Let us consider the mapping Γ_1. The meta images $\mathfrak{M}_1(f)$ and $\mathfrak{M}_1(g)$ are shown in Figure 33. The mapping Γ_1 can be defined in two ways. We illustrate two mappings Γ_1' and Γ_1'' in Table 3. Similarly, the mappings Γ_2 and Γ_3 can be defined. Since there exist an identity mapping for all i $\in$ [3], we conclude that image $\mathcal{A}$ and $\mathcal{B}$ are not equivalent irregular images.

Armed with these definitions, we are now able to state the analog of Proposition 5.1 for **D**-dimensional images.

Theorem 7.5. Consider two images $\mathcal{A}$ and $\mathcal{B}$ such that $\mathbf{H}_i(\mathcal{A}\cdot) = \mathbf{H}_i(\mathcal{B},\cdot)$ for all $i \in [\mathbf{D}]$. The voxels of $\mathcal{A}$ satisfy the following property:

$$x \in \mathcal{A} \text{ iff } \Gamma_{\alpha_1}^{-1}\left[\Gamma_{\alpha_2}\left[\Gamma_{\alpha_3}^{-1}[\cdots \Gamma_{\alpha_{2k}}(x)\cdots]\right]\right] \in \mathcal{A}$$

where $k \geq 1$ and, for all $j \in [2k]$, $\alpha_j \in [\mathbf{D}]$.

Proof. Each Γ_i maps a voxel in $\mathcal{A}$ to a voxel in $\mathcal{B}$. If we wish to map a voxel in $\mathcal{A}$ to another voxel in $\mathcal{A}$ (as was done in the proof of Proposition 5.1), then we need to employ a pair of mappings $\langle\Gamma_p,\Gamma_q^{-1}\rangle$. Since $\Gamma_p \bigcirc \Gamma_q^{-1}$ is not well defined, a fortiori we must use $\Gamma_q^{-1} \bigcirc \Gamma_p : \mathcal{A} \rightarrow \mathcal{A}$ for this purpose. Now, note that

$$x \in \mathcal{A} \text{ iff } \Gamma_{\alpha_p}(x) \in \mathcal{B} \text{ iff } \Gamma_{\alpha_q}(x) \in \mathcal{B}$$

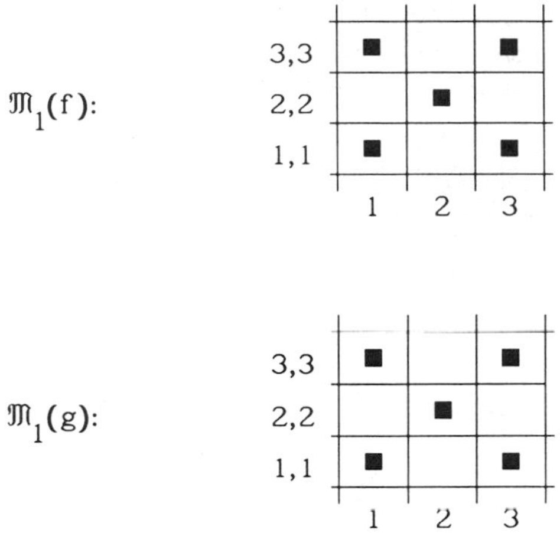

Figure 33. Meta images correspond to image F and G.

Table 3. Two Possible Γ_1 Mappings for the Images f and g of Example 7.13

x	$\Gamma_1'(x)$	$\Gamma_1''(x)$
⟨1,1,1⟩	⟨1,1,1⟩	⟨1,3,3⟩
⟨3,1,1⟩	⟨3,1,1⟩	⟨2,3,3⟩
⟨2,2,2⟩	⟨2,2,2⟩	⟨2,2,2⟩
⟨1,3,3⟩	⟨1,3,3⟩	⟨1,1,1⟩
⟨3,3,3⟩	⟨3,3,3⟩	⟨3,1,1⟩

Hence,

$$x \in \mathcal{A} \text{ iff } \Gamma_{\alpha_q}^{-1}[\Gamma_{\alpha_p}(x)] \in \mathcal{A}$$

By induction, one can easily verify that

$$\Gamma_{\alpha_1}^{-1} \bigcirc \Gamma_{\alpha_2} \bigcirc \Gamma_{\alpha_3}^{-1} \bigcirc \bigcirc \cdots \bigcirc \Gamma_{\alpha_{2k}} : \mathcal{A} \to \mathcal{A} \tag{11.44}$$

where each $\alpha_i \in [\mathbf{D}]$.

Remarks. If $\Gamma_q \equiv \Gamma_p$, then we obtain $\Gamma_q^{-1} \bigcirc \Gamma_p$ as an identity mapping. Hence, if one wishes to have (11.44) in the most compact form, then an additional requirement must be placed: $\Gamma_{\alpha_j} \neq \Gamma_{\alpha_{j+1}}$ for all $j \in [2k - 1]$. At the very least, we must require that $\alpha_j \neq \alpha_{j+1}$.

In Theorem 7.5, if all Γ_i are identity mappings, then the two images $\mathcal{A}$ and $\mathcal{B}$ are identical. As in Section V, one can show that if one Γ_j is a nonidentity mapping for some $j \in [\mathbf{D}]$, then all Γ_i are nonidentity mappings. If none of the Γ_i is an identity mapping, then we call $\mathcal{A}$ and $\mathcal{B}$ *equivalent irregular images*.

Definition 7.14. A **D**-dimensional image $\mathcal{A}$ will be called irregular image iff there exist permutations Γ_i and Γ_j, $i,j \in [\mathbf{D}]$ and $i \neq j$, defined in Definition 7.13 such that

$$x \in \mathcal{A} \text{ iff } \Gamma_j^{-1}(\Gamma_i(x)) \in \mathcal{A}$$

and Γ_i and Γ_j are not identity mappings.

Corollary 7.1 (Proposition 4.1). Consider two 2-dimensional images $\mathcal{A}$ and $\mathcal{B}$ such that $\mathbf{H}_1(\mathcal{A},\cdot) = \mathbf{H}_1(\mathcal{B},\cdot)$ and $\mathbf{H}_2(\mathcal{A},\cdot) = \mathbf{H}_2(\mathcal{B},\cdot)$. Then the pixels of $\mathcal{A}$ satisfy the following property: for all integers k $\neq$ 0,

$$\langle x, y\rangle \in \mathcal{A} \text{ iff } \langle \varphi_2^k(x), \varphi_1^{-k}(y)\rangle \in \mathcal{A}$$

Example 7.14. Consider the following two 2-dimensional images $\mathcal{A}$ and $\mathcal{B}$. (see Figure 21):

$$\mathcal{A} = \{\langle 1,1\rangle, \langle 2,2\rangle, \langle 3,3\rangle\}$$
$$\mathcal{B} = \{\langle 1,3\rangle, \langle 2,2\rangle, \langle 3,1\rangle\}$$

The two mappings, Γ_1 and Γ_2, are defined in Table 4. The application of Theorem 7.5 to the pixels in $\mathcal{A}$, results in

$$\langle 1,1\rangle \in \mathcal{A} \Leftrightarrow \Gamma_1^{-1}[\Gamma_2(1,1)] = \Gamma_1^{-1}(3,1) = \langle 3,3\rangle \in \mathcal{A}$$
$$\langle 2,2\rangle \in \mathcal{A} \Leftrightarrow \Gamma_1^{-1}[\Gamma_2(2,2)] = \Gamma_1^{-1}(2,2) = \langle 2,2\rangle \in \mathcal{A}$$
$$\langle 3,3\rangle \in \mathcal{A} \Leftrightarrow \Gamma_1^{-1}[\Gamma_2(3,3)] = \Gamma_1^{-1}(1,3) = \langle 1,1\rangle \in \mathcal{A}$$

Arriving at a similar result from **D**-dimensional image, $\mathbf{D} \geq 3$, results in a combinatorial explosion. The form (11.44) gives rise to at most $\mathbf{D}\cdot(\mathbf{D} - 1)^{2k-1}$ nonredundant mappings for every value of k; however, not all these mappings are necessarily distinct. The total number of distinct nonredundant mappings is $\eta!$ where η is the cardinality of image $\mathcal{A}$.

Table 4. Γ_1 and Γ_2 Mappings for the Images of Example 7.14

x	$\Gamma_1(x)$	$\Gamma_2(x)$
$\langle 1,1\rangle$	$\langle 1,3\rangle$	$\langle 3,1\rangle$
$\langle 2,2\rangle$	$\langle 2,2\rangle$	$\langle 2,2\rangle$
$\langle 3,3\rangle$	$\langle 3,1\rangle$	$\langle 1,3\rangle$

We now illustrate Theorem 7.5 with the help of a few three-dimensional equivalent irregular images.

Example 7.15. Consider the following two 3-dimensional images $\mathcal{A}$ and $\mathcal{B}$, such that $\forall i \in [3]$, $\mathbf{H}_i^+(\mathcal{A},\cdot) = \mathbf{H}_i^+(\mathcal{B},\cdot)$.

$$\mathcal{A} = \{\langle 1,1,1\rangle, \langle 2,2,2\rangle, \langle 3,3,3\rangle\}$$
$$\mathcal{B} = \{\langle 3,1,1\rangle, \langle 2,2,2\rangle, \langle 1,3,3\rangle\}$$

The mappings Γ_i are defined in Table 5. Now, we apply Theorem 7.5 to the voxels in $\mathcal{A}$. As

$$\Gamma_2^{-1}[\Gamma_1(1,1,1)] = \Gamma_2^{-1}(1,3,3) = \langle 3,3,3\rangle$$
$$\Gamma_2^{-1}[\Gamma_1(2,2,2)] = \Gamma_2^{-1}(2,2,2) = \langle 2,2,2\rangle$$

we conclude that

$$\langle 1,1,1\rangle \in \mathcal{A} \Leftrightarrow \langle 3,3,3\rangle \in \mathcal{A}$$
$$\langle 2,2,2\rangle \in \mathcal{A} \Leftrightarrow \langle 2,2,2\rangle \in \mathcal{A}$$

Finally, we provide an example of a disguised identity function:

$$\Gamma_3^{-1} \bigcirc \Gamma_1 \bigcirc \Gamma_2^{-1} \bigcirc \Gamma_1 \equiv \left[\Gamma_3^{-1} \bigcirc \Gamma_1\right]^2$$

Example 7.15 dealt with simple equivalent irregular images. Our next example shows a pair of nontrivial equivalent irregular images.

Example 7.16. Consider the two Chinese characters of Figure 3. As observed there, these two images have different $\mathbf{V}^+$ functions and hence cannot be irregular. However, one can construct two 3-dimensional images based on these characters, which are irregular. For example, consider the two images $\mathcal{A}$ and $\mathcal{B}$ whose meta images $\mathfrak{M}_1(\mathcal{A})$ and $\mathfrak{M}_1(\mathcal{B})$ are shown in Figure 34. The mappings Γ_i, $i \in [3]$, are given in Table 6. Note that Γ_1 and Γ_2 are identity mappings. We obtain the following equivalences:

Table 5. The Γ_i Mappings for the Images of Example 7.15

x	$\Gamma_1(x)$	$\Gamma_2(x)$	$\Gamma_3(x)$
$\langle 1,1,1\rangle$	$\langle 1,3,3\rangle$	$\langle 3,1,1\rangle$	$\langle 3,1,1\rangle$
$\langle 2,2,2\rangle$	$\langle 2,2,2\rangle$	$\langle 2,2,2\rangle$	$\langle 2,2,2\rangle$
$\langle 3,3,3\rangle$	$\langle 3,1,1\rangle$	$\langle 1,3,3\rangle$	$\langle 1,3,3\rangle$

$\mathfrak{M}_1(\mathscr{A})$:

$\mathfrak{M}_1(\mathscr{B})$:

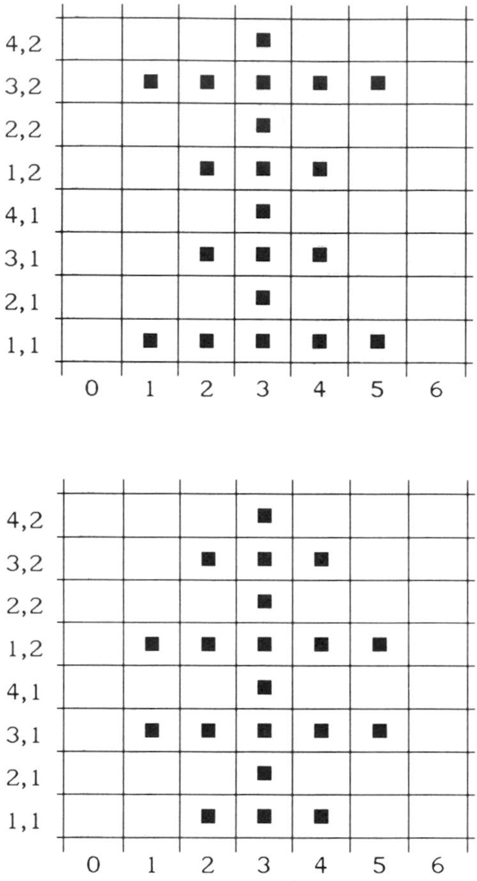

Figure 34. Two 3-dimensional images that have been constructed from the Chinese characters introduced in Figure 30. The images of $\mathscr{A}$ and $\mathscr{B}$ in 1st dimension were defined as the union of these two characters.

$$\langle 1,1,1\rangle \in \mathscr{A} \Leftrightarrow \Gamma_1^{-1}\,[\Gamma_3(1,1,1)] = \langle 1,3,2\rangle \in \mathscr{A}$$

$$\langle 2,1,1\rangle \in \mathscr{A} \Leftrightarrow \Gamma_1^{-1}\,[\Gamma_3(2,1,1)] = \langle 2,3,2\rangle \in \mathscr{A}$$

$$\langle 3,1,1\rangle \in \mathscr{A} \Leftrightarrow \Gamma_1^{-1}\,[\Gamma_3(3,1,1)] = \langle 3,3,2\rangle \in \mathscr{A}$$

$$\langle 4,1,1\rangle \in \mathscr{A} \Leftrightarrow \Gamma_1^{-1}[\Gamma_3\,(4,1,1)] = \langle 4,3,2\rangle \in \mathscr{A}$$

$$\langle 5,1,1\rangle \in \mathscr{A} \Leftrightarrow \Gamma_1^{-1}\,[\Gamma_3(5,1,1)] = \langle 5,3,2\rangle \in \mathscr{A}$$

$$\langle 3,2,1\rangle \in \mathscr{A} \Leftrightarrow \Gamma_1^{-1}\,[\Gamma_3(3,2,1)] = \langle 3,4,2\rangle \in \mathscr{A}$$

Table 6. The Three Different Mappings Between the Two Images of Example 7.16

x	$\Gamma_1(x)$	$\Gamma_2(x)$	$\Gamma_3(x)$
⟨1,1,1⟩	⟨1,1,2⟩	⟨1,1,2⟩	⟨1,3,1⟩
⟨2,1,1⟩	⟨2,1,2⟩	⟨2,1,2⟩	⟨2,3,1⟩
⟨3,1,1⟩	⟨3,1,2⟩	⟨3,1,2⟩	⟨3,3,1⟩
⟨4,1,1⟩	⟨4,1,2⟩	⟨4,1,2⟩	⟨4,3,1⟩
⟨5,1,1⟩	⟨5,1,2⟩	⟨5,1,2⟩	⟨5,3,1⟩
⟨3,2,1⟩	⟨3,2,2⟩	⟨3,2,2⟩	⟨3,4,1⟩
⟨2,3,1⟩	⟨2,3,2⟩	⟨2,3,2⟩	⟨2,1,1⟩
⟨3,3,1⟩	⟨3,3,2⟩	⟨3,3,2⟩	⟨3,1,1⟩
⟨4,3,1⟩	⟨4,3,2⟩	⟨4,3,2⟩	⟨4,1,1⟩
⟨3,4,1⟩	⟨3,4,2⟩	⟨3,4,2⟩	⟨3,2,1⟩
⟨2,1,2⟩	⟨2,1,1⟩	⟨2,1,1⟩	⟨2,3,2⟩
⟨3,1,2⟩	⟨3,1,1⟩	⟨3,1,1⟩	⟨3,3,2⟩
⟨4,1,2⟩	⟨4,1,1⟩	⟨4,1,1⟩	⟨4,3,2⟩
⟨3,2,2⟩	⟨3,2,1⟩	⟨3,2,1⟩	⟨3,4,2⟩
⟨1,3,2⟩	⟨1,3,1⟩	⟨1,3,1⟩	⟨1,1,2⟩
⟨2,3,2⟩	⟨2,3,1⟩	⟨2,3,1⟩	⟨2,1,2⟩
⟨3,3,2⟩	⟨3,3,1⟩	⟨3,3,1⟩	⟨3,1,2⟩
⟨4,3,2⟩	⟨4,3,1⟩	⟨4,3,1⟩	⟨4,1,2⟩
⟨5,3,2⟩	⟨5,3,1⟩	⟨5,3,1⟩	⟨5,1,2⟩
⟨3,4,2⟩	⟨3,4,1⟩	⟨3,4,1⟩	⟨3,2,2⟩

$$\langle 2,3,1\rangle \in \mathcal{A} \Leftrightarrow \Gamma_1^{-1}[\Gamma_3(2,3,1)] = \langle 2,1,2\rangle \in \mathcal{A}$$

$$\langle 3,3,1\rangle \in \mathcal{A} \Leftrightarrow \Gamma_1^{-1}[\Gamma_3(3,3,1)] = \langle 3,1,2\rangle \in \mathcal{A}$$

$$\langle 4,3,1\rangle \in \mathcal{A} \Leftrightarrow \Gamma_1^{-1}[\Gamma_3(4,3,1)] = \langle 4,1,2\rangle \in \mathcal{A}$$

$$\langle 3,4,1\rangle \in \mathcal{A} \Leftrightarrow \Gamma_1^{-1}[\Gamma_3(3,4,1)] = \langle 3,2,2\rangle \in \mathcal{A}$$

To conclude, we state that the result of Theorem 7.5 cannot be improved on to obtain a result similar to Proposition 7.1. In other words, we do not know of a mechanical scheme for the construction of irregular images. Herein lies the deficiency of this particular generalization.

E. Concluding Remarks

Since we have proposed two pseudocharacterizations, one may wonder which one is better. The following theorem answers that question.

Theorem 7.6. Consider two images $\mathcal{A}$ and $\mathcal{B}$ for which

$$\forall i \in [\mathbf{D}], \qquad \mathrm{F}_i(\mathcal{A};\cdot) = \mathbf{F}_i(\mathcal{B};\cdot)$$

Then we must have

$$\forall i \in [\mathbf{D}], \qquad \mathbf{H}_i(\mathcal{A},\cdot) = \mathbf{H}_i(\mathcal{B},\cdot)$$

Proof. For all $i \in [\mathbf{D}]$, $\mathbf{F}_i(\mathcal{A},\cdot) = \mathbf{F}_i(\mathcal{B};\cdot)$, there exist bijective mappings $\omega_i : C_{\mathcal{A},i} \rightarrow C_{\mathcal{B},i}$ between columns of $\mathcal{A}$ and $\mathcal{B}$. Now, all we have to show is that there exist bijective mappings between rows of $\mathcal{A}$ and $\mathcal{B}$:

$$\forall i \in [\mathbf{D}], \qquad \exists \varphi_i : R_{\mathcal{A},i} \rightarrow R_{\mathcal{B},i} \ni \varphi_i \text{ is bijective}$$

Fix any $i \in [\mathbf{D}]$. Consider any $\langle \alpha_1,\alpha_2, \ldots ;\alpha_{\mathbf{D}}\rangle \in \mathcal{A}$. Then

$$\langle \alpha_1, \ldots \alpha_{i-1},\alpha_{i+1}, \ldots ,\alpha_{\mathbf{D}}\rangle \in R_{\mathcal{A},i}$$

Define φ_i as follows:

$$\varphi_i(\alpha_1, \ldots \alpha_{i-1},\alpha_{i+1}, \ldots ,\alpha_{\mathbf{D}})$$
$$= \langle \omega_1(\alpha_1), \ldots ,\omega_{i-1}(\alpha_{i-1}),\omega_{i+1}(\alpha_{i+1}), \ldots ,\omega_{\mathbf{D}}(\alpha_{\mathbf{D}})\rangle$$

Then φ_i is clearly well defined. Now, since ω_i's are bijective mappings, φ_i must also be a bijective mapping.

To show that the converse of the above result is not true, consider the following example.

Example 7.17. Consider two images $\mathcal{A}$ and $\mathcal{B}$ of Example 7.15. They had same $\mathbf{H}_1$, $\mathbf{H}_2$, and $\mathbf{H}_3$ functions. However, $\mathbf{F}_2(\mathcal{A};3,3) = 1 \neq 0 = \mathbf{F}_2(\mathcal{B};3,3)$.

As we did in Section III for the two-dimensional case, we can express all the functions introduced earlier in a morphological framework. We again assume the lattice $\mathbf{L}$ and the transformation $\mathfrak{L}$ of Section III. We define three parameterized families of $\mathbf{D}$-dimensional structuring elements

$$\{H_{i,n}\}_{i\in[\mathbf{D}],\ n\in\mathfrak{R}^+}, \qquad \{F_{i,m}\}_{i\in[\mathbf{D}],m\in\mathcal{T}_{\mathbf{D}}-1}, \qquad \{C_N\}_{N\in\mathcal{T}_{\mathbf{D}}}$$

as follows: For all $v \in \mathfrak{R}^{\mathbf{D}}$,

$$H_{i,n}(v) = \begin{cases} 1 & \text{if } v = \langle 0,0, \ldots ,0\rangle \\ 0 & \text{if } \Omega_i(v) = \langle 0,0, \ldots ,0\rangle \text{ and } \Pi_i(v) = n \\ \alpha & \text{otherwise} \end{cases}$$

$$F_{i,m}(v) = \begin{cases} 1 & \text{if } v = \langle 0,0, \ldots ,0\rangle \\ 0 & \text{if } \Omega_i(v) = m \text{ and } \pi_i(v) = 0 \\ \alpha & \text{otherwise} \end{cases}$$

$$C_N(v) = \begin{cases} 1 & \text{if } v = \langle 0,0, \ldots ,0\rangle \\ 0 & \text{if } v = N \\ \alpha & \text{otherwise} \end{cases}$$

Theorem 7.7. For any observed image $\mathcal{A}$, $i \in [\mathbf{D}]$, $n \in \mathfrak{R}^+$, $m \in \mathfrak{T}_{\mathbf{D}-1}$, and $N \in \mathfrak{T}_{\mathbf{D}}$, we have

$$\mathbf{H}_i(\mathcal{A};n) = \delta[\mathbf{OPEN}(\mathfrak{L}(\mathcal{A});H_{i,n})]$$

$$F_i(\mathcal{A};m) = \delta[\mathbf{OPEN}(\mathfrak{L}(\mathcal{A});F_{i,m})]$$

$$\mathbf{C}_{\mathbf{D}}(\mathcal{A};N) = \delta[\mathbf{OPEN}(\mathfrak{L}(\mathcal{A});C_N)]$$

VIII. CONCLUSIONS

In this chapter we have presented a few morphologically derived features for complete as well as pseudocharacterization of multidimensional objects irrespective of their position, orientation, and size. A salient feature of these algorithms is that they are implementable on an as-is basis (unlike the joint central moments and Fourier descriptors, where one is theoretically required to store infinitely many values). Our characterizations, as opposed to existing characterizations in the literature, are based on algorithms that can be implemented. All the algorithms for pseudocharacterization lend themselves to parallel implementation. Moreover, their computation can be carried out very easily.

The algorithms presented in this chapter and those presented in [5–7] can be employed for a variety of tasks. In references [12] and [16,17] one can find the applications of the **HOR**, **VER**, $\mathbf{H}^+$, $\mathbf{H}^-$, $\mathbf{V}^+$, and $\mathbf{V}^-$ functions in printed character recognition and in the recognition of "complex" objects that are encountered in an industrial environment.

The function $\mathbf{C}_2$ can be useful for lossless image coding provided one can arrive at an image reconstruction algorithm based on these functions. As mentioned in [5], such algorithm may not be straightforward. So far, we have only obtained a partial reconstruction algorithm for the λ function [12]. The reconstruction leads to messy (to say the very least) combinatorial problems which, beyond a point, become insurmountable.

ACKNOWLEDGMENT

Thanks are due to Professor Charles R. Giardina, who suggested that the analysis for the two-dimensional case can be and should be extended to the multidimensional case. He patiently went over most of the proofs. This work formed a significant portion of Hanjin Lee's Ph.D. thesis at the Stevens Institute of Technology.

REFERENCES

1. Alt, H., et al., Congruences, similarity, and symmetries of geometric objects, *Discrete Comput. Geom.*, *3*, 237–256 (1988).

2. Atallah, M. J., On symmetry detection, *IEEE Trans. Comput.*, *C-34*, 663–666 (1985).
3. Reingold, E. M., On the optimality of some set algorithms, *J. ACM*, *19*, 649–659 (1972).
4. Atkinson, M. D., An optimal algorithm for geometrical congruence, *J. Algorithms*, *8*, 159–172 (1987).
5. Sinha, D., and Giardina, C. R., Discrete black and white object recognition via morphological functions, *IEEE Trans. Pattern Anal. Machine Intell.*, *PAMI-12*, 275–293 (1990).
6. Sinha, D., On the performance of pattern recognition, algorithms that work only with integers, in *Electronic Imaging East '89*, pp. 415–420.
7. Sinha, D., Geometry of discrete sets with application to pattern recognition, in *SPIE Conference on Intelligent Robots and Computer Vision VIII: Algorithms and Techniques*, vol. 1192, paper 17, Philadelphia, Nov. 1989.
8. Serra, J., *Image Analysis and Mathematical Morphology*, vol. 2. *Theoretical Advance*, Academic Press, London, 1988.
9. Giardina, C. R., Wrap-around morphological algebra, private communication.
10. MacLane, S., and Birkhoff, G., *Algebra*, Macmillan, New York, 1967, pp. 73–78, 88, 98.
11. Giardina, C. R., and Sinha, D., Pointed fuzzy sets, in *SPIE Conference on Intelligent Robots and Computer Vision VIII: Algorithms and Techniques*, vol. 1192, Philadelphia, Nov. 1989, pp. 659–668.
12. Lee, H., Characterization of finite multi-dimensional discrete sets from the perspective of invariant pattern recognition, Ph.D. Thesis, Stevens Institute of Technology, 1990.
13. Hu, M. K., Visual pattern recognition by moment invariants, *IRE Trans. Inform. Theory*, *8*, 179–187 (1962).
14. Childs, L., *A Concrete Introduction to Higher Algebra*, Springer-Verlag, New York, 1979, pp. 22, 52.
15. Pavel, M., *Fundamental of Pattern Recognition*, Marcel Dekker, New York, 1989, p. 69.
16. Lee, H., and Sinha, D., Efficient, parallel, and invariant shape recognition algorithms for complex industrial tasks, in *IEEE International Conference on Systems, Man, and Cybernetics*, Los Angeles, Nov. 1990, pp. 361–366.
17. Lee, H., Printed international character recognition, *IEEE Region 10 International Conference TENCON'91*, New Delhi, Aug., 1991.

Chapter 12

The Morphological Approach to Segmentation: The Watershed Transformation

S. Beucher and F. Meyer

Centre de Morphologie Mathématique
Ecole des Mines de Paris
Fontainebleau, France

I. INTRODUCTION

Segmentation is one of the key problems in image processing. In fact, one should say segmentations because there exist as many techniques as there are specific situations. Among them, gray-tone images segmentation is very important and the relative techniques may be divided into two groups: the techniques based on contour detection and those involving region growing. Many authors have tried to define general schemes of contour detection using low-level tools [1,2]. Unfortunately, because they work at a very primitive level, a great number of algorithms must be used to emphasize their results.

An original method of segmentation based on the use of watershed lines has been developed in the framework of mathematical morphology. This technique, which may appear to be close to the region-growing methods, leads in fact to a general methodology of segmentation and has been applied with success in many different situations.

In this chapter, the principles of morphological segmentation will be presented and illustrated by means of examples, starting from the simplest ones and introducing step by step more complex segmentation tools.

In Section II, we shall review briefly various morphological tools which are used throughout this chapter. These basic transformations are useful for the description of some algorithms used in morphological segmentation. We shall not introduce the basic notions of mathematical morphology; the reader not familiar with them is invited to refer to [3,4].

Section III will be devoted to the presentation of the watershed lines and to their use in segmentation through a very simple didactic example. A simple watershed algorithm will be described.

A real segmentation problem will be presented in Section IV. The problems which arise will be discussed and solved by means of the second great morphological tool used in segmentation: homotopy modification.

In Section V, some algorithms for watershed construction and for homotopy modification will be described. However, the computational cost is the major drawback of the method. Hence the optimality and speed of the algorithms become a critical issue.

In Section VI, various examples of segmentations taken in many domains of image analysis will be discussed.

At this point, a general scheme for segmentation using mathematical morphology will be introduced. More complex algorithms based on a hierarchical approach to the segmentation will be presented. Then examples of complex segmentation will be given.

Finally, the advantages and drawbacks of this methodology will be discussed.

Although we will try in this chapter to be as complete as possible, it is not possible to give an extensive presentation of all the existing techniques of morphological segmentation. Such a review may be found in [5] or in an introductory paper by the authors [6].

II. A REVIEW OF SOME BASIC TOOLS

A. Notation

For simplicity, we will mainly present the segmentation tools in the framework of digital pictures. In this representation, a graytone image can be represented by a function $f : \mathbf{Z}^2 \rightarrow \mathbf{Z}$. $f(x)$ is the gray value of the image at point x. The points of the space $\mathbf{Z}^2$ may be the vertices of a square or of a hexagonal grid.

A section of f at level i is a set $X_i(f)$ defined as

$$X_i(f) = \{x \in \mathbf{Z}^2 : f(x) \geq i\}$$

In the same way, we may define the set $Z_i(f)$:

$$Z_i(f) = \{x \in \mathbf{Z}^2 : f(x) \leq i\}$$

We have obviously

$$X_i(f) = Z^c_{i+1}(f)$$

We shall denote by $X \oplus B$ (resp. $f \oplus B$) the dilation of a set X (resp. a function f) by an elementary disk B (square or hexagon) and by $X \ominus B$ (resp. $f \ominus B$) the elementary erosion. The corresponding opening and closing by the same structuring element are denoted respectively by X_B and X^B. We shall also denote by γ and φ some general morphological openings and closings.

B. Definition of Some Basic Transformations

In this section, some useful morphological tools for segmentation are described: gradient, top-hat transform, distance function, geodesic distance function, and more generally the geodesic reconstructions. Then the notion of homotopy and homotopic transformations are introduced.

The gradient image (or the top-hat transform) is often used in the watershed transformation, because the main criterion for the segmentation in many applications is the homogeneity of the gray values of the objects present in the image. But other criteria may be relevant and other functions may be used. In particular, when the segmentation is based on the shape of the objects, the distance function is very helpful.

The geodesic transformations are of primary importance for the explanation of both the watershed and the homotopy modification algorithms. Among these transformations, the geodesic skeleton by zones of influence and the reconstruction (both for sets and for functions) are fundamental approaches.

1. Morphological Gradient

The morphological gradient [5] of a picture is defined as

$$g(f) = (f \oplus B) - (f \ominus B)$$

When f is continuously differentiable, this gradient is equal to the modulus of the gradient of f (Figure 1):

$$g(f) = \left[\left(\frac{\partial f}{\partial x}\right)^2 + \left(\frac{\partial f}{\partial y}\right)^2\right]^{1/2}$$

The simplest way to approximate this modulus is to assign to each point x the difference between the highest and the lowest pixels within a given neighborhood of x. In other words, for a function f, it is the difference between the dilated function $f \oplus B$ and the eroded function $f \ominus B$.

2. The Top-Hat Transformation

The top-hat transform WTH(f) of a function f is defined as the difference between the function and its morphological opening [7]:

$$\mathrm{WTH}(f) = f - \gamma(f)$$

This transformation is a very good contrast detector suitable for enhancing the white and narrow objects in the image (Figure 2). Different sizes and shapes may be chosen for the structuring element used in the opening and this leads to very

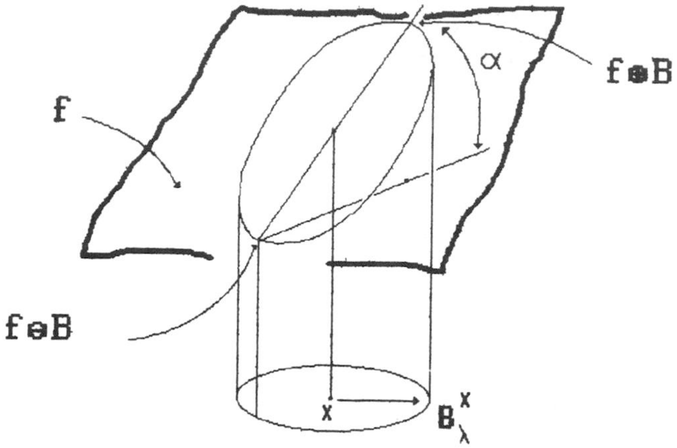

Figure 1. Construction of the morphological gradient.

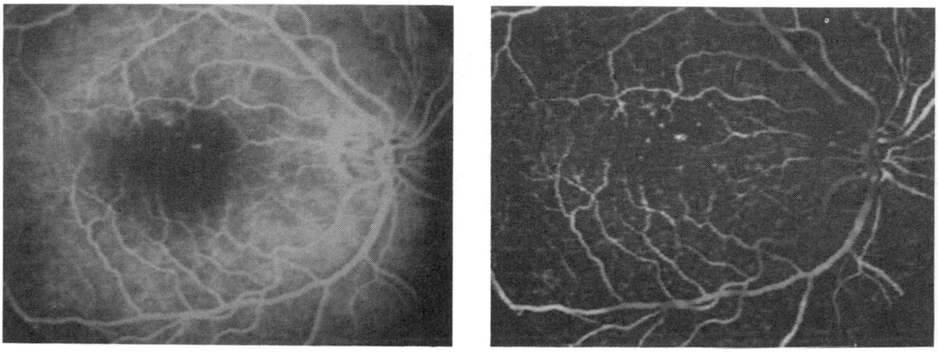

(a) (b)

Figure 2. White top-hat transform (b) of image (a).

efficient filters [8]. A similar definition called black top-hat BTH(f) uses closing to enhance the black and narrow features:

$$\text{BTH}(f) = \varphi(f) - f$$

3. Distance Function

Let Y be a set of $\mathbf{Z}^2$. For every point y of Y, define the distance $d(y)$ of y to the complementary set Y^c (Figure 3):

$$\forall\, y \in Y, \qquad d(y) = \text{dist}(y, Y^c)$$

where dist(y,Y^c) is the distance of y to the nearest point of Y^c.

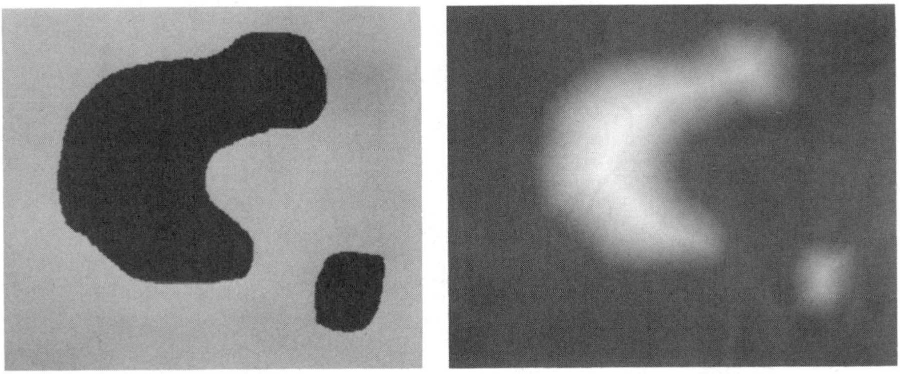

(a) (b)

Figure 3. Distance function (b) of a set (a).

It can easily be shown that a section of d at level i is given by

$$X_i(d) = \{y : d(y) \geq i\} = Y \ominus B_i$$

where B_i is a disk of radius i.

This distance function is very helpful for segmenting binary objects, as shown later on.

4. Geodesy, Geodesic Distance

The geodesic transformations are very efficient in mathematical morphology. Starting from the notion of geodesic distance, one may define geodesic dilations and erosion and consequently, in geodesic spaces, the majority of the morphological transformations [9]. Here we introduce only the geodesic distance and two basic operators linked to this distance: the geodesic SKIZ (skeleton by zones of influence) and the reconstruction of a set from a marker.

Let $X \subset \mathbf{Z}^2$ be a set, x and y two points of X. We define the geodesic distance $d_X(x,y)$ between x and y as the length of the shortest path (if any) included in X and linking x and y (Figure 4a).

Let Y be any set included in X. We can compute the set of all points of X that are at a finite geodesic distance from Y:

$$R_X(Y) = \{x \in X : \exists\, y \in Y, d_X(x,y) \text{ finite}\}$$

$R_X(Y)$ is called the X-reconstructed set by the market set Y. It is made of all the connected components of X that are marked by Y.

Suppose that Y is composed of n connected components Y_i. The geodesic zone of influence $z_X(Y_i)$ of Y_i is the set of points of X at a finite geodesic distance from Y_i and closer to Y_i than to any other Y_j (Figure 4b):

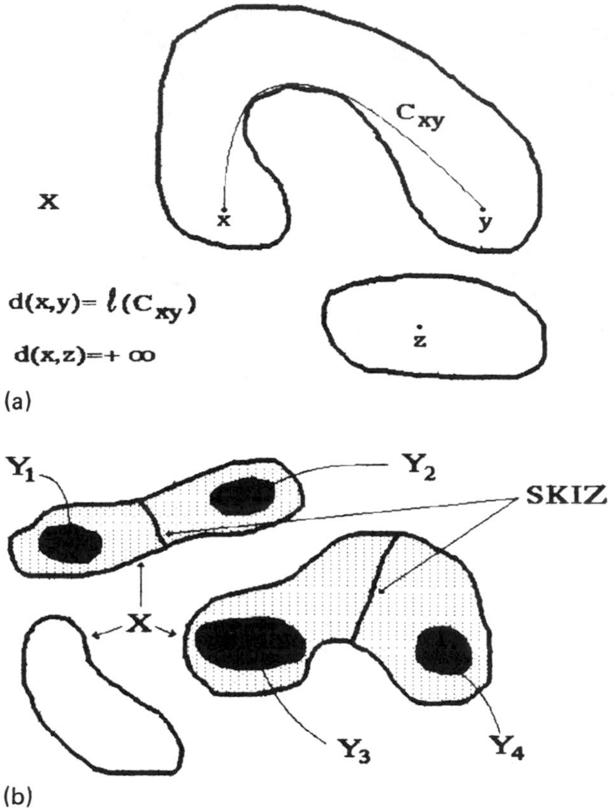

Figure 4. (a) Geodesic distance and shortest paths; (b) geodesic SKIZ of a set Y included in X.

$$z_X(Y_i) = \left\{ x \in X : \begin{array}{l} d_X(x,Y_i) \text{ finite} \\ \forall j \neq i,\ d_X(x,Y_i) < d_X(x,Y_j) \end{array} \right\}$$

The boundaries between the various zones of influence give the geodesic skeleton by zones of influence of Y in X, $\mathrm{SKIZ}_X(Y)$.

We shall write

$$\mathrm{IZ}_X(Y) = \bigcup_i z_X(Y_i)$$

and

$$\mathrm{SKIZ}_X(Y) = X / \mathrm{IZ}_X(Y)$$

where / stands for the set difference.

5. Geodesy for Functions: Reconstruction, Regional Extrema

Reconstruction. Introducing the geodesic transformations for the functions is not so easy because, on the one hand, there are many possible extensions of the binary operators and, on the other hand, the underlying geodesic distance is not obvious. Nevertheless, there exists a trick for extending the binary reconstruction to gray-tone images: it consists in using the sections of the functions. Indeed, any gray-tone picture may be considered either as a function f or as a pile of sections $X_i(f)$ (or $Z_i(f)$ as previously defined). Giving all the possible sections of a function f allows one to know for any point x the corresponding value $f(x)$:

$$f(x) = \max(i : x \in X_i(f))$$

or

$$f(x) = \min(i : x \in Z_i(f))$$

Consider two functions g and f and suppose that $f \leq g$. The corresponding sections of these two functions at level i are $X_i(g)$ and $X_i(f)$. This latter set is obviously included in the former one. For every level i, define a new set obtained by reconstructing $X_i(g)$ using $X_i(f)$ as a marker. It can be shown [5] that the new sets $R_{X_i(g)}(X_i(f))$ define a pile of embedded sections of a new function called the reconstruction of g by f (Figure 5) and denoted $R_g(f)$. In a similar way, the dual reconstruction of a function g by a function f (with $f \geq g$), denoted $R^*_g(f)$ is obtained by reconstructing the sections $Z_i(g)$ using $Z_i(f)$ as a marker (Figure 6).

As illustrated in Figure 6, the function f can be considered as a "wrap-up film" which packs the function g considered as a "parcel." The wrap-up film is of a type which contracts when heated. This contraction, however, occurs only in a horizontal direction, never in a vertical direction. The reconstruction and its dual transformation are clearly increasing. The reconstruction of g is always below the original function. Hence the transformation is antiextensive. Furthermore, the result remains unchanged if the reconstruction is repeated: the transformation is idempotent. It follows that the reconstruction is in fact a morphological opening [10]. The dual operation is a closing.

Minima, maxima of a function. Among the various features that can be extracted from an image, the minima and the maxima are of primary importance in the watershed transformation.

The set of all the points $\{x,f(x)\}$ belonging to $\mathbf{Z}^2 \times \mathbf{Z}$ can be seen as a topographic surface S. The lighter the gray value of f at point x, the higher the altitude of the corresponding point $\{x,f(x)\}$ on the surface.

The minima of f, also called regional minima, are defined as follows.

Consider two points s_1 and s_2 of this surface S. A path between $s_1(x_1,f(x_1))$ and $s_2(x_2,f(x_2))$ is any sequence $\{s_i\}$ of points of S, with s_i adjacent to s_{i+1}. A nonascending path is a path where

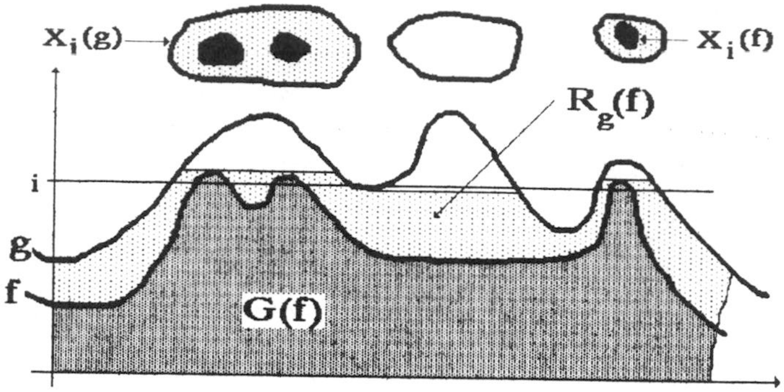

Figure 5. Reconstruction of a function g by a marker function f.

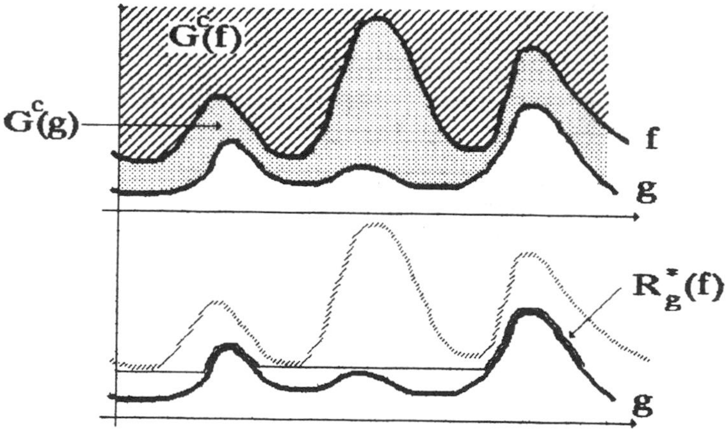

Figure 6. Dual reconstruction.

$$\forall\ s_i(x_i, f(x_i)),\ s_j(x_j, f(x_j)) \qquad i \geq j \Leftrightarrow \leq f(x_j)$$

This path is made of the concatenation of horizontal portions and of strictly descending ones.

A point $s \in S$ belongs to a minimum iff there exists no nonascending path stating from $s(x,f(x))$ and joining any point $s'(x', f(x'))$ of S such that $f(x') < f(x)$. A minimum can be considered as a sink of the topographic surface (Figure 7). The set $m(f)$ of all the minima of f is made of various connected components $m_i(f)$. A similar definition holds for the maxima $M(f)$.

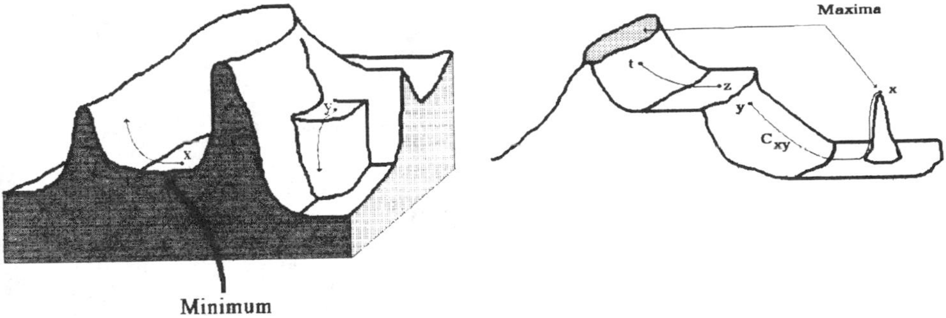

Figure 7. Minima and maxima of a function.

There exist various techniques for extracting the extrema of a function f. The most common one (but unfortunately one of the slowest ones) consists in using the reconstruction. It can be shown that [5]

$$k_{m(f)} = f - R_f(f - 1)$$

and

$$k_{M(f)} = R_f^*(f + 1) - f$$

where $k_{m(f)}$ and $k_{M(f)}$ are respectively the indicator functions of the minima and the maxima of f (Figure 8):

$$k_{m(f)}(x) = 1 \qquad \text{iff } x \in m(f)$$
$$k_{m(f)}(x) = 0 \qquad \text{if not}$$

6. Homotopy and Homotopic Transformations

Homotopy is a topological property of sets. Instead of defining homotopy in pure mathematical terms, let us simply give a practical definition: two sets X and Y are said to be homotopic if the first one can be superimposed onto the second one by means of continuous deformations. A transformation Φ is said to be homotopic if it transforms any set X into an homotopic set $\Phi(X)$ (Figure 9). A simply connected set will be transformed into a simply connected set, a set with one hole into a set with one hole, and so on. A typical example of homotopic transform is given by the skeleton of a set [11,12].

The extension of homotopy to functions is more difficult. In that case, it can be shown [11] that a homotopic transformation $\Phi(f)$ of a function f preserves the number and the relative positions of the extrema of f.

Another definition of homotopy for functions, more restrictive but easier to manipulate, can be used. Two functions f and g are said to be homotopic if, for any level i, the sections $X_i(f)$ and $X_i(g)$ are homotopic sets (Figure 10).

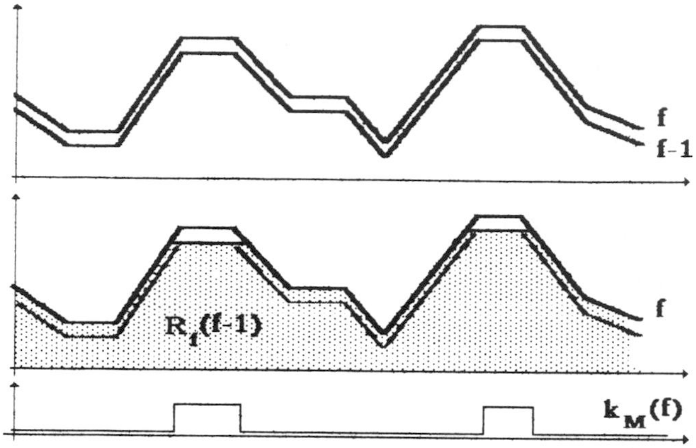

Figure 8. Maxima and reconstruction of a function f by $(f - 1)$.

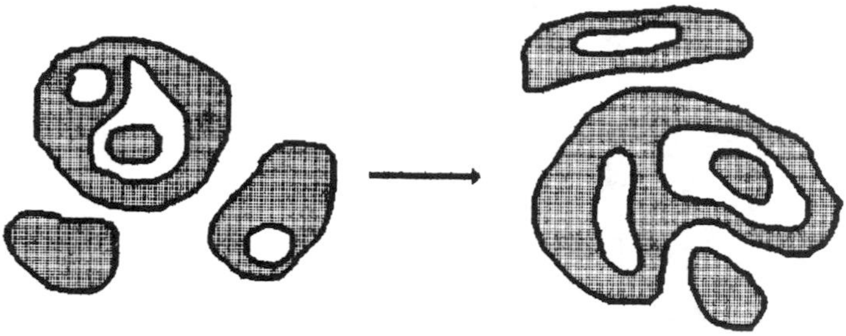

Figure 9. Example of homotopic transformation.

III. THE WATERSHED TRANSFORMATION

Let us introduce now one of the main tools used for segmentation in mathematical morphology: the watershed transformation [13]. After a didactic presentation as a flooding process, we shall explain its use for segmentation on a very simple example.

A. The Watershed Transformation

Consider again an image f as a topographic surface and define the catchment basins of f and the watershed lines by means of a flooding process. Imagine that we pierce each minimum $m_i(f)$ of the topographic surface S and that we plunge this surface into a lake with a constant vertical speed. The water entering through the holes floods the surface S. During the flooding, two or more floods coming

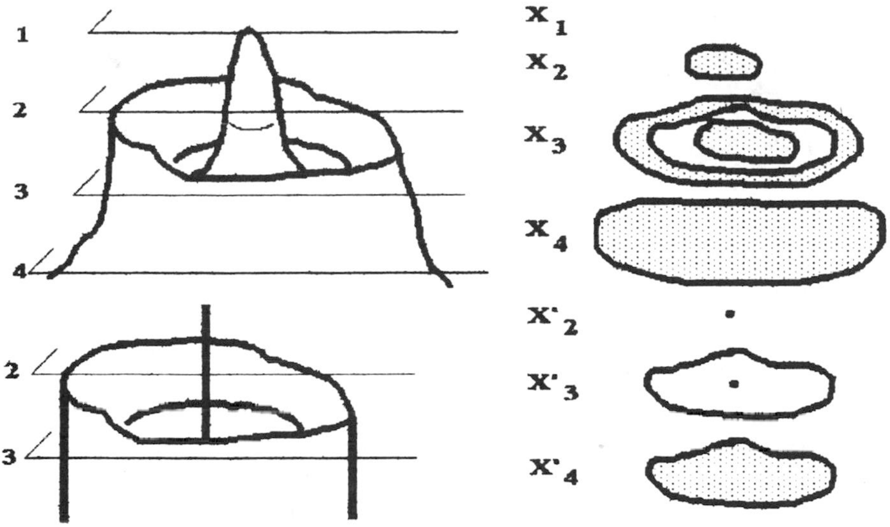

Figure 10. A restrictive definition of the homotopy for functions.

from different minima may merge. We want to avoid this event and we build a dam on the points of the surface S where the floods would merge. At the end of the process, only the dams emerge. These dams define the watershed of the function f. They separate the various catchment basins $CB_i(f)$, each one containing one and only one minimum $m_i(f)$ (Figure 11).

B. Use of the Watershed in Segmentation: A (Too) Simple Example

The application of the watershed to image segmentation will be shown through a very simple example: the segmentation of single dots in an image (radon gas bubbles in a radioactive material).

The dots in Figure 12a draw a topographic surface made of hollows with a roundish bottom. Each hollow has a unique bottom. The segmentation problem lies in finding the best contour of the bubbles.

A solution consisting of simply using a threshold is not sufficient because with a high threshold, the highest hollows are correctly detected, but the deepest ones are much too large. A lower threshold, while detecting correctly the deepest hollows, misses the higher.

Since absolute values cannot be used, we may try instead the variation of the gray-tone function, that is, its gradient (Figure 12c). The corresponding gradient image should present a volcano-type topography as depicted in Figure 12b. The contours of the radon bubbles therefore correspond to the watershed lines of the

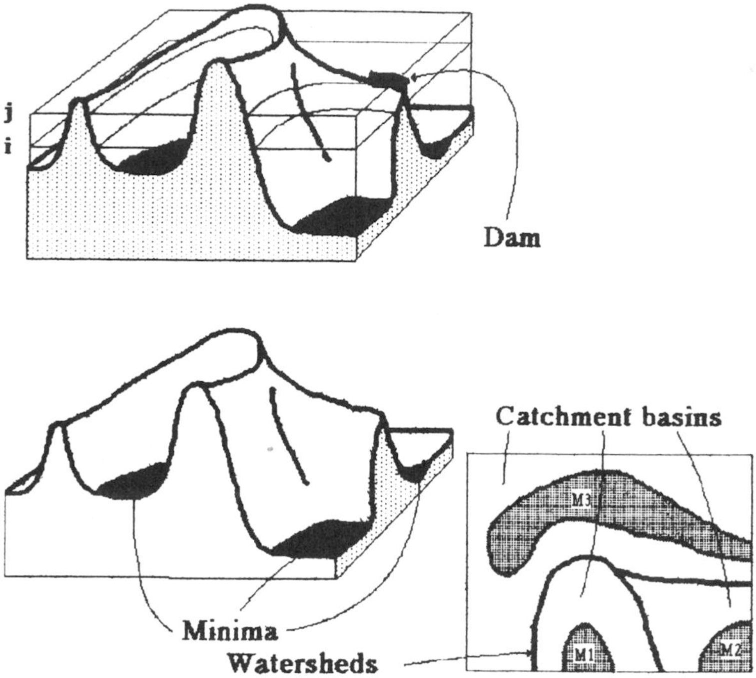

Figure 11. (a) Flooding of the relief and dam building; (b) catchment basins and watershed lines.

gradient image $g(f)$ (Figure 12d). In this gradient image, each dot of the original picture becomes a regional minimum surrounded by a closed chain of mountains. The bubble itself corresponds to a catchment basin of the gradient function, and the varying altitude of the chain of mountains expresses the contrast variation along the contour of the original dot.

C. Building the Watershed

Let us conclude this introductory example by a simple watershed algorithm which uses the basic morphological operators described in the first part.

The definition of the watershed transformation by flooding may be directly transposed by using the sections of the function f.

Consider (Figure 13) a section $Z_i(f)$ of f at level i, and suppose that the flood has reached this height. Consider now the section $Z_{i+1}(f)$. We see immediately that the flooding of $Z_{i+1}(f)$ is performed in the zones of influence of the connected components of $Z_i(f)$ in $Z_{i+1}(f)$. Some connected components of $Z_{i+1}(f)$ which are not reached by the flood are, by definition, minima at level $i + 1$.

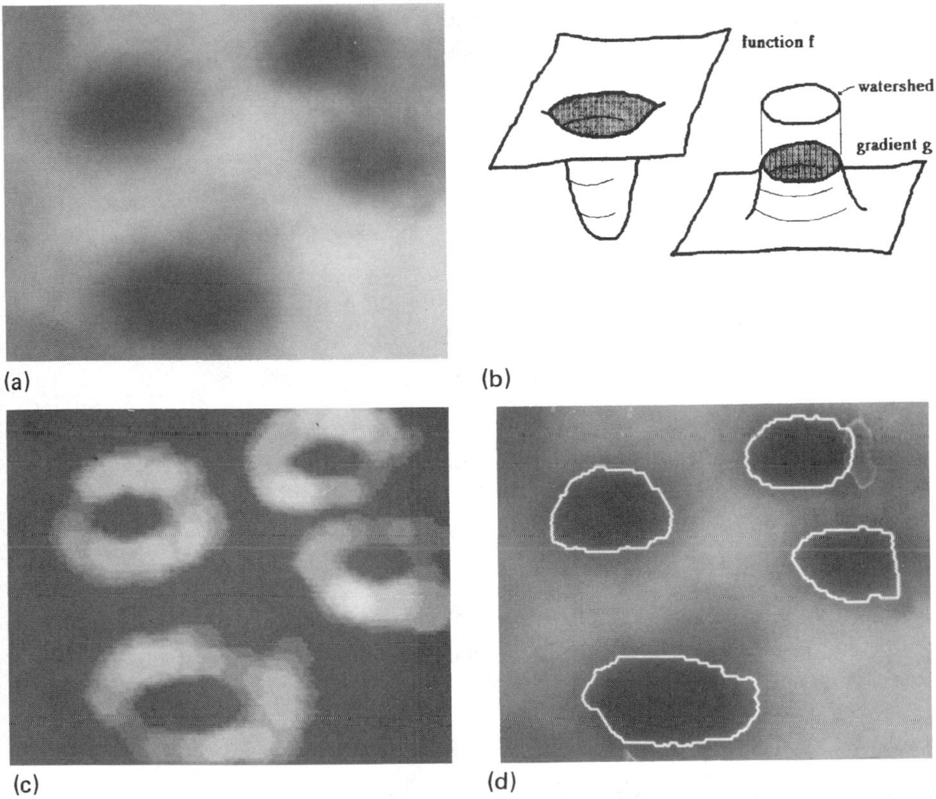

Figure 12. (a) Bubbles of gas in a radioactive material; (b) corresponding topographic surface of the initial function and of the gradient image; (c) morphological gradient; (d) watershed transform of the gradient image.

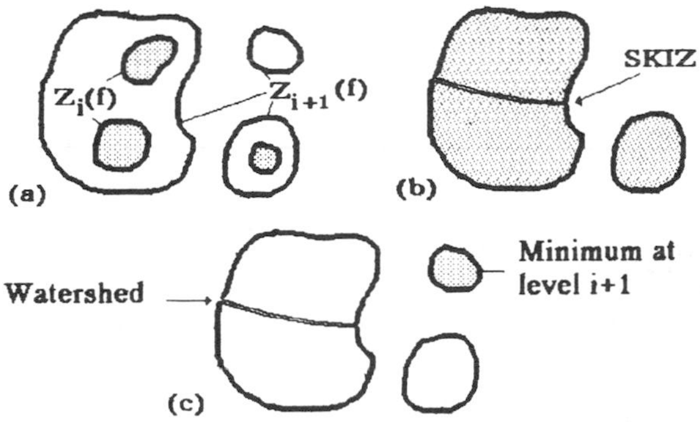

Figure 13. Watershed construction using a geodesic SKIZ.

These minima must therefore be added to the flooded area. Denoting by $W_i(f)$ the section at level i of the catchment basins of f and by $m_{i+1}(f)$ the minima of the function at height $i + 1$, we have

$$W_{i+1}(f) = \left[\mathrm{IZ}_{Z_{i+1}(f)}(X_i(f))\right] \cup m_{i+1}(f)$$

The minima at level $i + 1$ are given by

$$m_{i+1}(f) = Z_{i+1}(f) / R_{Z_{i+1}(f)}(Z_i(f))$$

This iterative algorithm is initiated with $W_{-1}(f) = \emptyset$. At the end of the process, the watershed line DL(f) is equal to

$$\mathrm{DL}(f) = W_N^c(f) \qquad (\text{with } \max(f) = N)$$

We shall discuss more deeply in the following the main groups of algorithms used for the watershed and focus our attention on some of them, but before doing so we must try to apply the previous method on a more complex example.

IV. SEGMENTATION IN THE REAL WORLD

The problems encountered in the segmentation process will be best illustrated by presenting a complete and typical segmentation problem in the field of automated cytology.

A. The Oversegmentation Problem

Figure 14a shows two overlapping nuclei. The inside of the nuclei is textured by the chromatin structure. Their outside is cytoplasm, which is textured itself. Any technique based on thresholding fails in this case. The importance of the nuclear texture obscures completely the gradient image (Figure 14b) and makes it difficult to discriminate between the contour lines of the nucleus and the chromatin patterns. For this reason the image has to undergo a filtering to smooth the inside texture while preserving the boundaries of the nucleus. We will use a morphological sequential filter in which a closing is followed by an opening. The openings and closings used here are based on reconstructions (Figure 14c and d). For a complete presentation of these filters, refer to [14].

Let g be the gradient of the sequentially filtered image f (Figure 14e). The construction of the watershed line associated to all regional minima of the function g is illustrated by Figure 14f. Surprisingly, many catchment basins have appeared. Each of them is generated by a different regional minimum of the image. And although the gradient image seems relatively clean, there are many regional minima, even in the background, which seemed homogeneous. This fact is general: the construction of the watershed line leads to severe oversegmentations. This may be amended by two types of methods. The first one [15] is presented below. The second [5] consists in a hierarchical segmentation of the image. This approach will be presented in the Section VII. Let us now see the first solution to the oversegmentation problem.

B. Flooding from Selected Sources

1. Description

The oversegmentation produced by direct construction of the watershed line is due to the fact that every regional minimum becomes the center of a catchment basin. Not all regional minima, however, have the same importance. Some of them are just produced by noise, others by minor structures in the image.

In order to avoid oversegmentation, we need some additional information: suppose we know before flooding which minima correspond to the centers of the nuclei and which to the background. If we come back to our flooding scheme, we will bore a hole only in these minima before immersing the relief. It is the only difference from the preceding algorithm; the flooding and the building of dams take place as previously. The catchment basins of the minima which are not pierced are filled up by overflowing of the neighboring catchment basin; as soon as the water reaches the saddle point between both basins, the water rushes through the pass and fills the previously empty basin (Figure 15). No dam is constructed between these two basins. A dam is constructed only for separating floods originating from different pierced minima. In the end, both spots and background will be covered by the flood, except for the divide line that separates them.

In fact, it is not even necessary to choose sources within the minima. Any region may be chosen. Nor is it necessary that the various markers be connected particles. It is sufficient that they share the same label. Two particles with the same label will be considered to belong to the same region and no dam will be erected between them, if their flooded areas happen to merge.

2. Searching the Markers

Until now the filtering done smoothed the inside texture of the nuclei, while preserving the outside contours. The detection of markers requires even more severe filtering. A first dilation reduces the sizes of the nuclei (a dilation enlarges the white parts). The result of a dilation is, in fact, an open set. A morphological closing does a further smoothing. As a result, one gets the function f, where each nucleus appears as a dark basin (Figure 14g). Such basins are easily detected by a top-hat transformation associated to a dual reconstruction. The marker function is simply the previous one to which a constant height h has been added. The reconstruction operation fills up completely the dark basins.

After subtraction of the initial function from the reconstructed one (in fact, it is a top-hat transformation, as the dual reconstruction is a closing), all nuclei appear as white domes. The final binary markers are obtained by thresholding

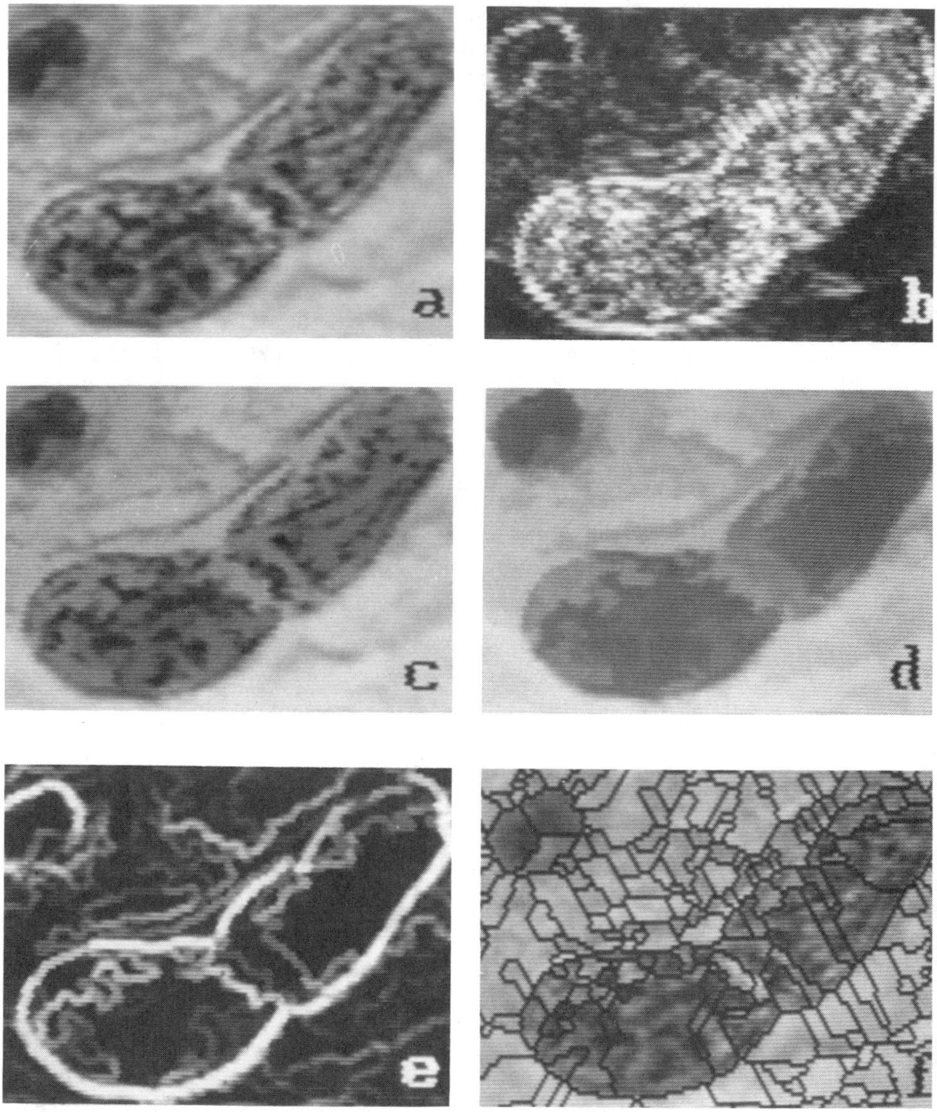

Figure 14. A typical sequence of segmentation. (a) Initial image of two overlapping nuclei. (b) Morphological gradient of the initial image. (c and d) Filtering of the original image. (e) Morphological gradient of the filtered image. (f) Watershed line of the gradient of the filtered image; the result is oversegmented. (g) After dilation and closing, each center of a nucleus appears as a dark basin. (h) Inside markers obtained by a top-hat transformation superimposed on the initial image. (i) Outside markers are the watershed lines of the initial image; the flooding sources are the inside markers. (j) Inside and outside markers superimposed on the gradient image. (k) Watershed of the gradient image with sources corresponding to the markers. (l) Resulting contour.

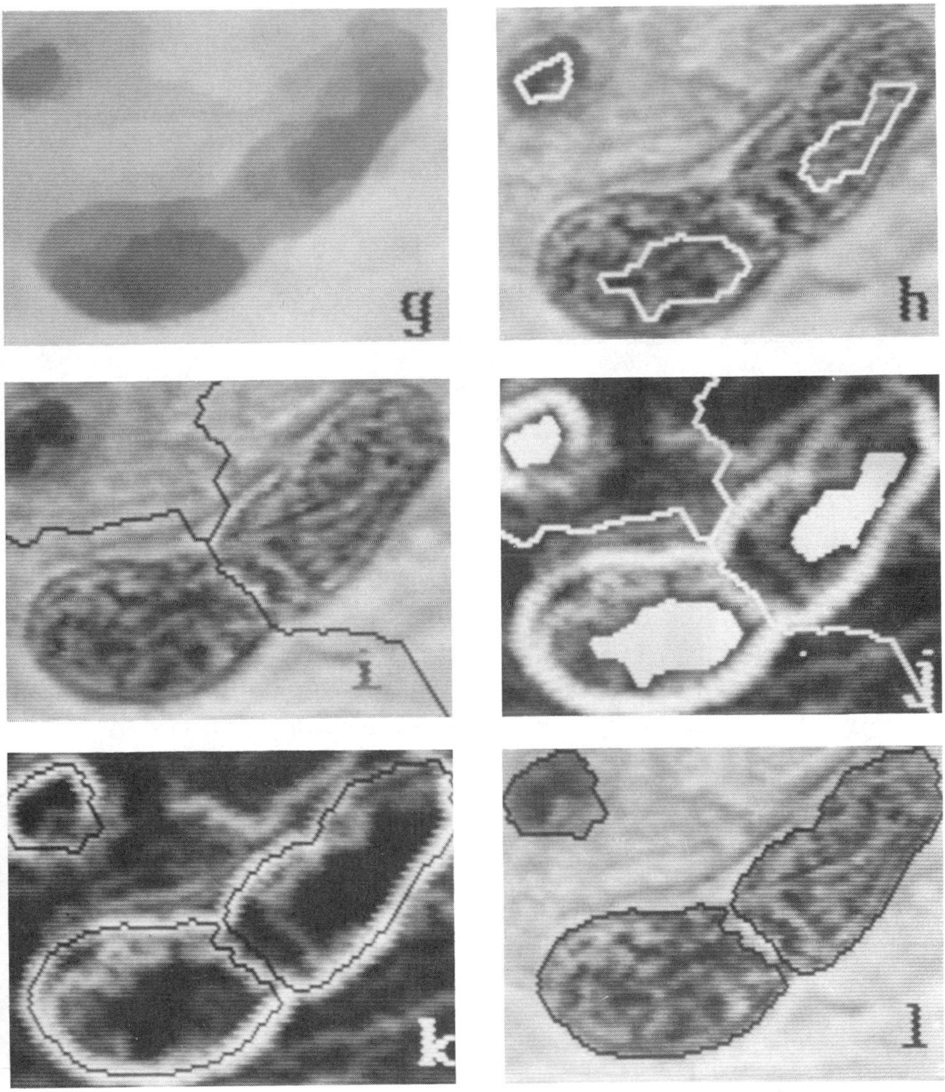
g
h
i
j
k
l

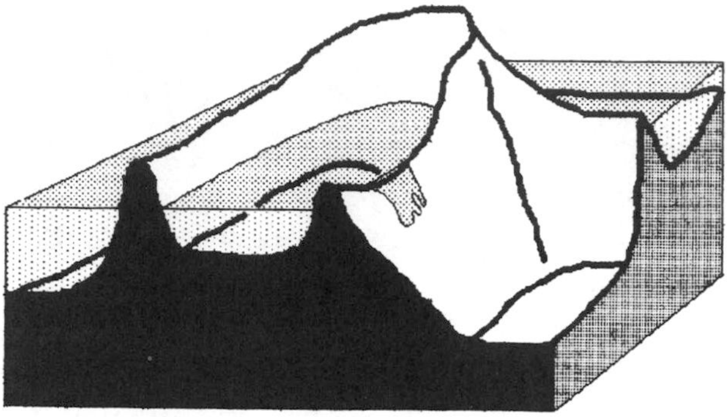

Figure 15. Overflow from a selected catchment basin to an adjacent one.

these domes; the threshold level is easy to choose, since it depends on the height h used in the wrapping algorithm. Moreover, the size and shape of the markers are not critical for the remaining treatment. Only their existence and location are critical. Figure 14h shows the inside markers superimposed on the original image.

3. Construction of the Outside Markers

The outside markers are nothing other than the watershed line of the original image, when it is flooded by sources corresponding to the inside markers. The result of the flooding is shown in figure 14i. To each inside marker corresponds a catchment basin. In this way, we are sure to select a connected outer marker.

4. Construction of the Final Contours

The contour of each nucleus is necessarily between its inside and its outside marker. Its detection is easy. One floods the gradient image obtained previously; the sources are the inside and outside markers detected above. Figure 14j presents the inside and outside markers superimposed on the gradient image. The catchment basins corresponding to the inside markers are the binary masks of the nuclei (Figure 14k and l).

Before describing more sophisticated algorithms, let us simply rewrite the watershed algorithm given above when we introduce this selection of markers. This algorithm can be written as follows.

If $W_i(g)$ is the section at level i of the new catchment basins of g, we have

$$W_{i+1}(g) = \mathrm{IZ}_{(Z_{i+1} \cup M)}(W_i(g))$$

with

$$W_{-1}(g) = M, \qquad \text{marker set}$$

Surprisingly, this algorithm is simpler than the pure watershed algorithm because we do not take the real minima of g into account.

C. Marker Selection and Homotopy Modification

The previous procedure can be decomposed in two steps. The first one consists in modifying the gradient function g in order to produce a new gradient g'. This new image is very similar to the original one, except that its initial minima have disappeared and have been replaced by the set M (Figure 16). This image modification is also called homotopy modification. In fact, we have replaced the old minima of the function g by new ones corresponding to the markers; in other words, we have changed the homotopy of the function g. Recalling the behavior of the dual reconstruction of a function, we can easily see from Figure 16 that this can be performed by reconstructing the sections of g with the markers M.

We have

$$\forall i, \qquad Z_i(g') = R_{Z_i(g) \cup M}(M)$$

If we denote by k_M the indicator function of the markers and by $i_{\max}$ the maximum value of g, we can write

$$k' = i_{\max}(1 - k_M)$$

and then

$$g' = R^*_{\mathrm{Inf}(g,k')}(k')$$

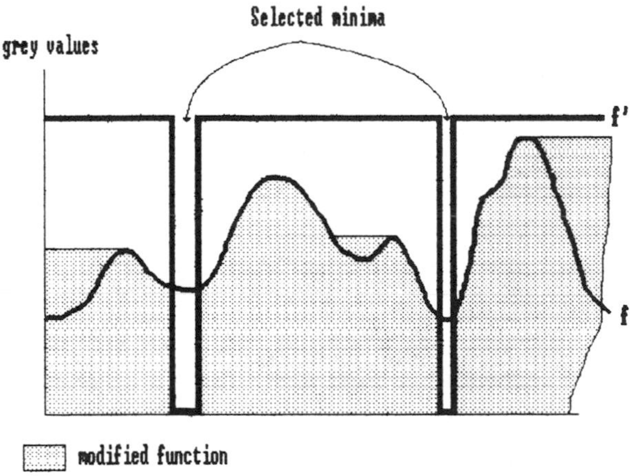

Figure 16. Principle of the homotopy modification of a function by a set of selected markers.

The second step simply consists in performing the watershed of the modified gradient g'.

V. ALGORITHMS OF WATERSHED

A. Review of the Different Classes of Algorithms

The watershed algorithms can be divided in two groups. The first group contains algorithms which simulate the flooding process. The second group is made of procedures aiming at direct detection of the watershed points. Each group of algorithms can subsequently be divided into three classes: parallel algorithms, sequential ones, or ordered algorithms.

An algorithm is parallel if the neighboring points of the point to be transformed take all their values in the original image. The algorithm is said to be parallel, because the result is independent of the order in which the points are transformed. As a matter of fact, all points could be transformed in parallel.

In a sequential (also called recursive) algorithm, the newly computed value of a point will serve as argument for the transformation of its not yet transformed neighbors. The result of the transformation depends completely on the scanning order. Generally, for simplifying the access to the image memories, one adapts forward and inverse raster scanning. Rosenfeld and Pfaltz showed the equivalence between parallel and sequential algorithms [16].

The sequential algorithms are generally much faster than the parallel algorithms. In a parallel algorithm, the value of a point has an influence only on its neighbors. In a sequential algorithm the value of a point may have an influence on the values of all points processed after it. As an example, the distance function of a binary set requires only two scans [17,18]. The reconstruction algorithm needs a variable number of scans depending on the shapes of the functions but not on their size.

The recursive algorithms will not be described here. The reader will find in [19] an extensive review of the main sequential algorithms of mathematical morphology.

The ordered algorithms are essentially the same as the sequential algorithms except for the scanning order of the points. The scanning order is made in such a way that each point is visited only once, at the very moment when its neighborhood is sufficiently well known to determine its value. Vincent has published many such algorithms in the binary case [20]. Verwer et al. described an ordered algorithm for the construction of the geodesic distance function [21] and, more recently, an algorithm for integrating functions [22]. Below we will introduce a data structure called an ordered queue [23] which makes it possible to implement in a quite natural way a series of ordered algorithms. We will describe the implementation for the reconstruction transformation and for the construction of the watershed line. The algorithms described in the previous section belong to the first group and are parallel; they simulate the flooding of the topographic surface drawn by f.

Before presenting the ordered algorithms which also belong to the first group, let us briefly describe another algorithm belonging to the second group and based on the arrows representation of a function f [5].

B. A Brief Introduction to a Second Group Algorithm

From $f : \mathbf{Z}^2 \rightarrow \mathbf{Z}$, we may define an oriented graph whose vertices are the points of $\mathbf{Z}^2$ and with edges or arrows from x to any adjacent point y iff $f(x) < f(y)$ (Figure 17).

The definition does not allow arrowing of the plateaus of the topographic surface. This arrowing can be performed by means of geodesic dilations. The operation is called the completion of the arrows graph. Moreover, in order to suppress problems due to the fact that a watershed line is not always of zero thickness, a more complicated procedure called overcompletion is used, which leads to double arrowing for some points. Then, starting from this complete graph (overcompleted), we may select some configurations which, locally, correspond to divide lines. These configurations are represented in Figure 18 for the 6-connectivity neighborhood of a point on a hexagonal grid (up to a rotation).

Any point receiving arrows from more than one connected component of its neighborhood may be flooded by different lakes. Consequently, this point may belong to a divide line. In a second step, the arrows starting from the selected points must be suppressed. These points, in fact, cannot be flooded, so they cannot propagate the flood. In doing so, we change the arrowing of the neighboring points and consequently the graph of arrows. Provided that the overcompletion of this new graph has been made, some new divide points may then appear. The procedure is rerun until no new divide point is selected (Figure 19).

This algorithm produces local watershed lines. The true divide lines can be extracted easily; they are the only ones which form closed curves.

The arrows representation is also very useful for detecting very quickly the extrema of a function. The minima, for instance, are the connected components of $\mathbf{Z}^2$ which do not receive any arrow. One detects all plateaus of the function on one side and the lower borders of the plateaus on the other side. Then, in a second phase, all plateaus having lower neighbors are reconstructed. The plateaus which could not be reconstructed are the regional minima of the image.

C. Ordered Algorithms

Leaving the algorithms of the second group, let us come back to those of the first group using an ordered queue. We shall describe mainly the implementation of the dual reconstruction and of the construction of the catchment basins. We shall first introduce a data structure called ordered queue (OQ). Its principal merit is

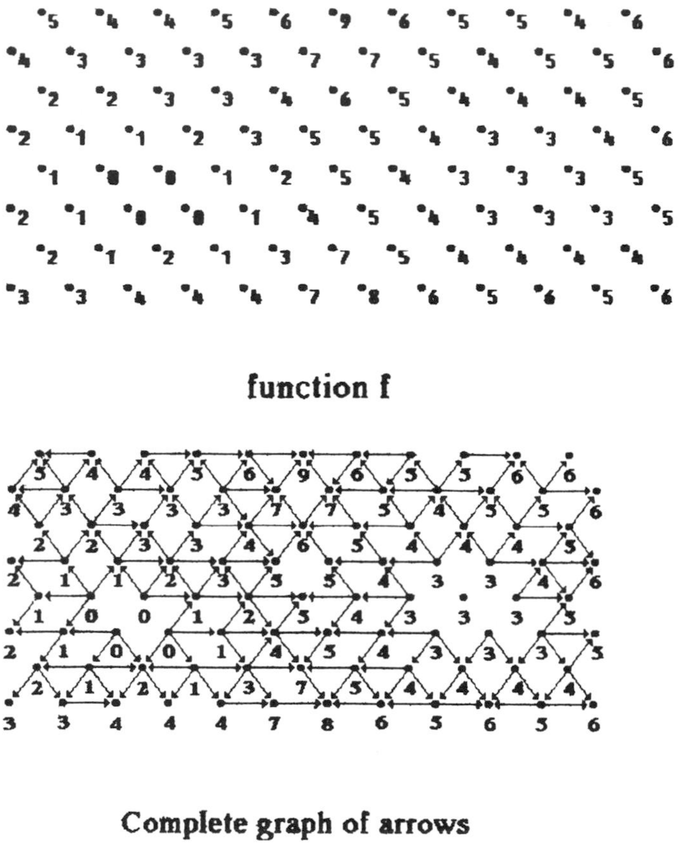

Figure 17. (a) Function f and (b) its corresponding graph of arrows.

to facilitate the storage of points in any order and their retrieval in the order of flooding. For this reason, this structure is at the base of an elegant optimal implementation of the reconstruction operation and watershed line.

1. The Ordered Queue

A hierarchical ordering relation in flooding. During the flooding of a topographic surface, there appears a dual order relation between the pixels (we consider here the flooding with sources placed at the regional minima of the function). It is clear that a point x is flooded before a point y if y is higher than x on the relief. This constitutes the first level of the hierarchy. It is simply the order relation between the gray values. A second order relation occurs on the plateaus. Let X be a plateau at an altitude h. Before X begins to be flooded, all neighboring

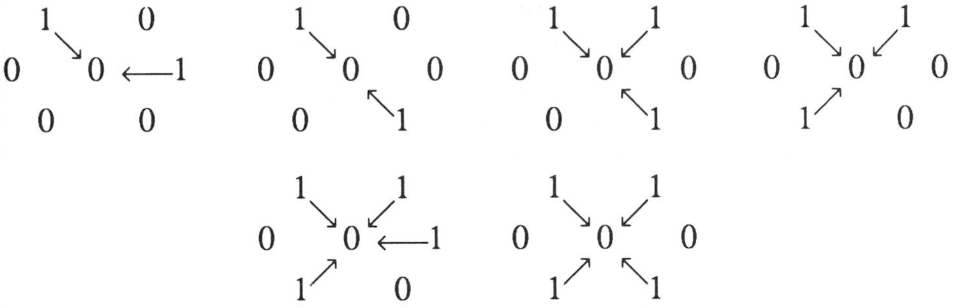

Figure 18. Configurations of arrows corresponding to possible divide points (hexagonal grid).

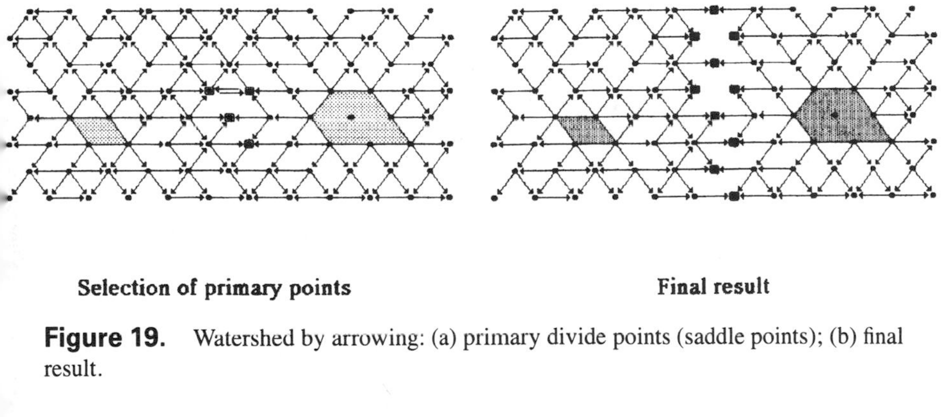

Selection of primary points **Final result**

Figure 19. Watershed by arrowing: (a) primary divide points (saddle points); (b) final result.

points of X with a lower altitude than h have been flooded. One supposes that the flooding of the plateau is not instantaneous but progressive. The flood progresses inward into the plateau with uniform speed. The first neighbors of already flooded points are flooded first. Second neighbors are flooded next, etc. This introduces a second order relation among points with the same altitude, corresponding to the time when they are reached by the flood. If two points x and y belong to the same plateau X, of height h, x will be reached by the flow before y if the geodesic distance within the plateau X to the points of lower altitude is smaller for x than for y. In the next section we show that an ordered queue naturally introduces this hierarchical order relation. Clients are served in the order in which they arrive, but, in consideration of its rank in society, each client is served

Before all clients of lower rank, even if they arrived before it
Before all clients with the same rank who arrived after it
After all clients with a higher rank, even if they arrived after it

Description of an ordered queue

Functional description: A hierarchical queue may be seen as a multiple queue. Clients arrive and will be served according to their order of priority. Each client is put at the end of the queue corresponding to its level of priority: it will be served after all clients with the same priority who arrived before it. Only one client may be served at a time. Once the queue of a given priority is empty, it is suppressed. If a client with high priority arrives after the suppression of the queue to which it belongs, it will be put in the queue of highest priority still existing.

The ordered queue is organized in such a way that it is possible to know whenever a client extracted from the queue has a lower priority than the previous client. The functional specification of an ordered queue is the following:

{creation} /* function without argument creating an ordered queue */
create ordered file: $\emptyset \rightarrow$ FILE_H
{destruction} /* disallocates the storage attributed to an ordered queue */
suppress ordered file: FILE_H $\rightarrow \emptyset$
{insertion} /* inserts an element at the end of the queue of corresponding priority */
insert: FILE_H x (client, priority) $\rightarrow$ FILE_H
{serve} /* gives the address and priority of the client with the highest priority, who arrived first */
serve: FILE_H $\rightarrow$ element

Illustration of the possible actions: Figure 20 shows how a simple queue works. We have represented the queue as a cylinder and the clients as coins in the cylinder. Each arriving client is put on the top of the cylinder. An opening at the base of the cylinder permits the removal of the client who arrived first.

Figure 21a–d show how an ordered queue works. It can be seen in Figure 21a that an ordered queue is in fact a series of simple queues. Each simple queue is assigned a level of priority. In the drawings the priority is represented as a gray value; the darkest gray values correspond to the highest priorities. In our example, we have five levels of priority. All cylinders are open at the top, which means that at any moment it is possible to introduce a client of any priority in the queue. On the contrary, only the queue with the highest priority has an opening at its basement. Figure 21b shows the extraction of an element of the structure: it is the client who arrived first in the queue of highest priority still existing in the

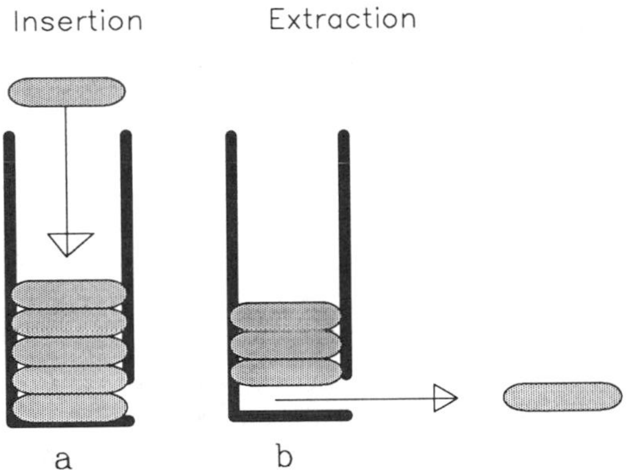

Figure 20. Mechanism of a simple queue.

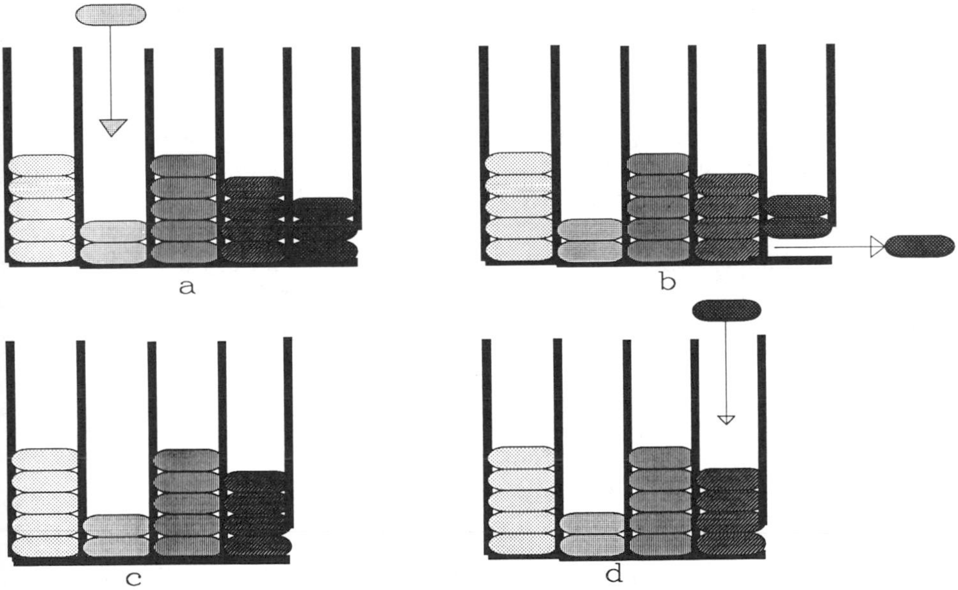

Figure 21. Principle of an ordered queue: (a) the ordered queue; (b) extraction of an element of highest priority; (c) the number of simple queues is reduced; (d) treatment of a highest-priority element when the corresponding queue has been suppressed.

structure. If the attempt to extract a new element of the current queue fails, the queue is suppressed and the queue with the priority immediately below is opened for extraction of the next elements (Figure 21c). Figure 21d illustrates the last feature of an ordered queue: a latecomer of high priority arrives, and the queue with the same priority has already been suppressed. Then the client is put at the end of the current queue of highest priority.

Implementation of an ordered queue. There are several possibilities for implementing an ordered queue. We have chosen an implementation sparing memory. One has to represent a number of queues equal to the number of gray levels in the image. Rather than allocating a different space to each of them, we allocate a common space to all of them and put them one after the other. A pointer identifies the end of each queue. The empty space is structured as a stack: the space necessary to store a newcomer is taken at the top of the stack; conversely, the space liberated by a client leaving the ordered queue is returned to the stack.

2. The Reconstruction Algorithm

The algorithm. We have presented above the dual reconstruction operation, which permits a filling up of the basins of a function. A parallel implementation has been presented in which the number of iterations before convergence is linear with the size of the objects. We will present here an optimal implementation using ordered queues. The algorithm has two phases: an initialization phase followed by the working phase.

Initialization phase. An ordered queue (OQ) is created with a number of queues equal to the number of gray values. The aim of the initialization is to store at least one point belonging to each regional minimum of the marker function in the ordered queue. Each point is stored in the queue with a priority equal to its gray value. There exist several optimal algorithms for finding the regional minima. One of them is based on arrowing and has been presented above.

Working phase. During the working phase all points of the image are taken into consideration and get the gray-tone value corresponding to the reconstruction operation. During the process, a pixel may have three different states; it must be possible to recognize the status of each point during the working phase. For this reason, we use three labels:

The pixel has its final value: label 2. This happens when the point leaves the ordered queue.

The pixel has been stored in the ordered queue but has not been assigned its final value yet: label 1.

The pixel has never been taken into consideration: no label.

f is the initial function (also called "parcel" in the illustration), g the function (above f) used as a marker for the dual reconstruction (named "film"). The treatment goes as follows:

Until the ordered queue is empty, repeat:

{ **Step 1**: A client x is extracted from the ordered queue. If the label of x indicates that x already has its final value, start again step 1. Otherwise give a label 2 to the pixel x, indicating that its value is now definitive.

Step 2: apply the following treatment to any neighbor y of x without label:

The new value v of the point y is computed: $v = \max(g(x), f(y))$. The pixel y is given this value in the image g. The address of the pixel y is stored in the ordered queue with the priority v. The pixel y is also given a label 1 indicating its presence in the queue. }

Illustration. A one-dimensional drawing will illustrate the way the algorithm works. Figure 22a represents a profile of the functions f (parcel) and g (film). The function f is hatched; the function g is above f and is indicated with bold lines. The regional minima of the function g are indicated below the function.

Initialization. The function f has six gray levels. Hence an ordered queue with six levels of priority is created. The first point met during a forward scanning of the image is stored in the ordered queue: point **c** with a priority 0, points **a** and **j** with a priority 1 (here low values mean high priorities). In contrast to the working phase, the points put in the ordered queue during the initialization are not given a label 1.

Working phase. Figure 22c to h show how the image g is modified during the treatment. The labels of the points are indicated under the corresponding profiles of the functions. At the beginning of the working phase, no point has a label.

Point **c** is the first to come out of the ordered queue. Point **c** has no label and can be treated. It is given a label 2, indicating that its value is definitive. The first neighbor of **c** to be examined is point **b**. **b** has no label and its new value is given by $\max(f(b), g(c)) = \max(0,0) = 0$. Point **b** is stored in the ordered queue with a priority 0 and is given a label 1. The new value of the function g at point **b** is now 0. The other neighbor of **c** is point **d**. Its treatment is similar to the treatment of **b**.

The next point coming out of the ordered queue is point **b**. Point **a**, as a neighbor of **b**, is introduced in the OQ with a priority 0. But **a**, as an initialization point, was already present in the OQ. Hence point **a** appears in the OQ at two different places. The first time **a** comes out of the OQ, it is given a label 2. The second time it will not be further processed.

The processing of point **d** introduces point **e** in the OQ with a priority 4. This is represented in Figure 22d.

The treatment of **a** consists in giving it the label 2. As a boundary point, point **a** introduces no other point in the OQ. The queue of priority 0 is now empty and is removed from the structure. The next point comes out of the queue of priority 1; it is point **a**, which has already a label 2. Point **a** is skipped and one proceeds

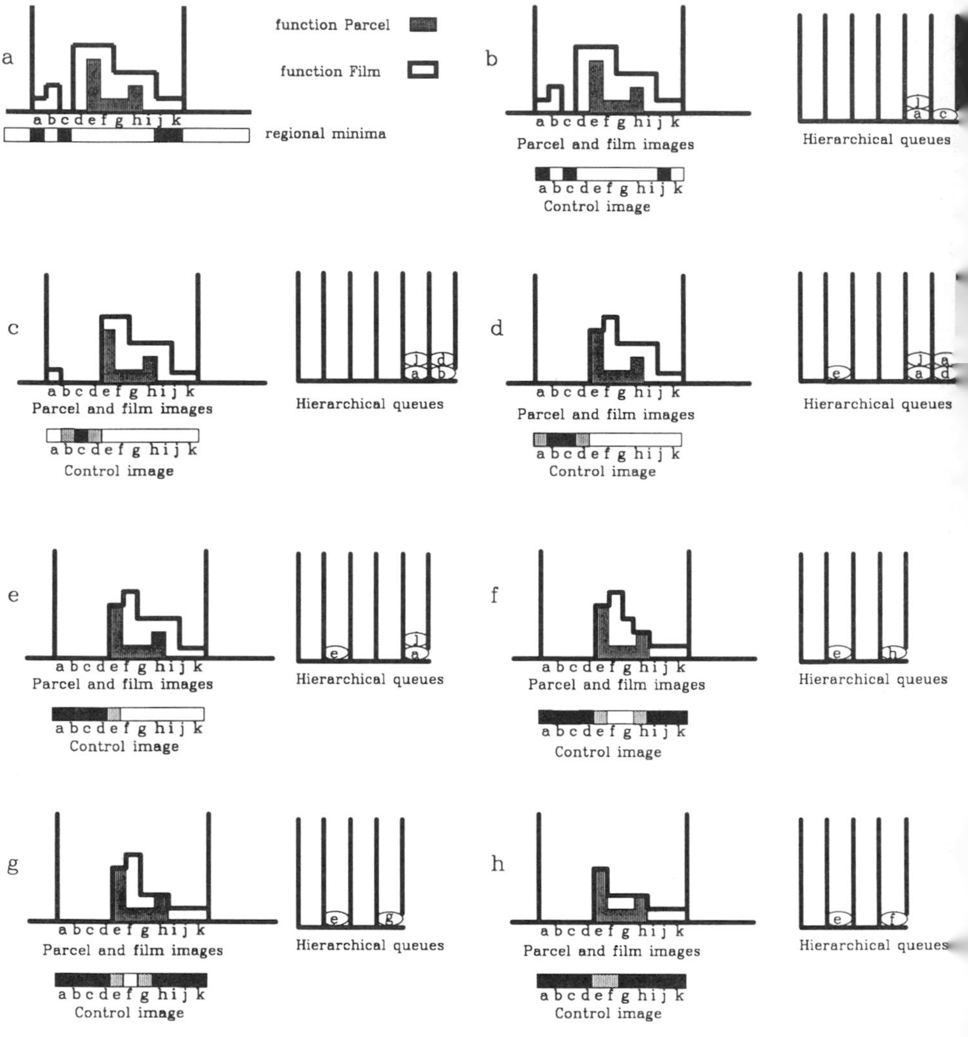

Figure 22. Illustration of the dual reconstruction by an OQ algorithm (see text).

by treating point **j**. The treatment of **j** introduces points **i** and **k** in the OQ with a priority 1. After the treatment of points **i** and **k**, point **h** is introduced in the OQ with a priority 2. The queue of priority 1, being empty, is suppressed. The function *g* has now the shape indicated in Figure 22f.

The treatment proceeds in this way (Figure 22g) until the result indicated in Figure 22h is obtained. The last two points **c** and **f** present in the OQ have all their neighbors labeled; they are not replaced when they leave the OQ. At this moment, the OQ is empty and the treatment is finished.

The points are treated in an order proportional to their gray values. Each point is processed only once. In this sense, the algorithm is optimal.

3. The Watershed Algorithm

General presentation. The input is now a gray-tone function f to be flooded and a set of markers M, which serve as sources for the flooding. If the markers are the regional minima of the image f, then the result is the plain watershed line associated with the relief f. If it is not the case the result is the watershed line of a function $f' = R^*_{\mathrm{Inf}(f,k')}$, k' being the function defined in Section IV.C.

The use of an ordered queue makes it possible to flood directly from a set of markers without doing the reconstruction operation. Indeed, the picturesque presentation of the flooding in Figure 15 will be faithfully simulated.

The markers are identified by labels. Each region will keep the label of the marker which has been the source of the flood. A marker may have several distinct connected components as long they share the same label.

There exist two versions of the algorithm. In the first version, the catchment basins touch each other, without any frontier between them. In the second version such frontiers are generated. Only the first version will be presented in detail.

The algorithm. An initialization phase is followed by a working phase.

Initialization. An ordered queue is created with as many priority levels as there are gray tones in the image f.

A boundary point of a marker belongs to a marker and has in its neighborhood a point outside a marker. All boundary points of the markers are entered in the ordered queue; the value of each point in the image f determines the priority level in the ordered queue.

Working phase. An image g is created by labeling the markers M. The treatment follows:

Until the ordered queue is empty, repeat:

{ A client x is extracted from the ordered queue. To each neighbor y of x having no label in the image g the same treatment is applied:
- the point y is given the same label as x in the image g.
- the point y is stored in the ordered queue; its value in the image f determines its priority level in the ordered queue. }

Illustration. The series of Figure 23a–h illustrate how the algorithm works. The left part of the figure presents on the top the topographic surface and below the zone which has been flooded. Each flooded zone bears the label of the source from which it has been flooded. The content of the ordered queue is represented in the right part of each figure.

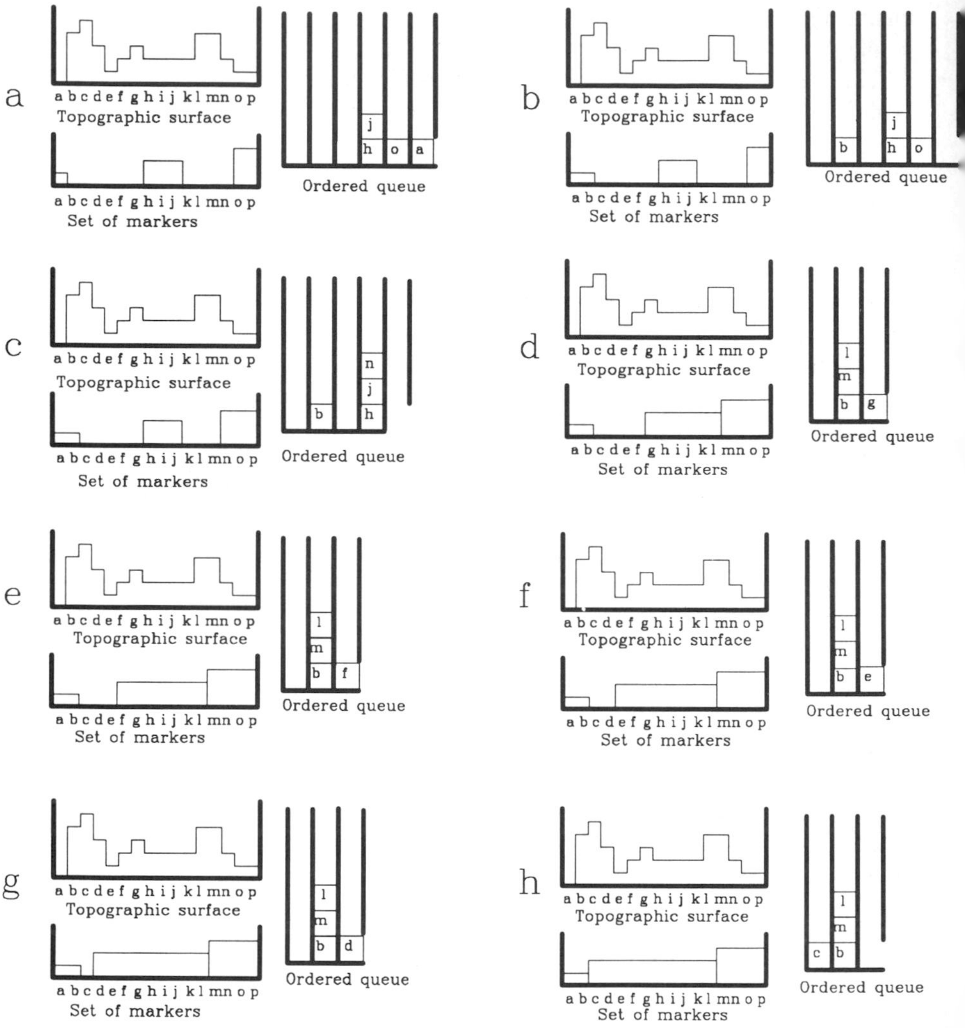

Figure 23. Watershed line construction using an OQ algorithm for simulating the flooding (see text).

Initialization. An ordered queue is created with six levels of priority corresponding to the five gray tones of the topographic surface.

The inside boundary points are stored in the ordered queue with a priority corresponding to their altitude on the topographic surface: point **a** with priority 0, point **o** with priority 1, and points **h** and **j** with priority 2. The resulting ordered queue is shown in Figure 23a.

Working phase. Point **a** is the first point to leave the ordered queue. Its only neighbor is **b**. Point **b**, having no label, takes the label 1 from **a** and is put into the ordered queue with a priority equal to its altitude, i.e., 4. The queue of level 0 is now empty and is suppressed. The state of images and queues is illustrated by Figure 23b.

The next point leaving the ordered queue is **o**. Its left neighbor having no label gets the label 3 from **o** and is stored in the queue of priority 2. The right neighbor of **o** is **p**; **p** already has a label and is not further processed (Figure 23c). The queue of priority 1, being empty, is suppressed.

The treatment of the points belonging to the queue of priority 2 proceeds as follows:

Point **h** gives its label 2 to its neighbor **g**: **g** enters the ordered queue with priority 3.
Point **j** gives its label 2 to its neighbor **k**: **k** enters ordered queue with priority 2.
Point **n** gives its label 3 to its neighbor **m**: **m** enters the ordered queue with priority 4.
Point **k** gives its label 2 to its neighbor **l**: **l** enters the ordered queue with priority 4.

The queue of priority 2 is now empty and is suppressed. The state of images and queues is illustrated in Figure 23d.

The next point to be treated is **g**. Its only neighbor without a label is **f**. But **f** has an altitude equal to 3 and should be put in the queue of priority 2. Yet, this queue has just been suppressed. Point **f** will then be put in the queue with the highest priority still existing, in our case the queue of priority 3. Simultaneously, point **f** is given the label 2. The flooding coming from the source (**hij**) will fill up the catchment basin associated with the minimum **e**, where no marker was placed.

The flooding of this neighboring catchment basin continues with the treatment of point **f**. Point **e** is introduced in queue 3, despite the fact that its altitude is 1 (Figure 23f). After the treatment of the points **e** and **d**, queue 3 is now empty and is suppressed (Figure 23g and h). The next points to leave the OQ are successively points **b**, **m**, **l**, and **c**. All their neighbors already having labels, their treatment introduces no new points in the OQ. Thus the queue is completely emptied when the last point, **c**, leaves it. This achieves the flooding. Each point of the image has been assigned to the region from which the flood came first.

Discussion. The implementation of the watershed line we have described is the simplest and fastest using an ordered queue. A slight modification of the rules which affect each point to the catchment basins leads to several variants. As may be seen in Figure 23, the algorithm does not produce frontiers between catchment basins: each point in the field belongs to a catchment basin.

Another version of the watershed ordered algorithm produces a frontier with a thickness of one pixel point. Both algorithms share the following features:

Each point being considered only once during the treatment phase, the algorithms are indeed optimal.

The flooding is done according the order relation induced by the ordered queue and analyzed above under "The Watershed Algorithms."

VI. EXAMPLES OF SEGMENTATIONS

Three examples of segmentation are described in this section. Each one has been chosen to illustrate a particular topic of this methodology. The first example, the electrophoresis gel segmentation, although not very complex shows that different choices of markers may lead to different results. The second example, the overlapping grains separation, is a binary application of the watershed segmentation. In this case, the criterion used for segmenting objects is based not on their gray values but on their relatively convex shapes. The third example, finally, is more complex. The objects to be segmented are facets in a cleavage fracture in steel. It shows that, despite the fact that the markers are difficult to obtain, once they are defined, the tasks consisting in comparing the facets or defining their neighborhood relationships become easier.

A. Segmentation of Electrophoresis Gels

This first example consists of contouring blobs of proteins in an electrophoresis gel (Figure 24a). This problem seems to be easier than the nuclei segmentation presented above and, in fact, a similar approach is used.

The initial image is filtered. An alternate sequential filter is applied. The minima of the filtered image are the markers of the blobs (Figure 24b). We must also define a marker for the background. In order to get a connected marker surrounding the blobs, we use, as we did for the cells, the watershed of the initial filtered image (Figure 24c). From this, we obtain our set of markers M (Figure 24d). Finally, the watershed of the modified gradient image is performed. The result is given in Figure 24e.

It is clear in this example that the final segmentation depends on the selection of the minima of the initial function as blob markers. If some blobs do not correspond to minima (as is sometimes the case), they will not be contoured correctly. Moreover, using a connected marker for the background induces, by construction, each detected blob to be surrounded by a simple closed arc and that there are no touching blobs.

But, if we use another marker for the background, the result will be different. To demonstrate this, let us choose as background marker the maxima of the initial filtered image (Figure 25a). This marker is not connected, and the watershed

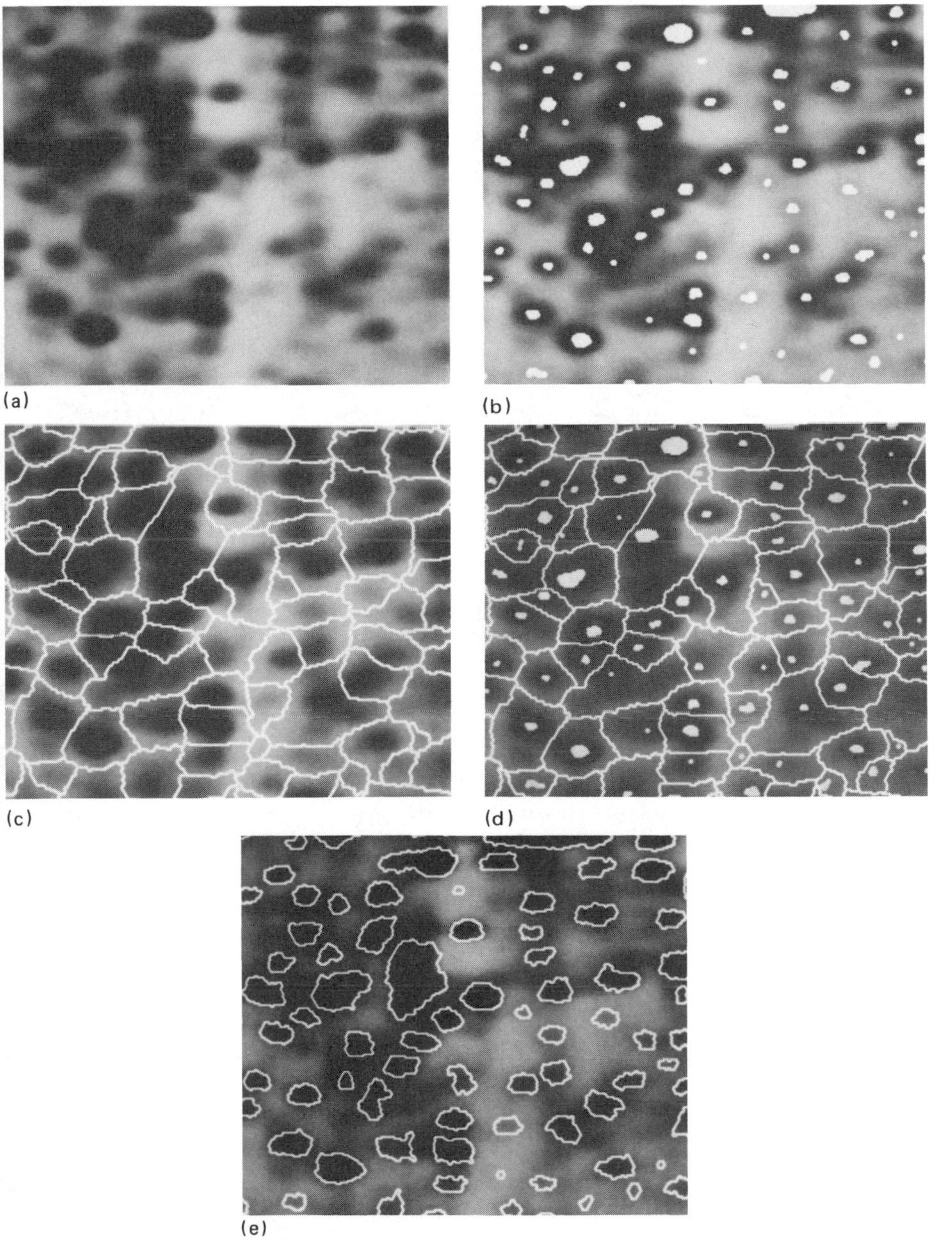

Figure 24. (a) Electrophoresis gel; (b) minima of the filtered image marking the blobs; (c) watershed of the filtered image used as background marker; (d) set of selected markers; (e) final segmentation.

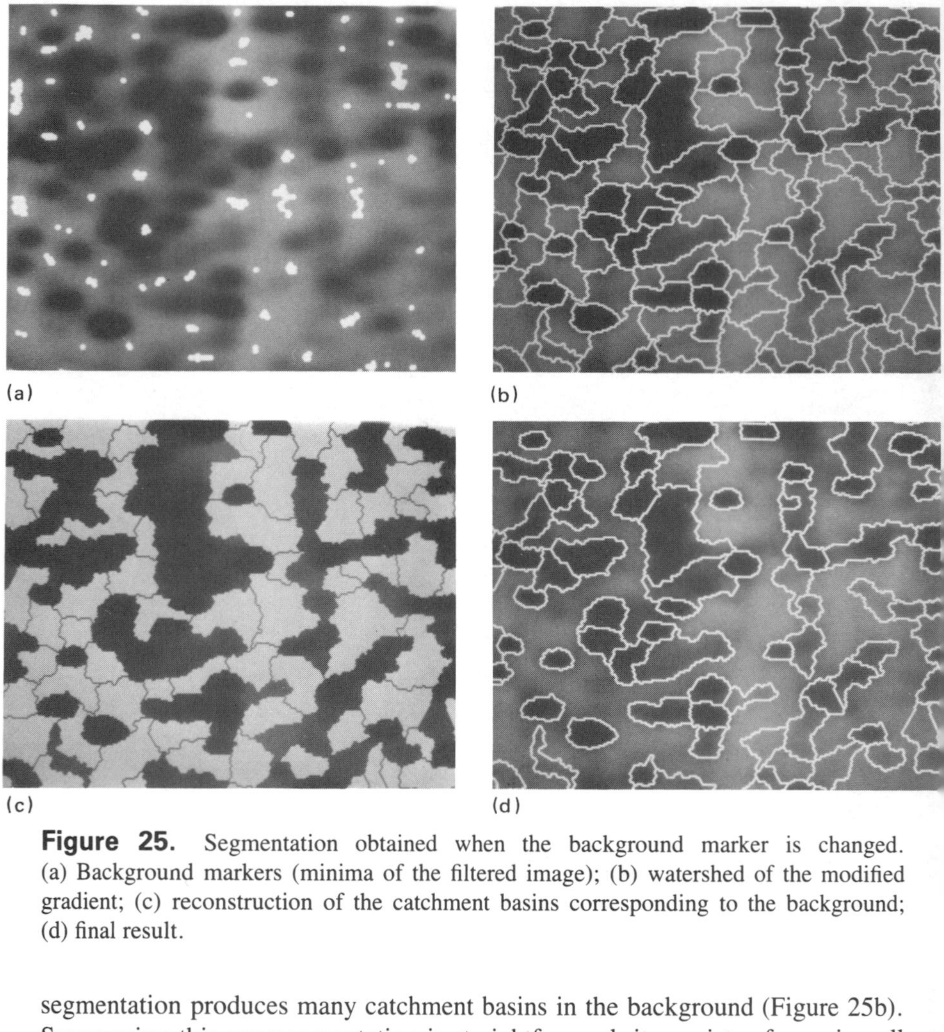

Figure 25. Segmentation obtained when the background marker is changed. (a) Background markers (minima of the filtered image); (b) watershed of the modified gradient; (c) reconstruction of the catchment basins corresponding to the background; (d) final result.

segmentation produces many catchment basins in the background (Figure 25b). Suppressing this oversegmentation is straightforward; it consists of merging all the basins marked by the maxima (Figure 25c). If the same is done with the catchment basins corresponding to the minima of the initial function, the final segmentation (Figure 25d) is rather different. In that case, the objects which have been detected are not the individualized blobs but the heaps of proteins.

Note that the use of a nonconnected marker is not a problem when an ordered algorithm is used. In such a case, all the connected components of the marker have the same label.

B. Segmentation of Overlapping Grains

This example presents another case where the watershed line is most useful: the separation of overlapping grains. The initial picture (Figure 26a) represents coffee grains. This picture can easily be thresholded and it may be seen from their shape that many grains overlap or touch each other. To segment them, no contrast criterion can be used because there is obviously no visible boundary between two overlapping grains.

The solution of the problem consists in using the distance function of the binary set (Figure 26b). The maxima of the distance function mark the different

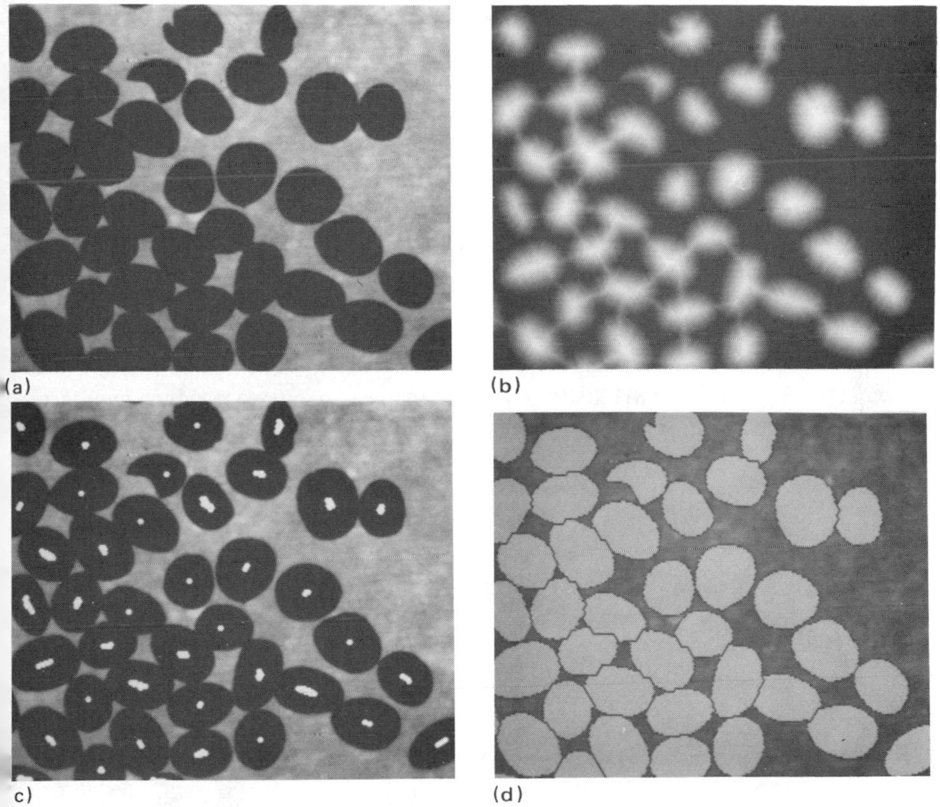

Figure 26. (a) Grains segmentation (coffee beans); (b) distance function; (c) maxima of the distance function; (d) result of the segmentation by thalweg lines of the distance function.

grains (Figure 26c). They can be used (after a slight filtering to solve some parity problems on the digitization grid) to build the talweg lines, defined as the watershed lines of the inverted distance function (Figure 26d).

C. Stereoscopic Analysis of a Fracture in Steel

This third example is a problem of segmentation of cleavage facets in a scanning electron micrograph of a steel fracture (figure 27). The marker selection in this case is more complex. A primary definition of a facet is used: a facet is supposed to be a more or less convex and homogeneous region of the image. That is why the functions used for the watershed along with the marker selection are built by combining a photometric criterion (contrast between facets due to variations in gray tones or to blazing ridges) and a shape criterion (facets are supposed to be more or less convex).

Two functions are defined: the first one, f_1, is the maximum of the gradient function of the initial image f and of the top-hat transformation. The top-hat transform WTH(f) is used for enhancing in the image the blazing zones while the gradient detects the contrast between adjacent facets (Figure 28a):

$$f_1 = \max(g(f), \mathrm{WTH}(f))$$

The second function f_2 is the distance function to the blazing zones and to the contours. It can be shown [5] that this function may be built by dilating the previous function f_1 by a cone (Figure 28b). This technique allows the combination of the two criteria depicted above. Using a gray-tone image instead of a binary one to compute the distance function is just an extension which avoids an arbitrary thresholding of f_1.

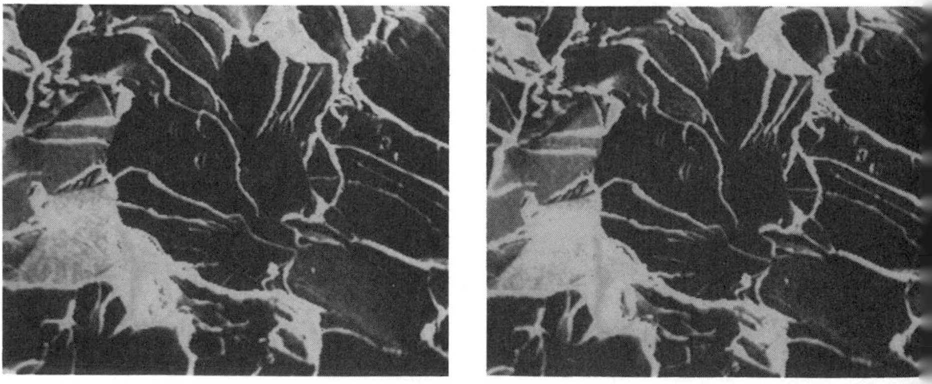

Figure 27. Stereo pair of a cleavage fracture in steel.

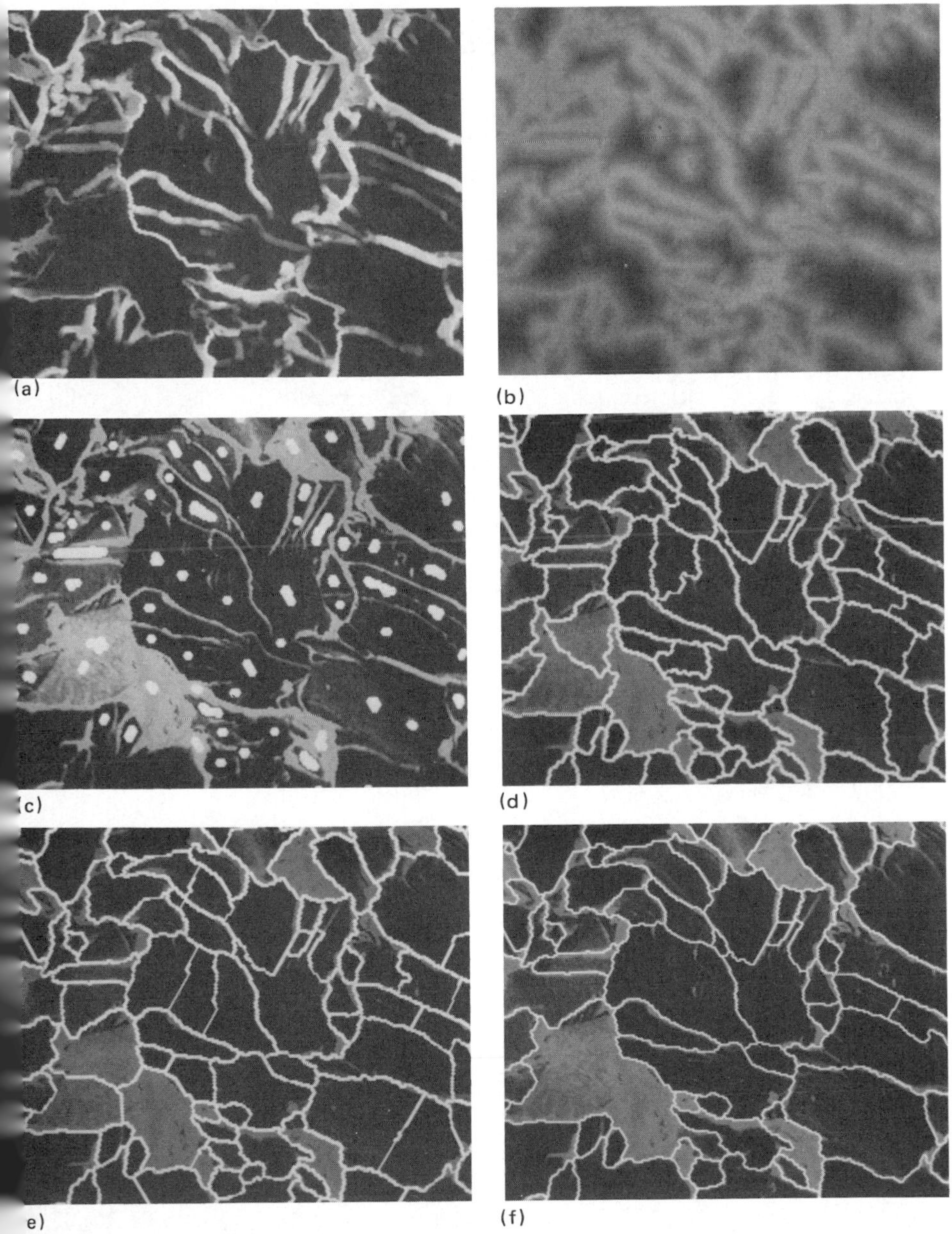

Figure 28. (a) First function used for marker selection; (b) second function; (c) markers of the facets; (d) watershed lines of the first function; (e) watershed lines of the second function; (f) final contours.

The markers of the facets are the minima of f_2 (Figure 28c). We can see that more than one marker may appear in regions which obviously correspond to simple facets. This multiple marking leads to an oversegmentation of the facets.

In order to eliminate this oversegmentation, the watershed transformations of the two functions f_1 and f_2 are performed (Figure 28d and e). These two functions have been modified by the same set of markers (that is, the minima of f_2), and only the divide lines which are superimposed in the two watershed transforms are kept (Figure 28f). This procedure allows one to distinguish between the watershed lines which do not follow the highly contrasted regions in the initial image.

The methodology of segmentation based on the primary definition of the markers of the objects to be extracted is particularly helpful here. Indeed, when the first picture of the stereoscopic pair has been segmented and the corresponding facets have been selected, the markers used in this first step can be used again to segment the homologous facets in the second picture of the stereo pair. The procedure is the following: the markers attached to a facet in the first image are "thrown" onto the second image f_2' corresponding for the second picture to the image f_2. These markers fall along the steepest slope of f_2' and each one reaches a unique minimum of f_2'. These minima are the markers of the homologous facet in the second picture (Figure 29). In this way, we establish a one-to-one correspondence between the markers of the two pictures of the stereo pair and, therefore, between the segmented facets (Figure 30).

As soon as the same facet (or part of a facet) has been segmented in the two pictures of the stereo pair, the computation of its size and orientation in space is relatively easy. By following the corresponding points in the two contours, it is

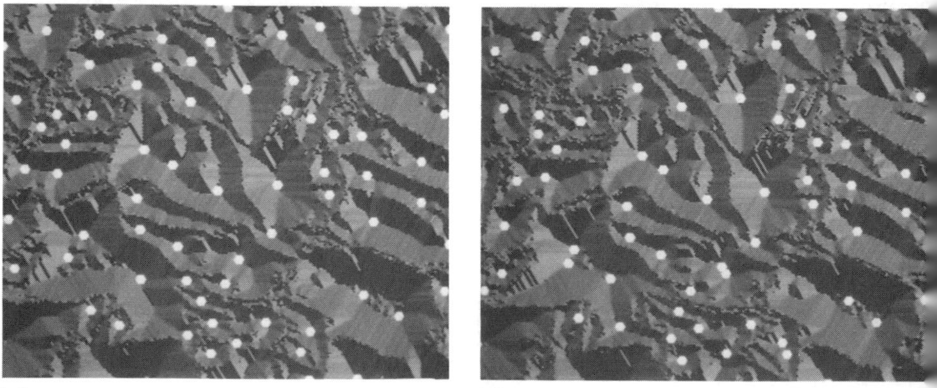

(a) (b)

Figure 29. (a) Markers of the first image; (b) corresponding markers in the second one.

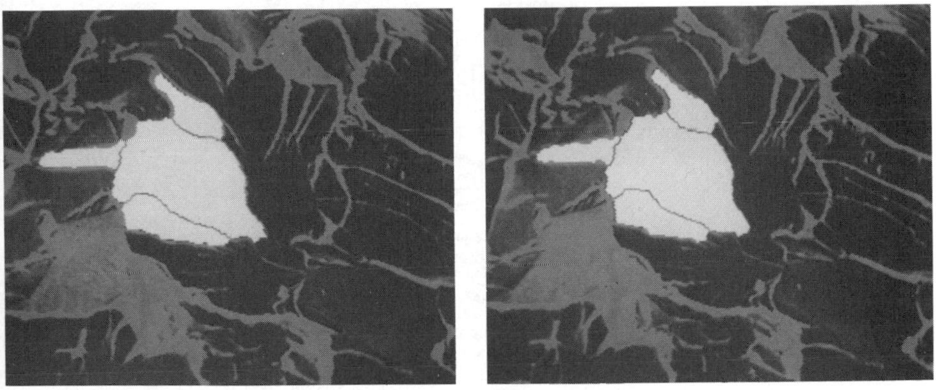

Figure 30. Homologous facets in the stereo pair.

possible to calculate the shift between them and hence their height. Assuming that a facet is almost a plane, its interpolation is performed. Finding the cleavage angle between two adjacent facets (which is in fact the required parameter) is immediate.

This approach to stereovision—segmenting the objects first instead of trying to find the homologous pixels in the two images immediately—is very powerful: the watershed transformation coupled with the marker selection allows us to find directly the corresponding objects in the stereo pair. Moreover, this topological approach allows us to control this correspondence very accurately (two adjacent objects in the scene are in most cases adjacent in both images of the stereo pair).

D. The Segmentation Paradigm

These examples of segmentation lead to a general scheme. Image segmentation consists in selecting first a marker set M pointing out the objects to be extracted and then a function f quantifying a segmentation criterion. This criterion can be, for instance, the changes in gray values, but as seen in the previous examples, other features can be used and even, as illustrated in the case of the cleavage fractures, a mixture of them. This function is modified to produce a new function f' having as minima the set of markers M. The segmentation of the initial image is performed by the watershed transform of f' (Figure 31).

The segmentation process is therefore divided into two steps: an "intelligent" part whose purpose is the determination of M and f and a "straightforward" part consisting in the use of the basic morphological tools, namely the watershed transform and image modification.

This technique has demonstrated its efficiency in various domains of image analysis for both binary and gray tone pictures. This methodology is also helpful

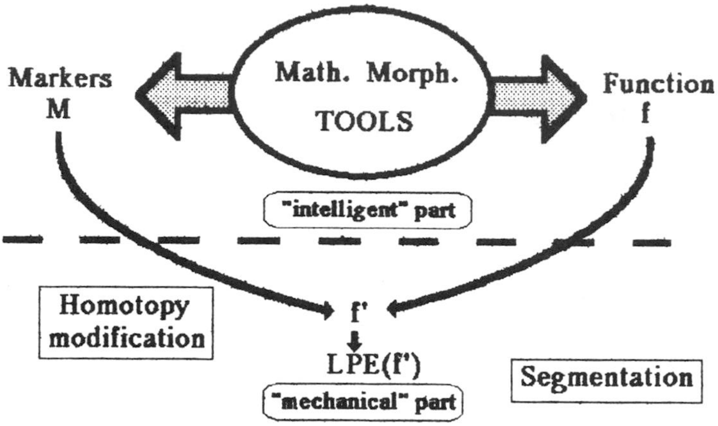

Figure 31. Synopsis of the morphological segmentation methodology.

in three-dimensional segmentation [24], in color image contouring [25], or for the extraction and tracking of objects in time sequences [26].

Many other examples of segmentation based on this scheme may be found in the literature [5,6,27–29].

VII. HIERARCHICAL SEGMENTATION

Until now, our aim was to prevent oversegmentation in selecting good markers and, by means of homotopy modification, to produce as many catchment basins in the watershed as we had selected objects.

This last part will be devoted to the description of a hierarchical segmentation technique which does not prevent oversegmentation but instead tries to suppress the irrelevant boundaries on the watershed transform. We shall see that this approach also uses a watershed transformation defined in this case on a simplified version of the initial image.

A. Introduction

The previous examples have proved that the marker extraction and the good choice of the function used in the watershed are the intelligent process of the segmentation, needing a great deal of effort and skill. The final result is therefore closely dependent on this first task.

Unfortunately, in some cases, marker selection and extraction are not easy. Some pictures are very noisy and image processing becomes more and more complex. In other cases, the objects to be detected may be so complex and so varied in shape, gray level, and size that it is very hard to find reliable algorithms enabling their extraction. For that reason, we need to go a step further in the segmentation.

When attempting to segment a gray-tone image, we know that the initial watershed transformation of the gradient image provides very unsatisfactory results: many apparently homogeneous regions are fragmented in small pieces. Fortunately, the watershed transformation itself, applied on another level, will help us to merge the fragmented regions. Indeed, if we look at the boundaries produced by the segmentation, they do not have the same weight. Those which are inside the almost homogeneous regions are less significant. In order to compare these boundaries, we need to introduce neighborhood relations between them through the definition of a new graph. This graph is built from a simplified version of the original image called a partition or mosaic image.

B. The Mosaic Image

Although the construction of the mosaic image is not necessary for defining the hierarchical segmentation, it will help for understanding the procedure.

Consider a gray-tone image f and its corresponding morphological gradient image $g(f)$. A simplified image can be computed in the following way:

1. We calculate the watershed of the gradient image.
2. We label every catchment basin of the watershed with the gray value in the initial image f corresponding to the minima of $g(f)$.

Figure 32 illustrates this operation.

We will describe the principle of the hierarchical segmentation by means of a simple example. The initial image is an X-ray photograph of metallic particles in the burst produced by a shaped-charge weapon (Figure 33a).

The result is a simplified image (Figure 33b), made of a mosaic of pieces (the catchment basins) of constant gray levels, where no information regarding the contours has been lost. This simplified image, also called a mosaic image, may then be used to define a valued graph, to which the morphological transforms, and in particular the watershed, can be extended.

C. Hierarchical Segmentation

1. An Introduction

When we look at the mosaic image, some regions seem to be almost homogeneous. In fact, they are made of a patchwork of pieces of constant gray levels, the variation in gray values (the step) between two adjacent tiles being low. On the contrary, when we cross a boundary separating two different regions, the step is much higher. In other words, the criterion used to decide whether we are inside an homogeneous region is the fact that the transitions between the different tiles of the mosaic image which partition this homogeneous region are lower than the

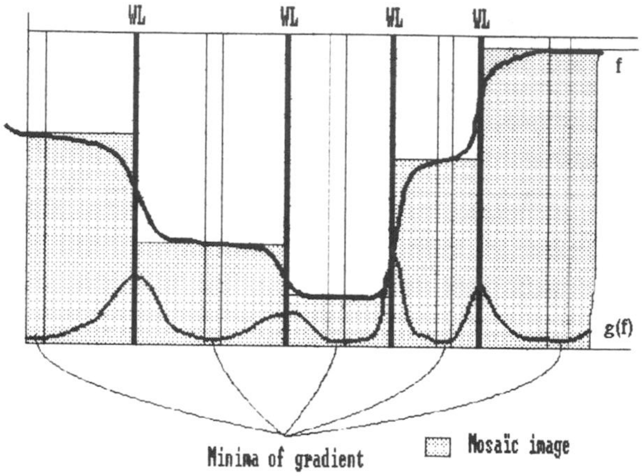

Figure 32. Computation of the mosaic image.

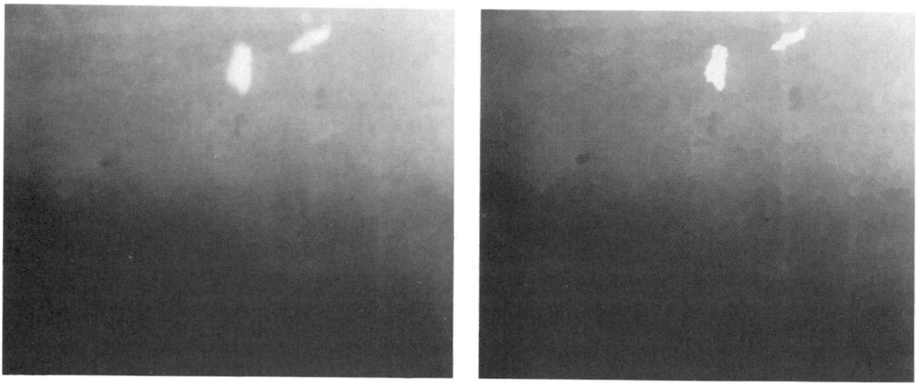

(a) (b)

Figure 33. (a) Initial and (b) mosaic image of an X-ray photograph of metallic particles.

transitions between tiles belonging to different homogeneous regions. Going a step further, one can say that a homogeneous region is marked by minimal transitions in the mosaic picture.

Figure 34a illustrates this notion in a very simple case. The step in gray values between the two tiles supposed to belong to the same homogeneous region is lower than the surrounding ones.

A hierarchical segmentation will then consist in merging adjacent tiles of the original mosaic image using a flooding process starting from the minimal transi-

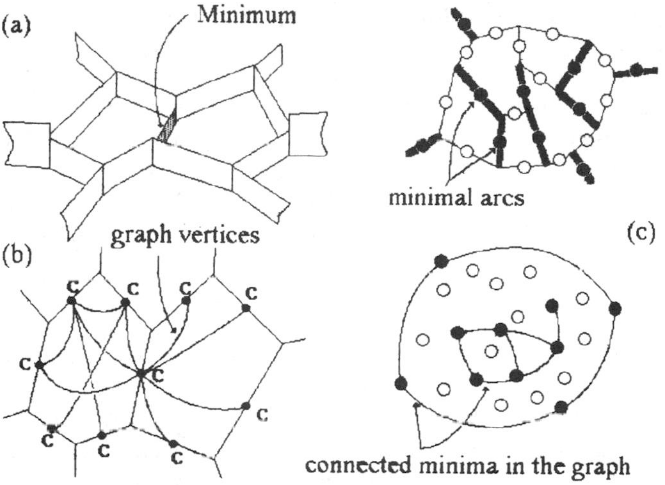

Figure 34. (a) Gradient of the mosaic image; (b) corresponding graph used in the hierarchical approach; (c) example of minimal arcs and their correspondence in the graph.

tions. But this watershed transformation needs, to be realized, the definition of a particular graph built from the mosaic.

2. Construction of a New Graph

Let us build a new valued graph from the mosaic image. First, the summits are made of the transitions between tiles. These summits are valued with the absolute value of the gray-tone difference (the step) between these tiles. Second, the vertices of this new graph, hence the neighborhood relationships between the transitions in the mosaic image, must be set. Two boundaries of the mosaic (two summits of the graph) are considered neighbors if they surround the same catchment basin. This rule simply means that any action on a boundary between two pieces of the mosaic will affect the pieces themselves and therefore the other boundaries which contour them.

The valuation of this graph may be calculated by means of the gradient of the mosaic image (Figure 34a). This morphological gradient is obtained by performing the difference between the dilation and the erosion of the mosaic image. The final graph (Figure 34b) is a nonplanar valued graph.

3. Watershed on the Graph and Hierarchy

All the morphological transformations can be extended to the graph defined above, where the summits correspond to the simple arcs of the primitive watershed transform and the vertices connect the boundaries surrounding the same primitive catchment basin. In particular, the notion of minimum as it has been defined using paths on the graph of a function can be applied to this valued graph. In our case, the weakest boundaries of the mosaic image correspond to regional minima of the new graph (Figure 34c). We may also flood the "relief" defined by this valued graph starting from these minima. This watershed transformation produces watershed lines composed of points of the new graph which correspond in fact to boundaries of the primitive segmentation of the initial image. Only these boundaries, corresponding to the divide lines of the graph, remain. We have thus suppressed the boundaries of the primitive watershed which are surrounded by more contrasted ones.

4. Illustration

The implementation of the algorithm is not difficult. It is possible to realize it with an ordered queue. We can also use more classical watershed algorithms. The fact that the graph used is not planar is not a real problem. Moreover, one can show [5] that this graph may be transformed into a planar one and in a second step that this planar graph may be used to build an image, called a hierarchical image, to which the classical watershed algorithms may be applied, producing as a result the hierarchical segmentation described above.

Let us show this algorithm on our simple example. To contour the metallic particles, we compute the mosaic image and we suppress some oversegmented regions by performing the first degree of this hierarchical process. But this procedure may be repeated, giving higher and higher levels of hierarchy. At the end of these iterations, the final image is obviously empty. Nevertheless, it is possible to label every boundary of the primitive watershed image with the higher level of hierarchy in which this boundary remains (Figure 35b). It is then easy to see that the particles correspond to the maxima of this new image.

The result of this hierarchical segmentation is given in Figure 35c. From that picture, the extraction of the particles is straightforward. They correspond to the new homogeneous regions that contain the maxima of the initial image (Figure 35d).

D. Another Example

This hierarchical segmentation can be used efficiently for extracting features from complex scenes [30]. Let us apply this technique for delineating the road in the scene represented in Figure 36.

The initial image having been very noisy, the result of the watershed transformation of the gradient image is oversegmented (Figure 36b). However, the road being a rather homogeneous region in the image, we hope that the hierarchical segmentation will help in extracting it from the image. The mosaic image is performed (Figure 36c), and then its gradient (Figure 36d). The result of the first level of hierarchy is given in Figure 36e. In the second step, the region in front of the scene may be extracted. This marker of the road may be used again in an

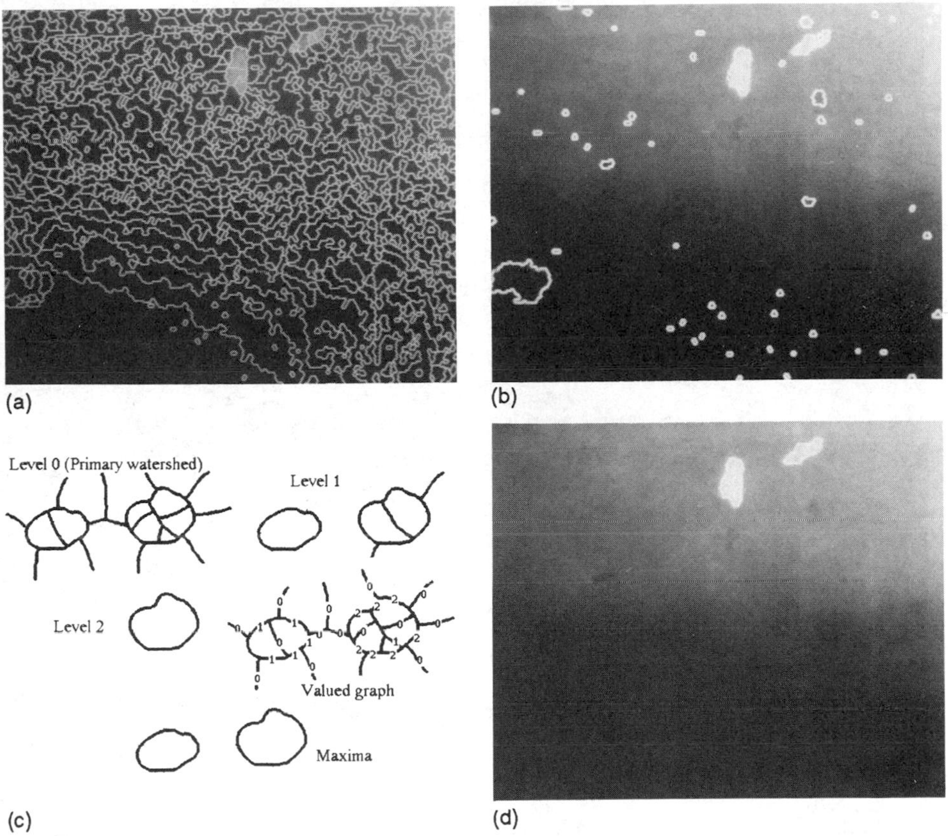

Figure 35. (a) Initial watershed; (b) hierarchical segmentation; (c) principle of labeling of the arcs of the watershed; (d) final result.

homotopy modification and watershed procedure to produce a more refined contouring of the road (Figure 36f).

E. Discussion

The result of the watershed transformation yields to a hierarchical segmentation of the image, as illustrated in the previous examples. The selection of some markers can be made at this level to segment features in the image (for example, the road in the last example). Further levels of hierarchy may also be defined by iterating this procedure (as shown in the introductory example).

Starting from a highly fragmented image, we have obtained after simplification a new mosaic. It is obviously possible to iterate this simplification process. By this means we get a hierarchy of simplification stages, the last always being a

(a) (b) (c) (d) (e) (f)

Figure 36. (a) Road scene viewed through the windscreen of a car; (b) result of the watershed applied to the gradient; (c) mosaic image; (d) gradient of the mosaic image; (e) first stage of the hierarchy; (f) extraction of the marker of the road.

uniform image. It is also possible to change the valuation of the graph between the different stages of hierarchization. Doing so, we can introduce various criteria of hierarchization. For instance, in the case of the fracture image (Section VI), one could use as a valuation the angle between two adjacent facets calculated from the measure of the altitude of the boundary points. Many alternative techniques are also possible, such as valuation according to the size, shape, or orientation of the tiles of the mosaic picture and calculation of the new gradient values we get when the tiles are merged.

VIII. CONCLUSION

The morphological approach to image segmentation problems by means of the watershed transformation is an efficient technique, in terms of the results it produces as well as the control kept by the user on every stage of the process. Let us briefly discuss some of these advantages.

1. The watershed transform provides closed contours, by construction. This fact is of primary importance because we do not have to worry about the contour closing of objects, which could be a problem when using contour detection methods. In other words, the watershed transform aims at extracting objects or regions in the image. This property is particularly helpful when there is no visible contour, as shown in the coffee grains example.
2. When computing the watersheds, the watershed lines always correspond to contours which appear in the image as obvious contours of objects, even when severe oversegmentation occurs. This explains, in particular, the good positioning of contours and the appropriate construction of the mosaic image used in the hierarchical approach. This property is very interesting because it gives to the watershed transformation a great advantage compared to the split-and-merge methods, where the first splitting is often a simple regular sectioning of the image leading sometimes to unstable results.
3. It is a general method which can be applied to many situations. We gave some examples of its use in various applications. But, in fact, these examples are only a small selection of the domains in image analysis where this technique has been used efficiently. Remember that this methodology can be applied to three-dimensional images where the contour detection techniques fail because the notion of contour in a three-dimensional image is not easy to define and to handle.
4. The great advantage of this methodology is that it splits the segmentation process into two separate steps. First, we have to detect what we want to extract: it is the marker selection. Then we have to define the criteria which are used to segment the image. These criteria may be photometric (contrast variations), or based on the shape of the objects, or a combination of both. This combination of different criteria is made easier through the use of powerful morphological tools (geodesic transforms, homotopy modification, and so on) as shown in the examples.

This last assertion means that image segmentation cannot be performed accurately and adequately if we do not construct the objects we want to detect. In this approach, the picture segmentation is not the primary step of image understanding. On the contrary, a fair segmentation can be obtained only if we know exactly what we are looking for in the image.

In fact, there is no general method available to achieve this marker detection and object selection. But why should such a general method exist? The everyday practice of image analysis shows, on the contrary, that in many problems you are not able to see the features you want to extract if you don't know a priori what you are looking for. For instance, we saw in the electrophoresis gel example that the blobs may be extracted individually or as heaps. These are two different ways of seeing the same image. That means that you must often build or construct the objects you want to detect. Image segmentation is not the primary step in image understanding; it is its consequence.

REFERENCES

1. Marr, D., *Vision,* Freeman, San Francisco, 1982.
2. Canny, J. F., Finding edges and lines in images, Artificial Intelligence Laboratory, MIT, Cambridge, Massachusetts, TR-720, 1983.
3. Serra, J., *Image Analysis and Mathematical Morphology*, Academic Press, London, 1982.
4. Coster, M., and Chermant, J. L., *Précis d'analyse d'images*, Editions CNRS, France, 1985.
5. Beucher, S., Segmentation d'images et morphologie mathématique, Doctorate thesis, School of Mines, Paris, 1990.
6. Meyer, F., and Beucher, S., Morphological segmentation. *J. Visual Commun. Image Repres.*, *1(1)*, 21–45 (1990).
7. Meyer, F., Cytologie quantitative et morphologie mathématique, Doctorate thesis, School of Mines, Paris, 1979.
8. Matheron, G., Filters and lattices, in *Image Analysis and Mathematical Morphology*, vol. 2, Academic Press, London, 1988, pp. 115–136.
9. Lantuejoul, C., and Beucher, S., On the use of the geodesic metric in image analysis, *J. Microsc.*, *121*(Pt 1) (1981).
10. Matheron, G., *Random Sets and Integral Geometry*, Wiley, New York, 1975.
11. Meyer, F., Skeletons and perceptual graphs, *Signal Process.*, *16(4)*, 335–363 (1989).
12. Matheron, G., Examples of topological properties of skeletons, in *Image Analysis and Mathematical Morphology*, vol. 2, Academic Press, London, 1988, pp. 217–233.
13. Beucher, S., and Lantuejoul, C., Use of watersheds in contour detection, in *Proceedings, International Workshop on Image Processing*, *CCETT/IRISA*, Rennes, France, 1979.
14. Serra, J., *Image Analysis and Mathematical Morphology*, vol. 2, *Theoretical Advances*, Academic Press, London, 1988.

15. Meyer, F., Sequential algorithms for cell segmentation: maximum efficiency? in *Proceedings, International Symposium on Clinical Cytometry and Histometry*, Schloss Elmau, 1986.
16. Rosenfeld, A., and Pfaltz, J. L., Sequential operations in digital picture processing, *J. ACM*, *13*, 471–494 (1966).
17. Borgefors, G., Distance transformations in digital spaces, *CVGIP*, *34*, 344–371 (1986).
18. Danielson, P. E., Euclidean distance mapping, *CVGIP*, *14*, 227–248 (1980).
19. Meyer, F., Algorithmes séquentiels, *Proceedings, Onzième Colloque GRETSI*, Nice, France, 1987, pp. 543–546.
20. Vincent, L., Algorithmes morphologiques à base de files d'attente et de lacets. Extension aux graphes, Doctorate thesis, School of Mines, Paris, 1990.
21. Verwer, B. J. H., Verbeek, P. W., and Dekker, S. T., An efficient uniform cost algorithm applied to distance transforms, *PAMI*, *11*, 425–429 (1989).
22. Verbeek, P. W., and Verwer, J. H., Shading from shape, the eikonal equation solved by grey-weighted distance transformation, *Pattern Recogn. Lett.*, *11*, 681–690 (1990).
23. Meyer, F., Un algorithme optimal de partage des eaux, in *Proceedings 8th Congress AFCET*, Lyon-Villeurbanne, France, 1992, vol. 2, pp. 847–859.
24. Gratin, C., and Meyer, F., Mathematical morphology in three dimensions, in *Proceedings 8th ICS*, Irvine, California, Aug. 25–30, 1991.
25. Meyer, F., Color image segmentation, in *Fourth International Conference on Image Processing and Its Applications*, Maastricht, April 1992.
26. Friedlander, F., Le traitement morphologique d'images de cardiologie nucléaire, Doctorate Thesis, School of Mines, Paris, 1989.
27. Beucher, S., Segmentation tools in mathematical morphology, in *Proceedings, SPIE Congress*, San Diego, 1990.
28. Vincent, L., and Soille, P., Watersheds in digital spaces: an efficient algorithm based on immersion simulations, *IEEE PAMI*, *1*(6), 583–597 (1990).
29. Beucher, S., The watershed transformation applied to image segmentation, in *Tenth Pfefferkorn Conference*, Cambridge, UK, Scanning Microscopy International, 1991.
30. Beucher, S., Bilodeau, M., and Yu, X., Road segmentation by watershed algorithms, in *Proceedings, Prometheus Workshop*, Sophia-Antipolis, France (1990).

Chapter 13

Anamorphoses and Function Lattices (Multivalued Morphology)

Jean Serra

Centre de Morphologie Mathématique, Ecole des Mines de Paris
Fontainebleau, France

I. INTRODUCTION

A. Gray-Tone Morphology and Anamorphoses

At the end of the 1970s mathematical morphology, which had been initially designed as a set theory [1–3], went through a wave of extensions toward numerical functions. Now, prompted by some kind of physical intuition, or perhaps by some internal consistency, all of the authors who took part in this generalization adopted a common behavior when transposing binary notions into numerical ones. It was implicitly admitted that the extension had to involve *uniquely* inequalities and min, or max, operators (i.e., the concepts that correspond to the set-oriented inclusion, intersection, and union), but to keep away any possible arithmetic or algebraic operation.

The constraint held for the generalization of dilation and increasing mappings [4,5], as well as for thinnings and thickenings [6,7] or for segmentation by watersheds [8]. One can even detect it in Matheron's pioneer work on hit-or-miss topology for numerical functions [9]. Better still, such a new notion as Meyer's "top hat" transformation, which involves an arithmetic difference, was introduced in an apparently pure set way by relation (13.4) below [10].

The Minkowski addition $f \oplus B$ of a numerical function $f : \mathbf{R}^n \rightarrow \overline{\mathbf{R}}$ by a structuring element $B \subset \mathbf{R}^2$ typically illustrates this gray-tone approach:

$$(f \oplus B)(x) = \text{Sup}\{f(y), y - x \in B, B \subset \mathbf{R}^2\} \tag{13.1}$$

However, this common way of thinking was unconscious. I personally realized its meaning when Sternberg proposed, in 1979 [11, 37], another generalization of the set-oriented Minkowski addition [11], namely the following relationship:

$$(f \oplus g)(x) = \text{Sup}\{f(y) + g(x - y), x \in \mathbf{R}^2\} \tag{13.2}$$

(f and g, numerical functions), which encompasses relation (13.1). In spite of some intrinsic qualities, a feature of this second-generation operator was that it lost the common denominator of the previous transformations.

Now, what does such a property mean? When a grey tone operator only involves sup $\bigvee$, inf $\bigwedge$, and ordering $\leq$, it becomes automatically transportable ot any other gray ordering whose sup and inf are homologous to those of the first one. For example, if we replace the grey scale by that of its logs for in relation (1-2), we have

$$\text{Log}(f \oplus B) = \text{Log}\, f \oplus B$$

This makes the spatisl processing independent of the gray ordering. From a more technical point of view, if we call anamorphosis [11] an isomorphism between lattices (here the two gray ranges), then the above property consists in *commuting* (or being compatible) *with anamorphosis,* and the mappings which satisfy it are said to be *flat*.

The physical relevance of this compatibility is obvious for television images, which involve logarithmic gamma corrections, and various display amendments (e.g. histogram equalizations). But it allows as well to process data which have a large range of variation, such as those encountered in demography [35], by means of standard image processors or software programs. Also, owing to this property an n-bit image is transformed into another n-bit image. When one remembers that the limit of deconvolution implementation lies in practice in the bit depth acceptable by computers, one appreciates the property at its true worth.

Surprisingly, a decade before, a similar consensus on compatibility under anamorphosis of a kind hiding another name, arose in a different domain of Applied Mathematics. It was the search for robust estimates in mathematical statistics. When J. Tukey [40] proposed for example to take the median of a sample, instead of the classical arithmetic average, to estimate the mean of a population, he suggested to make the mean estimate independent of possible spurious tails in frequency distributions. Stated in our own words, this *exactly* coincides with anamorphosis compatibility. Besides, a large number of robust estimators, e.g. local histograms, local mode abscissa, adapted windows, and so on, can be interpreted in terms of flat structuring elements.

B. Function Lattices

If the concept is deep, it deserves to apply to other function lattices. In imagery, the two most important ones are multivalued images, obtained from a series of sensors, with color images in particular and (color) motion image sequences. But this goal immediately raises a number of nontrivial questions. Consider the relatively simple case of still color images in two dimensions. Practically, how do we define cross sections on them? By one set, by three sets? In the first case, does a decreasing stack of sets generate a function as for gray-tone images? Moreover, what can a "decreasing" stack of sets mean? How do we extend the very useful distinction between smaller and strictly smaller? What to do when the green is darker and the red lighter? Do minima and maxima exist? Can we generalize the rank operators, especially the median one? What does a top-hat mapping mean? Do we have to process the three channels separately? If not, in which way are they allowed to interact? Do these processing modes leave the color bit depth unchanged? If so, do they create additional colors or not? Etc.

Beyond all these questions a distinction between gray-tone and color images stands out. In the former case, the model of numerical functions $f : \mathbf{R}^2 \to \mathbf{R}$ is a matter for the following three structures:

1. A complete lattice
2. A chain (i.e., a totally ordered lattice)
3. A vector space (at least on $\mathbf{R}$, if not on $\overline{\mathbf{R}}$)

Point 3 refers to the class of linear anamorphoses, i.e., to changes of units on the gray scale. Commuting with linear changes obviously represents the minimum requirement as soon as we want to give, or to keep, a physical dimension to images through transformations [12]. When this requirement is not satisfied, the choices of units interact with image operations in an inextricable manner (e.g., in relation (13.2)). Fortunately, it is fulfilled by all linear mappings (convolution, kriging, Fourier transformation, etc.) and in morphology by the linear combinations of flat operators allowed by the lattice structure (e.g., top hat transforms for functions $f : \mathbf{R}^2 \to [0,1]$). The extension to a multidimensional vector space does not set basic problems, and we will not particularly focus on this in the following.

On the other hand, the distinction between partial (1) and total (2) orderings turns out to be essential for our purpose: a property of the gray-tone mapping cannot be extended to color cases if it requires the total ordering of $\overline{\mathbf{R}}$; otherwise it can. Furthermore, the color case remains somehow ambiguous because this non-totally ordered lattice is the product of three totally ordered ones. For this reason, it seems wiser to set the above series of questions in the general lattice framework (anyway, the proofs will not be more complicated than for the color case).

The organization of the chapter derives from these comments. After a bibliographic survey (Section II), Sections III to V are devoted to function lattices, flat operators, and anamorphic primitives, respectively. In Section VI, we come back to gray-tone functions, not only because they represent the simplest illustration but also because not all has been said about them (e.g., "Dolby" filters). The last section applies to color images the major results of the theoretical part. We have tried to multiply spots on various possible kinds of color processing, rather than a pseudoexhaustive approach.

This chapter is possible because of another text, due to Matheron [13], which is probably the most important, although unpublished, contribution to theoretical morphology of the last 3 years. The two definitions and the associated characterizations of a cross section and of an anamorphosis on function lattices given in [13] open a way that the present chapter continues. However, unlike ref. [13], our point of view remains purely algebraic, and none of the theorems proved below needs topological conditions.

II. FLAT OPERATORS: STATE OF THE ART

In this section, we recall a few results about gray-tone functions. The notation has been chosen to keep as close as possible to the major previous works and, in the meantime, to prepare the extension to more general lattices.

A. Notation

The space on which the functions are defined will be given the generic symbol E, which designates indifferently $\mathbf{R}^n$ or $\mathbf{Z}^n$, $n \geq 1$, or subparts of them. The set into which these functions map is denoted by T, which may be the completed real line $\overline{\mathbf{R}}$ (respectively $\overline{\mathbf{Z}}$) or any bounded or not closed interval of $\overline{\mathbf{R}}$ (resp. of $\overline{\mathbf{Z}}$). A numerical function, also called a gray-tone function, is a mapping f from E into T. We shall denote by T^E the class of these numerical mappings. Whereas T is supposed to be a complete lattice, for the usual ordering, no assumption holds on E.

The subgraph, also called umbra (resp. supergraph and antiumbra), of f is by definition the subset of the product space $E \otimes T$, which is generated by the pairs (x,t), $x \in E$, $t \in T$, such that $f(x) \geq t$ (resp. $f(x) \leq t$). For a given $t \in T$, the section $X_f(t)$ of the subgraph of f, or more briefly the subsection of f, is the set of points of E such that $f(x) \geq t$:

$$X_f(t) = \{x \in E : f(x) \geq t\}$$

Similarly, the supersection $S_f(t)$ at level t is

$$S_f(t) = \{x \in E : f(x) \leq t\}$$

As a general rule, ψ designates a set operator and Ψ a function operator.

B. Historical Review on Flat Operators

The search for gray-tone operators that are compatible under anamorphosis occurred, these past 15 years, in connection with the operators based on function thresholds. After two first papers which just extended Minkowski addition and subtraction to gray-tone functions [4,5], the first series of substantial algorithms was proposed by Meyer on the one hand (gradient by erosion, top hat transform [10]) and by Beucher and Lantuéjoul on the other (watersheds [8,14]). All these approaches were digital and based on the idea that a *set* transformation $\psi : \mathcal{P}(E) \rightarrow \mathcal{P}(E)$, acting on the threshold $X_f(i)$ of the initial function, could generate the threshold at the same level i of a transform function $\Psi(f)$:

$$X_{\Psi(f)}(i) = \psi(X_f(i)) \tag{13.3}$$

More generally, the starting set operator ψ incorporated more than one set input (two for gradient and top hats, all the thresholds lower than t for the watershed at level t). But in any case the output was the section of the transform $\Psi(f)$ at level t. For example, Meyer's initial algorithm for the top hat transform was

$$X_{\Psi(f)}(i) = \bigcup_{i=0}^{i\max -i} \left[X_f\,(i + i')\backslash([X_f(i + 1)] \ominus B) \oplus B \right] \tag{13.4}$$

which actually corresponds to the threshold at level i of function f minus its (flat) morphological opening by B. Starting from these digital approaches, in 1982 Serra [7] extended them to the Euclidean case of functions $f : \mathbf{R}^n \rightarrow \overline{\mathbf{R}}$ by treating separately the one- and the two-operand set mappings ψ. He had to assume semicontinuity for function f and operators ψ to keep valid relations such as (13.3) or (13.4). In addition, the first one demanded ψ to be increasing. For the non-increasing case, he replaced relation (13.3) by

$$X_{\Psi(f)}(i) = \bigcap_{j \leq i} \psi(X_f(i)) \tag{13.5}$$

in the digital case. This allowed him to extend thinnings and thickenings from binary to gray-tone images. For the first time, the requirement of commutativity under anamorphosis was introduced. It was proved that the class of gray operations generated by inf and finite composition of Minkowski additions satisfies the commutation (or compatibility) requirement.

Further on, various studies focused on the single-operand operators. Wendt et al. [15], in 1986, called them "stack filters" and analyzed their digital properties. Following the same direction, Zeng et al. [16] concentrated on the self-duality (median filters). From a more theoretical point of view, Maragos and Schafer in 1987 [17] based a theory of morphological operations on semicontinuity and increasingness. By adopting the opposite point of view, Heijmans in 1991 [18] extended some results by Serra [7,19] and by Janovitz [36] but without any increasingness or semicontinuity assumption. He extended relation (13.5) to the continuous case, generating the so-called "flat operators." A number of strong results are derived. Among others, as far as the present study is concerned, he proved the compatibility under anamorphosis for (increasing or not) flat operators Ψ_1: $\overline{\mathbf{R}}^E \to \overline{\mathbf{R}}^E$, and for increasing flat operators Ψ: $T^E \to T^E$, when T is a finite lattice (in the latter case "anamorphosis" is no longer a bijection).

C. Comments on the Historical Review

(1) The idea of putting together cross sections and anamorphoses seems very natural as soon as one notices that

$$X_{\alpha f}(t) = X_f(\alpha^{-1}(t)) \qquad \text{and} \qquad S_{\alpha f}(t) = S_f(\alpha^{-1}(t))$$

for any anamorphosis α on the gray-tone axis. In a sense, it is because we have already these two commutation relations that the flat operators Ψ are compatible with anamorphoses.

(2) Independently of anamorphoses, formula (13.5) remains a tremendous tool to generate function transformations from set transformations and vice versa. How to extend it?

(3) Relation (13.5) or its t-continuous version, hides an implicit choice made by Serra, Janovitz, and Heijmans. Starting from the same function f and the same set mapping ψ, we can generate another function, at least, by putting

$$S_{\Psi f} = \bigcap_{t'>t} \psi(S_f(t'))$$

and probably more than one. What about them?

(4) Curiously, the second wave of studies, corresponding to the last 8 years, has ignored the multioperand operators, whereas they were within the scope of the first wave of studies (e.g., Meyer's top hats or Beucher's watersheds).

(5) Last, but not least, nothing in the previous studies tells which properties are due to the total ordering of T and which to its lattice structure only.

III. CROSS SECTIONS AND ANAMORPHOSES ON FUNCTION LATTICES

Beyond gray-tone functions, image processing deals more and more with multispectral pictures and with motion pictures. These two cases turn out to cover sufficiently different situations to prompt us to base the approach on general function lattices.

Let E be a set, such as a Euclidean or a digital space, and let T be a complete lattice where inequality, sup, and inf are denoted by $<$, $\wedge$, and $\vee$, respectively. Then the set T^E of the mappings from E into T, when it is equipped with product ordering (i.e., $g < f$ if $g(x) < f(x)$ for all $x \in E$) becomes in turn a complete lattice, where the sup and inf, still denoted by $\bigvee$ and $\bigwedge$, are defined by the relations

$$(\bigvee f_i)(x) = \bigvee f_i(x) \qquad \text{and} \qquad (\bigwedge f_i(x) = \bigwedge f_i(x) \tag{13.6}$$

We saw that, for gray-tone functions, dimensionality questions are tied up with cross sections and anamorphoses. But can we still define these two notions on general function lattices? If so, how?

A. Anamorphoses

This old term dates back to the Renaissance, when painting art discovered the law of perspective. It cassically designates monotonous and continuous deformations of the coordinates of an image. In Figure 1, which illustrates the notion, the anamorphosis holds on the space domain D in which the image is defined. In image processing, anamorphoses more often hold on the intensity distortions than on the geometric ones, i.e., on lattices T and, by extension, T^E. However, Figure 1 suggests that an anamorphosis should link two lattices T and T' in an isomorphic way, i.e., by preserving ordering, sup, and inf in both senses. This led Matheron [13, Section 1] to propose the following definition.

Definition 3.1. A mapping α from a lattice T into a lattice T' is an anamorphosis when it is an isomorphism for the complete lattice structure, i.e.,

1. α is a bijection (one-to-one and onto mapping)
2. α and α^{-1} are both dilations and erosions

The two conditions are not redundant: one can see, for example, in Figure 2 a dilation-erosion α that is not bijective. The two following criteria illustrate the need for both properties 1 and 2.

Criterion 3.2. A bijection $\alpha : T \to T'$ is an anamorphosis if and only if α and α^{-1} are increasing.

Proof. If α and α^{-1} are increasing, then the equivalences

$$\alpha(x) < x' \Leftrightarrow x < \alpha^{-1}(x'), \qquad x \in T, x' \in T'$$
$$x < \alpha(x') \Leftrightarrow \alpha^{-1}(x) < x'$$

show that α is a dilation and an erosion. The converse proof is trivial.

Criterion 3.3 (from [13]). A mapping $\alpha : T \to T'$ is an anamorphosis if and only if it is *onto* and satisfies the following condition:

$$x < y \Leftrightarrow \alpha(x) < \alpha(y), \qquad x,y \in T \tag{13.7}$$

The proof is easy; see [13, Section 1].

We see from this last criterion that the definition of an anamorphosis already introduced for the gray-tone case (α strictly increasing mapping) is a particular case of the general one. Nevertheless, when T and T' are not totally ordered

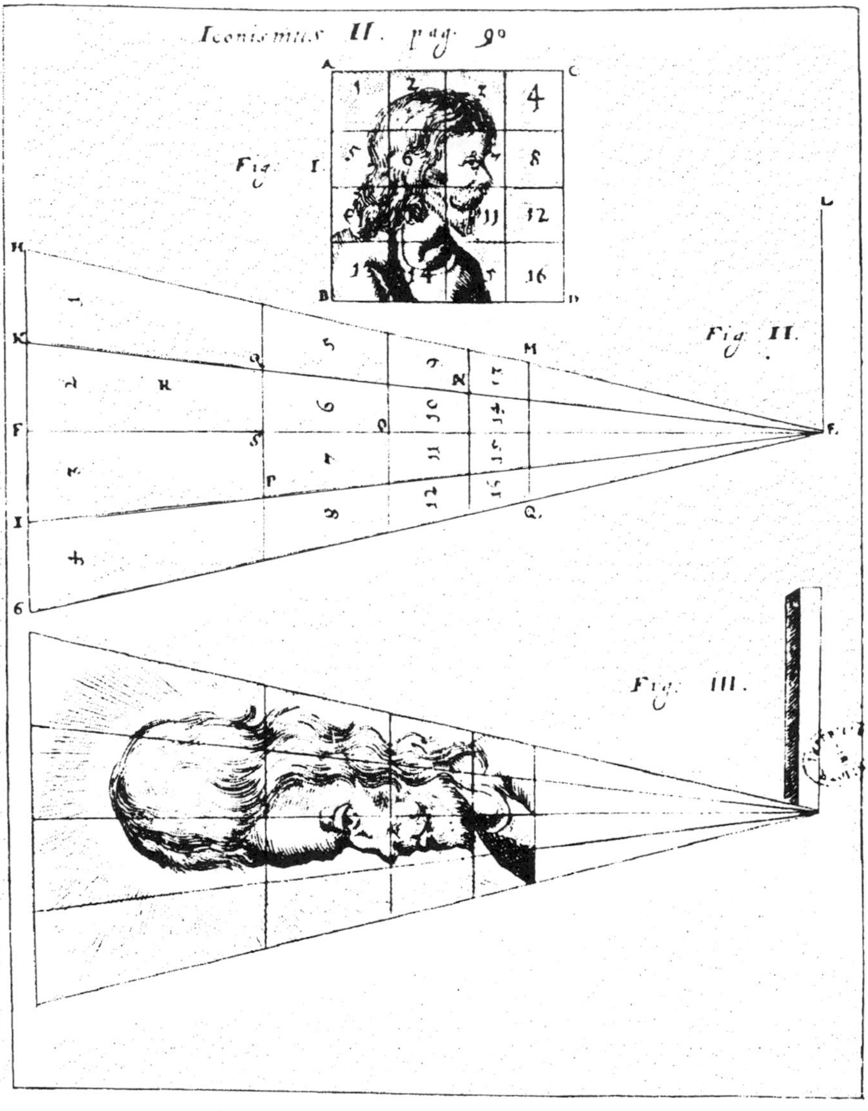

Figure 1. Anamorphoses of the portrait of King Louis XIII by P. du Breuil (1649), from Balturiatis [33]. The square domain D is changed into a parallelogram P in such a way that the two lattices $\mathcal{P}(D)$ and $\mathcal{P}(D)$ of their respective subsets are isomorphic, i.e., such that if A, $B \in \mathcal{P}(D)$ become A', $B' \in \mathcal{P}(P)$, then $A' \cap B'$ and $A' \cup B'$ are the transforms of $A \cap B$ and $A \cup B$, respectively.

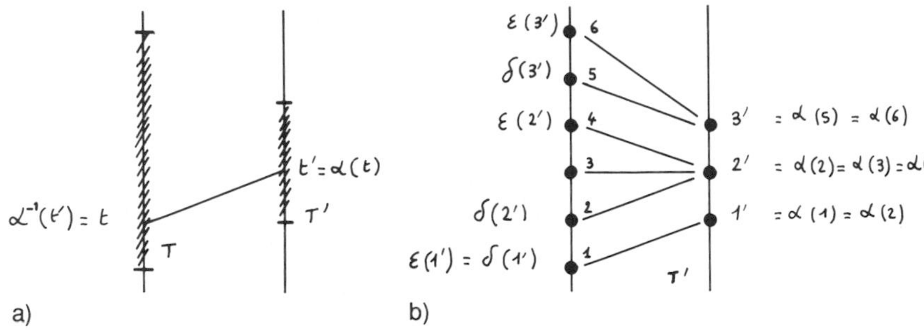

Figure 2. In the continuous case (a), the strict increasingness of the bijection $\alpha : T \to T'$ implies that $\alpha(s) \geq t \Leftrightarrow s \leq \alpha^{-1}(t)$ and that $\alpha(s) \leq t \Leftrightarrow s \leq \alpha^{-1}(t)$; hence α is both a dilation and an erosion, with α^{-1} as adjoined erosion and dilation. In the discrete case (b), since T and T' are totally ordered, α is still a dilation and an erosion, but the adjoined erosion and dilation (from T' into T) namely $\varepsilon(t') = \max\{s : \alpha(s) = t'\}$ and $\delta(t') = \min\{s : \alpha(s) = t'\}$, are distinct from each other.

(e.g., multispectral or motion images), the strictly increasing mappings from T onto T' are generally not anamorphoses.

It is clear that anamorphoses, as any operator acting on lattices T or T', admit immediate extensions as product operators on T^E or T'^E. It suffices to put for every $x \in E$

$$(\alpha \circ f)(x) = \alpha[f(x)]$$

to define α as a mapping from T^E into T'^E (e.g., the log of a gray-tone image is generated by taking the log of the intensity at each point). For simplicity, we keep the same notation for both operators $\alpha : T \to T'$ and $\alpha : T^E \to T'^E$. The context is always explicit enough to make the distinction clear.

Another aspect of Definition 3.1 concerns the adjunction of erosions and dilations it involves. Although the reader probably already knows these notions, otherwise he would not dive into such a chapter, we recall them briefly. (This basic morphology on complete lattices has been introduced and developed by Matheron [20] and by Serra [21] and taken up again with additional results by Heijmans and Ronse in [22].) Given two complete lattices T and T', any operator from T into T' that preserves the sup (resp. the inf) is called a dilation (resp. an erosion). Denoting these mappings by δ and ε, respectively, we have

$$\delta(\bigvee t_i) = \bigvee \delta(t_i), \qquad \varepsilon(\bigwedge t_i) = \bigwedge \varepsilon(t_i)$$

where the symbols $\bigwedge$ and $\bigvee$ concern lattice T in the left parts and lattice T' in the right ones.

Associated with every dilation $\delta : T \to T'$ there exists a unique erosion $\varepsilon : T' \to T$ that satisfies the equivalence

$$\delta(t) < t' \Leftrightarrow t < \varepsilon(t'), \qquad t \in T, t' \in T' \tag{13.8}$$

Moreover, if this equivalence holds between two operators, from $T \to T'$ and $T' \to T$, then they turn out to be a dilation and an erosion, respectively, and they are said to be adjoined or reciprocal. More precisely we have, given $\delta : T \to T'$

$$\varepsilon(t') = \bigvee\{t, t \in T, \delta(t) < t'\}$$

and given $\varepsilon : T \to T'$,

$$\delta(t') = \bigwedge\{t, t \in T, \varepsilon(t) > t'\} \tag{13.9}$$

In particular, in case of anamorphoses, α^{-1} is both the adjoined erosion and dilation of α, considered itself as a dilation and as an erosion.

Let us conclude this section with a brief comment on duality. Starting from the set of the elements of a complete lattice T, reorder them by stating that t_1 is smaller than t_2 iff $t_1 > t_2$ in T. Obviously, this new order generates a new lattice T^* on the elements of T, where $\text{Sup}\{x_i\} = \bigwedge x_i$ and $\inf\{x_i\} = \bigvee x_i$ for any family $\{x_i\}$ in T. The lattice T^* is said to be *dual* of T and every proposition on T may be associated with a dual version on T^* by inverting $>$ and $<$, on the one hand, and $\bigvee$ and $\bigwedge$ on the other hand. Strictly speaking, the mapping $T \to T^*$ is not an anamorphosis, though it preserves the lattice structure, but rather an involution.

B. Supersections, Subsections

The following generalization and Proposition 3.4 are due to Matheron [13, Section 1]. Let us recall them. The supergraph, or antiumbra, Γ of a mapping $f : E \to T$ is defined as the subset of the product space $E \otimes T$ which is made of all the pairs (x,t), $x \in E$, $t \in T$, such that $t > f(x)$.

For a given $x \in E$, the set of $t \in T$ such that (x,t) belong to the supergraph Γ is the set of majorants of $f(x)$. Conversely, given $t \in T$, the set $S(t)$ of points $x \in E$ such that $(x,y) \in \Gamma$ defines the cross section of the supergraph $\Gamma(t)$ at level t. We shall call them, more briefly, the supersections of f. Hence we have

$$S_f(t) = \{x\, ; x \in E, f(x) < t\}$$

Similarly, if the subgraph $\Gamma^* \subset E \otimes T$ is the set of pairs (x,t) such that $f(x) > t$, then the cross sections of Γ^*, also called subsections of f, are given by the relationship

$X_f(t) = \{x\, ; x \in E, f(x) > t\}$; furthermore, we have

$$S_f(\bigwedge t_i) = \bigcap S_f(t_i) \qquad \text{and} \qquad X_f(\bigvee t_i) = \bigcap X_f(t_i)$$

Hence the mapping $t \to S_f(t)$ from T into $\mathcal{P}(E)$ is an erosion as well as (changing the sense of the T axis) the mapping $t \to X_f(t)$. According to relation (13.9), the associated dilations δ_S and δ_X from $\mathcal{P}(t)$ into t are nothing but

$$\delta_S(Z) = \wedge\{t : Z \subset S_f(t)\} = \wedge\{t : x \in Z \Rightarrow f(x) \leq t\} = \vee\{f(x), x \in Z\}$$
$$\delta_X(Z) = \vee\{t : Z \subset X_f(t)\} = \vee\{t : x \in Z \Rightarrow f(x) \geq t\} = \wedge\{f(x), x \in Z\}$$

In particular, when Z is reduced to the point $\{x\}$, considered as a set, $\delta_X(\{x\}) = \delta_S(\{x\}) = f(x)$. Hence any erosion $\varepsilon : T \to \mathcal{P}(E)$ generates a unique function $f : E \to T$ via its adjoined dilation δ, by putting

$$f(x) = \delta(\{x\}), \qquad x \in E$$

As the image $\varepsilon(T)$ is considered to be the family of supersections or subsections of f, the adjoined dilation δ will be the above δ_S or δ_X. But f remains the same in both cases, and we may state

Proposition 3.4 (from [13], Section 1). A mapping $t \to A(t)$ from T into $\mathcal{P}(E)$ is the supersection (resp. the subsection) of a certain function $f : E \to T$ if and only if A is an erosion from T (resp. from the dual lattice T^*) into $\mathcal{P}(E)$.

Note that we had to replace T by its dual T^*, when dealing with subsections, in order to make the mapping $t \to X(t)$ increasing. This explains why δ_X is a dilation in spite of its expression in terms of inf.

Figure 3 gives an elementary example of such a situation. The space E is constituted by six aligned points. The lattice T equals $\mathcal{P}(B)$, where B is the two-element family {(.); (˙)}, so T possesses the four following elements, including the empty set $\emptyset$:

$$T = \{(:); (.); (\dot{}); (\emptyset)\}$$

As a consequence of the proposition, when a same dilation δ is performed on each supersection S, we do not necessarily generate the series of supersections of a hypothetical transformed function $\delta(f)$. This was done in Figure 3d; the intersection of the second line with the third one should correspond to the set of those points $x \in E$ where $(\delta f)(x) < (.) \wedge (\dot{}) = \emptyset$. However, the last line, which is also supposed to represent the points $x \in E$ for which $(\delta f)(x) < (\emptyset)$, is different!

Hence the flat approach does not apply to general function lattices. Even in the digital case, we cannot implement an increasing (set) operator level by level and expect to obtain the corresponding levels of a hypothetical transformed function. We must find another way.

IV. FLAT OPERATORS ON LATTICES

To achieve a meaningful generalization to function lattices of the classical notion of a flat gray-tone operator, we have to reach the following three goals:

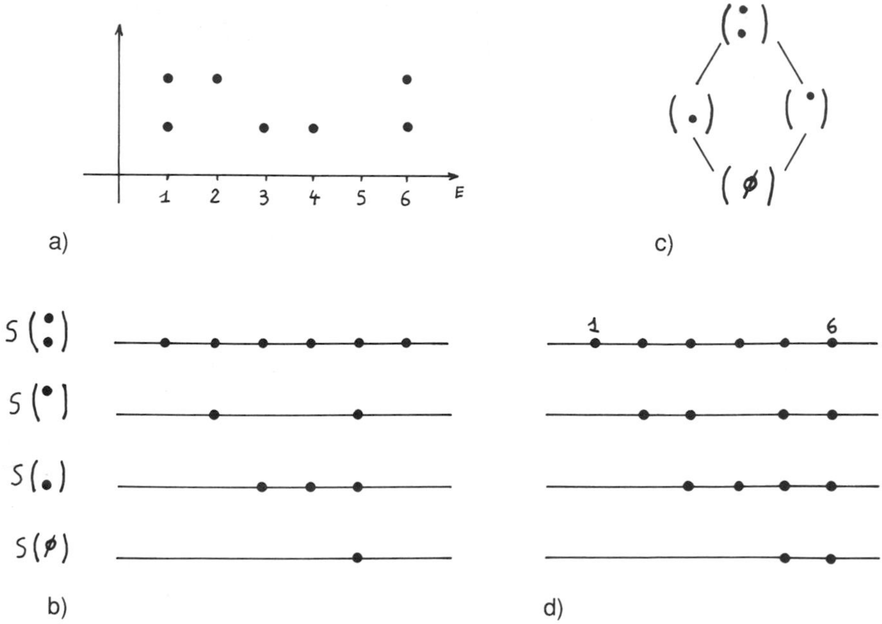

Figure 3. (a) Function f from E into $T = \mathcal{P}(B)$ where B consists of two elements $(\cdot)$ and $(.)$; (b) supersections S of f; (c) ordering relation scheme on T; (d) the dilations of the supersections S by a pair A of consecutive points in E.

1. To extend correctly the operations "union (or intersection) of a set family, $\{A(t)\}$ as $t > t_0$." What concept does the *strictly* larger symbol correspond to?
2. To define a class of set primitives whose associated flat (function) operators cover in particular watersheds, summit extractors, etc.
3. To introduce some properties on the set primitives that make the associated function operators compatible under anamorphosis.

We will now examine these three points successively.

A. Generalization of the Strict Inequality

In the totally ordered lattice $\overline{\mathbf{R}}$, the only ordering relation is $\leq$. The "strictly larger (or smaller)" relations $>$ (or $<$) are introduced by duality with respect to the complement in $\overline{\mathbf{R}}$. Since

$$\{t : t \leq t_0\}^c = \{t : t > t_0\}$$

Similarly, in any complete lattice, equipped with ordering $<$, the complement of the proposition "y is smaller than x, that is, $y < x$" will be "y is not smaller

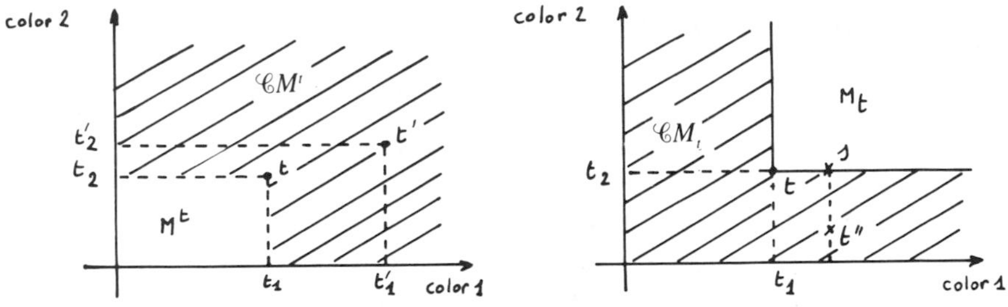

a) b)

Figure 4. Lattice T is the set of the pairs $(t_1 t_2)$ of intensities for bicolor images. Here $(t_1 t_2) < (t'_1 t'_2)$ iff $t_1 \leq t_2$ and $t'_1 \leq t'_2$. The four basic sets M_t, M^t, $\mathscr{C}M_t$, $\mathscr{C}M^t$ are represented. Note that the max s of t and t'' (b) is different from both of them.

than x, that is, $y \nless x$." This last proposition differs notably from "y is larger than x, and different from x," that is, "$y > x$, $y \neq x$," as can be seen in Figure 4.

In other words, the four basic sets associated with $x \in T$ are M^x of its minorants, M_x of its majorants, and their two complements $\mathscr{C}M^x$ and $\mathscr{C}M_x$ in T:

$$M_x = \{y : y \in T,\ y > x\}, \qquad M^x = \{y : y \in T,\ y < x\}$$

$$\mathscr{C}M_x = \{y : y \in T,\ y \ngtr x\}, \qquad \mathscr{C}M^x = \{y : y \in T,\ y \nless x\}$$

Finally, whereas the relation $\ngtr$ does not induce an ordering, we have the following implication:

$$s \ngtr t \text{ and } t_1 > t \Rightarrow s \ngtr t_1 \tag{13.10}$$

A notation that makes a distinction between $y \leq x$ and $y < x$ becomes irrelevant as soon as the lattice under study is not totally ordered and has to be replaced by $y < x$ (y smaller than or equal to x) and $y \nless x$. In particular, when sets are concerned, for this reason the notation $Y \subset X$ is more adequate than $Y \subseteq X$. Moreover, it has been proved [13,23] that when the lattice T is provided with a topology adapted to its ordered structure, the sets M^x and M_x are closed in T, hence $\mathscr{C}M_x$ and $\mathscr{C}M^x$ are open, whereas the set $M^x \backslash \{x\}$ is generally not open, apart from the totally ordered lattices.

B. Set Primitives and Flat Operators

Definition 4.1. Let T be a complete lattice, E be an arbitrary space, and f be a mapping from E into T, i.e, an element of T^E. Any mapping $(t,f) \rightarrow \psi_f(t)$ from $T \otimes T^E$ into $\mathscr{P}(E)$ is called a *set primitive*.

The interest in such primitives lies in their ability to generate function mappings Ψ from T^E into itself. Indeed, given a primitive ψ_f, the mapping

$$\begin{cases} Z(t) = \cap \{\psi_f(s),\ s \not< t\}, & \text{if } t \neq t^m \\ Z(t) = E, & \text{if } t = t^m \end{cases} \tag{13.11}$$

where t^m denotes the maximum element of lattice T, is always an erosion. To prove it, consider a family $\{t_i \in T\}$, $i \in I$, in T, whose inf is $t = \bigwedge_i t_i$. We have

$$Z(t) = \cap\{\psi_f(s),\ s \not< \wedge\ t_i\} \tag{13.12}$$

Since T is a complete lattice, to say "$s < \wedge\ t_i$" is equivalent to saying "$s < t_i$, for all $i \in I$" or again "there is no $i_0 \in I$ such that $s \not< t_{i_0}$" or else

$$\{s : s \not< \wedge\ t_i\} = \bigcup_{i \in I}\{s : s \not< t_i\} \tag{13.13}$$

Now the intersection (13.12) of the $\psi_f(s)$'s over the union (13.13) of the domains equals the intersection, with respect to i, of the intersections of the $\psi_f(s)$'s in each domain $\{s : s \not< t_i\}$, i.e.,

$$Z(t) = \bigcap_i \bigcap_{s \not< t_i} \psi_f(s) \tag{13.14}$$

or again $Z(\wedge\ t_i) = \cap\{Z(t_i),\ i \in I\}$; therefore, the mapping $t \to Z(t)$ is an erosion, and, according to Proposition 3.4, there exists a unique function $\Psi_1(f)$ whose $Z(t)$ are the supersections

$$S_{\Psi_1(f)}(t) = \bigcap_{s \not< t} \psi_f(t) \tag{13.15}$$

According to relation (13.9), this function is

$$\Psi_1(f) = \inf\{t : x \in \bigcap_{s \not< t} \psi_f(t)\} \tag{13.16}$$

Now, if we focus on subsections X instead of supersections S, a similar approach allows us to associate with the primitive ψ a mapping $\Psi_0 : T^E \to T^E$ as follows:

$$X_{\Psi_0(f)}(t) = \bigcap_{s \not> t} \psi_f(t)$$

or

$$\Psi_0(f) = \sup\{t : x \in \bigcap_{s \not> t} \psi_f(s)\} \tag{13.17}$$

which is equivalent. These first results are summarized in the following theorem.

Theorem 4.2. Given an arbitrary set primitive, in the sense of Definition 4.1, there exists a unique mapping $\Psi_0 : T^E \to T^E$ such that for every pair $(t,f) \in T \otimes T^E$ the subsection of $\Psi_0(f)$ at level t and at the minimum level t_m is

$$X_{\Psi_0(f)}(t) = \bigcap_{s \not> t} \psi_f(t); \qquad X_{\Psi_0(f)}(t_m) = E$$

Similarly, there exists a unique mapping $\Psi_1 : T^E \to T^E$ such that for every pair $(t,f) \in T \otimes T^E$, the subsection of $\Psi_1(f)$ at level t and at the maximum level t^m is

$$X_{\Psi_1(f)}(t) = \bigcap_{s \not< t} \psi_f(t); \qquad X_{\Psi_1(f)}(t^m) = E$$

The expressions for $\Psi_0(f)$ and $\Psi_1(f)$ are given by relations (13.16) and (13.17), respectively.

Comments

1. Ψ_0 and Ψ_1 are not the only function mappings that one can draw from the primitive ψ. It is also possible, for example, to define a middle element Ψ between Ψ_0 and Ψ_1 by putting

   $$\Psi = (\mathrm{Id} \wedge \Psi_1) \vee \Psi_0$$

   (Id = identity operator) among a number of other possible mappings. This said, Ψ_0 and Ψ_1 are the most direct derivations from primitive ψ.
2. By playing with the duality with respect to complement in $\mathcal{P}(E)$, we could have introduced Ψ_0 and Ψ_1 from unions performed on $\psi_f(t)$. Indeed, by taking the complements in relation (13.15), for example, we obtain

   $$\bigcup_{s \not< t} \Psi_f^c(S) = [S_{\Psi_1(f)}(t)]^c = \{x : (\Psi_1 f)\,(x) \not< t\}$$

However we shall keep the logic of the intersection for the next derivations, and in particular to prove the following representation:

Theorem 4.3. Given a set primitive $\psi : T \otimes T^E \to \mathcal{P}(E)$, its two function extensions Ψ_0 and Ψ_1 admit the following representations:

$$(\Psi_0 f)(x) = \bigvee\{t : x \in \bigcap_{s \not> t} \psi_f(s)\} = \bigwedge\{t : x \in \psi_f(t)\}$$

$$(\Psi_1 f)(x) = \bigwedge\{t : x{:} \in \bigcap_{s \not< t} \psi_f(s)\} = \bigvee\{t : x \notin \psi_f(t)\}$$

In particular, Ψ_0 is always smaller than Ψ_1.

Proof. We shall prove the representation associated with Ψ_0. Given $x \in E$, put

$$t_1 = \bigwedge \{t : x \notin \psi_f(t)\}$$

From "$x \notin \psi_f(s) \Rightarrow s > t_1$," we draw "$s \not> t_1 \Rightarrow x \in \psi_f(s)$." Now let $t < t_1$. For all $s \not> t$, we have, according to relation (13.10), $s \not> t_1$, hence $x \in \psi_f(s)$, and

$$x \in \bigcap_{s \not> t} \psi_f(s) = X_{\psi_0(f)}(t)$$

that is, $(\Psi_0 f)(x) > t$, and finally

$$(\Psi_0 f)(x) > \bigvee \{t,\, t < t_1\} = t_1, \qquad \text{that is, } (\Psi_0 f)(x) \in M_{t_1} \tag{13.18}$$

To prove the reverse inequality, consider t such that $t > t_1$ and $t \neq t_1$. In $\mathscr{C}M_t = \{s : s \not> t\}$ there necessarily exists an element s such that $x \notin \psi_f(s)$. If not, all s such that $x \notin \psi_f(s)$ should be included in M_t, which would result in $t < \wedge \{s: x \notin \psi_f(s)\} = t_1$, which is impossible by hypothesis. Therefore, for every $t > t_1$ and $t \neq t_1$, we may write

$$x \notin \bigcap_{s \not> t} \psi_f(s) \Leftrightarrow x \in [X_{\Psi_0(f)}(t)]^c \Leftrightarrow (\Psi_0 f)(x) \in \mathscr{C}M_t$$

Hence we have

$$(\Psi_0 f)(x) \in \cap\{\mathscr{C}M_t, t > t_1, t \neq t_1\} = \mathscr{C}\,[\cup\,\{M_t, t > t_1, t \neq t_1\}]$$

Now, since $\cup\,\{M_t, t > t_1, t \neq t_1\} = M_{t_1}/\{t_1\}$, we have

$$(\Psi_0 f(x) \in \mathscr{C}\,[M_{t_1} \cap \{t_1\}^c] = M_{t_1}^c \cup \{t_1\}$$

The last relation, combined with relation (13.18), results in $(\Psi_0 f)(x) = t_1$, which achieves the proof.

Comments

1. Owing to the theorem, the two function mappings Ψ_0 and Ψ_1 are now directly linked to the primitive ψ, and no longer via intersections.
2. Also, the theorem generalizes to complete lattices a classical result valid for $T = \overline{\mathbf{R}}$, according to which, for any set family $\{\psi_f(t)\}$, $t \in \overline{\mathbf{R}}$, we have

   $$\text{Sup}\{t : x \in \bigcap_{s<t} \psi_f(s)\} = \inf\{t : x \notin \psi_f(t)\}$$

   (here, the symbol $<$ means "strictly smaller").
3. We started from set primitives to reach function mappings. The theorem shows a converse route. Indeed, given a function mapping $\Psi : T^E \Rightarrow T^E$, put

   $$\psi_f(t) = X^c_{\Psi f}(t) \tag{13.19}$$

   and consider the two function mappings Ψ_0 and Ψ_1 generated by ψ_f. As for Ψ_1 we have, according to the theorem,

   $$(\Psi_1 f)(x) = \vee\,\{t : x \notin \psi^f(t)\} = \vee\,\{t : x \in X_{\Psi f}(t)\} = (\Psi f)(x)$$

   that is, $\Psi_1 = \Psi$. Moreover, the correspondence between ψ and Ψ_1 is a bijection. As for Ψ_0, the result is just trivial, since we have

   $$(\Psi_0 f)(x) = \wedge\,\{t : x \in X_{\Psi f}(t)\} = \wedge\,\{t : (\Psi f)(x)\} > t\} = t_m$$

   where t_m is the minimal element of T. Better results concerning Ψ_0 will be obtained by considering $[S_{\Psi f} t)]^c$ as a primitive at level t. Then Ψ is identified with the lower extension Ψ_0 of the primitive. These results can be summed up as follows:

   Corollary. For any function mapping $\Psi : T^E \to T^E$, there exists one and only on set primitive $\psi : T \otimes T^E \to \mathscr{P}(E)$ whose Ψ is the upper (resp. the lower) extension, namely

$$\psi_f(t) = X^c_{\Psi_f}(t) \text{ (resp. } \psi_f(t) = S^c_{\Psi_f}(t)), \qquad t \in T, f \in T^E$$

4. *Empty families*. In relation (13.11) as well as in Theorem 4.2, we were led to introduce families of indices that may be empty. Then, for consistency, we have to put the two following conditions:

$$\cap \{X_i, \ i \in \text{the empty family}\} = E \tag{13.20}$$

$$\cup \{X_i, \ i \in \text{the empty family}\} = \emptyset$$

An example of such families is given by $s \not< t^m$ or $s \not> t_m$. Indeed, each time they occurred, the two cases were treated according to the above logical rule (see relation (13.11) or Theorem 4.2).

We will now illustrate the important Theorem 4.3 by the two cases of an increasing or of a decreasing primitive $\psi_f : T \to \mathcal{P}(E)$. What about Ψ_0 in such cases?

Let $\psi : T \to \mathcal{P}(E)$ be increasing (see Figure 5a). If moreover $\psi(t_m) = \emptyset$, then no x belongs to $\psi(t_m)$ and

$$\Psi_0(x) = \wedge \{t : x \notin \psi(t)\} = t_m$$

If $\psi(t_m) = A \neq \emptyset$, then $x \in A$ implies that $\{t : x \notin \psi_f(t)\}$ is the empty family, hence $\Psi_0(x) = t^m$. If $x \notin A$, then as previously $\Psi(x) = t_m$.

If $\psi : T \ ;\to \mathcal{P}(E)$ is now decreasing, the result is directly derived from $\Psi_0(x) = \wedge \{t : x \notin \psi(t)\}$, as shown in Figure 5.

V. FLAT PRIMITIVES AND ANAMORPHOSES

A rather natural approach to examining the behavior of a function mapping $\Psi : T^E \to T^E$ under anamorphosis consists of introducing this isomorphism via a set primitive ψ which generates Ψ.

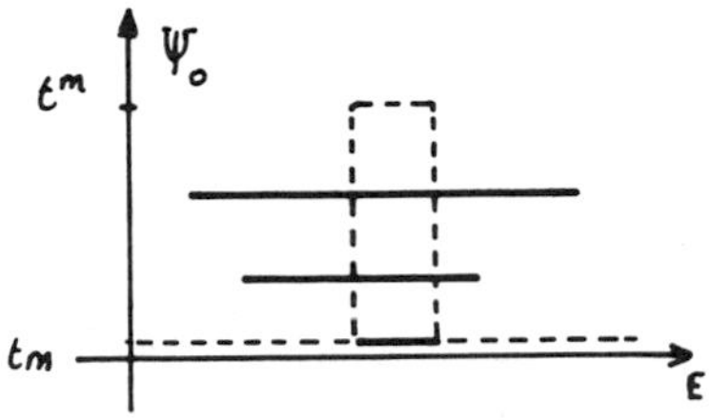

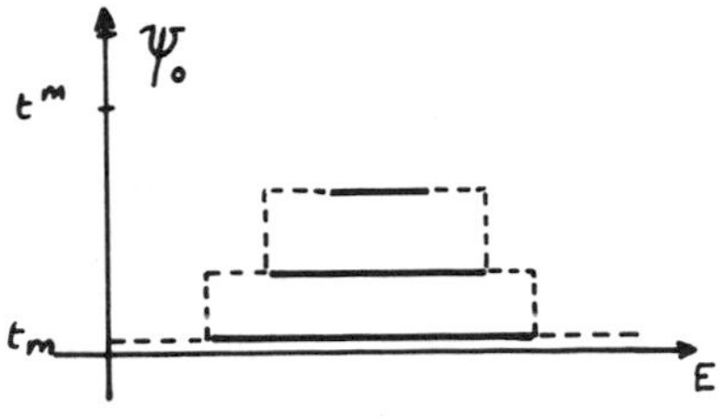

Figure 5. (Solid lines) The increasing (a) or decreasing (b) set primitive ψ. (Dashed lines) The resulting function Ψ_0.

From now on, unless otherwise stated, T' is a complete sublattice of T, so that every mapping Ψ on T^E is also a mapping on T'^E. It is within this framework that the notion of a flat primitive is introduced as follows.

Definition 5.1. Let T be a complete lattice, T' be a complete sublattice of T, and E be an arbitrary space. A set primitive $\psi : T \quad T^E \to \mathcal{P}(E)$ is said to be *flat* when, for every anamorphosis α on (T,T'), one has for all $t' \in T'$

$$\psi_{\alpha f}(t') = \psi_f(\alpha^{-1}t'), \qquad t' \in T', f \in T^E \tag{13.21}$$

This definition generalizes the usual notion of a flat (or stack) mapping for gray-tone functions (see Section II). Indeed, in this case $\psi_f(t)$ takes either form $\psi(X_f(t))$ or $\psi(S_f(t))$. Thn, owing to the two following relations (which are true because α is an increasing bijection):

$$S_{\alpha f}(t) = \{x : \alpha f(x) < t\} = \{bx : f(x) < \alpha^{-1}t\} = S_f(\alpha^{-1}t) \tag{13.22}$$
$$X_{\alpha f}(t) = X_f(\alpha^{-1}t)$$

we immediately obtain relation (13.21). But one can imagine many different operators, in the sense of the above definition (see Section VI).

A. Compatibility Under Anamorphosis

Consider now the two extensions Ψ_1 and Ψ'_1 of the same set primitive ψ, according to whether we work in T or in T', that is,

$$(\Psi_1 f)(x) = \bigvee_{\text{in } T} \{t;\ t \in T,\ x \notin \psi_f(t)\}$$
$$(\Psi'_1 f)(x) = \bigvee_{\text{in } T'} \{t';\ t' \in T',\ x \notin \psi_f(t')\} \tag{13.23}$$

These two function mappings, which are generally independent, become *equivalent* when ψ is flat, and only in this case. We can state this key property more precisely as follows.

Theorem 5.2. Let $\psi : T \otimes T^E \to \mathcal{P}(E)$ be a set primitive and Ψ_1 and Ψ'_1 its upper extensions to the lattice T and to the sublattice T', respectively. The pair (Ψ_1, Ψ'_1) is compatible with the anamorphoses $\alpha : T \to T'$, that is,

$$\alpha\Psi_1 = \Psi'_1\alpha \tag{13.24}$$

if and only if the set primitive ψ is flat.

Proof. We begin with the "if" part of the proof. Given $t' \in T'$, the family of indexes s' of T' such that $s' \nless t'$ is transformed by α^{-1} into the family $\alpha^{-1}(s') \in T$, $\alpha^{-1}(s') \nless \alpha^{-1}(t')$, and this is a bijection since α is an anamorphosis. Therefore, the flatness assumption (13.21) implies the equality

$$\cap \{\psi_f(\alpha^{-1}s'),\ \alpha^{-1}(s') \in T,\ \alpha^{-1}(s') \nless \alpha^{-1}(t')\}$$
$$= \cap \{\psi_{\alpha f}(s'),\ s' \in T',\ s' \nless t'\}$$

where the indexes for intersection vary in T in the left member and in T' in the right one. Now the right member is nothing but $S_{\Psi'_1 \alpha f}(t')$, and the left one equals $S_{\alpha\Psi_1 f}(t')$. Hence we obtain $(\alpha\Psi_1)(f) = (\Psi'_1\alpha)(f)$ via the identity of their upper sections. By duality, the proof extends to Ψ_0 and Ψ'_0 via the lower sections.

Conversely, suppose that $\alpha\Psi_1 = \Psi'_1\alpha$ for the two mappings Ψ_1 and Ψ'_1 which derive from a common primitive ψ we have;

$$X_{\Psi'_1 \alpha f}(t') = X_{\alpha\Psi_1 f}(t') = X_{\psi_1 f}(\alpha^{-1}t'), \qquad t' \in T'$$

Now, we know from the corollary of Theorem 4.2 that we have $\psi_{\alpha f}(t) = X^c_{\Psi'_1 \alpha f}(t')$ on the one hand and $\psi_f(\alpha^{-1}t) = X^c_{\Psi_1 f}(\alpha^{-1}t)$ on the other hand; hence ψ is flat.

Comments

1. Clearly, an equivalent theorem can be based on the two *lower* extensions Ψ_0 and ψ'_0 of a given ψ to lattices T^E and T'^E respectively. Then we have

   $$\alpha\Psi_0 = \Psi'_0\alpha$$

   for all anamorphoses $\alpha : T \rightarrow T'$ if and only if ψ is flat.
2. The proof also remains valid when we consider an arbitrary class A of anamorphoses $\alpha : T \rightarrow T'$. We just have to restrict to class A both flatness requirement (Definition 5.1) and anamorphosis commutation. In particular, if relation (13.21) is true for one anamorphosis α_0 only, then Ψ_1 and Ψ'_1 (or Ψ_0 and Ψ'_0) are compatible with α_0 only. For example, when $T = \overline{\mathbf{R}}$, or a product of completed lines, or else $\mathcal{P}(\mathbf{R}^n)$, then an important class of anamorphoses is generated by linear changes of coordinates, such as affinities and homothetics. Actually, the function mappings Ψ that commute with those anamorphoses are the only ones to hold a physical meaning, i.e., to be independent of the choice of units on the axes [12].
3. In the theorem, we have assumed the existence of a common primitive Ψ. By weakening this condition, we are led to the following corollary (stated here for upper extension):

   Corollary. Let T and T' be two complete lattices and E an arbitrary space. Given two primitives ψ and ψ', on T and T', respectively, to which correspond the two upper extensions Ψ_1 and Ψ'_1, and given a class A of anamorphoses $\alpha : T \rightarrow T'$, we have the following equivalence:

   $$\alpha\Psi_1 = \Psi'_1\alpha \Leftrightarrow \psi'_{\alpha f}(t') = \psi_f(\alpha^{-1}t')$$

   for all $\alpha \in A$, $t' \in T'$, and $f \in T^E$.
   (Proof similar to that of the theorem.)
4. We will conclude this section by a theorem of structure which contains one of the very reasons for the quality and robustness of anamorphosis compatibility.

Proposition 5.3. Given an anamorphosis $\alpha : T \to T'$, where T' is a complete sublattice of T, the class of mappings Ψ_1 (resp. Ψ'_1), defined as in Theorem 5.2, such that the pairs (Ψ_1, Ψ'_1) are compatible for α, is closed under sup in T (resp. in T') and under composition.

Proof. Let $\{\psi_i\}$ be a family of primitives. We see from relation (13.23) that the mappings Ψ_1 and Ψ'_1 associated with $\psi = \cup\{\psi_i\}$ are nothing but $\bigvee \Psi_{A,1}$ and $\bigvee \Psi'_{i,1}$, respectively. Now, since α is a dilation, we have

$$\alpha\Psi_1 = \alpha(\bigvee \Psi_{i,1}) = \bigvee \alpha\Psi_{i,1} = \bigvee \Psi'_{i,1}\alpha = (\bigvee \Psi'_{i,1})\alpha = \Psi'_1\alpha$$

The closure under composition is also immediate, since

$$\alpha\Psi_1\Psi_2 = \Psi'_1\alpha\Psi_2 = \Psi'_1\Psi_2\alpha$$

(Dual statement for extensions of the type Ψ_0.)

B. Finite Lattices

Up to now, we have implicitly admitted that a mapping $\alpha : T \to T'$ could be simultaneously a dilation, an erosion, and a bijection. Now, although such conjunction of properties can occur when T and T' are equivalently infinite, it becomes impossible when T and a fortiori its sublattice T' are *finite*. Then one can still construct mappings that are dilations and erosions (see, e.g., Figure 2), but surely not bijections. Therefore, a complementary analysis of what remains and what is lost of Theorem 5.2, when bijectivity is dropped out, may be instructive.

We will keep the same background and the same notation as previously (T and T' may be infinite or not), except for α, which is now only a dilation-erosion from T *onto* T'. As a dilation its reciprocal is the erosion $\varepsilon : T' \to T$, and as an erosion its reciprocal is the dilation $\delta : T' \to T$. The splitting of the former mapping α^{-1} into the pair (ε, δ) turns out to transpire into dissymmetric properties of the cross sections S and X. Indeed we have

$$S_{\alpha\Psi f}(t') = \{x \in E : (\alpha\Psi f)(x) < t'\} = \{x : (\Psi f)(x) < \varepsilon(t')\}$$

that is,

$$S_{\alpha\Psi f}(t') = S_{\Psi f}(\varepsilon(t')) \tag{13.25}$$

and similarly

$$X_{\alpha\Psi f}(t') = X_{\Psi f}(\delta(t')) \tag{13.26}$$

Also, this splitting of α^{-1} into (ε, δ) suggests we duplicate the definition of the flatness as follows:

Definition 5.3. A set primitive $\psi : T \otimes T^E \to \mathcal{P}(E)$ is said to be upper flat (resp. lower flat) for a given class A of dilations (resp. erosions) $\alpha : T \to T'$ when, for all $\alpha \in A$, $t \in T'$, and $f \in T^E$,

$$\psi_{\alpha f}(t') = \psi_f(\varepsilon\, t') \qquad \text{resp. } \psi_{\alpha f}(t') = \psi_f(\delta\, t') \tag{13.27}$$

Consider then the upper extension Ψ_1 and Ψ'_1 of an upper primitive ψ to the function lattices T^E and T'^E, respectively. For all $t' \in T'$ and $f \in T^E$, we have

$$\begin{aligned} S_{\Psi'_1 \alpha f}(t') &= \cap \{\psi_{\alpha f}(s'),\ s' \in T',\ s' \not< t'\} \\ &= \cap \{\psi_f(\varepsilon s'),\ s' \in T',\ s' \not< t'\} \end{aligned}$$

On the other hand, we may write, since $T' \subset T$ and $\varepsilon s' \not< \varepsilon t' \Rightarrow s' \not< t'$:

$$S_{\Psi_1 f}(\varepsilon t') = \cap \{\psi_f(s),\ s \in T,\ s \not< \varepsilon t'\}$$

i.e., by putting $s = \varepsilon s'$

$$S_{\Psi_1 f}(\varepsilon t') = \cap \{\psi_f(\varepsilon s'),\ \varepsilon s' \in T,\ \varepsilon s' \not< \varepsilon\, t'\} \supset S_{\Psi'_1 \alpha f}(t')$$

hence, by taking relation (13.25) into account, $S_{\Psi'_1 \alpha f}(t') \subset S_{\alpha \Psi_1 f}(t')$, that is, finally $\Psi'_1\alpha > \alpha\Psi_1$.

The dual result, on lower primitives and lower extensions, is proved in a similar way, allowing us to state the following theorem.

Theorem 5.4. Let T be a complete lattice, $T' \subset T$ be a complete sublattice of T, and E be an arbitrary space. If a set primitive $\psi : T \otimes T^E \to \mathcal{P}(E)$ is upper flat (resp. lower flat) for a family A of dilations (resp. erosions) from T onto T', then $\Psi'_1\alpha > \alpha\Psi_1$ (resp. $\Psi'_0\alpha < \alpha\Psi_0$) for all $\alpha \in A$.

It can be noted that the result concerning the upper extensions is not improved by the additional assumption that α is an erosion, nor that concerning the lower extensions by the assumption that α is a dilation. Nevertheless, the inequality of the theorem indicates a trend: the more degraded T is, the larger Ψ becomes. More precisely, let T'' be a complete sublattice of T_1, and α' a dilation from T' onto T'', of adjoined erosion ε'. The three extensions of the upper flat primitive ψ to T, T', and T'' are Ψ_1, Ψ'_1, and Ψ''_1, respectively. Then the theorem implies that

$$\Psi_1 < \varepsilon\Psi'_1\alpha < \varepsilon\varepsilon'\Psi'_2\alpha\alpha'$$

where $\alpha\alpha'$ turns out to be nothing but the upper flat dilation from T onto T'' and $\varepsilon\varepsilon'$ the adjoined erosion.

This theorem enlightens the bias that can be introduced by lookup tables when nonincreasing transformations are implemented. However, there exists one case in which the equality $\Psi'_1\alpha = \alpha\Psi_1$ of Theorem 5.2 is recovered, namely when ψ is increasing and upper semicontinuous; we know [13] then that $\cap \{\psi(s),\ s < t,\ s \in T\}$ = reduces to $\psi(t)$. Hence, assuming that ψ is upper flat, we obtain

$$S_{\Psi'_1 \alpha f}(t') = \psi_{\alpha f}(t') = \psi_f(\varepsilon t') = S_{\psi_1 f}(\varepsilon t') = S_{\alpha \Psi_1 f}(t')$$

that is, $\alpha\Psi_1 = \Psi'_1\alpha$, and we can state the following theorem.

Theorem 5.5. Given $f \in T^E$, let $\psi_f : T \to \mathcal{P}(E)$ be an increasing and upper semi-continous primitive, which is also upper flat for a family A of dilations from T onto T'. Then for every $\alpha \in A$, we have $\alpha\Psi_1 = \Psi_1'\alpha$. Similarly, if ψ is lower semi-continuous increasing, and lower flat for a family A of erosions α from T onto T', then $\alpha\Psi_0 = \Psi_0'\alpha$.

Corollary. When the lattice T is finite and $\Psi : T \to E$ is increasing, then for every dilation (resp. erosion) α from T onto a sub-lattice T' of T, we have $\alpha\,\Psi_1 = \Psi_1'\,\alpha$ (resp. $\alpha\,\Psi_0 = \Psi_0'\,\alpha$).

This corollary generalizes the result established by Heijmans when T is a finite chain [18], which was presented in Section II.

C. Involution and Duality

We saw in the introduction that the notion of a rank operator could not be extended from totally ordered lattices to non-totally ordered ones. This said, there are in practice only three useful rank operators, namely the sup, the inf, and the median. By definition, the first two extend to any type of complete lattices, hence they are not really based on the rank approach. Should it be the same for median operators?

When we order a finite sequence of numbers, of odd size, the median is, by construction, the element of the series whose value always remains invariant when the sense of ordering is changed from increasing to decreasing. This property implies that

$$\mathrm{med}(f \mid K) = -\,\mathrm{med}(-f \mid K) \tag{13.28}$$

where $f \in \overline{\mathbf{R}}^E$ and K is the finite old neighborhood. Conversely, the median is the only one of the rank operators to satisfy relation (13.28). However, this equality does not characterize median operators, since it is also fulfilled by all unbiased operators such as convolutions (e.g., $f * g = -(-f * g)$). Also, it is not unique, and for any $\lambda \in \overline{\mathbf{R}}$ we have as well

$$\mathrm{med}(f \mid K) = \lambda - \mathrm{med}(\lambda - f \mid K)$$

However, all operators $t \to -t$; $t \to \lambda - t$, from $\overline{\mathbf{R}}$ into itself share two typical properties:

1. If $\overline{\mathbf{R}}^*$ stands for the dual lattice of $\overline{\mathbf{R}}$, they are anamorphoses from $\overline{\mathbf{R}}$ into $\overline{\mathbf{R}}^*$.
2. They are identical to their inverse, or, as well, they give the identity by iteration.

Arising from this, definitions of involution, duality, and self-duality for complete lattices may be stated as follows.

Definition 5.7. An anamorphosis α from a complete lattice T onto its dual T^*, such as $\alpha = \alpha^{-1}$, is called an *involution on T*.

Definition 5.8. Given a complete lattice T and an involution α on T, a pair of mappings (Ψ, Ψ^*) on T is said to be *dual* with respect to α when $\alpha\Psi = \Psi^*\alpha$, and a mapping Ψ is *self-dual* with respect to α when $\alpha\Psi = \Psi\alpha$.

It may be instructive to consider the impact of self-duality and involution on primitives and cross sections.

Proposition 5–9. A mapping α on T is an involution if and only if, for all $f \in T^E$, and all $t \in T$, we have

$$S_{\alpha f}(t) = X_f(\alpha t) \qquad \text{and} \qquad S_f(\alpha t) = X_{\alpha f}(t) \tag{13.29}$$

Proof. If α is an involution, we may write

$$S_{\alpha f}(t) = \{x : x \in E,\ \alpha f(x) < t\} = \{x : f(x) > \alpha t)\} = X_f(\alpha t)$$

Conversely, if the two extreme members are equal for all f and all t, then $\alpha t < t' \Leftrightarrow t > \alpha t'$, hence the pair (α,α) is an adjunction from T onto T^*; that is, α is an involution.

Coming back to the duality, we see from the structure of the equation $\alpha\Psi^* = \Psi\alpha$ that Ψ^* is nothing but the *same algorithm* as Ψ, but applied to the dual lattice T^*, with its reverse order. In particular, we have

Proposition 5.10. The two upper and lower extensions to T^E of a primitive ψ are dual with respect to an involution α if and only if ψ is flat for α.

Proof. From Proposition 5.8 and the corollary of Theorem 4.3, we may write for any primitive ψ and any involution α

$$\psi^c_f(\alpha t) = S_{\Psi_0 f}(\alpha t) = X_{\alpha\Psi_0 f}(t)$$

$$\psi^c_{\alpha f}(t) = X_{\Psi_1 \alpha f}(t)$$

hence the equality of the left members in the two lines is equivalent to that of the right ones.

This proposition shows, among other things, that the compatibility of Ψ under anamorphosis and that of its dual go hand in hand.

Corollary. Let Ψ be a mapping fromT^E into itself, α an involution on T, and β an anamorphosis from T into a complete sublattice $T' \subset T$. If Ψ commutes with α, then it is equivalent that Ψ, or that its dual Ψ^*, be compatible under β.

Proof. One can always suppose Ψ to be of the type Ψ_1 for some primitive ψ. Since Ψ commutes with α, ψ is flat for α and, according to the proposition, $\Psi^* = \alpha\psi\alpha$ is equal to the smaller extension Ψ_0 of ψ. Moreover, if Ψ is compatible under anamorphosis β, then (Theorem 5.2) ψ is flat for β, hence Ψ_0 also commutes with β. The converse proof is derived in the same way.

Proposition 5.10 also shows that, when Ψ is considered of the type Ψ_1 in T, for a certain primitive ψ, flat for involution α, then Ψ is as well of the type Ψ_0 in T^*, for the primitive, which derives from ψ by changing the $X_f(t)$ into S_f and vice versa. This means in particular that if these permutations do not modify ψ, then Ψ is self-dual for α, and conversely. Hence we may state:

Proposition 5.11. A mapping Ψ on T^E is self-dual for an involution α if and only if the representation of its upper (or of its lower) primitive ψ in terms of cross sections $X_f(s)$ and $S_f(s)$ is a symmetrical function of X_f and S_f.

The following is a criterion easier to handle, since it allows us to generate a self-dual mapping from any family of mappings in T^E by means of the morphological center [19, Chapter 8].

Criterion 5.12. Let $\{\Psi_i\}$, $i \in I$, be a family of mappings in T^E that is self-dual under a given involution α. Then the morphological center

$$\nu = (\mathrm{Id} \wedge (\vee \Psi_i)) \vee (\wedge \Psi_i)$$

where Id denotes the identity mapping, is self-dual under α.

Proof. Put $\Psi = \vee \Psi_i$. Since $\{\Psi_i\}$ is self-dual under α—i.e., since for every $i \in I$ there exists a $j \in I$ such that $\Psi_i^* = \Psi_j$—we have $\psi^* = (\vee \Psi_i)^* = \wedge \Psi_i$, consequently $\Psi > \Psi^*$, and

$$\nu = (\mathrm{Id} \wedge \Psi) \vee \Psi^* = (\mathrm{Id} \vee \Psi^*) \wedge \Psi$$

Since α inverts sup and inf, we derive from the former equation

$$\alpha\nu = (\alpha \vee \alpha\Psi) \wedge \alpha\Psi^* = (\alpha \vee \Psi^*\alpha) \wedge \Psi\alpha$$

and from the latter $\nu\alpha = (\alpha \vee \psi^*\alpha) \wedge \Psi\alpha$, i.e. $\alpha\nu = \nu\alpha$.

In practice, Criterion 5.12 is applied for the families $\{\gamma\varphi,\varphi\gamma\}$ or $\{\gamma\varphi\gamma,\varphi\gamma\varphi\}$, when γ is the opening dual of an arbitrary closing φ. It is notable that the self-dual operators of type ν remain defined when the median and more generally the rank operator approach fails, by lack of total ordering on T. Moreover, the behavior of the centers ν, under iteration, is more satisfactory than that of the rank ones (when they exist) since there is no point $x \in E$ for which the sequence $f(\alpha)$, $(\nu f)(x)$, $(\nu \circ \nu)(f)(x)$, etc. may oscillate. This last property of pointwise monotony is true not only when ν is based on one of the two families ($\{\gamma\varphi,\gamma\varphi\}$ or $\{\gamma\varphi\gamma,\varphi\gamma\varphi\}$ but also under more general conditions [19, Chapter 8].

VI. APPLICATION TO GRAY-TONE IMAGES

This section must be considered as an illustration of the above general theory, in the simplest and the most familiar case when T, totally ordered, is a closed segment of $\overline{\mathbf{R}}$ or of $\overline{\mathbf{Z}}$. We will review a few already known gray-tone transformations and comment on their more or less partial commutativity under anamorphosis. The only original algorithm proposed below concerns Dolby filters. (In this section, we come back to the classical notation: $>$ for "strictly larger" and $\geqslant$ for "ordered.")

A. Threshold Combinations

When introducing Definition 5.1 for flatness for a primitive ψ, we saw that the classical stack or flat operators, i.e., the mappings ψ of the type $X_{\Psi(f)}(t) =$

$\psi(X_f(t))$ where $\psi : \mathcal{P}(E) \to \mathcal{P}(E)$ is fixed, were flat in the general sense of Definition 5.1.

However, they are far from being the only possible ones. Alternatively, we may start from an arbitrary set mapping ψ_0 depending on 2, 3, . . . , n operands. In case of two operands, for example, when ψ_0 maps $\mathcal{P}(E) \times \mathcal{P}(E)$ into $\mathcal{P}(E)$, all the combinations such as

$$\bigcup_{s>t} \psi_0[X_f(t), X_f(s)]; \qquad \bigcap_{s_1>t} \bigcap_{s_2>t} \psi_0 [X_f(s_1), X_f(s_2)]$$

or those obtained by changing some $\cap$ into $\cup$, X_f into S_f, $>$ into $<$, etc. generate flat primitives ψ. More generally, any algorithm ψ that involves the lattice structure of T *only*, and that does not comprise fixed values, is flat for all anamorphoses on $\overline{\mathbf{R}}$. This covers complex algorithms such as watersheds but rules out the weighted dilations of the type $\text{Sup}\{f(x - y) + g(y)\}$ for numerical functions. Actually, addition is irrelevant to the lattice structure of $\overline{\mathbf{R}}$

B. Summit Detection, Thickenings, Watersheds

These three classical mappings belong to the previous type.

1. Summit Detection

The function $f : \mathbf{R}^n \to [0,1]$ will be taken upper semicontinuous to ensure the existence of maxima. Then the set of its regional maxima at altitude t is given by

$$Y_f(t) = X_f(t) \cap [\bigcup_{s>t} r(X_f(s); X_f(t))]^c = \bigcap_{s>t} [X_f(t) \cap r^c(x_f(s); X_f(t))]$$

where $r(A;B)$ designates the "reconstructed opening," i.e., the set of those connected components of B that hit A. Hence, the transformed function $\Psi(f)$ that equals t at point x when x belongs to a maximum of level t, and 0 when $f(x)$ is not a maximum value, has for subsection at level t

$$X_{\Psi(f)}(t) = \bigcup_{s \geq t} Y_f(s)$$

i.e., a threshold combination that commutes under anamorphosis.

2. Thickenings

The classical gray-tone thickening algorithm [6] is the following:

$$(\Psi f)(x) = \inf\{f(y), y \in T_x\}$$
$$\text{when } \text{Sup}\{f(y), y \in T'_x\} \leq f(x) < \text{Inf}\{f(y), y \in T_x\}$$
$$(\Psi f)(x) = f(x) \qquad \text{when not}$$

where T_x and T'_x are two disjoined structuring elements in $\mathbf{R}^n$, implanted at point x. It is shown in [7, p. 451] that this operator admits the following decomposition:

$$X_{\Psi(f)}(t) = \bigcap_{s<t} [X_f(s) \cup (X_f(s) \ominus T) \cap (X_f^c(s) \ominus T')]$$

In other words, the thickening Ψ is exactly the extension of type Ψ_0 associated with the expression between brackets as primitive (see Theorem 4.2). Note that the second thickening of type Ψ_1 based on the same primitive and whose expression is given by Theorem 4.3 is trivially equal to $+\infty$ everywhere (if we want to obtain two thickenings here, we should consider the two upper and lower extensions of the *hit or miss* $(X_f(s) \ominus T) \cap (X_f^c(s) \ominus T')$ and take the sup of each of them with f).

3. Watersheds

The now classical watershed is a transformation due to Beucher and Lantuéjoul [8,14] that segments a function around each of its minima. From a function it generates a *set*, whose connected components segment the function. It is mentioned here as an example of an implicit expression of threshold combination.

Denote by $e(A;B)$, $A,B \subset \mathbf{R}^n$, the complement in B of the geodesic SKIZ of $A^c \cap B$, and by $m_f(t)$ the set of minima of a l.s.c. continuous function f at level t. Then put

$$Y_f(t) = e[K_f(t)\ ;\ S_f(t)) \cup m_f(t)]$$

with

$$K_f(t) = \bigcup_{t'<t} e[Y_f(t')\ ;\ \bigcap_{\rho>t'} S_f(\rho)]$$

Then the watersheds $W(f)$ of f are given by

$$W(f) = \cup \{Y_f(t),\ t \in [-\infty, \infty]\}$$

Whereas the above expression for $Y_f(t)$ is not explicitly solved, we see clearly from $Y_f(t)$ and $K_f(t)$ that the watershed is *invariant* under anamorphosis, as well as its complement called "divide lines"). One can also define a divide line function $\Psi(f)$ by putting

$$(\Psi f)(x) = f(x) \qquad \text{when } x \in W^c(f)$$
$$(\Psi f)(x = 0 \qquad \text{when not}$$

Then, the divide line function is compatible under any anamorphosis on $\overline{\mathbf{R}}$

C. Dolby Openings and Filters

We have just described transformations that commute with all anamorphoses. Others, such as convolution, top hat transforms [7,10], and dimensional measurements [12], commute with linear anamorphoses only (i.e., $t' = at + b$, $t \in \overline{\mathbf{R}}$). We would like, in this section, to show an example of compatibility with anamorphoses that keep unchanged one given value of $\overline{\mathbf{R}}$.

In audio signal processing, the Dolby technique (named after its inventor) applies a severe filter to the high frequencies when the amplitude is low and a weak one when it is high. Though the operator proposed below is based on an algorithm different from actual Dolby, we borrow the terminology because of the similarity of purposes.

With every function $f : \mathbf{R}^n \to [0,1]$, associate its lower-thresholded version f_e for a given value t_0 as follows:

$$\begin{cases} f_1(x) = f(x) & \text{when } f(x) \geq t_0 \\ f_1(x) = 0 & \text{when } f(x) < t_0 \end{cases}$$

Consider two flat openings γ_1 and γ_2, with $\gamma_1 < \gamma_2$ (i.e., γ_1 is more severe than γ_2). Apply γ_1 to the initial image f, γ_2 to the lower-thresholded function f_1, and take their max $\gamma(f)$:

$$\gamma(f) = \gamma_1(f) \vee \gamma_2(f_1)$$

It is easy to see that γ is still an opening. The same detail will be removed by γ if it is darker than t_0 and kept if not. Since γ_1 and γ_2 are flat for all anamorphoses, we have

$$X_{\gamma(f)}(t) = X_{\gamma_2(f)}(t) \qquad \text{if } t \geqslant t_0$$

$$X_{\gamma(f)}(t) = X_{\gamma_1(f)}(t) \cup X_{\gamma_2(f)}(t_0) \qquad \text{if } t < t_0$$

This implies that γ is still flat but uniquely for the anamorphoses α such that $\alpha(t_0) = t_0$.

By duality, if we start from the upper-thresholded version

$$\begin{cases} f_s(x) = 1 & \text{when } f(x) > t_0 \\ f_s(x) = f(x) & \text{when } f(x) \leq t_0 \end{cases}$$

of f, and from two flat closings φ_1 and φ_2, with $\varphi_1 > \varphi_2$, we generate the "Dolby" closing

$$\varphi(f) = \varphi_2(f_s) \wedge \varphi_1(f)$$

The operator φ acts *less* on the higher values than on the lower ones and closes coarse valleys and holes above t_0, whereas their dark equivalents are less modified. Now introduce these two other lower- and upper-thresholded variants f_1' and f_s' of f:

$$\begin{cases} f_1'(x) = f(x) & \text{when } f(x) > t_0 \\ f_1'(x) = f(t_0) & \text{when } f(x) \leq t_0 \end{cases}$$

$$\begin{cases} f_s'(x) = f(t_0) & \text{when } f(x) \leq t_0 \\ f_s'(x) = f(x) & \text{when } f(x) > t_0 \end{cases}$$

Clearly, the two mappings

$$\gamma'(f) = \gamma_1(f) \vee \gamma_2(f_1'), \qquad \varphi'(f) = \varphi_1(f) \wedge \varphi_2(f_s')$$

are openings and closings, respectively, but act in the opposite way to γ and φ. For example, the action of φ' is fine above t_0 and coarse below. Hence the composition products $\varphi'\gamma$ and $\gamma\varphi'$, as well as the corresponding center, or again the alternating sequential filters that may derive from families $\{\varphi'^j\}$ and $\{\gamma^j\}$, systematically preserve more details above t_0 than below. And all of them commute with the anamorphoses α for which $\alpha(t_0) = t_0$.

Finally, by now taking the products $\varphi\gamma$ and $\gamma\varphi$ and all their derivatives, we shall keep thin reliefs when they are light but also thin cracks when dark. A variant of these operations consists in taking two different values, t_0 for γ and t_1 for φ, with $t_0 > t_1$. An illustration of this variant is given in Figure 8, about the processing of hue in color images.

VII. APPLICATION TO MULTIVALUED IMAGES

Given n totally ordered complete lattices $T_1, \ldots, T_n$ of product T (i.e., $t \in T \Leftrightarrow t = \{t_1, \ldots, t_n; t_1 \in T_1, \ldots, t_n \in T_n\}$, we call multispectral or multivalued functions from E into T all mappings $f : E \to T$. Sometimes, one also speaks of vector functions, since at each point $x \in E$, $f(x)$ is the series of n components $\{f_1(x), f_2(x), \ldots f_n(x)\}$. In the present chapter we prefer to avoid this terminology, which refers implicitly to addition and to multiplication by a scalar (see the introduction). In contrast, it is more convenient to distinguish clearly between

1. The set of all possible anamorphoses on T
2. Some preferential subsets (e.g., anamorphoses with a fixed point)
3. When T, or its restriction to finite coordinates, is a vector space (or an algebraic module), the family of linear anamorphoses on T

Typically, color images are multivalued functions, where the three components R, G, B are also vector spaces [38]. By changing this system of coordinates into a sort of spherical one, we find the H, L, S (hue, lightness, saturation) representation, which is also a lattice. But is it the same as the R, G, B one? Is it still a vector space?

In other situations, the components of f may pertain to different worlds. It is the case in economic geography when one studies the triple {population density, slope of the relief, number of cars per family} [39]; we still have a complete, but partly discrete, partly continuous, lattice, and surely not a vector space.

A. Product Lattices T and T^E

Starting from any family of (totally ordered or not) lattices T_i there always exists a canonic lattice structure on the product space $T = T_1 \otimes T_2 \ldots T_n$. In this structure, t is smaller than t' (in lattice T) iff for *all* components $i \in \{1,n\}$, t_i is smaller than t'_i. Then the sup (resp. the inf) of a family $\{t^j, j \in J\}$ is the element of T where each component i is the sup (resp. the inf) of the $\{t_i^j\}$.

When each T_i is totally ordered, there are applied the classical numerical notation ($<$ = strictly smaller; $\leq$ lattice order) and curved symbols for the product lattice T:

$$t \prec t' \Leftrightarrow t_i \leq t'_i$$

$$i \in [1, \ldots, n] \qquad (13.30)$$

$$(\curlyvee t)_i = \bigvee t)_i, \qquad (\curlywedge t)_i = \bigwedge t_i$$

The lattice T^E of the multivalued functions from E onto T, where E is an arbitrary space, derives from T as usual. Lattices T and T^E are midway between the totally ordered case and the general one. For example, as a consequence of the loss of total ordering, the supremum $t \curlyvee t'$ generally differs from both t and t'. If t_1 and t_2 represent a two-spectrum color decomposition, as in Figure 6, then the color of $t \curlyvee t'$ is different from those of t and t'. In other words, by performing morphological operations we may generate (spurious) new colors. However, since the T_i are totally ordered we shall not generate new color *components* by using $\curlyvee$ and $\curlywedge$ (see Figure 6). Ultimately, this first result can be considered as a proposition and presented as follows:

Proposition 7.1. Let $T = T_1 \otimes T_2 \otimes \cdots T_n$ be a multivalued lattice and T^E the class of the multivalued functions from E into T. Then the subclass of T^E made of the functions that take k_1 given values only in T_1, k_2 in T_2, . . . , k_n in T_n, is closed under $\curlyvee$, $\curlywedge$, and under compositions of these two operators.

The supersections $S_f(t)$ of a function $f : E \to T$, that is, the sets

$$S_f(t) = \{x \in E : f(x) \prec t\}$$

are obtained from the partial sections $S_i(t)$ of the components

$$S_{f_i}(t_i) = \{x \in E : f_i(x) \leq t_i\}$$

Since $f \quad t$ is equivalent to $f_i(x) \leq t_i$ for all i, we have

$$S_f(t) = \bigcap_{i=1}^{n} S_{f_i}(t_i) \qquad (13.31)$$

Similarly, we have for the subsections

$$X_f(t) = \bigcap_{i=1}^{n} X_{f_i}(t_i)$$

It may be noticed that *both* relations $S_f(t)$ and $X_f(t)$ involve intersections.

A notion that vanishes with the total ordering is that of a rank operator. The only "rank" operators that remain are those that do not depend on the possibility of totally ordering any sample of data, namely the sup and the inf. Now, if we want to extend the median because this operator is flat and self-dual, and not

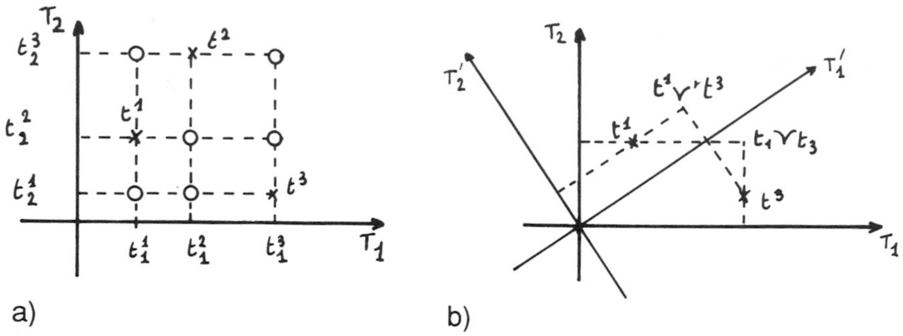

Figure 6. (a) From the three colors t^1, t^2, t^3 we can generate the six supplementary colors indicated with rings, by means of and . However, their projections are those of t^1, t^2, t^3. (b) The sup of t^1 and t^3 is not preserved under rotation of the axes.

because a rank is affected to it, the "midway status" of T allows the extension. It just consists of taking the median of each component in the same window. Flatness and self-duality are obviously transferred to the resulting color operator. The other rank operators may be extended similarly via the components (e.g., given a window of seven elements, take the last but one of each coordinate). The technique is heuristically acceptable, but, what should it mean? Either the existence of a rank was really needed in one dimension and it is now lost, or it was not, and then why introduce it in the one-dimensional case?

Another major feature of the multivalued lattice T ties in with *uniqueness* questions. When the components T_i do not share dimensional characteristics, which is the case in the above example of economic geography, the product lattice T still remains valid, with no alternative (except for possible changes in direction on some T_i). The situation is basically different when the T_i represent a system of coordinates in a vector space. Then there exists an infinite number of other equivalent systems of coordinates for the same vector space; they derive from the first one by rotations, similarities, passages to spherical, cylindric, or polar coordinates, etc. However, the product lattices T and T' associated with two equivalent systems of coordinates are not equivalent at all (see Figure 6b). Therefore, the same three-valued processing results in different color images according to whether it is applied to the R, G, B or the H, L, S representation.

B. Trace

Since the directions of the T_i are mutually independent, one can generate several product lattices T by taking the opposite order in one or more of the T_i. Formally speaking, these changes do not affect the properties of T, such as Proposition 7.1; but practically speaking, they lead to different results.

However, given a population of multivalued functions $f^j : E \to T$ consider the multivariate statistics of the scatter plot $\{f_i^j(x),\ i \in 1, N\}$ for all j and all $x \in E$. In this display the axes are the T_i and the origin is taken to be the point where each coordinate T_i is the median value of the ith histogram. There always exists at least one set of orderings for the T_i that maximizes the number of pixels falling in the positive ($1/2^{n}$th sector of the $\{T_i\}$ space (if the distribution admits a spherical symmetry in some subspaces, then there exists more than one solution). The resulting ordering on T will be called standard. In R, G, B color imagery, a standard order is typically given by maximization of the lightness, since the white is produced by adding colors (in the physical sense, if not in the electronic one).

The standard ordering allows us to condense image information into two scalar simplified versions by taking the numerical max and min of the f_i at each point x:

$$s(x) = \bigvee f_i(x), \qquad l(x) = \bigwedge f_i(x)$$

The two upper and lower scalar functions thus generated often carry enough information to serve as markers for gradient and extrema. They are called the traces of f.

1. Gradients, Extrema

As a matter of fact, in image processing gradients are used as contrast descriptors and extrema as dome or blob markers, that is, they are led astray from their initial roles in calculus of variations. Indeed, the gradient is more often reduced to its scalar part (e.g., Beucher's model), as the angular part can be treated separately. Similarly, the strict extractions of extrema generate a confusing noise and are replaced by dome extractors [24] or preceded by flat filters, which both suppress noise and enlarge the significant extrema.

Holding to the same point of view, a significant extension of the gradient module to multivalued images is provided by the trace s^* of the modules of the i derivations:

$$s^* = \bigvee |f_i'|$$

The same principle may also be used for top hat transforms or extrema extractors; the traces of these operations, performed separately on each T_i, generate (scalar) function markers of interest. It may be noticed that the actual definition of a multivalued maximum at level t, namely a connected component of $X_f(t)$ such that for all $t' \nless t$, $X_f(t') = \emptyset$, does not coincide with that obtained from the trace of the maxima of the T_i. Actually, the very definition of a maximum can be implemented. It leads to the intersection of the components' maxima and is likely to be often empty. In contrast, the intersection of Grimaud's domes extractors [24] seems more realistic.

2. Lattice of Angles, or Directions, on the Circle

What to do when one of the parameters under study is a direction $\omega \in [0,2\pi]$? If the distribution of this parameter around the disk is isotropic, the choice of a certain diameter is arbitrary. If not, there exists a maximum mode. One will take for sup one of the angles, say ω_0, of this mode, and for inf the opposite direction. Change the angular origin to have $\omega_0 = \pi$. Then the following total ordering leads to complete lattice structure

$$\omega \prec \omega' \Leftrightarrow \cos\omega < \cos\omega'$$
$$\text{or } \cos\omega = \cos\omega' \quad \text{and} \quad \sin\omega \leq \sin\omega'$$

If the parameter under study is defined modulo π (i.e., on a direction), then the previous analysis reduces to

$$\omega \prec \omega' \Leftrightarrow \cos\omega \leq \cos\omega'$$

This latter relation, which amounts to assimilating the semicircle to a segment, works as well when the range of the angles under study is smaller than the semicircle.

C. Flat Operators and Anamorphoses

In the gray-tone case, the total ordering allows us to identify anamorphoses with strictly increasing mappings from T onto a sublattice $T' \subset T$ (Criterion 3.3) and dilations/erosions with increasing mappings, as shown in Figure 2. This nice simplicity is now lost, and we can see from Figure 7a that an increasing mapping from T into its diagonal T' may not be a dilation.

However, let T' be a complete sublattice of T, of components $\{T_i'\}$, and let $\{\alpha_i\}$ be a family of anamorphoses, each α_i acting on the corresponding component T_i'. Consider the operator

$$\alpha(t) = \{\alpha_1(t_1), \alpha_2(t_2), \ldots, \alpha_n(t_n)\}$$

when $t = \{t_1, \ldots, t_n\} \in T$. α is obviously a bijection such that it is increasing as well as its inverse. Hence, according to Criterion 3.2, it designates an anamorphosis from T onto T':

Criterion 7.2. Let $T = T_1 \otimes T_2 \cdots T_n$ be a complete n-valued lattice and $T' = T_1' \otimes \cdots T_n'$ be a complete sublattice of T. Then each family $\{\alpha_i\}$ of anamorphoses from the components T_i onto their homologous T_i' generates a product anamorphosis α from T onto T' where

$$\alpha(t) = \{\alpha_i(t_i)\} \qquad (13.32)$$

The converse is obviously false: the product of three anamorphoses on H, L, S, respectively, generates an anamorphosis on color images (from Criterion 7.2), but one cannot decompose it into three independent anamorphoses on R, G, and B channels. In the following, we shall not demand more than commutation under

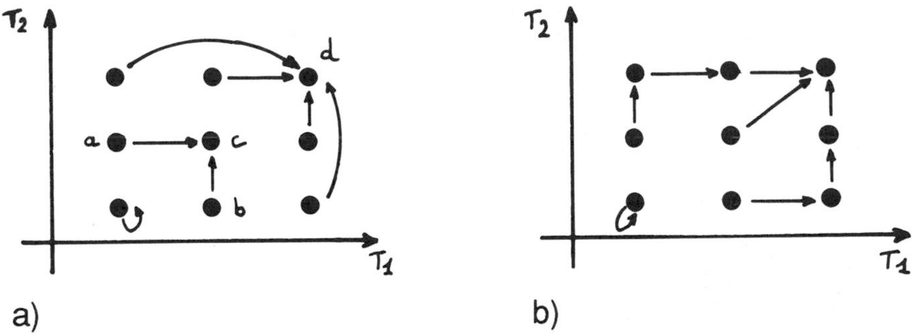

Figure 7. (a) T is a two-valued lattice of nine elements (the dots). T' is the diagonal of T; α maps T onto T' as indicated by the arrows. Although α is increasing, $C = \alpha(a) \vee \alpha(f) \neq \alpha(a \vee b) = d$, hence α is not a dilation. (b) Example of a dilation of T onto a sublattice of T that is not reducible to the product of increasing mappings on each of the two axes.

product anamorphoses, except for the linear anamorphoses, for which the converse of Criterion 7.2 is true.

The canonic product lattices that we have introduced in Section VII.A are not the only ones, though they are the simplest ones. One can also imagine cross conditions between the orderings of the components. The resulting structures will be called conditional lattices. Moreover in the product lattices themselves, one can also impose cross constraints that make a mapping on one component dependent of the other ones. Such operations are said to be conditional. In the three following subsections, we successively illustrate the three cases.

D. Product Lattices, Independent Processing

We will begin by taking for T the product lattice that is generated from the components T_i by relation (13.30).

1. Identical Processing

When the components play symmetrical roles (e.g., R, G, B channels), the simplest way to approach the product T consists in providing each component with the same processing. Indeed, the few color morphological treatments that are currently published proceed in this manner (see, for example, 25,26) and still do it even when the components are not symmetrical, such as the triplet lightness-hue-contast.

2. Adaptive Processing

However, in this latter case, it is more consistent to adapt the processing of each component to its specification. Consider, for example, the comparative implementations of Figure 8.

(a)

(b)

Figure 8. (a) Santa Anna Guiding the Virgin, by Stella. (b) Processing I. (c) Processing II.

(c)

The first processing holds identically on R, G, and B channels. One implements the center ν of Criterion 5.12 for the family $\{\varphi\gamma, \gamma\varphi\}$, where φ is the hexagonal (flat) closing of size two and γ the dual opening, and obtains Figure 8b (processing I).

The second processing holds on the H, L, S representation. Lightness is not modified, whereas saturation is strongly amended by alternating sequential filters (ASFs) of size six (flat hexagons). Finally the hue is approached by an ASF of size three. But in the latter case the opening and closing primitives are of the Dolby type and keep the small details when they belong to the first and the last fourth of the hue range.

The second processing reduces the information more than the first one. However, it yields better quality, just because of a more adapted choice of the parameters.

E. Conditional Lattices

The latter experiment suggests that priorities be introduced on the components. Not only is this always possible, but it leads to lattices that involve genetically, so to speak, functional relationships between components. For simplicity, we present the method in the case of two components and when a linear relationship is involved.

From the point of view of the priorities, the two extreme situations we may encounter are indicated in Figure 9a and c. On the one hand, in $T = T_1 \otimes T_2$, no dissymmetry between T_1 and T_2 is introduced; on the other hand, in the "scanning" lattice of a TV beam, if T_1 stands for the vertical direction, upside down, and T_2 for the horizontal one, oriented from left to right, a complete priority is given to T_1 since then

$$\underset{\text{(in scanning lattice)}}{t \prec t'} \Leftrightarrow \begin{cases} t_1 < t'_1 \\ \text{or } t_1 = t'_1 \text{ and } t_1 \le t'_2 \end{cases}$$

We will insert a sort of compromise between these situations by relaxing the ordering condition on t_2 as follows:

$$t \prec t' \Leftrightarrow \begin{cases} t_1 \le t'_1 \\ t_2 \le t'_2 + k(t'_1 - t_1) \end{cases} \tag{13.33}$$

Geometrically speaking, the parameter k corresponds to the opposite of the slope of the oblique lines in Figure 9b. Clearly, relation (13.31) defines an ordering on which we can build the following lattice:

$$\curlyvee \{t^j\} = (t_1,t_2) \qquad \text{with } \begin{cases} t_1 = \sup \{t^j_1\} \\ t_2 = \sup\{t^j_2 + ht^j_1\} - kt_1 \end{cases}$$

(A similar expression is obtained for $\curlywedge$ by changing sup into inf.) We see in Figure 9b that $a = t \curlyvee t'$ is closer to the point with the highest T_1 coordinate (i.e., t) than in the product lattice $T_1 \otimes T_2$. But this point a remains distinct from t, contrary to what happens in the scanning lattice.

The extension of ordering (13.33) to 3, . . . , n components may be achieved in several ways. For $n = 3$, for example, one can keep or not keep the symmetry between two components. This yields the following two respective orderings:

$$(1)\ t \prec t' \Leftrightarrow \begin{cases} t_1 \le t'_1,\ t_2 \le t'_2 \\ t_3 \le t'_3 + k_1(t'_1 - t_1) + k_2(t'_2 - t_2) \end{cases}$$

$$(2)\ t \prec t' \Leftrightarrow \begin{cases} t_1 \le t'_1,\ t_2 \le t'_2 + k_0(t'_1 - t_1) \\ t_3 \le t'_3 + k_1(t'_1 - t_1) + k_2(t'_2 - t_2) \end{cases}$$

In these relationships the additional term(s) $k(t'_i - t_i)$ may be replaced by any strictly decreasing function $f : \mathbf{R}^* \to \mathbf{R}^*$, such as $\exp\{-(t'_i - t_i)\}$. Finally, we could also modify the vertical side of the angle of vertex t (see Figure 9b). If it becomes oblique, with a positive slope, we generate a lattice more severe than $T_1 \otimes T_2$, where the points larger than t occupy the acute angle of vertex t.

F. Conditional Processing

Instead of combining the components in the genesis of the multivalued lattice itself, we may alternatively, whereas not exclusively, keep for T the simple product lattice and resort to cross-combinations at the level of the processing. Here the possibilities are infinite. For example, any dilation δ on T that is generated by n dilations δ_i on theT_i ($i = 1, . . . ,n$), may be replaced by the geodesic dilation [27] where each δ_i is treated geodesically inside the function $f = \bigvee f_i$. This function is similar to a lightness function, when the f_i $\{i = 1,2,3\}$ are the three colored images. By duality, the erosions may become geodesic ones with

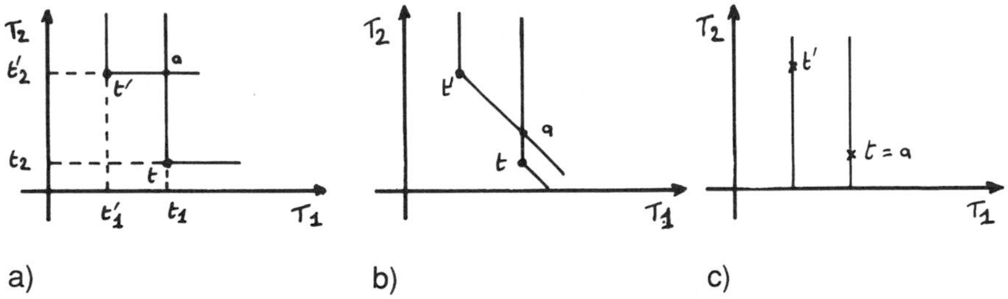

Figure 9. (a) $T = T_1 \otimes T_2$ (independence of the components). (b) Partial dependence of the components, via a linear relationship; the points larger than t are those of the obtuse angle of vertex t. (c) Scanning lattice, where a maximum priority is given to T_1.

respect to $f' = \bigwedge f_i$. Such a technique would result in making colors more uniform.

One can easily multiply the examples of this type. Instead we shall focus on two peculiar processing modes whose efficiency has been proved in real applications. The first one proposes a solution to the problem of halos; the second designs watersheds in color images.

1. Adaptive Color Edge Detection [28]

Consider, in a color image, two homogeneous zones A and B that share the same frontier. In the gray-tone case, every threshold through the edge generates a separation, or a segmentation of the two zones. It can be more or less appropriate, but it does exist. In our case, the thresholds of the three channels result in X_r, X_g, and X_b. If, by convention the zone A corresponds to higher values for the three channels (up to changes in direction on the axes), then according to relation (13.30)

$$A = X_r \cap X_g \cap X_b \qquad \text{and} \qquad B = X_r^c \cap X_g^c \cap X_b^c$$

and as a general rule, the halo $A^c \cap B = B^c \cap A$ will not be empty, even if the thresholds have been chosen accurately. This is a consequence of the lack of total ordering on T. Obviously, we always have the possibility of using a (binary) skeleton to reduce the halo $A^c \cap B$, but the local color perturbations are still there and appear as spurious colorations near the edge (Figure 10a).

We will reduce this noise by means of an adapted contrast mapping κ [29], whose primitives are closings and openings on the channels. Indeed, in the case of Figure 10a, an opening γ resets the phases between r and g by suppressing a peak on g. In another case, a closing φ should be better.

A criterion has to be fixed, in order to allow the choice between $(\gamma g)_x$ and $(\varphi g)_x$ at the point x. Take a scalar indicator of the spatial variation based on the trace or the sum l, and involving the derivatives of the signal, i.e.,

$$d = \inf\{|(\varphi g)' - l'|; |(\gamma g)' - l'|\} \tag{13.34}$$

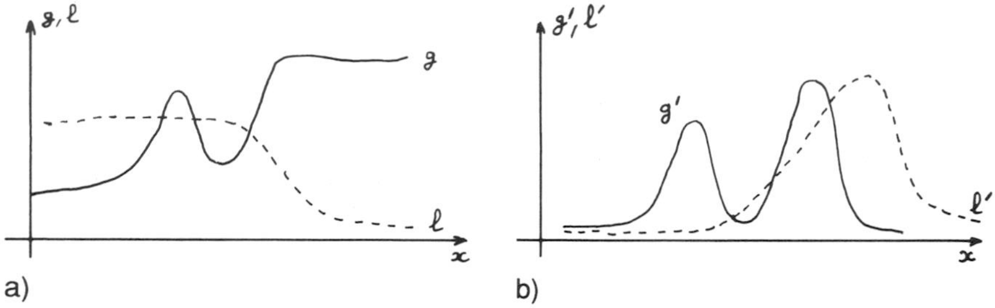

Figure 10. (a) Parasitic noise in the green channel near a transition. (b) Introduction of a reference signal l' for the gradient.

This allows us to construct the contrast operator κ as follows:

$$(\kappa g)_x = \begin{cases} (\varphi g)_x & \text{if } d(x) = |(\varphi g)' - l'|_x \\ (\gamma g)_x & \text{if } d(x) \neq |(\varphi g)' - l'|_x \end{cases}$$

In the same way, one can take more aggressive primitives such as dilations and erosions and also adapt the channel without introducing explicitly the reference l' [28]. Clearly, when the primitive γ and φ are flat, κ commutes with linear anamorphoses on T. Vitria [30] has applied this color edge cleaning to electronic circuit detection, where it proved useful.

2. Watersheds for Color Images

The algorithm that follows was proposed by F. Meyer to segment seven paintings, from the 16th to the 19th century, in a comparative study of the styles of the artists [31,32]. The procedure directly derives from two remarks:

1. The marked watersheds [25] are mappings with two independent inputs, the set of the markers and one set output.
2. The progression of each catchment basin from its marker requires only a metric system, to perform geodesic SKIZ.

From the first remark, we infer that marking may not depend on problematic color minima (which do not exist in general). It suffices to enter the marker in an interactive manner or by taking the minima of any of the scalar intensity indicators proposed in relation (13.34), or other ones.

From the second remark, we are led to the Euclidean or to the city block distance between each pair of pixels in the three-color domain.

We are then brought back to a scalar watershed segmentation. It results in disjoined connected catchment basins. By taking the color histogram in each of them one can extract the coordinates r_0, g_0, b_0 of the main mode (or of its center of gravity if the mode is not reduced to a point) and assign this dominant color

r_0, g_0, b_0 to all pixels of the catchment basins. It is remarkable that if the distance chosen is the city block, all the steps of the algorithm commute with the anamorphoses on the components.

ACKNOWLEDGMENTS

The author wishes to express his gratitude to T. Jochems, who implemented color algorithms, and to G. Matheron for his useful comments. We also gratefully acknowledge our stimulating discussions with S. Beucher, F. Meyer, and J. Vitria about color processing. This work has been supported by grant JMP-767 of the Franco-Spanish Mercurio Program.

REFERENCES

1. Matheron, G., *Eléments pour une théorie des milieux poreux*, Masson, Paris, 1967.
2. Haas, H., Matheron, G., and Serra, J., Mophologie mathématique et granulométries en place, *Ann. Mines*, *11*, 736–753, *12*, 767–782.
3. Serra, J., Introduction à la morphologie mathématique, fascicule 3, Cahiers du C.M.M., Ecole des Mines, Paris, 1969.
4. Serra, J., Morphologie mathématique pour les fonctions "à peu près en tout ou rien," Technical Report C.M.M., Ecole des Mines, Paris, 1975.
5. Rosenfeld, A., and Kak, A., *Digital Picture Processing*, Academic Press, San Diego, 1976.
6. Goetcharian, F. N., From binary to grey level tone image processing by using fuzzy logic concepts, *Pattern Recogn. 12*, 7–15 (1980).
7. Serra, J., *Image Analysis and Mathematical Morphology*, Academic Press, London, 1982.
8. Beucher, S., and Lantuéjoul, C., Use of watersheds in contour detection, in *Proceedings, International Workshop on Image Processing*, CCETT/IRISA, Rennes, France, September 1979.
9. Matheron, G., Théorie des ensembls aléatoies, fascicule 4, Cahiers du C.M.M., Ecole des Mines, Paris, 1969.
10. Meyer, F., Contrast features extraction, in *Proceedings, 2nd European Symposium on Quantitative Analysis of Microstructures in Material Sciences, Biology and Medicine*, Caen, October 1977, pp. 374–380.
11. Sternberg, S. R., Cellular computers and biomedical image processing, in *Biomedical Images and Computers* (J. Sklansky and J. C. Bisconte, eds.), Lecture Notes in Medical Informatics, vol. 17, Springer-Verlag, Berlin, 1982, pp. 204–319.
12. Rivest, J. F., Serra, J., and Soille, P., Dimensionality in Image Analysis Technical Report, Ecole des Mines de Paris, September 1991; *JVCIR*, in press.
13. Matheron, G., Les treillis compacts, Technical Report, Paris School of Mines, N-23/90/G, November 1990.
14. Beucher, S., Segmentation d'images et morphologie mathématique, doctorate thesis, School of Mines, Paris, June 1990.

15. Wendt, P. D., Coyle, U. J., and Gallagher, N. C., Stack filters, *IEEE Trans. Acoust. Speech Signal Process.*, *34*, 898–911 (1986).
16. Zeng, B., Zou, H., and Neuvo, Y., F.I.R. Stack hybrid filters apt, *Engine*, *30*(7), 965–975 (1991).
17. Maragos, P., and Schafer, R. W., Morphological filters. Part I. Their set-theoretic analysis and relations to linear shift-invariant filters, *IEEE Trans. Acoust. Speech Signal Process.*, *35*, 1153–1169 (1987).
18. Heijmans, H. J. A. M., Theoretical aspects of grey-level morphology, *IEEE Trans. Pattern Anal. Machine Itell.*, *13*(6), 568–592 (1991).
19. Serra, J., ed., *Image Analysis and Mathematical Morphology*, vol. 2, *Theoretical Advances*, Academic Press, London, 1988.
20. Matheron, G., in *Image Analysis and Mathematical Morphology*, vol. 2, *Theoretical Advances* (J. Serra, ed.), Academic Press, London, 1988, chapters 3–6.
21. Serra, J., Mathematical morphology for complete lattices, in [19], chapter 1.
22. Heijmans, H. J. A. M., and Ronse, C., The algebraic basis of mathematical morphology. Part I. Dilations and erosions, *Comput. Vision Graphics Image Process.*, in press.
23. Giers, G., Hofmann, K. H., Keimel, K., Lawson, J. D., Mislove, M., and Scott, D. S., *A Compendium of Continuous Lattices*, Springer, Berlin, 1980.
24. Grimaud, M., La géodésie numérique en Morphologie Mathématique. Application à la détection automatique des microcalcifications, thesis, 1991.
25. Meyer, F., and Beucher, S., Morphological segmentation, *JVCIR*, *1*, 21 (1990).
26. Pei, S. C., and Chen, F. C., Subband decomposition of monochrome and color images by mathematical morphology, 1991.
27. Lantuéjoul, Ch., and Beucher, S., On the use of geodesic metric in image analysis. *J. Microsc.*, *121*, 39–49 (1981).
28. Serra, J., Sept mini études, Technical Report, Ecole des Mines de Paris, no. 33/91/MM, 1991.
29. Meyer, F., and Serra, J., Contrasts and activity lattice, *Signal Process.*, *16*, 303–317 (1989).
30. Vitria, J., Estudi sobre l'automatizacio de l'analisi visual de circuits integrats mitjançant morfologia matematica, Ph.D. thesis, UAB, Barcelona, 1990.
31. Meyer, F., Color image segmentation, in *Proceedings*, *Image Processing and Its Applications*, *IEE Conf.*, Maastricht, April 1992.
32. Gourdon, F., Segmentation couleur de tableaux appliquée à l'étude du rapport des objets entre eux, Technical Report, Ecole des Mines de Paris, no. S-2/90/MM, July 1990.
33. Balturiatis, *Anamorphoses*, Flammarion, Paris, 1957.
34. Giardina, C. R., and Dougherty, E. R., *Morphological Methods in Image and Signal Processing*, Prentice-Hall, Englewood Cliffs, N.J., 1988.
35. Haralick, R. M., Sternberg, S. R., and Zhuang, X., Image analysis using mathematical morphology, *IEEE Trans. Pattern Anal. Machine Intell.*, *PAMI-9*, 532–550 (1987).
36. Janowitz, M. F., A model for ordinal filtering of digital images, In *Statistical Image Processing and Graphics* (E. J. Wegman and D. J. DePriest, eds.), Marcel Dekker, New York, 1986.

37. Sternberg, S. R., Grayscale morphology, *Comput. Vision Graphics Image Process.*, *35*, 333–355 (1986).
38. Torres, L., Lleida E., Apuntes de television, Department of Signal Theory and Telecommunications, UPC, 1991.
39. Voiron-Canicio, C., A morphological approach to the spatial analysis of district data, Technical Report, University of Nice, France (proposed for publication to *JVCIR*), 1991.
40. Tukey, J., *Exploratory Data Analysis*, Addison-Wesley, London, 1977.

Index